Point Processes and Their Statistical Inference

PROBABILITY: PURE AND APPLIED

A Series of Textbooks and Reference Books

Editor

MARCEL F. NEUTS

University of Arizona
Tucson, Arizona

Point Processes and Their Statistical Inference

Second Edition
Revised and Expanded

Alan F. Karr

Department of Mathematical Sciences
The Johns Hopkins University
Baltimore, Maryland

CRC Press
Taylor & Francis Group
Boca Raton London New York

CRC Press is an imprint of the
Taylor & Francis Group, an **informa** business

First published 1991 by Marcel Dekker, Inc.

Published 2020 by CRC Press
Taylor & Francis Group
6000 Broken Sound Parkway NW, Suite 300
Boca Raton, FL 33487-2742

First issued in paperback 2020

ISBN 13: 978-0-367-58003-2 (pbk)
ISBN 13: 978-0-8247-8532-1 (hbk)

Visit the Taylor & Francis Web site at
http://www.taylorandfrancis.com

and the CRC Press Web site at
http://www.crcpress.com

Library of Congress Cataloging-in-Publication Data

Karr, Alan F.
Point processes and their statistical inference / Alan F. Karr --
2nd ed., rev. and expanded.
        p. cm. -- (Probability, pure and applied; 7)
Includes bibliographical references and index.
ISBN 0-8247-8532-0 (acid-free paper)
1. Point processes. I. Title. II. Series.
QA274. 42. K37 1991   90-28376
519.2--dc20   CIP

To Betsy and Cathy

Preface to the Second Edition

As I have worked on this second edition, I have recalled frequently Benjamin Franklin's epitaph on himself, written sixty-two years before his death:

> The body of Benjamin Franklin, Printer, like the cover of an old book, its contents torn out and stripped of its lettering and gilding, lies here, food for worms; but the work shall not be lost, for it will, as he believed, appear once more in a new and more elegant edition, revised and corrected by the Author.

The second edition is, without doubt, new, revised and corrected; I do hope that it is more elegant.

The main changes are rather complete reorganization and rewriting of material pertaining to the multiplicative intensity model and stationary point processes, additional material concerning the Cox regression model, and an expanded, updated bibliography. Also, I have provided expanded explanations of many fundamental statistical concepts.

Computer text processing is both boon and anathema to perfectionists. Literally on every page, consequently, there are changes that correct the embarrassingly large number of minor errors in the first edition, and that (are intended to) enhance readability. In particular, I have adopted dual symbols for indicating ends of results and proofs: □ marks the end of an unproved result or example, while proofs terminate with ■.

On the other hand, there must (and did) come a point at which the temptation to revise the text "one more time" must be refused; I hope that it was neither too soon nor too late.

Other changes are arguably unnecessary, but not, I think, gratuitous. One's taste changes over time, and what seemed well-turned phrases five years ago now appear affected or simply obscure, so the prose is, I hope, leaner and more natural. To those who dislike my penchant for "complicated" sentences and "unusual" words, however, I remain unapologetic: mathematics may not be literature, yet even here a distinctive style and precise, picturesque vocabulary are essential.

I thank my many friends and colleagues who have read and cited the first edition, especially those who provided lists of errors and various other forms of criticism. Their response has been most gratifying, and has been the principal stimulus for my undertaking a task I once vowed never to do. Also, and belatedly, I thank Marilyn Karr for a heroic proofreading job for the first edition.

Alan F. Karr

Preface to the First Edition

Inference for stochastic processes seems to have become a distinct subfield of probability and statistics. No longer viewed as falling between probability and statistics but part of neither, it is now recognized as spanning both. Questions of inference are current important concerns — not only in classical senses but also in state estimation — for the traditionally important stochastic processes (Markov chains, diffusions, and point processes).

For point processes, progress has been especially rapid because of simple qualitative structure, the wealth of tractable but broad special cases, and, above all, persistent, stimulating contact with applications. This contact provides the impetus for the solution of problems at the same time as, in consonance with most of the history of mathematics, it confirms that important physical problems lead to the most challenging and interesting mathematical questions. Yet as recently as the mid-1970s inference for point processes was more a collection of disparate techniques than a coherent field of scientific investigation. The introduction of martingale methods for statistical estimation, hypothesis testing, and state estimation revolutionized the then inchoate field, at once expanding the family of models amenable to meaningful analysis and unifying, albeit conceptually more than practically, the entire subject. Developments since then have been numerous and diverse, but without engendering fragmentation.

The goal of this book is to present a unified (but necessarily, of course, incomplete) description of inference for point processes before the field becomes so broad that such a goal is hopeless. By depicting the probabilistic and statistical heart of the subject, as well as several of its boundaries, I have endeavored to convey simultaneously the unity and diversity whose interplay and tension I find so alluring.

Although conceived primarily as a research monograph, the book is nevertheless suitable as a text for graduate-level courses and seminars. The style of presentation is meant to be accessible to probabilists and statisticians alike, but (reflecting my own development as a probabilist who became — and remains – very interested in statistics) more background is presumed in probability than in statistics. Therefore, most probabilistic ideas are introduced without explanation while comparably difficult statistical concepts are usually described briefly at least. To attempt to read the book without a semester graduate course in probability (at the level, for example, of Chung, 1975), including an introduction to discrete time martingales, and a corresponding course in mathematical statistics (e.g., from Bickel and Doksum, 1977) would be unrealistic and ineffective, but this background should suffice. No deep prior knowledge of point processes is presumed; even such basic processes as Poisson processes are treated in some detail.

The most difficult decision during writing was how to treat continuous time martingale theory. A full development was deemed beyond the scope of the book; instead I have summarized relevant major results in Appendix B, and throughout most of the text have imposed assumptions sufficiently strong to render arcane theoretical details unnecessary. The slight loss of generality is more than offset by the gain in clarity and cogency; one is able to concentrate on the really important questions. A reader familiar with discrete time martingales can understand the key ideas by analogy.

I have resisted, in most cases, the temptation to generalize beyond recognition. In Freeman Dyson's beautiful metaphor I am more part of Rutherford's Manchester than Plato's Athens, in spirit a diversifier and a seeker of unity in the small. But unfettered diversity is chaos and a book without unifying ideas merely a dictionary. There are important themes: maximum likelihood, empirical averages (in several guises) as estimators, martingale representations, and, above all, exploitation of special structure (whether that structure be broadly or narrowly prescribed) that do pervade the entire book, although sometimes in the minor mode and often in variations. This somewhat schizophrenic but, I think, inescapable, viewpoint surfaces not only in the large scale structure of the book, but also in the structure of individual chapters and sections. Chapters 1–5 concern broad subjects while Chapters 6–10 investigate in more detail correspondingly narrower topics. In Chapters 6–9 rather stringent structural assumptions are stipulated but once the context is established we impose few additional restrictions; hence our problems and methods are distinctly "nonparametric."

The book can be used in a variety of advanced graduate-level courses or research seminars. Chapters 1 and 2 could, perhaps with supplementation, be used for a one-quarter or one-semester introduction to point processes. Together with Chapters 4 and 5, they present the two main lines of theory and inference: distributional theory/empirical inference and intensity theory/martingale inference; with judicious omissions they could be covered in one semester, but a more leisurely pace is also feasible. Chapter 3 (which was surprisingly difficult to write) is meant less for detailed study than as

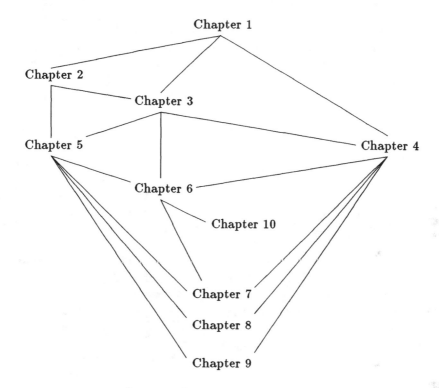

guide and appetizer for what follows. To omit it seems a mistake; to devour it is unnecessary. Each of Chapters 6–10 presents diversity in the form of inference for an important special class of point processes; they are largely independent of one another and are probably best covered selectively as interests of instructor and students dictate. Any of them could form the basis for a single set of lectures.

The accompanying figure shows the chapter interrelationships.

Every chapter contains a number of exercises, whose purpose is to extend and enrich the main body of material. No essential points have been relegated to exercises, but a reading of the book that ignores them would be incomplete. They are important agents of diversity as well as the principal means whereby applications are presented. The level of difficulty varies: some are computations, some are improvements pertaining to special cases of results in the text, still others are theorems taken from the literature that are not central to our development but are of importance and interest notwithstanding.

An author should also, I think, say what a book is not. This is not a book about data analysis, nor does it pretend to present applications in depth or breadth even though the text and exercises taken together do provide a reasonable sample. Bayesian ideas are crucial to state estimation, but Bayesian inference *per se* is absent. There is much more emphasis on estimation than on hypothesis testing; the argument that every estimator implicitly defines a test statistic is somewhat disingenuous because our estimators are often estimating infinite-dimensional objects and their distributional properties are typically too poorly understood to permit analysis of power, let alone optimality. Among important classes of point processes clustered point processes have received, relatively, the shortest shrift; such choices are inevitable, and this is one class of processes for which alternative treatments do exist. Cox processes may seem overrepresented, but I cannot remain dispassionate on a subject which has been a personal special interest for so long.

Much of the book was written during an itinerant year in 1983. M. R. Leadbetter and S. Cambanis of the University of North Carolina, where I spent a sabbatical semester, were especially hospitable, as were, for shorter periods, P. Franken of Humboldt University and E. Arjas of the University of Oulu. Extremely helpful readings of the manuscript were generously provided by R. Gill (Amsterdam), S. Johansen (Copenhagen), O. Kallenberg (Göteborg), M. R. Leadbetter (Chapel Hill) and J. Smith (Washington, D.C.); I hope they can recognize how valuable their comments were. Errors that remain, of course, are mine. I also wish to thank R. Serfling of The Johns Hopkins University for his constant encouragement and friendship, and the Air Force Office of Scientific Research for their support of my research, leading to a number of results that appear here for the first time.

Alan F. Karr

Contents

**10 Inference for Stochastic Processes Based on Poisson
   Process Samples**   **382**

Point Processes
and Their
Statistical Inference

1

Point Processes: Distribution Theory

The origins of our subject are assuredly ancient, dating from speculation about the distribution of stars in a sky once thought spherical and from records of floods, earthquakes and other natural events with exhibiting no evident periodic structure. By contrast, the modern mathematical theory is less lengthy. It stems from work of Palm and later Khintchine on problems of teletraffic theory and work, notably of Gibbs, on statistical mechanics; and from increased understanding of the most fundamental of point processes, the Poisson process. Since, there has developed on the one hand a general, powerful and elegant corpus of mathematical theory having both a life of its own and important consequences in other branches of probability, and on the other hand a plethora of applications to such diverse fields as biology, geography, meteorology, physics, operations research and engineering. Following the best mathematical tradition, the two aspects have nurtured one another and, despite some recent signs of divergence, remain in healthy contact.

A point process is a model of points randomly distributed in some space; they are indistinguishable except for their locations. Points may represent times of events, locations of objects or even, as elements of function spaces, paths followed by a stochastic system. Here we emphasize point processes on general spaces. Thus events or objects having secondary characteristics associated with them are viewed simply as points in a larger space; for example, a point of mass u located at $x \in \mathbf{R}^3$ becomes simply a point (x, u) in $\mathbf{R}^3 \times \mathbf{R}_+$. Such marked point processes, then, are nothing more than point processes on product spaces and can be treated using the same methods.

As random processes, point processes are subject to *statistical inference*, one principal theme of this book. For example, the probability law of a point process N may be unknown, and need to be determined from observations, for example of independent, identically distributed (i.i.d.) realizations of N, or from just one realization (which may be impossible). We emphasize the case that the law of N is either entirely unknown or of known structure but with an unknown infinite-dimensional "parameter." Two general approaches to inference will be developed. In Chapter 4 we treat empirical estimators of objects such as the Laplace functional, which determine the law of N and from which additional quantities of interest, for example mean and covariance measures, can be calculated. The strength of these methods is generality in regard to the underlying space and the structure of the point process, but they require multiple realizations. Our second broad class of techniques, the martingale methods presented in Chapter 5, applies only to point processes on $\mathbf{R}_+$ and only if certain structural assumptions are imposed, but within this context their power is extraordinary.

Alternatively, a point process may be stipulated to admit particular qualitative structure, for example, to be a Poisson process (Chapter 6) or Cox process (Chapter 7), rendering the estimation problem somewhat simpler in form although not necessarily easier in substance since typically the unknown parameter remains infinite-dimensional. Other classes of point processes with special structure treated here include renewal processes (Chapter 8) and stationary point processes (Chapter 9).

Our second salient theme is *state estimation* for point processes that are observed only partially. The observations are represented by a σ-algebra $\mathcal{H}$ ("observability" means measurability with respect to $\mathcal{H}$), and the objective is to reconstruct, with minimum error, unobservable functionals of the point process N, i.e., random variables $X \in \sigma(N)$ not measurable with respect to $\mathcal{H}$. Often observations correspond to complete knowledge of N over a subset A of the underlying space. It is required that the law of N be known. This realization by realization reconstruction of unobserved aspects of N entails estimation of random variables, leading to the term "state estimation," as opposed to "parameter" or "law" estimation. Throughout we employ the nearly universal criterion of mean squared error, with respect to which the optimal state estimator of a random variable X — with minimum mean squared error (MMSE) — is the conditional expectation $\hat{X} = E[X|\mathcal{H}]$: provided that $E[X^2] < \infty$, $E[(\hat{X} - X)^2] \leq E[(Y - X)^2]$ for all $Y \in L^2(\mathcal{H})$.

Were one to stop here, however, the problem would lack independent content. The key issues involve computation of MMSE state estimators as explicit functions (nonlinear in general) of the observations and, especially, recursive computation of state estimators.

A recursive representation enables one to update state estimators as additional data becomes available, rather than compute from scratch a revised estimator that incorporates both old and new observations. We illustrate in stylized form. Suppose that observations $\mathcal{H}_1$ have been obtained previously and that $\hat{X}_1 = E[X|\mathcal{H}_1]$ has been calculated (by whatever method). At a subsequent time, further observations $\mathcal{H}_2$ are obtained. A recursive representation avoids computing the revised estimator $\hat{X}_{12} = E[X|\mathcal{H}_1 \vee \mathcal{H}_2]$ directly, by substituting an expression of the generic form $\hat{X}_{12} - \hat{X}_1 = F(\hat{X}_1, Y_2)$, where $Y_2 \in \mathcal{H}_2$ is a suitable random element and F is some function. Two things are accomplished here: only the increment $\hat{X}_{12} - \hat{X}_1$ in the estimator is computed ($\hat{X}_1$ is known already), and the contributions of the old and new observations are separated. The ultimate in recursive representations are stochastic differential equations for state estimators for processes on $\mathbf{R}_+$, as discussed in Chapters 5, 6 and 7. There the linear order structure of the real line is essential because it determines a natural sequence in which observations arise.

Finally, the problems of statistical inference and state estimation may be concatenated. Suppose that there are i.i.d. copies N_i of a point process with unknown probability law $\mathcal{L}$ and that $N_1, \ldots, N_n$ have been observed previously. Suppose that N_{n+1} has just been observed over the set A (mathematically the observations are a σ-algebra denoted by $\mathcal{F}^{N_{n+1}}(A)$) and that it is desired to perform state estimation for a functional X_{n+1} of the unobserved part of N_{n+1} (over A^c). Without knowledge of $\mathcal{L}$ the "true" state estimator $\hat{X}_{n+1} = E[X_{n+1}|\mathcal{F}^{N_{n+1}}(A)]$ cannot be computed. If, however, one knew the functional form $\hat{X}_{n+1} = G(\mathcal{L}, N_{n+1}|_A)$, where $N_{n+1}|_A$ is the restriction of N_{n+1} to A (i.e., the observations), then a "pseudo-state estimator" $\hat{X}_{n+1}^*$ can be formed by invoking the *principle of separation*, which dictates that one utilize $N_1, \ldots, N_n$ (and in practice N_{n+1} as well) to form an estimator $\hat{\mathcal{L}}$ of $\mathcal{L}$, then take as the pseudo-state estimator $\hat{X}_{n+1}^* = G(\hat{\mathcal{L}}, N_{n+1}|_A)$. The key to the problem is to impose qualitative assumptions that allow computation of G without knowledge of $\mathcal{L}$. Mixed Poisson processes admit the most satisfactory solution, as described in Chapter 7; other cases are examined in Chapters 6, 8 and 10.

More generally, one may envision state estimation for random variables that are not $\mathcal{F}^N$-measurable but are related to N notwithstanding. The prime example is Cox processes, which have specified structural properties conditional on a directing random measure that is unobservable in all important applications. The random variable X may then be a functional of the directing measure. Questions and issues outlined above remain germane.

The remainder of this chapter introduces point processes and their distributional theory. Chapter 2 treats the martingale theory of point processes on $\mathbf{R}_+$. Chapter 3 formulates in some detail the kinds of problems of statistical inference and state estimation treated in this book and by way of illustration examines the most important point processes, ordinary Poisson processes on $\mathbf{R}_+$. Chapters 1 and 4, and 2 and 5, form parallel pairs. Chapter 4 addresses distribution-based inference using empirical estimators, with focus on large-sample properties, corresponding to observation of independent realizations. Chapter 5 presents an analogously broad description of intensity-based inference for point processes on the line. Important special classes of point processes — Poisson processes, Cox processes, renewal processes and stationary point processes — are the subjects of Chapters 6–9, which are unified by our exploiting special structure as effectively as possible in order to derive particularly precise results. A slightly tangential topic is introduced in Chapter 10: inference for stochastic processes sampled according to a point process.

1.1   Random Measures and Point Processes

Let E be a locally compact Hausdorff space whose topology has a countable base (hereafter abbreviated LCCB), with Borel σ-algebra $\mathcal{E}$. As in Appendix A, let $\mathcal{B}$ denote the ring of bounded Borel sets, and recall that a measure μ on E is a Radon measure if $\mu(A) < \infty$ for every $A \in \mathcal{B}$. With $\mathbf{M}$ the set of Radon measures, elements of the subsets $\mathbf{M}_p = \{\mu \in \mathbf{M} : \mu(A) \in \mathbf{N} \text{ for all } A \in \mathcal{B}\}$, $\mathbf{M}_s = \{\mu \in \mathbf{M}_p : \mu(\{x\}) \leq 1 \text{ for all } x \in E\}$, $\mathbf{M}_a = \{\mu \in \mathbf{M} : \mu \text{ is purely atomic}\}$, and $\mathbf{M}_d = \{\mu \in \mathbf{M} : \mu \text{ is diffuse}\}$ are point measures, simple point measures, purely atomic measures, and diffuse measures. On $\mathbf{M}$ we define the σ-algebra $\mathcal{M}$ given by (A.2): $\mathcal{M}$ is generated by the coordinate mappings $\mu \to \mu(f) = \int f d\mu$, where f ranges over the set C_K of continuous functions on E whose support is compact. By $\mathcal{M}_p$, $\mathcal{M}_s$, $\mathcal{M}_a$, $\mathcal{M}_d$ we denote the trace σ-algebras on $\mathbf{M}_p$, $\mathbf{M}_s$, $\mathbf{M}_a$, $\mathbf{M}_d$. Convergence in $\mathbf{M}$ is vague convergence: $\mu_n \to \mu$ if and only if $\mu_n(f) \to \mu(f)$ for every $f \in C_K$. The resultant Borel σ-algebra is $\mathcal{M}$. For additional details and results, see Appendix A.

We come now to the principal definitions.

Definition 1.1. Let $(\Omega, \mathcal{F}, P)$ be a probability space.

a) A *random measure* on E is a measurable mapping M of $(\Omega, \mathcal{F})$ into $(\mathbf{M}, \mathcal{M})$;

b) A *point process* on E is a measurable mapping N of $(\Omega, \mathcal{F})$ into $(\mathbf{M}_p, \mathcal{M}_p)$.

For a random measure M and $A \in \mathcal{E}$, the random variable $M(A)$ is termed the measure or mass of A and the integral $M(f) = \int_E f(x)M(dx)$, a well-defined random variable for f either nonnegative and $\mathcal{E}$-measurable or in C_K, is called the integral of f with respect to M. A random measure is interpreted as a random distribution of mass or charge (but only of one sign) in E.

A point process N is a random distribution of indistinguishable points in E: $N(A)$ is the number of points in the set A. There exists a representation

$$N = \sum_{i=1}^{K} \varepsilon_{X_i}, \tag{1.1}$$

where ε_x is the point mass at x [$\varepsilon_x(A)$ is 1 or 0 according as $x \in A$ or not], where the X_i are measurable mappings of $(\Omega, \mathcal{F})$ into $(E, \mathcal{E})$, termed the points or atoms of N, and where K is a random variable with values in $\bar{\mathbf{N}} = \{0, 1, \ldots, \infty\}$. The X_i need not be distinct and the order of their indices is not unique. We employ the representation (1.1) throughout; it follows, for example, that $N(f) = \sum_{i=1}^{K} f(X_i)$.

When $E = \mathbf{R}_+$, the X_i are customarily replaced by random times T_i chosen so that $0 \leq T_1 \leq T_2 \leq \cdots$ and we adopt the customary terminology: the T_i are the times of events or *arrival times*, their differences $U_i = T_i - T_{i-1}$ are *interarrival times*, and $N_t = N([0, t])$ is the number of arrivals in the time interval $[0, t]$. The fundamental relation between the *counting process* (N_t) and arrival time sequence (T_n) is that for each n and t, $\{N_t \geq n\} = \{T_n \leq t\}$; thus each process contains information sufficient to reconstruct the other.

A point process N is *simple* if $P\{N \in \mathbf{M}_s\} = 1$. In the case the points X_i in (1.1) are distinct [almost surely (a.s.)] and on $\mathbf{R}_+$ we may take $0 \leq T_1 < T_2 < \cdots$, with the interpretation that there are no simultaneous arrivals. For simple point processes on $\mathbf{R}$, especially stationary point processes, the most convenient representation is $N = \sum_{i=-K_1}^{K_2} \varepsilon_{T_i}$, where K_1 and K_2 are random variables taking values in $\bar{\mathbf{N}}$ and $\cdots < T_0 \leq 0 < T_1 < T_2 < \cdots$.

Similarly, a random measure M is *purely atomic* if $P\{M \in \mathbf{M}_a\} = 1$, in which case (1.1) generalizes [see Theorem A.4] to become

$$M = \sum_{i=1}^{K} U_i \varepsilon_{X_i}, \tag{1.2}$$

with K and the X_i as in (1.1) and the U_i nonnegative random variables. For many purposes one can identify M with the (marked) point process

$$N = \sum_{i=1}^{K} \varepsilon_{(X_i, U_i)}$$

on $E \times \mathbf{R}_+$; we use this device to extend point process results and techniques to atomic random measures (see, e.g., Section 6.5).

Finally, a random measure M is *diffuse* if $P\{M \in \mathbf{M}_d\} = 1$.

Associated with a random measure M are σ-algebras representing various forms of observation. The most important correspond to complete knowledge of M over a set A:

$$\mathcal{F}^M(A) = \sigma(M(B) : B \in \mathcal{E}, B \subseteq A).$$

For brevity, we put $\mathcal{F}^M = \mathcal{F}^M(E)$, which represents complete observation of M. In Chapters 3 and 6 we also consider integral data corresponding to σ-algebras $\mathcal{F}^M(g_1, \ldots, g_m) = \sigma(M(g_1), \ldots, M(g_m))$, which depict knowledge of M limited to the integrals $M(g_1), \ldots, M(g_m)$. When $E = \mathbf{R}_+$ we write $\mathcal{F}_t^M$ for $\mathcal{F}^M([0, t])$. In this case particularly, but also to some extent in general, we refer to the $\mathcal{F}^M(\cdot)$ individually and *en masse* as the *internal history* of M. Chapters 2 and 5 examine larger filtrations $\mathcal{G}(\cdot)$ such that $\mathcal{F}^M(A) \subseteq \mathcal{G}(A)$ for each A, in which case M is said to be *adapted* to the $\mathcal{G}(A)$.

We now introduce several important classes of point processes. Of these, Poisson processes are truly fundamental, while Cox processes, because of their role in state estimation and their being on the one hand rather general, yet on the other hand closely linked to Poisson processes, play nearly an equal role.

Definition 1.2. Let μ be an element of **M**. A point process N on E is a *Poisson process* with mean measure μ if

a) Whenever the sets $A_1, \ldots, A_k \in \mathcal{E}$ are disjoint, the random variables $N(A_1), \ldots, N(A_k)$ are independent;

b) For each $A \in \mathcal{E}$ and $k \geq 0$,

$$P\{N(A) = k\} = e^{-\mu(A)}\mu(A)^k/k!. \tag{1.3}$$

[By convention $N(A) = \infty$ a.s. if $\mu(A) = \infty$.]

Thus, N has *independent increments* and for each A the random number of points in A has a Poisson distribution with mean $\mu(A)$. (The term "mean measure" is introduced formally in Section 2.) Each of these properties nearly implies the other (see Section 5 and the chapter notes for details). The Poisson processes earliest to be studied were the ordinary (or homogeneous) Poisson processes on $\mathbf{R}_+$, for which $\mu(A) = \lambda|A|$, where $|A|$ is the Lebesgue measure of A and $\lambda > 0$ is known as the *rate* or *intensity* of N, terminology justified in Sections 8 and 3.5. Other Poisson processes on $\mathbf{R}_+$ are sometimes called nonhomogeneous Poisson processes. Inference for Poisson processes is the topic of Chapter 6.

A Cox process is a Poisson process with the mean measure made random.

Definition 1.3. Let M be a random measure on E and let N be a point process on E, defined over the same probability space. Then N is a *Cox process directed by M* if conditional on M, N is a Poisson process with mean measure M:

a) When $A_1, \ldots, A_k$ are disjoint, the random variables $N(A_1), \ldots, N(A_k)$ are conditionally independent given M;

b) For each A and each $k \geq 0$,

$$P\{N(A) = k|M\} = e^{-M(A)}M(A)^k/k!.$$

We refer to (N, M) as a *Cox pair* and to M as the *directing measure* of N. The terms "doubly stochastic Poisson process" and "Poisson process in a random environment" are synonymous with Cox process.

The most elementary Cox processes are the Poisson processes, for which $M \equiv \mu$ is deterministic. Next are the *mixed Poisson processes*, with directing measures $M = Y\nu$, where Y is a nonnegative random variable and ν is a fixed element of $\mathbf{M}$; thus a mixed ordinary Poisson process on $\mathbf{R}_+$ has rate $\lambda(\omega)$ that is random but not a function of time. Other specific classes of Cox processes are discussed in later chapters. The mapping of probability laws of random measures into laws of Cox processes directed by them is a bijection; this property is useful in establishing theoretical results as well as in applications (see, e.g., Exercise 1.7). The principal characterization of Cox processes, as limits of thinned point processes, appears in Section 5, along with the associated invariance property. Cox processes also present the challenging state estimation problem of reconstructing the directing measure from observation of the Cox process, which we solve in Chapter 7.

Empirical processes are not treated in complete detail in this book, but several results do appear in Chapter 4, and they are in some ways the most basic point processes, especially since Poisson processes can be constructed as mixed empirical processes (Exercise 1.1), a construction central to some of our inference procedures.

Definition 1.4. Given i.i.d. random elements $X_1, \ldots, X_n$ of E, the point process $N = \sum_{i=1}^{n} \varepsilon_{X_i}$ is an (n-sample) *empirical process* on E.

Thus N represents the positions of n points in E chosen independently and with the same distribution.

In a mixed empirical process the number of points also becomes random, but is independent of their locations.

Definition 1.5. Let $X_1, X_2, \ldots$ be i.i.d. random elements of E and let K be a nonnegative, integer-valued random variable independent of the X_i. Then the point process $N = \sum_{i=1}^{K} \varepsilon_{X_i}$ is a *mixed empirical process*.

On the real line, point processes can be defined via the interarrival times — the "interval" approach, as opposed to the "counts" approach based on the counting process (N_t); the simplest structural assumption is that the interarrival times be i.i.d.

Definition 1.6. A point process $\sum_{n=1}^{\infty} \varepsilon_{T_n}$ on $\mathbf{R}_+$ is a *renewal process* with interarrival distribution F (a probability on $\mathbf{R}_+$) if $T_0 = 0$ and if the interarrival times U_i are i.i.d. with distribution F.

Renewal processes are significant because many stochastic processes contain embedded renewal processes, whose points typically are times at which the underlying process (called a regenerative process) renews itself probabilistically. Important properties of renewal processes are of two types: those associated with the renewal equation and asymptotic properties following from the renewal theorem. Some of each are discussed in Chapter 8.

The remaining class of point processes accorded chapter-length treatment (in Chapter 9) is stationary point processes. Some machinery is needed to formulate the most appropriate definition, which is postponed to Section 8, but for completeness here is a provisional version.

Definition 1.7. A point process $N = \sum \varepsilon_{X_i}$ on $\mathbf{R}^d$ is *stationary* if for each x the translated point process $N_{(x)} = \sum \varepsilon_{X_i - x}$ has the same distribution as N.

Additional classes — marked point processes, cluster processes, infinitely divisible point processes, and symmetrically distributed point processes — are introduced elsewhere in the chapter.

1.2   Distributional Descriptors and Uniqueness

The distribution of a point process (more generally, of a random measure) can be specified in alternative forms, some of which — particularly the Laplace functional — are relatively more amenable to explicit calculation, computational manipulation, and derivation of characterization and convergence properties. Incomplete but useful descriptions in the form of first-, second- and higher-order moment measures are also available. Definitions and results in this section are formulated for point processes, but we indicate as well how they generalize to random measures. We begin with objects that determine uniquely (possibly with mild additional assumptions) the distribution of a point process.

Definition 1.8. Given a point process N defined over $(\Omega, \mathcal{F}, P)$, the *distribution* of N is the probability measure $\mathcal{L}_N = PN^{-1}$ on $(\mathbf{M}_p, \mathcal{M}_p)$.

Definition 1.9. Point processes N_1 and N_2 (not necessarily defined over the same probability space) are *identically distributed* if $\mathcal{L}_{N_1} = \mathcal{L}_{N_2}$.

Equality in distribution is written as $N_1 \overset{d}{=} N_2$.

Definition 1.10. The *Laplace functional* of a point process N is the mapping L_N defined by

$$L_N(f) = E[e^{-N(f)}] = E[e^{-\sum_i f(X_i)}],$$

where f is nonnegative and $\mathcal{E}$-measurable (which we write as $f \in \mathcal{E}^+$).

Definition 1.11. The *zero-probability functional* of N is the mapping $z_N : \mathcal{E} \to [0,1]$ defined by $z_N(A) = P\{N(A) = 0\}$.

All four definitions extend verbatim to random measures, except that in Definition 1.8, $(\mathbf{M}_p, \mathcal{M}_p)$ is replaced by $(\mathbf{M}, \mathcal{M})$; however, the zero-probability functional is useful only for simple point processes. The Laplace functional is evidently a generalized Laplace transform and possesses the usual properties: uniqueness, tractability in the sense of its being computable in key special cases and for transformations of point processes, and determination of convergence in distribution (Section 3). It is a fundamental tool throughout this book.

The main uniqueness theorem can now be given.

Theorem 1.12. For point processes N_1 and N_2 the following assertions are equivalent:

a) $N_1 \overset{d}{=} N_2$;

b) $(N_1(A_1), \ldots, N_1(A_k)) \overset{d}{=} (N_2(A_1), \ldots, N_2(A_k))$ for all k and all A_1, $\ldots, A_k \in \mathcal{B}$;

c) $N_1(f) \overset{d}{=} N_2(f)$ for all $f \in \mathcal{E}^+$;

d) $L_{N_1}(f) = L_{N_2}(f)$ for all $f \in \mathcal{E}^+$.

If in addition N_1 and N_2 are simple, then each of a) – d) is equivalent to

e) $z_{N_1}(A) = z_{N_2}(A)$ for all $A \in \mathcal{E}$.

Proof: (Sketch). Equivalence of a) and b) is standard. If b) holds and $f = \sum_{i=1}^{k} a_i 1_{A_i}$ is a simple, nonnegative function, then

$$L_{N_1}(f) = E\left[e^{-\sum_{i=1}^{k} a_i N_1(A_i)}\right] = E\left[e^{-\sum_{i=1}^{k} a_i N_2(A_i)}\right] = L_{N_2}(f); \qquad (1.4)$$

extension to general f is by approximation. Conversely, if c) holds, then in particular (1.4) does, from which b) follows the uniqueness theorem for

Laplace transforms. Equivalence of c) and d) is apparent. That a) entails e) is evident; the reverse implication comes from the property that the events $\{\mu : \mu(A) = 0\}$ constitute a π-system on $\mathbf{M}_s$ generating $\mathcal{M}_s$, so that in the simple case a) follows from e) by the monotone class theorem. ∎

Equivalence of a)–d) remains valid for random measures. Although e) has no direct counterpart, certain results derived by means of Cox processes exist for diffuse random measures; for details of these and several formally weaker versions of Theorem 1.12, see the exercises and chapter notes.

Moments of a point process are described in the following manner.

Definition 1.13. The *mean measure* or *first moment measure* of a point process N is the measure μ_N on E given by $\mu_N(A) = E[N(A)]$; N is *integrable* if $\mu_N \in \mathbf{M}$. More generally, for each k let N^k be the point process given by

$$N^k(dx_1, \ldots, dx_k) = N(dx_1) \cdots N(dx_k);$$

then the *moment measure of order* k is the measure $\mu_N^k = E[N^k]$.

Definition 1.14. The *covariance measure* of N is the signed measure

$$\rho_N(A \times B) = \mathrm{Cov}(N(A), N(B)) = \mu_N^2(A \times B) - \mu_N(A)\mu_N(B) \qquad (1.5)$$

on $E \times E$. [To avoid complications we use the covariance measure only when the second moment measure belongs to $\mathbf{M}(E^2)$.]

Moment and covariance measures for a random measure are defined exactly as in Definitions 1.13 and 1.14.

For many purposes (e.g., definition of Palm distributions), ordinary moment measures of point processes are unsatisfactory because the products N^k contain points on various diagonals. Without contributing information not already present, such points engender spurious singularity of moment measures. Their removal is effected by replacing N^k by the point process

$$N^{(k)}(dx_1, \ldots, dx_k)$$
$$= N(dx_1)(N - \varepsilon_{x_1})(dx_2) \cdots \left(N - \sum_{i=1}^{k-1} \varepsilon_{x_i}\right)(dx_k), \qquad (1.6)$$

which, provided that N is simple, consists only of k-tuples of points whose coordinates are distinct. In terms of the representation (1.1) we have

$$N^k = \sum_{i_1} \cdots \sum_{i_k} \varepsilon_{(X_{i_1}, \ldots, X_{i_k})},$$

while

$$N^{(k)} = \sum_{i_1} \cdots \sum_{i_k} \varepsilon_{(X_{i_1}, \ldots, X_{i_k})}.$$
$$\scriptstyle i_1 \neq \cdots \neq i_k$$

We refer to the measures $\mu_N^{(k)} = E[N^{(k)}]$ as *factorial moment measures* of N.

Moment measures can be computed from the Laplace functional:

$$\mu_N(f) = -\frac{d}{d\alpha} L_N(\alpha f)\Big|_{\alpha=0},$$

while from

$$E[N(f)^2] = -\frac{d^2}{d\alpha^2} L_N(\alpha f)\Big|_{\alpha=0}$$

we obtain

$$E[N(f)N(g)] = (1/2)\{E[N(f+g)^2] - E[N(f)^2] - E[N(g)^2]\},$$

from which the covariance measure can be calculated. Since in addition $z_N(A) = \lim_{t\to\infty} L_N(t1_A)$, the Laplace functional truly is the fundamentally useful descriptor of $\mathcal{L}_N$.

Here are the key examples.

Example 1.15. (Poisson process). Let N be a Poisson process with mean measure μ. Then

$$L_N(f) = \exp[-\textstyle\int_E(1 - e^{-f})d\mu]; \tag{1.7}$$

consequently (or directly from (1.3))

$$z_N(A) = e^{-\mu(A)}, \tag{1.8}$$
$$\mu_N(A) = \mu(A), \tag{1.9}$$
$$\rho_N(A \times B) = \mu(A \cap B). \tag{1.10}$$

[That $N(A), N(B)$ can only be positively correlated follows from the independent increments property of N.] Furthermore, N is simple if and only if μ is diffuse. The difference between ordinary and factorial moment measures can be inferred from the formulas

$$\mu_N^2(f) = E\left[\textstyle\sum_{i,j}f(X_i, X_j)\right] = \int f(x,x)\mu(dx) + \int\int f(x,y)\mu(dx)\mu(dy),$$

and

$$\mu_N^{(2)}(f) = E\left[\textstyle\sum_{i\neq j}f(X_i, X_j)\right] = \int\int_{\{x\neq y\}} f(x,y)\mu(dx)\mu(dy);$$

the latter measure is absolutely continuous with respect to $\mu \times \mu$ but the former is not. □

Example 1.16. (Cox process). Let N be a Cox process directed by the random measure M. Then

$$L_N(f) = L_M(1 - e^{-f}), \tag{1.11}$$

$$z_N(A) = L_M(1_A), \tag{1.12}$$

$$\mu_N = \mu_M, \tag{1.13}$$

$$\rho_N(A \times B) = \rho_M(A \times B) + \mu_M(A \cap B). \tag{1.14}$$

The Cox process is simple if and only if the directing measure is diffuse.

To derive (1.11), for example, one conditions on M and applies (1.7):

$$L_N(f) = E[E[e^{-M(f)}|M]] = E\left[e^{-\int(1 - e^{-f})dM}\right] = L_M(1 - e^{-f}).$$

Application of (1.8)–(1.10) to deduce (1.12)–(1.14) is analogous. ▯

A useful consequence reduces some random measure results, particularly for diffuse random measures, to properties of point processes (see Exercise 1.7 for an application).

Proposition 1.17. Let N be a Cox process directed by M. Then $\mathcal{L}_N$ and $\mathcal{L}_M$ determine each other uniquely. ▯

We conclude the section by discussing independence: point processes N_1 and N_2 defined over the same probability space are *independent* if the σ-algebras $\mathcal{F}^{N_1}$ and $\mathcal{F}^{N_2}$ are independent in the usual sense. Theorem 1.12 furnishes sufficient conditions.

Proposition 1.18. For point processes N_1 and N_2 the following are equivalent:

    a) N_1 and N_2 are independent;

    b) $E[e^{-N_1(f)-N_2(g)}] = L_{N_1}(f)L_{N_2}(g)$ for all f and g.

If N_1 and N_2 are simple, these are equivalent to

    c) $P\{N_1(A) = 0, N_2(B) = 0\} = z_{N_1}(A)z_{N_2}(B)$ for all A and B. ▯

Neither of the conditions $E[e^{-N_1(f)-N_2(f)}] = L_{N_1}(f)L_{N_2}(f)$ for all f nor $P\{N_1(A) = 0, N_2(A) = 0\} = z_{N_1}(A)z_{N_2}(A)$ for all A suffices for independence; they fail even for Poisson processes on a singleton set.

The preceding discussion extends to more than two point processes as well as to general random measures.

1.3   Convergence in Distribution

Several important classes of point processes, notably Cox processes and infinitely divisible point processes, admit characterizations as the set of all possible limits in distribution of sequences of point processes satisfying some sort of rarefaction condition; see Sections 4 and 5 for details. Here we formulate convergence in distribution for point processes and derive criteria for it. Recall that convergence in $\mathbf{M}_p$ is vague convergence (see Appendix A): $\mu_n \to \mu$ if and only if $\mu_n(f) \to \mu(f)$ for all $f \in C_K$. The vague topology is metrizable in such a way that $\mathbf{M}_p$ becomes a complete, separable metric space, and since $\mathcal{M}_p$ is the Borel σ-algebra engendered by the vague topology one can then speak of weak convergence of probability measures on $(\mathbf{M}_p, \mathcal{M}_p)$. This leads to the following definition.

Definition 1.19. Let $N, N_1, N_2, \ldots$ be point processes on E. Then (N_n) *converges in distribution* to N, written $N_n \overset{d}{\to} N$, provided that $\mathcal{L}_{N_n} \to \mathcal{L}_N$ weakly as probability measures on $(\mathbf{M}_p, \mathcal{M}_p)$, that is, $E[H(N_n)] \to E[H(N)]$ for every bounded, continuous function $H : \mathbf{M}_p \to \mathbf{R}$.

Of course, these point processes need not be defined over the same probability space, but for notational simplicity we treat them as if they were.

In proving convergence in distribution the crucial and difficult step is usually to establish relative compactness, which by Prohorov's theorem (Billingsley, 1968, Theorems 6.1 and 6.2) is equivalent to tightness. Recall that a sequence (P_n) of probabilities on a (complete, separable) metric space is *tight* if for every $\varepsilon > 0$ there is a compact set K such that $P_n(K) > 1 - \varepsilon$ for each n. According to Theorem A.10, a subset Γ of $\mathbf{M}_p$ is relatively compact if and only if

$$\sup_{\mu \in \Gamma} |\mu(f)| < \infty \tag{1.15}$$

for every $f \in C_K$ or, equivalently, if and only if

$$\sup_{\mu \in \Gamma} \mu(A) < \infty \tag{1.16}$$

for each $A \in \mathcal{B}$.

Using these characterizations we derive the fundamental property that a sequence of point processes is tight if and only if all of its one-dimensional distributions are tight.

Lemma 1.20. For a sequence (N_n) of point processes the following are equivalent:

a) (N_n) is tight [i.e, $(\mathcal{L}_{N_n})$ is tight];
b) $(N_n(f))$ is tight for every $f \in C_K$;
c) $(N_n(A))$ is tight for every $A \in \mathcal{B}$.

Proof: (Sketch). Equivalence of b) and c) is clear, and they are deduced from a) via (1.15) and (1.16). To show that b) implies a), let f_k be nonnegative functions in C_K with $f_k \uparrow 1$; then given $\varepsilon > 0$ for each k there is c_k

such that $P\{N_n(f_k) > c_k\} < \varepsilon/2^k$ for all n. The set $\Gamma = \bigcap_k\{\mu : \mu(f_k) \leq c_k\}$ is relatively compact in $\mathbf{M}_p$ and (with $\bar{\Gamma}$ the closure of Γ) $P\{N_n \notin \bar{\Gamma}\} \leq \sum_k P\{N_n(f_k) > c_k\} \leq \varepsilon$ for all n, and consequently a) holds. ∎

There ensues the major theorem on convergence in distribution for point processes.

Theorem 1.21. Let $(N_n), N$ be point processes on E. Then the following assertions are equivalent:

a) $N_n \overset{d}{\to} N$;

b) $(N_n(A_1), \ldots, N_n(A_k)) \overset{d}{\to} (N(A_1), \ldots, N(A_k))$ for all k and all $A_1, \ldots, A_k \in \mathcal{B}$ such that $P\{N(\partial A_i) = 0\} = 1$ for each i;

c) $N_n(f) \overset{d}{\to} N(f)$ for all $f \in C_K$;

d) $L_{N_n}(f) \to L_N(f)$ for all nonnegative $f \in C_K$.

Proof: (Sketch). Equivalence of c) and d) is essentially immediate, while the implications a) $\Rightarrow$ b) and a) $\Rightarrow$ c) are consequences of the continuous mapping theorem (Billingsley, 1968). If c) is fulfilled, then (N_n) is tight by Lemma 1.20, so that every subsequence $(N_{n'})$ admits a further subsequence $(N_{n''})$ converging in distribution to some point process $\tilde{N}$. By the implication a) $\Rightarrow$ c) and Theorem 1.12, we have that $\tilde{N} \overset{d}{=} N$; hence $N_{n''} \overset{d}{\to} N$. Since convergence in distribution is metrizable, $N_n \overset{d}{\to} N$. That b) entails a) is shown in a similar manner. ∎

For simple point processes, the zero-probability functional yields another criterion for convergence in distribution. However, convergence of zero-probability functionals does not alone imply tightness, and additional assumptions are necessary.

Proposition 1.22. Let $(N_n), N$ be point processes with N simple, and assume that (N_n) is tight. Then $N_n \overset{d}{\to} N$ if and only if $z_{N_n}(A) \to z_N(A)$ for all $A \in \mathcal{B}$ with $P\{N(\partial A) = 0\} = 1$. ☐

The proof is a straightforward modification of that of Theorem 1.21. No improvement results from assuming that the N_n are also simple. Tightness of (N_n) can be established using Lemma 1.20 or other criteria stated next.

Proposition 1.23. A sequence (N_n) of point processes is tight if either of the following conditions is fulfilled:

a) For each $A \in \mathcal{B}$ and each m there is a finite partition $A = \sum_{j=1}^{k_m} A_{mj}$ such that

$$\lim_{m\to\infty} \limsup_{n\to\infty} \sum_j P\{N_n(A_{mj}) \geq 2\} = 0;$$

b) For each $A \in \mathcal{B}$ with $P\{N(\partial A) = 0\} = 1$,

$$\limsup_{n \to \infty} E[N_n(A)] \leq E[N(A)] < \infty. \square$$

The definition of convergence in distribution and Theorem 1.21 extend to random measures with no substantive modification. Using Proposition 1.17 one can reduce random measure versions of Theorem 1.21 to point process versions, and can obtain specialized results applicable when the limit is diffuse. For example (compare Exercise 1.7), if (M_n) is a tight sequence of random measures and M is diffuse then $M_n \xrightarrow{d} M$ if and only if there is $t \in (0, \infty)$ such that $E[e^{-tM_n(A)}] \to E[e^{-tM(A)}]$ for all $A \in \mathcal{B}$. Sufficiency of this condition (necessity is apparent) is demonstrated by constructing Cox processes N_n, N directed by tM_n, tM, respectively; then together with (1.12), the condition implies that $z_{N_n}(A) \to z_N(A)$ for all $A \in \mathcal{B}$, and since tightness of (M_n) is equivalent to that of (N_n), Proposition 1.22 yields $N_n \xrightarrow{d} N$. Finally, (1.11) and the random measure version of Theorem 1.21 provide the desired conclusion.

Similar reasoning shows that the mapping associating the law of a Cox process and that of the directing random measure in not only a bijection (Proposition 1.17) but also continuous in both directions.

For point processes on $\mathbf{R}$ or $\mathbf{R}_+$ specialized results hold. Specifically, let $N_n = \sum_i \varepsilon_{T_{ni}}$ and $N = \sum \varepsilon_{T_i}$ be point processes on $\mathbf{R}_+$ with $0 \leq T_{n1} \leq T_{n2} \leq \cdots$ for each n (and similarly for the T_i), and let (U_{ni}), (U_i) be the associated interarrival time sequences.

Theorem 1.21 (bis). The following are equivalent for point processes on $\mathbf{R}_+$:

a) $N_n \xrightarrow{d} N$;

b) $(T_{n1}, \ldots, T_{nk}) \xrightarrow{d} (T_1, \ldots, T_k)$ for each k;

b) $(U_{n1}, \ldots, U_{nk}) \xrightarrow{d} (U_1, \ldots, U_k)$ for each k. $\square$

An analogous result holds for point processes on $\mathbf{R}$.

We conclude with two examples; others are given in the exercises.

Example 1.24. (Poisson process). Using the equivalence of a) and b) in Theorem 1.21, one sees via (1.7) that for Poisson processes N_n and N, with mean measures μ_n and μ, $N_n \xrightarrow{d} N$ if and only if $\mu \to \mu$ (vaguely). $\square$

Example 1.25. (Renewal process). The law of a renewal process depends continuously on the interarrival distribution: $N_n \xrightarrow{d} N$ if and only if $F_n \to F$ weakly as probability measures on $\mathbf{R}_+$. Moreover, if (N_n) is a sequence of renewal processes and if $N_n \overset{d}{\to} N$, then (since independence can be created but not destroyed in the limit) N is also a renewal process. $\square$

1.4   Marked Point Processes; Cluster Processes

Marked point processes are a fundamental tool in the study of thinned and compound point processes, Palm distributions, stationary point processes and cluster processes, as well as in applications, for example queueing and reliability theory. We introduce them in this section, along with several important constructions.

Definition 1.26. Let $N = \sum \varepsilon_{X_i}$ be a point process on E and let E' be a second LCCB space. A *marked point process* with underlying process N is any point process

$$\bar{N} = \sum_i \varepsilon_{(X_i, Z_i)} \tag{1.17}$$

on $E \times E'$. The random element Z_i of E' is called the *mark* associated to X_i; E' is the *mark space*.

In one sense a marked point process in nothing more than a point process on the product space $E \times E'$, so that there is little theory of marked point processes *per se*, but interpretation and usage show that the concept does have a life of its own. In most cases, the X_i are primary characteristics of the objects under study and the marks Z_i are secondary attributes. Often the distinction is the stochastic structure of the marks is simpler than that of the X_i or dependent on the X_i, as for processes with position-dependent marks (Example 1.28). For example, the X_i may be the locations of unit charge ions, each marked by $Z_i = 1$ or $Z_i = -1$ according as its charge is positive or negative, or the arrival times T_i of customers at a queueing system may be marked by the associated service or waiting times. Classical compound Poisson processes on $\mathbf{R}_+$ constitute another example.

More generally, a purely atomic random measure $M = \sum U_i \varepsilon_{X_i}$ on E may be viewed instead as the marked point process $\sum \varepsilon_{(X_i, U_i)}$ on $E \times \mathbf{R}_+$. Hence point process methods and results are applicable more widely that is apparent at first; see, for example, the discussion of infinitely divisible random measures with independent increments in this section, as well as the exercises.

In the most elementary model, the marks are independent of the point process N.

Example 1.27. (Independent marking). Let $N = \sum \varepsilon_{X_i}$ be a point process on E, let Z_i be i.i.d. random elements of E', and suppose that the

Z_i are independent of N as well. Then the marked point process $\bar{N}$ of (1.17) is said to be obtained from N by *independent marking*. For f a function on $E \times E'$,

$$L_{\bar{N}}(f) = E\left[E[e^{-\sum f(X_i, Z_i)}|N]\right] = L_N(H),$$

where $H(x) = -\log[\int_{E'} e^{-f(x,z)}\rho(dz)]$, with ρ the distribution of the X_i. The associated purely atomic random measure $M = \sum U_i \varepsilon_{X_i}$ is studied in Section 5 under the slight misnomer "compound point process." $\square$

More general, albeit equally tractable, is the case that the distribution of the mark Z_i can depend (only) on X_i and the marks are conditionally independent given N.

Example 1.28. (Position-dependent marking). Given a transition probability K from E to E', suppose that the Z_i are conditionally independent given N with $P\{Z_i \in B|N\} = K(X_i, B)$. Then N is said to be obtained from N by *position-dependent marking*, and we have $L_{\bar{N}}(f) = L_N(H)$ for $H(x) = -\log[\int_{E'} e^{-f(x,z)}K(x, dz)]$. $\square$

For an example from hydrology, where the X_i are times of precipitation events and the Z_i are amounts, see Exercise 1.12.

On the other hand, the marks can be complicated. In study of Palm distributions for stationary point processes on $\mathbf{R}^d$ (Section 8) one deals with marked point processes in which X_i is marked by the translate of N by X_i:

$$\bar{N} = \sum_i \varepsilon_{(X_i, \sum_j \varepsilon_{x_j - x_i})}$$

on $\mathbf{R}^d \times \mathbf{M}_p$. For regenerative processes (Chapter 8) the point process of regeneration times is marked by the paths followed by the process between them.

Finally, we view clustered point processes as marked point processes, but first we need a concept of addition for point processes. Given point processes $N_1, \ldots, N_k$ defined on the same space E and over the same probability space, their *superposition* is the point process N satisfying

$$N(A) = \sum_{i=1}^k N_i(A) \tag{1.18}$$

for each set A. Superposition is of particular interest when the components N_i are independent, for then the Laplace functionals multiply:

$$L_N(f) = \prod_{i=1}^k L_{N_i}(f), \tag{1.19}$$

which not only is very useful computationally but also furthers the analogy between Laplace functionals and Laplace transforms. More generally, if

$N_1, N_2, \ldots$ are point processes, then we define their superposition N in the same way as in (1.18), i.e.,

$$N(A) = \sum_{i=1}^{\infty} N_i(A),$$

but only provided that N actually defines point process, in other words that $N(B) < \infty$ for each bounded set B. In this case (1.19) extends as well.

We proceed to clustered point processes.

Definition 1.29. Let $N = \sum \varepsilon_{X_i}$ be a point process on E and let $E' = \mathbf{M}_p(\tilde{E})$, where $\tilde{E}$ is also LCCB. Let $\bar{N} = \sum \varepsilon_{(X_i, \tilde{N}_i)}$ be a marked point process with underlying process N and mark space E'. Then provided it takes values in $\mathbf{M}_p$, the point process $\tilde{N} = \sum_i \tilde{N}_i$ on $\tilde{E}$ is a *clustered point process*, or simply *cluster process*.

When $\tilde{E} = E$ the interpretation is particularly precise, but it makes sense in other cases as well. One should think of each X_i as a primary point, or *cluster center*, to which is associated the *cluster* of secondary points (*cluster members*) in $\tilde{E}$ comprised by the point process $\tilde{N}_i$. The clustered point process $\tilde{N}$ consists of all the cluster members. However, although term "cluster" is picturesque and suggestive of intuitive notions of clustering, it should not be taken too literally: even when $\tilde{E} = E$ the points of $\tilde{N}_i$ need not be near either each other or X_i.

Among clustered point processes the most important are the Poisson cluster processes.

Definition 1.30. Let $\tilde{N}$ be a clustered point process on $\tilde{E}$ with underlying marked point process $\bar{N} = \sum \varepsilon_{(X_i, \tilde{N}_i)}$. Then $\tilde{N}$ is a *Poisson cluster process* if

a) $N = \sum \varepsilon_{X_i}$ is a Poisson process on E;

b) $\bar{N}$ is obtained from N by position-dependent marking in the manner of Example 1.28.

That is, there is a transition kernel K from E to $E' = \mathbf{M}_p(\tilde{E})$ such that the $\tilde{N}_i$ are conditionally independent given N, with $P\{\tilde{N}_i \in \Gamma | N\} = K(X_i, \Gamma)$, $\Gamma \in \mathcal{M}_p(\tilde{E})$. The Laplace functional $L_{\tilde{N}}$ is given by $L_{\tilde{N}}(f) = L_{\bar{N}}(H)$, where $H(x, \nu) = \nu(f)$, and consequently

$$L_{\tilde{N}}(f) = \exp\left[-\int \mu(dx) \int \{1 - e^{-\nu(f)}\} K(x, d\nu)\right], \qquad (1.20)$$

where μ is the mean measure of N. This expression will be considered further in Section 5.

Every Poisson process is a Poisson cluster process: with $\tilde{E} = E$ and $K(x, \cdot) = \varepsilon_{\varepsilon_x}$, we have $\tilde{N}_i = \varepsilon_{X_i}$, and hence $\tilde{N} = N$. For a Poisson cluster process $\tilde{N}$, both the marked point process $\bar{N}$ and the point process $\sum \varepsilon_{\tilde{N}_i}$ are Poisson (Exercise 1.11) and therefore Poisson processes are the key to study of Poisson cluster processes. The question then arises: How broad is the class of Poisson cluster processes? Theorem 1.32 contains the answer that a point process is a Poisson cluster process if and only if it is infinitely divisible, a property we now define.

Definition 1.31. A point process N is *infinitely divisible* if for every n there exist i.i.d. point processes $N_1, \ldots, N_n$ such that $N \stackrel{d}{=} \sum_{i=1}^{n} N_i$.

If N is Poisson with mean measure μ, then taking $N_1, \ldots, N_n$ to be Poisson with mean μ/n gives

$$L_N(f) = \left(\exp[-(1/n) \int (1 - e^{-f}) d\mu] \right)^n = \prod_{i=1}^{n} L_{N_i}(f);$$

consequently, N is infinitely divisible.

The infinitely divisible point processes carry a rich, elegant theory extending the classical theory of infinitely divisible probability measures on $\mathbf{R}_+$ (or that of increasing Lévy processes). In Section 5 we develop results analogous to those in the classical case; however, the following characterization is particular to infinitely divisible point processes. For compatibility with Definition 1.30 we use the same notation.

Theorem 1.32. Assume that $\tilde{E}$ is unbounded, and let $\tilde{N}$ be a point process on $\tilde{E}$ with $E[\tilde{N}(B)] < \infty$ for each bounded set B. Then the following statements are equivalent:

a) $\tilde{N}$ is infinitely divisible;

b) $\tilde{N}$ is a Poisson cluster process (one can take $E = \tilde{E}$ in Definition 1.30).

Proof: (Partial). We defer the proof that a) implies b) to Section 5.

For the converse, as observed above, the marked point process $\bar{N} = \sum \varepsilon_{(X_i, \tilde{N}_i)}$ is Poisson and hence for each n can be represented as

$$\bar{N} \stackrel{d}{=} \sum_{j=1}^{n} \sum_i \varepsilon_{(X_{ji}, \tilde{N}_{ji})},$$

where the point processes $\bar{N}_j = \sum_i \varepsilon_{(X_{ji}, \tilde{N}_{ji})}$, $j = 1, \ldots, n$, are i.i.d. Poisson processes. Then,

$$\tilde{N} \stackrel{d}{=} \sum_{j=1}^{n} \sum_i \tilde{N}_{ji},$$

which shows not only that $\tilde{N}$ is infinitely divisible but also that the components may be taken to be Poisson cluster processes. (The other implication shows that they *must* be Poisson cluster processes.) ∎

Additional discussion of infinitely divisible point processes appears in Section 5.

Infinite divisibility for random measures is defined in precise analogy with Definition 1.31, and versions of the results here and elsewhere in the chapter are valid. One must take care, however, to maintain the distinction between point process infinite divisibility and random measure infinite divisibility, since there are point processes (e.g., nonzero deterministic point processes) that are infinitely divisible as random measures but not as point processes. We use the term "infinitely divisible point process" only in the sense of Definition 1.31.

One class of infinitely divisible random measures, those with independent increments, is particularly amenable to analysis using point process methods.

Definition 1.33. A random measure M has *independent increments* if $M(A_1), \ldots, M(A_k)$ are independent whenever $A_1, \ldots, A_k$ are disjoint.

Concerning random measures with independent increments, the following is the principal characterization.

Theorem 1.34. A random measure M with independent increments admits can be decomposed as

$$M = \mu_0 + \sum_{j=1}^{k} W_j \varepsilon_{x_j} + \sum_i U_i \varepsilon_{X_i}, \qquad (1.21)$$

where

i) μ_0, the deterministic part of M, is a fixed element of **M**;

ii) $0 \le k \le \infty$, (x_j) is a sequence in E without accumulation points, and the W_j are independent random variables with $P\{W_j > 0\} > 0$ for each j (the x_j are *fixed atoms* of M);

iii) The marked point process $\bar{N} = \sum \varepsilon_{(X_i, U_i)}$ is a Poisson process on $E \times (0, \infty)$ and is independent of (W_j).

Thus, the random measure $\sum U_i \varepsilon_{X_i}$ is infinitely divisible with independent increments.

Proof: (Sketch). Let M' be the component of M remaining after all fixed atoms are removed. Given $A \in \mathcal{B}$, let (A_{nj}) be a null array of partitions of A (i.e., $A = \sum_j A_{nj}$ for each n, the partitions are successive refinements, and $\max_j \operatorname{diam} A_{nj} \to 0$). Since M' has no fixed atoms, the triangular array $M'(A_{nj})$ is a null array of random variables [independent in j for each n and satisfying $\max_j P\{M'(A_{nj}) > \varepsilon\} \to 0$ for every $\varepsilon > 0$]. Moreover, $M'(A) = \sum_j M'(A_{nj})$ for each n; hence by the classical theory of infinitely divisible distributions on $\mathbf{R}_+$ (see, e.g., Chung, 1974) $M'(A)$ is infinitely divisible. The independent increments property of M' implies, then, that all finite-dimensional distributions — of random vectors $(M'(A_1), \ldots, M'(A_k))$ — are infinitely divisible. Consequently, M' is itself infinitely divisible. We now remove from M' any deterministic component, after which we need only show that the residual random measure M'' has the purely atomic form indicated by (1.29). By the random measure version of Theorem 1.32, M'' can be represented as $M'' = \sum M_i$, where $\tilde{M} = \sum \varepsilon_{M_i}$ is a Poisson process on $\mathbf{M}$, whose mean measure we denote by η. It follows that

$$L_{M''}(f) = \exp[-\int \{1 - e^{-\mu(f)}\} \eta(d\mu)], \qquad (1.22)$$

a version of the important expression (1.31). The key consequence of (1.22) is that the independent increments property of M'' means that η is concentrated on the set of measures $\{u\varepsilon_x : x \in E, u > 0\}$ (the "degenerate" measures). Indeed, if $A \cap B = \emptyset$, then from $L_{M''}(1_{A \cup B}) = L_{M''}(1_B)L_{M''}(1_A)$, we infer that

$$\int \left\{ (1 - e^{-\mu(A \cup B)}) - [(1 - e^{-\mu(A)}) + (1 - e^{-\mu(B)})] \right\} \eta(d\mu) = 0,$$

which yields $\max\{\mu(A), \mu(B)\} = 0$ almost everywhere with respect to η. By induction, for disjoint $A_1, \ldots, A_\ell$, almost everywhere with respect to η at most one of $\mu(A_1), \ldots, \mu(A_\ell)$ is nonzero. Thus η is concentrated on the degenerate measures. Using the obvious identification $u\varepsilon_x \leftrightarrow (x, u)$, (1.22) becomes

$$L_{M''}(f) = \exp[-\int \{1 - e^{-uf(x)}\} \gamma(dx, du)] \qquad (1.23)$$

for some measure γ on $E \times (0, \infty)$, so that (1.21) holds as asserted. ∎

We conclude with a familiar special case.

Example 1.35. (Poisson process). By definition a Poisson process has independent increments and we have already noted its infinite divisibility. Conversely, let N be a simple point process without fixed atoms that simultaneously is infinitely divisible and has independent increments. Then in the representation (1.21) the U_i must all be one and hence N is Poisson. ☐

1.5 Transformations of Point Processes

Two important operations, marking and clustering, were discussed in Section 4. In this section we introduce additional ways in which point processes are transformed or combined to produce further point processes. We characterize infinitely divisible point processes in one more way, as limits of superpositions of null arrays of point processes. Analogous "limit" characterizations of Poisson and Cox processes are also described. The simplest transformation is mapping.

Definition 1.36. Let $N = \sum \epsilon_{X_i}$, be a point process on E and let E' be a second LCCB space. Given a measurable mapping $f : E \to E'$, provided that it is locally finite, the point process $N f^{-1} = \sum \epsilon_{f(X_i)}$ is the *mapping* of N by f.

In Section 8 and Chapter 9 we consider in detail the case that $E = \mathbf{R}^d$ for some $d \geq 1$ and f is a translation: $f(y) = y - x$ for fixed $x \in \mathbf{R}^d$. Point processes with translation-invariant distributional properties are termed stationary. A second special case of mapping, used in Section 4 and again in Section 6.3, is projection, whereby one restricts attention to a single component of a point process on a product space. Since mapping constitutes a change of variable, it follows by standard results that $L_{Nf^{-1}}(g) = L_N(g \circ f)$, from which other computations (e.g., pertaining to the mean measure and the zero-probability functional) ensue. If N is Poisson with mean μ and if μf^{-1} belongs to $\mathbf{M}(E')$, then $N f^{-1}$ is Poisson with mean μf^{-1}. Applications of this property are given in Exercise 1.19.

Our next class of transformations involves additional randomness and yields compound point processes, which are not point processes at all in general, but rather purely atomic random measures; however, the terminology is well established and we use it despite its not being precisely correct.

Definition 1.37. Let $N = \sum \epsilon_{X_i}$ be a point process on E and let $\bar{N} = \sum \epsilon_{(X_i, U_i)}$ be a marked point process with mark space $\mathbf{R}_+$, obtained from N by position-dependent marking using a transition kernel K (Example 1.28). Then the purely atomic random measure

$$M = \sum_i U_i \epsilon_{X_i} \tag{1.24}$$

is the *K-compound* of N, or *compound point process* defined by N and K.

In the special case that $K(x, \cdot) \equiv \rho(\cdot)$ for a probability ρ on $\mathbf{R}_+$, the U_i are i.i.d. and independent of N and we call M a ρ-compound of N, or *classical* compound point process. Classical compound Poisson processes on $\mathbf{R}_+$ are the best known examples; they have independent increments in general and stationary increments as well if N is a homogeneous Poisson process. Theorem 1.34 demonstrates that, except for fixed atoms and a deterministic part, every random measure with independent increments has the form (1.24), with the marked point process $\bar{N}$ a Poisson process on $E \times \mathbf{R}_+$.

For study of a compound point process $M = \sum U_i \epsilon_{X_i}$ it is often convenient to deal instead with the marked point process $\bar{N} = \sum \epsilon_{(X_i, U_i)}$, for then point process theories and computational methods can be brought to bear. In inference situations, though, one must take care if the U_i can be zero, for then $\bar{N}$ cannot be reconstructed from observation of M alone. (This is true in particular for thinned point processes, defined momentarily, but they *are* point processes, and can be dealt with directly.) Of course, if $U_i > 0$ for each i, then observation of M is equivalent to that of $\bar{N}$ and the two may be used interchangeably.

When the U_i take only the values zero and one, the compound point process $N' = \sum U_i \varepsilon_{X_i}$ is a point process satisfying $N'(A) \leq N(A)$ for every A; such a process we call a *thinning* of N. In Definition 1.37, the mark space is $\{0,1\}$ and we may identify the kernel K with the function $p(x) = K(x, \{1\})$ from E into $[0,1]$.

Definition 1.38. Let $N = \sum \varepsilon_{X_i}$ be a point process on E, let $p : E \to [0,1]$ be measurable, and let the U_i be random variables that are conditionally independent given N with $P\{U_i = 1|N\} = 1 - P\{U_i = 0|N\} = p(X_i)$. Then the point process $N' = \sum U_i \varepsilon_{X_i}$ is a p *thinning* of N, or *thinned point process*.

The interpretation is that, starting with N, one forms N' by, for each i, retaining the point X_i in precisely the same location with probability $p(X_i)$ and deleting it with the complementary probability $1 - p(X_i)$. Different points are retained or deleted independently. It is important to remember that *thinning function p* represents *retention* probabilities.

The principal computational properties, established by conditioning, are the following.

Proposition 1.39. Let N' be the p-thinning of N; then

$$L_{N'}(f) = L_N(-\log\{1 - p - pe^{-f}\}), \tag{1.25}$$

$$z_{N'}(A) = L_N(1_A \cdot \{-\log(1-p)\}), \tag{1.26}$$

$$\mu_{N'}(dx) = p(x)\mu_N(dx). \,\square \tag{1.27}$$

In the context of thinning the most important point processes are the Cox processes, whose fundamental role we now explain. If N is a Cox process directed by the random measure M, and if N' is the p-thinning of N, then from (1.11) and (1.25)

$$
\begin{aligned}
L_{N'}(f) &= L_N(-\log\{1 - p - pe^{-f}\}) \\
&= L_M(p(1 - e^{-f})) = L_{pM}(1 - e^{-f}), \tag{1.28}
\end{aligned}
$$

where $pM(dx) = p(x)M(dx)$. Consequently, N' is a Cox process directed by pM, so that the family of Cox processes is closed under thinning, but more is true: Cox processes are characterized by their being p-thinnings for every p.

Theorem 1.40. For a point process N on E, the following statements are equivalent:

a) N is a Cox process;

b) For every function $p : E \to (0,1]$ there is a point process N_p such that $N \overset{d}{=} N'_p$, where N'_p is the p-thinning of N_p.

Proof: To show that a) implies b), with M the directing measure, it suffices to take N_p to be the Cox process directed by $p^{-1}M$ and to apply (1.28). Conversely, for each n, N has the same distribution as the $p_n \equiv 1/n$-thinning of some N_n, and that b) implies a) will follow once Theorem 1.41 is proved. ∎

Nearly every invariance property is a manifestation of a deeper limiting property, as is true here.

Theorem 1.41. Let N_n be point processes on E and let p_n be functions from E to $(0,1]$ such that $p_n \to 0$ uniformly. For each n, let N'_n be the p_n-thinning of N_n. Then the following assertions are equivalent:

a) There is a point process N' such that $N'_n \overset{d}{\to} N'$;

b) There is a random measure M such that $p_n N_n \overset{d}{\to} M$.

When a) and b) hold, N is (equal in distribution to) a Cox process directed by M.

Proof: In view of the convergence assumptions on (p_n), for each f

$$
\begin{aligned}
L_{N'_n}(f) &= L_{N_n}(-\log\{1 - p_n - p_n e^{-f}\}) \\
&\cong L_{N_n}(p_n(1 - e^{-f})) = L_{p_n N_n}(1 - e^{-f}).
\end{aligned}
\tag{1.29}
$$

Since by Theorem 1.21 (N'_n) converges if and only if $\lim L_{N'_n}(f)$ exists for all f, and similarly for $(p_n N_n)$, equivalence of a) and b) is established. Taking limits in (1.29) yields $L_{N'}(f) = L_M(1 - e^{-f})$, which confirms the last statement. ∎

Cox processes, therefore, are those point processes that can arise as limits in distribution of sequences of point processes subjected to successively more severe thinning. Another class of point processes characterized as the set of all limits under a rarefaction mechanism is the infinitely divisible point processes. In this case the mechanism is uniform sparsity of summands in a triangular array.

Definition 1.42. Let $\{N_{nk} : n \geq 1, 1 \leq k \leq k_n\}$ be a triangular array of point processes, all on the same space E; then (N_{nk}) is a *null array* provided that

    a) For each n, $N_{n1}, \ldots, N_{nk_n}$ are independent;
    b) For every $B \in \mathcal{B}$, $\lim_{n\to\infty} \max_{k \le k_n} P\{N_{nk}(B) \ge 1\} = 0$.

Thus we arrive at one final characterization of infinitely divisible point processes.

Theorem 1.43. For a null array (N_{nk}) of point processes, the following statements are equivalent:
    a) There is a point process N such that $\sum_{k=1}^{k_n} N_{nk} \xrightarrow{d} N$;
    b) There is a measure λ on $\mathbf{M}_p \setminus \{0\}$ satisfying

$$\int_{\mathbf{M}_p}[1 - e^{-\mu(B)}]\lambda(d\mu) < \infty, \qquad B \in \mathcal{B},$$

such that for all $f \in C_K^+$,

$$\sum_{k=1}^{k_n}[1 - L_{N_{nk}}(f)] \to \int_{\mathbf{M}_p}[1 - e^{-\mu(f)}]\lambda(d\mu). \tag{1.30}$$

When a) and b) hold, then N is infinitely divisible with

$$L_N(f) = \exp\left[-\int_{\mathbf{M}_p}\{1 - e^{-\mu(f)}\}\lambda(d\mu)\right]. \tag{1.31}$$

Proof: (Sketch). If a) holds, then $\sum_k \log L_{N_{nk}}(f) \to \log L_N(f)$ for each f, and together with the null array assumption this implies that

$$\sum_k[1 - L_{N_{nk}}(f)] \to -\log L_N(f) \tag{1.32}$$

for $f \in C_K^+$. Let

$$\lambda = \lim_{n\to\infty} \sum_k \mathcal{L}_{N_{nk}}, \tag{1.33}$$

where $\mathcal{L}_{N_{nk}}$ is the probability law of N_{nk} and the limit is in the sense of vague convergence on $\mathbf{M}_p$. Existence of λ is the incompletely resolved issue in this argument. Combined, (1.32) and (1.33) yield

$$-\log L_N(f) = \lim_{n\to\infty} \sum_k \int[1 - e^{-\mu(f)}]\mathcal{L}_{N_{nk}}(d\mu) = \int[1 - e^{-\mu(f)}]\lambda(d\mu),$$

which demonstrates both (1.30) and (1.31).

Conversely, that (1.31) defines the Laplace functional of a Poisson cluster process is straightforward. Under the null array assumption, (1.30) implies that for $f \in C_K^+$,

$$\sum_k \log L_{N_{nk}}(f) \cong -\sum_k[1 - L_{N_{nk}}(f)]$$
$$\to -\int[1 - e^{-\mu(f)}]\lambda(d\mu) = \log L_N(f);$$

consequently, a) holds. ∎

Since every infinitely divisible point process is the limit of the null array of Definition 1.31, its Laplace functional has the form (1.31). Furthermore, it is evident that a point process with Laplace functional (1.31) is a Poisson cluster process. These two observations constitute the proof that a) entails b) in Theorem 1.32.

To recapitulate, we now have three characterizations of infinitely divisible point processes: as Poisson cluster processes, as limits of row superpositions of null arrays, and as having Laplace functionals of the form (1.31). The latter two are analogous to classical results for infinitely divisible probability measures on $\mathbf{R}_+$, to which, indeed, they reduce when E consists of but one point. Note also that (1.20) is also of the form (1.31).

Theorem 1.43 admits random measure analogues; see Kallenberg (1983) for details. Briefly, the null array characterization holds under the condition that $\lim_n \max_j P\{M_{nj}(B) > \varepsilon\} = 0$ for every $\varepsilon > 0$. In the representation (1.31), λ becomes a measure on $\mathbf{M} \setminus \{0\}$ and a deterministic component may be present. Sharper forms are available when the limit has independent increments. For point processes, such a limit is a Poisson process, a case we treat at some length in the next section.

1.6 Approximation of Point Processes

Because Poisson and Cox processes are so fundamental, it is of interest to investigate whether and how well a given point process of sequence of point processes (for example, row superpositions in a triangular array) can be approximated by a Poisson or Cox process. Limit theorems provide some information, of course, but are incomplete without rates of convergence. In this section we describe a paradigm for approximation of point processes, with approximation error as the key concept, and carry out the steps of the paradigm for an important special case: approximation of Bernoulli point processes and their superpositions by Poisson processes.

The elements of the paradigm are

- A *distance* between (distributions of) point processes;

- An *inequality* providing an upper bound on the distance between two point processes;

- *Limit theorems with rates of convergence*, derived by application of the inequality.

Additional questions such as optimality of particular approximations are also of interest.

Throughout the section we assume that E is a *compact* metric space, with metric d^*. We consider two distances between the probability laws of point processes N_1 and N_2, both dependent on an underlying metric on $\mathbf{M}_p$. The latter is *Prohorov metric*, which metrizes the vague topology, and is given by

$$d(\mu_1, \mu_2) = \inf\{\varepsilon > 0 : \mu_1(A) \le \mu_2(A^\varepsilon) + \varepsilon \text{ for all } A \in \mathcal{E}\} \wedge 1,$$

where $A^\varepsilon = \{x : d^*(x, A) < \varepsilon\}$ and $d^*(x, A) = \inf\{d^*(x, y) : y \in A\}$.

The two distances between $\mathcal{L}_{N_1}$ and $\mathcal{L}_{N_2}$, which for notational simplicity we write as distances between N_1 and N_2, are then

a) The *variation distance*

$$\alpha(N_1, N_2) = \sup\{|P\{N_1 \in \Gamma\} - P\{N_2 \in \Gamma\}| : \Gamma \in \mathcal{M}_p\};$$

b) The *Prohorov distance*

$$\rho(N_1, N_2) = \inf\{\varepsilon > 0 : P\{N_1 \in \Gamma\} \le P\{N_2 \in \Gamma^\varepsilon\} + \varepsilon, \Gamma \in \mathcal{M}_p\}.$$

Evidently $\rho(N_1, N_2) \le \alpha(N_1, N_2)$; we concentrate on bounds for the latter. The Prohorov distance has the disadvantage of being almost impossible to calculate, but is significant nevertheless because it metrizes convergence in distribution: $N_n \overset{d}{\to} N$ if and only if $\rho(N_n, N) \to 0$ (see Billingsley, 1968).

An important technique for dealing with the variation distance is coupling. By a *coupling* of point processes N_1 and N_2 we mean a construction of them on a single probability space. Given a coupling (N_1, N_2), we have

$$\alpha(N_1, N_2) \le P\{N_1 \ne N_2\} = P\{d(N_1, N_2) > 0\}, \qquad (1.34)$$

$$\rho(N_1, N_2) \le \inf\{\varepsilon > 0 : P\{d(N_1, N_2) \ge \varepsilon\} < \varepsilon\}; \qquad (1.35)$$

see Strassen (1965) for details. A *maximal coupling* achieves equality in (1.34) or (1.35), and hence is useful for analysis of the distance between N_1 and N_2. Maximal couplings exist for every choice of N_1 and N_2, in both the α- and ρ-senses, but neither is unique and in general they cannot be realized simultaneously. Moreover, only relatively few are known explicitly.

To illustrate the paradigm we consider the problem of approximating Bernoulli point processes and their superpositions by Poisson processes; as a consequence of our inequalities we obtain the Poisson limit theorem of Grigelionis (1963) (see also Franken, 1963).

Given $p \in (0, 1]$ and a probability measure F on E, a *Bernoulli point process* with parameters (p, F) is a point process $N = U\varepsilon_X$, where X is a random element of E with distribution F and U is a Bernoulli random variable independent of X with $p = P\{U = 1\} (= 1 - P\{U = 0\})$. Thus with probability $1 - p$, N has no points at all, and with probability p, one point with distribution F. For sufficiently small p the optimal Poisson approximation to N, with respect to the α-distance, is given by the next result.

Theorem 1.44. Let N be a Bernoulli process on E with parameters (p, F) and let N^* be Poisson with mean measure $\mu^* = [-\log(1-p)]F$. Then

$$\alpha(N, N^*) = p + (1 - p)\log(1 - p), \qquad (1.36)$$

and if $p < 1 - 1/e$, then $\alpha(N, N^*) < \alpha(N, N')$ for any Poisson process with mean $\mu' \neq \mu^*$, so that N^* is the optimal Poisson approximation to N.

Proof: We construct a maximal coupling (N, N^*) as follows. Starting with N^*, let N be obtained by choosing one point of N^* at random. That is, $N = 0$ on $\{N^*(E) = 0\}$, while for $k \geq 1$, on $\{N^*(E) = k\}$, assuming that on this event N^* has the representation $N^* = \sum_{i=1}^{k} \varepsilon_{X_i}$ (see Exercise 1.1), where the X_i are i.i.d. with distribution F, then $N = \sum_{i=1}^{k} V_i \varepsilon_{X_i}$, where $V = (V_1, \ldots, V_k)$ is independent of N^* with $P\{V = (1, 0, \ldots, 0)\} = \cdots = P\{V = (0, \ldots, 0, 1)\} = 1/k$. Then by conditional uniformity, N is indeed Bernoulli with parameters (p, F) and, moreover,

$$P\{N \neq N^*\} = P\{N^*(E) \geq 2\} = p + (1 - p)\log(1 - p).$$

Maximality is apparent: for any coupling of N and N^*, $P\{N \neq N^*\} \geq P\{N^*(E) \geq 2\}$ since $N(E)$ can be only zero or one.

To prove optimality, suppose that N' is Poisson with mean measure $\mu' = \lambda G$, where $\lambda > 0$ and G is a probability on E. Then first of all, since N is Bernoulli, for the maximal coupling (N, N'),

$$\alpha(N, N') = P\{N \neq N'\} \geq P\{N'(E) \geq 2\} = 1 - e^{-\lambda} - \lambda e^{-\lambda}.$$

The latter function increases in λ and hence we may restrict attention to values satisfying $\lambda \leq \lambda^* = -\log(1 - p)$. With $\lambda \leq \lambda^*$ and N' Poisson with mean measure λF, the maximal coupling of N and N' satisfies $P\{N \neq N'\} = p - \lambda e^{-\lambda}$; consequently with λ fixed, the optimal choice of G is the obvious choice $G = F$, since for the maximal coupling of N and a Poisson process N'' with mean λG,

$$\begin{aligned}
P\{N \neq N''\} &= P\{N''(E) \geq 2\} + P\{N'' = 0, N \neq 0\} \\
&\quad + P\{N''(E) = 1, N \neq N''\} \\
&= 1 - e^{-\lambda} - \lambda e^{-\lambda} + P\{N'' = 0, N \neq 0\} \\
&\quad + P\{N''(E) = 1, N \neq N''\} \\
&\geq p - \lambda e^{-\lambda} = P\{N \neq N'\},
\end{aligned}$$

with the inequality strict unless $G = F$. The problem is thus reduced to a one-dimensional minimization over λ, the solution of which is easily seen to be $\lambda = \lambda^*$. ∎

Approximation of the superposition of independent Bernoulli point processes is then straightforward, but the following result does not yield an optimal approximation even when E is a singleton (in which case point process results reduce to results for Poisson approximation of Bernoulli and binomial random variables).

Corollary 1.45. Let $N_1, \ldots, N_k$ be independent Bernoulli point processes with parameters $(p_1, F_1), \ldots, (p_k, F_k)$, let $N = \sum_{i=1}^k N_i$, and let N' be Poisson with mean measure $\mu = \sum_{i=1}^k [-\log(1 - p_i)]F_i$. Then

$$\alpha(N, N') < 1 - \prod_{i=1}^k [(1 - p_i)(1 - \log(1 - p_i))]. \square$$

From Corollary 1.45 one can deduce the Poisson limit theorem of Grigelionis (1963). However, a somewhat stronger version can be obtained from the nest result, which embeds a triangular array of Bernoulli point processes in a single Poisson process.

Let $\{N_{ni} : n \geq 1, 1 \leq i \leq k_n\}$ be a triangular array of Bernoulli point processes with parameters (p_{ni}, F_{ni}), with the N_{ni} independent as i varies for each n, and let $N_n = \sum_{i=1}^{k_n} N_{ni}$ be the nth row superposition.

Theorem 1.46. Assume that there exists a finite measure μ on E such that

$$-\sum_{i=1}^{k_n} [\log(1 - p_{ni})]F_{ni} \ll \mu$$

for each n, and let r_n be the Radon-Nikodym derivative. Then there exists a probability space on which are defined a Poisson process N' with mean measure μ and a version of the triangular array (N_{ni}) such that for each n,

$$P\{N_n \neq N'\} \leq 1 - e^{-\int |1 - r_n| d\mu} \prod_{i=1}^{k_n} \{(1 - p_{ni})[1 - \log(1 - p_{ni})]\}.$$

Consequently, if $\max_i p_{ni} \to 0$ and $\int |1 - r_n| d\mu \to 0$, then as $n \to \infty$

$$\alpha(N_n, N') = O(\max_i p_{ni} + \int |1 - r_n| d\mu). \tag{1.37}$$

Proof: Let N' be Poisson with mean m, and by repeated thinning (see Exercise 1.16) decompose N' as $N' = \hat{N}_n + \sum_{i=1}^{k_n} \hat{N}_{ni}$, where
  i) $\hat{N}_n$ and the $\hat{N}_{ni}$ are mutually independent Poisson processes;
  ii) $\hat{N}_n$ has mean measure $[1 - r_n(x) \wedge 1]\mu(dx)$;
  iii) $\hat{N}_{ni}$ has mean measure $[(r_n(x) \wedge 1)/r_n(x)][-\log(1 - p_{ni})]F_{ni}(dx)$.

Also, let $\bar{N}_{ni}$ be Poisson with mean measure $[r_n(x) - 1]^+ \mu(dx)$ and independent of N', with decomposition $\bar{N}_n = \sum_{i=1}^{k_n} \bar{N}_{ni}$, where each $\bar{N}_{ni}$ is Poisson
with mean measure $\{1 - [r_n(x) \wedge 1]/r_n(x)\}[-\log(1 - p_{ni})]F_{ni}(dx)$. Then for
each n and i, $N'_{ni} = \hat{N}_{ni} + \bar{N}_{ni}$ is Poisson with mean $[-\log(1 - p_{ni})]F_{ni}$, and
to it couple N_{ni} as described in Theorem 1.44, preserving independence of
the N_{ni} in the process. Then

$$\begin{aligned} P\{N_n \neq N'\} &\geq P\{\hat{N}_n = 0, \bar{N}_n = 0, N_{ni} = N'_{ni}, i = 1, \ldots, k_n\} \\ &= P\{\hat{N}_n = 0\}P\{\bar{N}_n = 0\}\prod_{i=1}^{k_n} P\{N_{ni} = N'_{ni}\}. \end{aligned}$$

Evaluation of the last expression using (1.36) yields (1.37). ∎

We illustrate with two examples.

Example 1.47. (Point process analogue of the Poisson limit theorem
for binomial distributions). Suppose that for each n the N_{ni}, $i = 1, \ldots, n$
are i.i.d. Bernoulli processes with parameters $(\lambda/n, F)$, where $\lambda > 0$ and
F is a fixed probability on E. Taking $\mu = \lambda F$ in Theorem 1.46 leads to
the convergence rate $\alpha(N_n, N') = O(n^{-1})$; observe that this applies to the
α-distance rather than just the ρ-distance. □

Example 1.48. (Empirical process). Let $X_1, X_2, \ldots$ be i.i.d. strictly
positive random variables whose distribution F admits a positive, continuous
density f on some interval $[0, \varepsilon]$. Let $E = [0, t]$, where $t > 0$ is fixed, and
put $N_{ni} = \mathbf{1}(X_i \leq t/n)\varepsilon_{nX_i}$, $1 \leq i \leq n$. Then N_{ni} is Bernoulli with parameters $(F(t/n), F_n)$, where $F_n(A) = F(n^{-1}A)/F(t/n)$ for $A \subseteq E$. Choosing
$\mu(dx) = \lambda\, dx$, where $\lambda = f(0)$, in Theorem 1.46 — so that N' is an ordinary
Poisson process with rate λ, and performing the necessary computations
yields

$$\alpha(N_n, N') = O(n^{-1}). \qquad \square \tag{1.38}$$

An alternative but related methodology for approximation of point processes on $\mathbf{R}_+$, based on martingales and compensators, is presented in Section 2.3. On $\mathbf{R}_+$ this method can be used to approximate more general point
processes than Bernoulli processes, but it requires the special structure of
the real line. Neither, however, are the methods of this section as restrictive
as they may appear at first.

Proposition 1.49. Given a point process N on a compact space E, let
$p = P\{N \neq 0\}$. Then there is an infinitely divisible point process N^* such
that

$$P\{N \neq N^*\} = p + (1 - p)\log(1 - p). \tag{1.39}$$

Proof: Let $G(\Gamma) = P\{N \in \Gamma | N \neq 0\}$, and consider the Bernoulli point process $\tilde{N} = \mathbf{1}(N \neq 0)\varepsilon_N$ on the space $\mathbf{M}_p$. Letting $\bar{N} = \sum \varepsilon_{N_j}$ be Poisson on $\mathbf{M}_p$ with mean $[-\log(1-p)]G$ and such that (1.36) holds, then $N^* = \sum_j N_j$ is infinitely divisible and satisfies (1.39). ∎

1.7 Palm Distributions

As will emerge, Palm distributions are a potent tool for inference for point processes, especially Cox processes (Chapter 7) and stationary point processes (Chapter 9). In addition many structural and conditioning properties are described best or most easily via Palm distributions. Heuristically, a Palm distribution of a simple point process N is the law, conditional on the presence of points of N at prescribed locations $x_1, \ldots, x_k$, of the *remainder* of N:

$$Q((x_1, \ldots, x_k), \Gamma) = P\{N - \sum_1^k \varepsilon_{x_i} \in \Gamma | N(\{x_i\}) = 1, i = 1, \ldots, k\}, \quad (1.40)$$

where the x_i are points in E. We must, however, define such distributions indirectly, because ordinarily $P\{N(\{x_i\}) = 1, i = 1, \ldots, k\} = 0$.

Before proceeding with the definitions, we refine our objective slightly. The order of the x_i is immaterial to the conditional distributions (1.40), which depend only on $\{x_1, \cdots, x_k\}$. Rather than parametrize the Palm distributions by points, therefore, it is more convenient to parametrize them by point measures $\mu = \sum_1^k \varepsilon_{x_i}$.

To begin, we recall the point processes $N^{(k)}$ of (1.6), whose means $\mu_N^{(k)} = E[N^{(k)}]$ are the factorial moment measures of N. For each k, let $\mathbf{M}_s(k) = \{\mu \in \mathbf{M}_s : \mu(E) = k\}$ be the set of simple point measures with mass k, and let $\mathbf{M}^* = \sum_{k=0}^{\infty} \mathbf{M}_s(k)$, the set of *finite* elements of $\mathbf{M}_s$.

Suppose now that $N = \sum \varepsilon_{X_i}$ is a simple point process on E. We first introduce certain measures on the product spaces $\mathbf{M}_s(k) \times \mathbf{M}_s$ and $\mathbf{M}^* \times \mathbf{M}_s$.

Definition 1.50. a) For each k, the measure

$$C_N^{(k)}(\Lambda \times \Gamma) = E\left[\int_{E^k} \mathbf{1}(\sum_1^k \varepsilon_{x_i} \in \Lambda, N - \sum_1^k \varepsilon_{x_i} \in \Gamma)N^{(k)}(dx)\right] \quad (1.41)$$

on $\mathbf{M}_s(k) \times \mathbf{M}_s$ is the kth order *Campbell measure* of N;

b) The measure

$$C_N(\Lambda \times \Gamma) = \sum_{k=0}^{\infty} (1/k!)C^{(k)}(\Lambda \times \Gamma)$$

on $\mathbf{M}^* \times \mathbf{M}_p$ is the *compound Campbell measure* of N.

For $k = 0$ we take $C_N^{(0)}(\Lambda \times \Gamma) = \mathbf{1}(0 \in \Lambda)P\{N \in \Gamma\}$; note also that, provided one identifies $\mathbf{M}_s(k)$ with $\mathbf{M}_s(E^k)$,

$$\mu_N^{(k)}(f) = \int \mu(f)C_N^{(k)}(d\mu \times \mathbf{M}_s). \tag{1.42}$$

Moreover, the $C_N^{(k)}$ are concentrated on disjoint subsets of $\mathbf{M}^*$, so that

$$C_N(\Lambda \times \Gamma) = \sum_{k=0}^{\infty}(1/k!)C_N^{(k)}(\Lambda \cap \mathbf{M}_s(k) \times \Gamma).$$

Finally we introduce the *compound factorial moment measure*

$$K_N(\Lambda) = C_N(\Lambda \times \mathbf{M}_s). \tag{1.43}$$

The reduced Palm distributions can now be defined.

Definition 1.51. The *reduced Palm distributions* of N are given by the transition probability

$$Q_N(\mu, d\nu) = \frac{dC_N(\cdot \times d\nu)}{dC_N(\cdot \times \mathbf{M}_s)}(\mu)$$

from $\mathbf{M}^*$ to $\mathbf{M}_s$.

Existence of Q follows by known theorems on disintegration of measures on product spaces. However, the construction is simple conceptually: for each Γ, $C(\cdot \times \Gamma) \ll C(\cdot \times \mathbf{M}_s)$, so there exists the Radon-Nikodym derivative $Q(\mu, \Gamma)$. One must then ensure that $Q_N(\mu, \cdot)$ is a probability measure for each μ. See Kallenberg (1983) for details.

Suppose now that $\mu \in \mathbf{M}_s(k)$, and note that

$$\int_{E^k}\mathbf{1}(\sum_{i=1}^k \epsilon_{x_i} = \mu)N^{(k)}(dx) = k!\,\mathbf{1}(N - \mu \geq 0).$$

Therefore,

$$
\begin{aligned}
Q_N(\mu, \Gamma) &= \frac{E\left[\int_{E^k} \mathbf{1}(\sum_{i=1}^k \epsilon_{x_i} = \mu, N - \sum_{i=1}^k \epsilon_{x_i} \in \Gamma)N^{(k)}(dx)\right]}{E\left[\int_{E^k} \mathbf{1}(\sum_{i=1}^k \epsilon_{x_i} = \mu)N^{(k)}(dx)\right]} \\[2mm]
&= \frac{E\left[\mathbf{1}(N - \mu \in \Gamma)\int_{E^k} \mathbf{1}(\sum_{i=1}^k \epsilon_{x_i} = \mu)N^{(k)}(dx)\right]}{E\left[\int_{E^k} \mathbf{1}(\sum_{i=1}^k \epsilon_{x_i} = \mu)N^{(k)}(dx)\right]} \\[2mm]
&= \frac{k!\,P\{N - \mu \in \Gamma, N - \mu \geq 0\}}{k!\,P\{N - \mu \geq 0\}} \\[2mm]
&= P\{N - \mu \in \Gamma | N - \mu \geq 0\},
\end{aligned}
$$

which confirms the interpretation of reduced Palm distributions.

A more rigorous interpretation can be established via limits. Let (A_{nj}) be a null array of partitions of E (see the proof of Theorem 1.34 and associated discussion), and for $\mu \in \mathbf{M}_s(k)$ and each n, let $A_n(\mu) = \bigcup \{A_{nj} : \mu(A_{nj}) > 0\}$, and observe that if $\mu \in \mathbf{M}_s(k)$, then for n sufficiently large, $A_n(\mu)$ is the union of precisely k of the A_{nj}; we denote these components by $B_{nj}(\mu)$, $j = 1, \ldots, k$.

Theorem 1.52. Almost everywhere with respect to the measure K_N of (1.43), for $\mu \in \mathbf{M}_s(k)$,

$$Q_N(\mu, \cdot) = \lim_{n \to \infty} P\{N_{A_n(\mu)^c} \in (\cdot) | N(B_{nj}(\mu)) = 1, j = 1, \ldots, k\}$$

in the sense of weak convergence of probability measures on $\mathbf{M}_s$, where $N_{A_n(\mu)^c}$ is the restriction of N to $A_n(\mu)^c$.

Proof: To simplify the notation, we assume that $\mu \in \mathbf{M}_s(1)$; to streamline the argument, we suppose that the mapping $\mu \to Q(\mu, \cdot)$ is continuous. Finally, let $\tilde{A}_n(\mu) = \{\varepsilon_y : y \in A_n(\mu)\}$ and let $\Lambda_n(\mu) = \{\nu : \nu(A_n(\mu)) = 0\}$. Then

$$P\{N_{A_n(\mu)^c} \in (\cdot) | N(B_{nj}(\mu)) = 1, j = 1, \ldots, \mu(E)\}$$

$$= \frac{P\{N_{A_n(\mu)^c} \in (\cdot), N(B_{nj}(\mu)) = 1, j = 1, \ldots, \mu(E)\}}{P\{N(B_{nj}(\mu)) = 1, j = 1, \ldots, \mu(E)\}}$$

$$= \frac{E\left[\int_{A_n(\mu)} \mathbf{1}(N - \varepsilon_x \in \Gamma \cap \Lambda_n(\mu)) N(dx)\right]}{E\left[\int_{A_n(\mu)} \mathbf{1}(N - \varepsilon_x \in \Lambda_n(\mu)) N(dx)\right]}$$

$$= \frac{C_N^{(1)}(\tilde{A}_n(\mu) \times [\Gamma \cap \Lambda_n(\mu)])}{C_N^{(1)}(\tilde{A}_n(\mu) \times \Lambda_n(\mu))}$$

$$= \frac{\int_{\tilde{A}_n(\mu)} K_N(d\nu) Q_N(\nu, \Gamma \cap \Lambda_n(\mu))}{\int_{\tilde{A}_n(\mu)} K_N(d\nu) Q_N(\nu, \Lambda_n(\mu))}$$

$$\cong \frac{Q_N(\mu, \Gamma \cap \Lambda_n(\mu))}{Q_N(\mu, \Lambda_n(\mu))}$$

by the continuity assumption. But since $A_n(\mu) \downarrow \operatorname{supp} \mu$ and since N is simple,

$$\int_{\mathbf{M}_s(1)} K_N(d\nu) Q_N(\nu, \Lambda_n(\mu)) = C_N^{(1)}(\mathbf{M}_s(1) \times \Lambda_n(\mu))$$

$$= E\left[\int \mathbf{1}(N - \varepsilon_x \in \Lambda_n(\mu)) N(dx)\right]$$

$$\to E[N(E)] = K_N(\mathbf{M}_s(1)),$$

where the last equality is by (1.42). Hence it follows that $Q_N(\cdot, \Lambda_n(\mu)) \to 1$ a.e., which suffices to complete the proof. ∎

In several contexts it is useful to work with point processes whose distributions are the reduced Palm distributions of N.

Definition 1.53. The *reduced Palm processes* of N are any point processes N_μ, $\mu \in \mathbf{M}^*$, with distributions $Q_N(\mu, \cdot)$.

Once more we illustrate with Poisson and Cox processes. The form of reduced Palm distributions for a Poisson process is yet another manifestation of the property of independent increments.

Example 1.54. (Poisson process). Let N be a Poisson process with diffuse mean measure μ. Then for almost every ν the reduced Palm process N_ν is also Poisson with mean μ. Indeed, if $\nu = \sum_{i=1}^{k} \epsilon_{x_i}$, then for disjoint neighborhoods V_i of the x_i, f a continuous function, and $U = (\bigcup_{i=1}^{k} V_i)^c$,

$$E\left[\prod_{i=1}^{k}\{N(V_i)e^{-N(f)+\sum f(x_i)}\}\right] / E\left[\prod_{i=1}^{k} N(V_i)\right]$$

$$= E\left[e^{-\int_U f dN}\prod_{i=1}^{k}[N(V_i)e^{-\int_{V_i} f dN + f(x_i)}/\mu(V_i)]\right]$$

$$= e^{-\int_U (1-e^{-f})d\mu}\prod_{i=1}^{k}\{e^{-\int_{V_i}(1-e^{-f})d\mu}[\int_{V_i} e^{-f}d\mu]e^{f(x_i)}/\mu(V_i)\}$$

$$\to e^{-\int(1-e^{-f})d\mu},$$

from which the assertion follows immediately. In fact, this property characterizes Poisson processes; see Exercise 1.36 for a related characterization of mixed Poisson processes. ☐

One can also define (unreduced, in effect) Palm distributions for random measures that are not necessarily point processes, but in general there is no longer a conditioning interpretation. The *compound Campbell measure* of a random measure M is the measure

$$C_M(\Lambda \times \Gamma) = \sum_{k=0}^{\infty}(1/k!)E\left[\int_{E^k} \mathbf{1}(\sum_1^k \epsilon_{x_i} \in \Lambda)M^k(dx)\mathbf{1}(M \in \Gamma)\right]$$

and the *compound moment measure* is

$$K_M(\Lambda) = C_M(\Lambda \times \mathbf{M}) = \sum_{k=0}^{\infty}(1/k!)E\left[\int_{E^k} \mathbf{1}(\sum_1^k \epsilon_{x_i} \in \Lambda)M^k(dx)\right].$$

The (unreduced) *Palm distributions* of M are given by the transition probability

$$Q_M^*(\mu, d\eta) = \frac{dC_M(\cdot \times d\eta)}{dC_M(\cdot \times \mathbf{M})}(\mu),$$

and the (unreduced) *Palm processes* of M are random measures M_μ with distributions $Q_M^*(\mu, \cdot)$.

Within this context, the following property holds for Cox processes. It is used in Section 7.2 for state estimation of the unobservable directing measure of a Cox process. Theorem 4.23 elucidates further the central role of Palm distributions in state estimation for point processes.

Proposition 1.55. Let (N, M) be a Cox pair. Then $K_N = K_M$ and, with respect to these measures, for almost every μ, the reduced Palm process N_μ has the distribution of a Cox process directed by the Palm process M_μ.

Proof: Because N is a Cox process directed by M, for each k and positive functions f and g

$$E\left[e^{-M(1-e^{-f})}\int e^{-\sum g(x_i)} M^k(dx)\right]$$
$$= E\left[e^{-N(f)}\int e^{-\sum g(x_i)} e^{\sum f(x_i)} N^{(k)}(dx)\right].$$

It follows that $K_N = K_M$, and hence that

$$\int K_M(d\mu)e^{-\mu(g)}\int Q_M(\mu, d\eta)e^{-\eta(1-e^{-f})}$$
$$= \int K_N(d\mu)e^{-\mu(g)}\int Q_N(\mu, d\nu)e^{-\nu(f)}$$
$$= \int K_M(d\mu)e^{-\mu(g)}\int Q_N(\mu, d\nu)e^{-\nu(f)}.$$

This implies that for fixed f, $L_{M_\mu}(1 - e^{-f}) = L_{N_\mu}(f)$ almost everywhere, and the proposition follows by standard reasoning and Theorem 1.12. ∎

There is a notational inconsistency, but we virtually never use unreduced Palm processes for point processes, so we let it stand. Indeed, there is really only one case in which unreduced Palm distributions for point processes are important: that of stationary point processes, in which a multiple of the unreduced Palm distribution $P\{N \in (\cdot)|N(\{0\}) = 1\}$ plays a central role, as we discuss in the next section.

1.8 Stationary Point Processes

Stationarity for point processes on $\mathbf{R}^d$ or $\mathbf{Z}^d$ means distributional invariance under translations. Among other properties, a stationary point process admits a single intensity parameter, which under ergodicity assumptions can be reconstructed from almost every realization. The intensity, moreover, carries a micro-level interpretation as well, given in Corollary 1.67.

Suppose now that E is either $\mathbf{R}^d$ or $\mathbf{Z}^d$, and for each x, let τ_x denote the translation operator $\tau_x y = y - x$. Then for μ a measure on E, we have

$\mu\tau_x^{-1}(A) = \mu(A + x) = \mu(\{z + x : z \in A\})$. Let λ be the normalized Haar measure on E, i.e., Lebesgue measure or the discrete uniform distribution.

Let N be a point process on E, defined on a probability space $(\Omega, \mathcal{F}, P)$ rich enough to support operators $(\theta_x)_{x \in E}$ satisfying

a) The mapping $(x, \omega) \to \theta_x \omega$ is jointly measurable;

b) $\theta_0 = I$, the identity on E;

c) $\theta_{x+y} = \theta_x \theta_y$ for each x and y.

This is really no restriction: on the canonical sample space $\Omega = \mathbf{M}_p$ one can take $\theta_x \mu = \mu\tau_x^{-1}$. We further define, for $N = \sum \epsilon_{X_i}$,

$$N_{(x)} = N\tau_x^{-1} = \sum \epsilon_{X_i - x}.$$

Here is the key definition.

Definition 1.56. a) The point process N is *stationary* with respect to (θ_x) if for each x, θ_x is measure-preserving: $P\theta_x^{-1} = P$, and

$$N_{(x)} = N \circ \theta_x; \tag{1.44}$$

b) N is *stationary in law* if $N_{(x)} \overset{d}{=} N$ for each x.

Obviously stationarity implies stationarity in law; conversely, a point process stationary in law admits a version that is stationary. In general we employ the more powerful formalism of stationarity.

The following examples exhibit diametrically different behaviors.

Example 1.57. (Poisson process). Let $\Omega = \mathbf{M}_p$, with $N(\omega) = \omega$. If under P, N is a (homogeneous) Poisson process with mean measure λ, then N is stationary with respect to the operators $\theta_x \omega = \omega\tau_x^{-1}$. That (1.44) holds is evident, while translation invariance of λ implies that

$$
\begin{aligned}
E[e^{-N \circ \theta_x(f)}] &= E[e^{-N(f \circ \tau_x)}] \\
&= \exp[-\int (1 - e^{-f \circ \tau_x}) d\lambda] \\
&= \exp[-\int (1 - e^{-f}) d\lambda] = E[e^{-N(f)}]. \square
\end{aligned}
$$

Example 1.58. (Lattice process). This process is of interest because of the explicitness of its construction and for devising counterexamples. Let $\Omega = [0, 1]$, with $\mathcal{F} = \mathcal{B}([0, 1])$ the Borel σ-algebra, and let P be the uniform distribution, so that $X(\omega) = \omega$ is uniformly distributed. Then the point process $N = \sum_{n=-\infty}^{\infty} \epsilon_{X+n}$, which is simply a random translation of the integer lattice, is stationary with respect to the operators $\theta_x \omega = \{\omega - x\}$, the fractional part of $\omega - x$. $\square$

All of the distributional descriptors in Section 2 are defined, of course, for stationary point processes, but moment measures play a relatively more significant role, particularly in inference. The first moment measure, if it exists, is translation invariant, and is hence a scalar multiple of λ.

Definition 1.59. Let N be a stationary point process. Then the *intensity* of N is the constant $\nu_N \in [0, \infty]$ satisfying

$$\mu_N = \nu_N \lambda. \tag{1.45}$$

Physically the intensity represents the mean spatial density of points of N; in Examples 1.57 and 1.58 the intensity is one. At least two other heuristic concepts of intensity can be formulated: a *local limit* $\lim E[N(V_n)]/\lambda(V_n)$ with $V_n \downarrow \{0\}$ and a *global limit* $\lim N(A_n)/\lambda(A_n)$ as $A_n \uparrow E$; these are made precise in Corollary 1.67 and Theorem 1.71. In nice cases, the three values coincide.

Generalizing (1.45), for $k > 1$ the moment measure μ^k is a $(k - 1)$-dimensional mixture of certain images of Lebesgue measure. For $z \in E^{k-1}$, let λ_z be the image of λ under the mapping $x \to (z_1 + x, \ldots, z_{k-1} + x, x)$; then we have the following result.

Proposition 1.60. Given a stationary point process N admitting a moment of order k, there exists a measure μ_*^k on E^{k-1} such that

$$\mu^k(\cdot) = \int_{E^{k-1}} \lambda_z(\cdot) \mu_*^k(dz). \quad \Box \tag{1.46}$$

Definition 1.61. The measure μ_*^k is the *reduced moment measure* of order k.

In particular, $E^0 = \{0\}$, so that $\mu_*^0 = \nu \varepsilon_0$, which exhibits (1.45) as a special case of (1.46).

Cumulant measures are needed to formulate some theorems in Chapter 9, so we introduce them at this point.

Definition 1.62. For a point process N admitting a moment of order k, the kth order *cumulant measure* γ^k is given by

$$\gamma^k(\otimes_{j=1}^k f_j) = \sum_{\mathcal{J} = \{J_\ell\}} (-1)^{|\mathcal{J}|-1}(|\mathcal{J}| - 1)! \prod_\ell \mu^{|J_\ell|}(\otimes_{j \in J_\ell} f_j),$$

where the sum is over all partitions $\mathcal{J}$ of $\{1, \ldots, k\}$. (For a discussion of the relationship to ordinary cumulants of random variables, see the proof of Theorem 9.7.)

Among cumulant measures, the most important is that for $k = 2$, the *covariance measure* [compare (1.5)]

$$\rho(f_1 \otimes f_2) = \gamma^2(f_1 \otimes f_2) = \mu^2(f_1 \otimes f_2) - \nu^2\lambda(f_1)\lambda(f_2).$$

Reduced cumulant measures γ_*^k satisfying

$$\gamma^k(\cdot) = \int_{E^{k-1}} \lambda_z(\cdot)\gamma_*^k(dz).$$

also exist, by virtue of stationarity.

The most important tool in analysis and inference for stationary point processes is the Palm measure. If N is a stationary point process with finite intensity ν_N, then each univariate *unreduced* Palm process $N_{\varepsilon_x}^*$ satisfies

$$N_{\varepsilon_x}^*\tau_x^{-1} \stackrel{d}{=} N_{\varepsilon_0}^*, \tag{1.47}$$

so that it suffices to study the unreduced Palm distribution $Q_N^*(\varepsilon_0, \cdot)$. We do so indirectly, in terms of the Palm measure P^*.

Definition 1.63. Let N be a stationary point process with finite intensity ν_N. The finite measure $P^*(\cdot) = \nu_N Q_N^*(\varepsilon_0, \cdot)$ on Ω is the *Palm measure* of N (or, depending on emphasis, of P).

Even though, as we see momentarily, P^* need not be a probability, integrals with respect to it are written as "expectations" E^*. The *Palm distribution* $P^*(\cdot)/P^*(\Omega)$ is the probability law of N conditional on there being a point at the origin.

We use the following characterization repeatedly.

Proposition 1.64. The Palm measure is the unique finite measure P^* on Ω such that

$$E\left[\int_E G(\theta_x\omega, x)N(\omega, dx)\right] = E^*\left[\int_E G(\omega, x)dx\right] \tag{1.48}$$

for each measurable function $G : \Omega \times E \to \mathbf{R}$.

Proof: It suffices to take $G(\omega, x) = \mathbf{1}(N(\omega) \in \Gamma)\mathbf{1}(x \in A)$ in (1.48), where $\Gamma \in \mathcal{M}_p$ and $A \in \mathcal{E}$. In this case,

$$
\begin{aligned}
E^*\left[\int G(\omega, x)\lambda(dx)\right] &= \lambda(A)P^*\{N \in \Gamma\} \\
&= \nu_N\lambda(A)Q_N^*(\varepsilon_0, \Gamma) \\
&= \nu_N\int_A\lambda(dx)Q_N^*(\varepsilon_0, \{\mu : \mu\tau_x^{-1} \in \Gamma\}) \\
&= \int_A\mu_N(dx)\int Q_N^*(\varepsilon_x, d\mu)\mathbf{1}(\mu\tau_x^{-1} \in \Gamma) \\
&= \int\int C_N^1(dx, d\mu)\mathbf{1}(\mu\tau_x^{-1} \in \Gamma)\mathbf{1}(x \in A) \\
&= E\left[\int_A\mathbf{1}(N_{(x)} \in \Gamma)N(dx)\right] \\
&= E\left[\int G(\theta_x\omega, x)N(dx)\right],
\end{aligned}
$$

where the third equality is by (1.47) and the next-to-last is by the analogue of (1.41) for unreduced Palm distributions. ∎

Note that in particular the intensity satisfies $\nu = P^*(\Omega)$.

In addition, (1.48) can be viewed in terms of marked point processes. Let $N = \sum \epsilon_{X_i}$ be stationary and note that for each i the translate $N_{(X_i)} = \sum_j \epsilon_{X_j - X_i}$ has a point at the origin. Then the marked point process

$$\bar{N} = \sum_i \epsilon_{(X_i, \sum_j \epsilon_{X_j - X_i})},$$

in which X_i is marked by $N_{(X_i)}$, has as its mean, by (1.48), the product measure $\lambda \times P^*$. This observation is crucial to the nonparametric estimation techniques for stationary point processes described in Chapter 9.

The following example continues Example 1.58.

Example 1.65. (Lattice process). For the process $N = \sum_{n=-\infty}^{\infty} \epsilon_{X+n}$ of Example 1.58, we have $P^* = \epsilon_0$, which is certainly plausible. To see this, apply (1.48) with $G(\omega, x) = e^{-\alpha\omega}\mathbf{1}(x \in [0,1])$, where $\alpha > 0$, to obtain

$$\int_\Omega P^*(d\omega)e^{-\alpha\omega} = E\left[\int_{[0,1]} e^{-\alpha\theta_x\omega}N(dx)\right] = E[e^{-\alpha\theta_{X(\omega)}\omega}] = 1.\square$$

The next result provides not only a direct conditioning interpretation of the Palm measure but also the micro-level interpretation of the intensity. It is analogous to Theorem 1.52.

Theorem 1.66. Let N be a stationary point process with finite intensity and Palm measure P^*.

a) If $E = \mathbf{Z}^d$, then $P^* \ll P$ with $(dP^*/dP)(\omega) = N(\omega, \{0\})$;

b) If $E = \mathbf{R}^d$ and Z is a positive random variable such that the mapping $x \to Z(\theta_x\omega)$ is continuous for each ω, and if V_n are neighborhoods of the origin such that $\overline{V_n} \downarrow \{0\}$ as $n \to \infty$, then

$$E^*[Z] = \lim_{n \to \infty} E[ZN(V_n)]/\lambda(V_n) \tag{1.49}$$

and

$$E^*[Z]/P^*(\Omega) = \lim_{n \to \infty} E[Z|N(V_n) > 0]. \tag{1.50}$$

Proof: a) This is straightforward and is omitted.

b) Taking $G(\omega, x) = Z(\theta_{-x}\omega)\mathbf{1}(x \in V_n)$ in (1.48) gives

$$E[ZN(V_n)] = E^*\left[\int_{V_n} Z(\theta_{-x}\omega)\lambda(dx)\right],$$

from which (1.49) follows by continuity of Z and the dominated convergence theorem.

To establish (1.50) we show first that

$$\lim_{n\to\infty} E[N(V_n)\mathbf{1}(N(V_n) \geq 2)] = 0.$$

Indeed, yet another application of (1.48) gives

$$
\begin{aligned}
E[N(V_n)\mathbf{1}(N(V_n) \neq 2)]/\lambda(V_n) &= E^*\left[\int_{V_n}\mathbf{1}(N(V_n - x) \geq 2)\lambda(dx)\right]\\
&\leq P^*\{N(V_n - V_n) \geq 2\}\\
&\to P^*\{N(\{0\}) \geq 2\} = 0,
\end{aligned}
$$

with the last equality holding because N is simple. Therefore,

$$\lim_{n\to\infty} \left|E[ZN(V_n)] - E[Z\mathbf{1}(N(V_n) > 0)]\right|/\lambda(V_n) = 0, \qquad (1.51)$$

and for $Z \equiv 1$,

$$\lim_{n\to\infty} P\{N(V_n) > 0\}/\lambda(V_n) = P^*(\Omega). \qquad (1.52)$$

Finally, (1.50) follows by combining (1.51) and (1.52). ∎

Here is the micro-level interpretation of the intensity.

Corollary 1.67. (Khintchine-Koroljuk theorem). Let N be a simple, stationary point process with intensity ν, and let V_n be neighborhoods of 0 such that $\overline{V_n} \downarrow \{0\}$. Then

$$\lim_{n\to\infty} P\{N(V_n) > 0\}/\lambda(V_n) = \lim_{n\to\infty} P\{N(V_n) = 1\}/\lambda(V_n) = \nu. \,\square \qquad (1.53)$$

Illustrative of the counterexamples yielded by the lattice process of Examples 1.58 and 1.65 is that for $V_n = (-1/n, 1/n)$ and $B = (1/2, 1)$, $P\{N(B) = 1 | N(V_n) \neq 0\} = 1/2$ for each n, while $P^*\{N(B) = 1\} = 0$. Thus, (1.49) can fail if the continuity hypothesis is not fulfilled.

The following result relates moment measures and reduced moment measures for stationary point processes.

Proposition 1.68. Let N be a stationary point process. Then for each k, the reduced moment measure μ_*^k of (1.46) satisfies $\mu_*^k = E^*[N^{k-1}]$. $\square$

The proof is left as Exercise 1.30.

Elsewhere in the book we present applications of Palm distributions to inference. We conclude the present discussion with a classical result from the theory of stationary point processes on $\mathbf{R}$.

Theorem 1.69. (Palm-Khintchine equations). Given a stationary point process N on $\mathbf{R}$ with Palm measure P^*, the *Palm functions*

$$p_k^*(t) = P^*\{N(0, t) = k\}$$

satisfy

$$P\{N(0,t) > k\} = \int_0^t p_k^*(s)ds. \qquad (1.54)$$

Proof: Note first that

$$\mathbf{1}(N(0,t) > k) = \int_{(0,t)} \mathbf{1}(N(s,t) = k)N(ds), \qquad (1.55)$$

i.e., $N(0,t)$ exceeds k if and only if there is some s at which N has a point, such that the interval (s,t) contains k additional points of N. We then apply (1.48) to the functional $G(N,s) = \mathbf{1}(N(0,t-s) = k)\mathbf{1}(0 < s < t)$:

$$\begin{aligned}
P\{N(0,t) > k\} &= E\left[\int_{(0,t)} \mathbf{1}(N(s,t) = k)N(ds)\right] \\
&= E\left[\int G(N\tau_s^{-1}, s)N(ds)\right] \\
&= E^*\left[\int G(N,s)ds\right] \\
&= \int_0^t P^*\{N(0,t-s) = k\}ds \\
&= \int_0^t P^*\{N(0,s) = k\}ds. \blacksquare
\end{aligned}$$

Moreover, (1.54) has a natural physical interpretation akin to that of (1.55): with ν denoting the intensity of N, νds is the rate at which points of N occur, while conditional on the presence of a point at s, $p_k^*(t-s)/\nu$ is the probability that the interval (s,t) contains k additional points.

We conclude by examining ergodicity for stationary point processes, which leads as well to the macro-level interpretation of the intensity. Suppose that N is stationary with respect to (θ_x).

Definition 1.70. a) An event Γ is *invariant* if $\theta_x^{-1}\Gamma = \Gamma$ for every x;

b) The point process N (or the probability P) is *ergodic* if $P\{\Gamma\}$ is 0 or 1 for every invariant event Γ.

We denote by $\mathcal{I}$ the σ-algebra of invariant events.

Here are the final results of the chapter, two ergodic theorems for point processes.

Theorem 1.71. Let N be a stationary point process with Palm measure P^*. Then

a) $P^* \ll P$ on $\mathcal{I}$ and for all $f \in C_K^+$,

$$E[N(f)|\mathcal{I}] = Y\lambda(f), \qquad (1.56)$$

where Y is the $\mathcal{I}$-measurable derivative dP^*/dP;

b) For $f \in C_K^+$,

$$\lim_{x \to \infty} \frac{\int_{\{0 \le y \le x\}} N \tau_y^{-1}(f) \lambda(dy)}{\lambda(\{y : 0 \le y \le x\})} = Y \lambda(f) \tag{1.57}$$

almost surely with respect to P and hence almost everywhere with respect to P^*. [In (1.57), inequalities and the limit are meant coordinatewise.]

Proof: a) Given $\Gamma \in \mathcal{I}$ and f, (1.48) implies that

$$E[E[N(f)|\mathcal{I}]; \Gamma] = E[N(f); \Gamma] = P^*(\Gamma) \lambda(f).$$

Choosing f_0 with $0 < \lambda(f_0) < \infty$ and putting $Y = E[N(f_0)|\mathcal{I}]/\lambda(f_0)$ thus gives a version of dP^*/dP on $\mathcal{I}$, and then (1.56) is a direct computation.

b) Existence of the almost sure limit and its equaling $E[N(f)|\mathcal{I}]$ ensue from general results in ergodic theory (see Dunford and Schwartz, 1958, and Matthes *et al.*, 1978). Identification of the limit is effected using (1.56). ∎

In particular, (1.57) implies that

$$\lim_{x \to \infty} N(\{y : 0 \le y \le x\}) / \lambda(\{y : 0 \le y \le x\}) = \nu_N$$

almost surely as well as in $L^1(P)$. Among other things, this provides a means of estimating ν_N; it and more general estimation procedures comprise the principal content of Chapter 9. However, in applications and theoretical analyses, it is of interest to allow more general sets K increasing to E or even just satisfying $\lambda(K) \to \infty$, along with more general functionals of N. The following spatial ergodic theorem of Nguyen and Zessin (1979a) does both of these and more.

Assume now that $E = \mathbf{R}^d$ for some d, and denote by $\mathcal{K}$ the family of bounded, convex subsets of E. For $K \in \mathcal{K}$, let $\delta(K)$ be the supremum of the radii of Euclidean balls contained in K. As $\delta(K) \to \infty$, K must expand to infinity in all directions.

We consider random measures

$$M(A) = \int_A N(dx) H(N \tau_x^{-1}) = \int_A N(dx) H(N \circ \theta_x) \tag{1.58}$$

where $H : \mathbf{M}_p \to \mathbf{R}$, which are covariant with respect to (θ_x) in the sense that for each A and y,

$$M(\tau_y A) \circ \theta_y = M(A). \tag{1.59}$$

The choice $H \equiv 1$ yields $M = N$, so that the next theorem extends Theorem 1.71.

Theorem 1.72. Let N be a stationary, ergodic point process with Palm measure P^* and let H be a function on $\mathbf{M}_p$ satisfying

$$E^*[H(N)] < \infty. \tag{1.60}$$

Then with M defined by (1.58),

$$\lim_{\delta(K)\to\infty} M(K)/\lambda(K) = E[M([0,1]^d)] \tag{1.61}$$

almost surely and in $L^1(P)$.

Proof: The underlying result in this case is the multi-dimensional ergodic theorem of Fritz (1970), which asserts that for $X \in L^1(P)$ and (K_α) a linearly ordered subclass of $\mathcal{K}$ with $\delta(K_\alpha) \to \infty$,

$$\lim(1/\lambda(K_\alpha))\textstyle\sum_{x\in K_\alpha} X \circ \theta_{-x} = E[X|\mathcal{I}]$$

almost surely. Here and below summations are over $\mathbf{Z}^d$. In particular, if P is ergodic, then

$$\lim(1/\lambda(K_\alpha))\textstyle\sum_{x\in K_\alpha} X \circ \theta_{-x} = E[X]. \tag{1.62}$$

Let $J = (0,1]^d$; then since M is a random measure and since as x varies over $\mathbf{Z}^d$ the sets $J + x$ constitute a partition of E,

$$
\begin{aligned}
M(K_\alpha) &= \textstyle\sum M(K_\alpha \cap (J+x)) \\
&= \sum_{x\in K_\alpha} M(J) \circ \theta_{-x} + \sum_{x\notin K_\alpha:(J+x)\cap K_\alpha\neq\emptyset} M((K_\alpha - x)\cap J) \circ \theta_{-x} \\
&\quad + \sum_{x\notin K_\alpha:(J+x)\cap K_\alpha^c\neq\emptyset} M((K_\alpha^c - x)\cap J) \circ \theta_{-x}, \tag{1.63}
\end{aligned}
$$

where we have also used (1.59). By (1.60) and (1.62),

$$
\begin{aligned}
\lim(1/\lambda(K_\alpha))\textstyle\sum_{x\in K_\alpha} M(J) \circ \theta_{-x} &= E[M(J)] \\
&= E\left[\textstyle\int_J N(dx)H(N\circ\theta_x)\right] = E^*[H(N)];
\end{aligned}
$$

this also confirms that (1.60) is the integrability hypothesis needed in order to apply (1.62).

The crucial property of convex sets satisfying $\delta(K) \to \infty$ is that their boundaries are small relative to their volumes, vanishing so in the limit. In (1.63) the second and third terms are, effectively, summations over the boundary of K_α and hence both, divided by $\lambda(K_\alpha)$, converge to zero. More precisely, the second term, which accounts for those $x \notin K_\alpha$ such that $J + x$ intersects K_α notwithstanding, may be bounded as follows:

$$\left|(1/\lambda(K_\alpha))\sum_{x \notin K_\alpha:(J+x)\cap K_\alpha \neq \emptyset}M((K_\alpha - x) \cap J) \circ \theta_{-x}\right|$$
$$\leq (1/\lambda(K_\alpha))\sum_{x \notin K_\alpha:d(x,\partial K_\alpha)\leq 1}|M(J)| \circ \theta_{-x} \to 0,$$

where $d(x, \partial K_\alpha)$ is the distance from x to the boundary of K_α. Here the convergence to zero occurs because $\lambda(\{x : d(x,\partial K_\alpha) \leq 1\})/\lambda(K_\alpha) \to 0$ (Exercise 1.35). The third term in (1.63) is handled analogously, which establishes almost sure convergence in (1.61). Convergence in $L^1(P)$ can either be shown directly (Nguyen and Zessin, 1979a) or deduced by combining almost sure convergence with uniform integrability. ∎

EXERCISES

1.1 a) Let N be a Poisson process with mean measure μ. Establish *conditional uniformity* of N: if $0 < \mu(A) < \infty$, then for each k, $P\{N_A \in \Gamma | N(A) = k\} = P\{\tilde{N}_k \in \Gamma\}$, where $\tilde{N}_k$ is the k-sample empirical process on A induced by the probability $\rho(B) = \mu(A \cap B)/\mu(A)$. (Given that N has k points in A, their locations are independent and identically distributed according to ρ.)
b) Show that if $\mu(E) < \infty$, then the Poisson process with mean measure μ is a mixed empirical process $N = \sum_{i=1}^K \varepsilon_{X_i}$, where the X_i are i.i.d. with distribution $\rho(B) = \mu(B)/\mu(E)$ and the random variable K is independent of the X_i and Poisson distributed with mean $\mu(E)$.

1.2 Let N be a Poisson process with mean measure μ. Prove that N is simple if and only if μ is diffuse.

1.3 Let N be a mixed Poisson process on E with diffuse mean measure μ. Prove that N is *symmetrically distributed* with respect to μ: whenever the sets $A_1, \ldots, A_k \in \mathcal{B}$ are disjoint with $\mu(A_i) = \mu(A_j) < \infty$ for each i and j, the random variables $N(A_1), \ldots, N(A_k)$ are interchangeable.

1.4 Let N be a Cox process on $\mathbf{R}_+$ directed by the diffuse random measure M, and let $N_t = N([0,t])$, with M_t defined analogously. Assume that $M_\infty = \infty$ almost surely, let $A_s = \inf\{t : M_t > s\}$, $s \in \mathbf{R}_+$, and define a point process $\tilde{N}$ by $\tilde{N}_s = N(A_s)$. Prove that $\tilde{N}$ is a homogeneous Poisson process with rate 1.

1.5 Let $N = \sum_{i=1}^\infty \varepsilon_{T_i}$ be a renewal process on $\mathbf{R}_+$ with interarrival distribution F satisfying $F(0) = 0$ (which implies that N is simple).
a) Let $\tilde{N}_t = 1 + N_t$. Prove that the mean measure of $\tilde{N}$ is the renewal measure $R = \sum_{n=0}^\infty F^n$, where F^n is the n-fold convolution of F with itself and $F^0 = \varepsilon_0$.
b) Show that for $f \geq 0$ the function $g(t) = E[\exp\{-\sum_{i=1}^\infty f(t+T_i)\}]$ satisfies the equation $g(t) = \int_0^\infty g(t+x)e^{-f(t+x)}F(dx)$.
c) Given $A \in \mathcal{B}$, derive an analogous equation for $z_N(A + t)$.

1.6 Let N be a simple point process on E such that $N(A)$ is Poisson distributed for every $A \in \mathcal{B}$. Prove that N is a Poisson process.

1.7 Let M_1 and M_2 be diffuse random measures on E. Prove that if there exists $t \in (0, \infty)$ such that $E[e^{-tM_1(A)}] = E[e^{-tM_2(A)}]$ for all $A \in \mathcal{E}$, then $M_1 \stackrel{d}{=} M_2$, [*Hint:* Use Cox processes.]

1.8 Prove that the zero-probability functional z_N of a point process N is *completely monotone* in the following sense. For a mapping Ψ of $\mathcal{B}$ into $\mathbf{R}$ and $A \in \mathcal{B}$, define $\Delta_A \Psi(B) = \Psi(A \cup B) - \Psi(B)$, and define iterates $\Delta_{A_1} \cdots \Delta_{A_n}$ recursively. Then Ψ is completely monotone if $(-1)^n \Delta_{A_1} \cdots \Delta_{A_n} \Psi(B) \geq 0$ for all choices of n, $A_1, \ldots, A_n$ and B.

1.9 a) Let N be the n-sample empirical process engendered by the probability ρ on E. Prove that $\mathcal{L}_N$ depends continuously on ρ.
b) Formulate and prove an analogous assertion for mixed empirical processes.

1.10 Let (N_n) be a sequence of point processes on E, all with the same mean measure $\mu \in \mathbf{M}$. Prove that (N_n) is tight.

1.11 Let $N = \sum \varepsilon_{X_i}$ be a Poisson process on E with mean measure μ, and let K be a transition probability from E to a second space E'.
a) Prove that there exist random elements Y_i of E' that are conditionally independent given N with $P\{Y_i \in B | N\} = K(X_i, B)$ for each i.
b) Prove that $\bar{N} = \sum \varepsilon_{(X_i, Y_i)}$ is a Poisson process on $E \times E'$ with mean measure $\mu(dx)K(x, dy)$.

1.12 This exercise applies the construction in Exercise 1.11 to a problem in hydrology. Let $E = E' = \mathbf{R}_+$, let $N = \sum \varepsilon_{T_i}$ represent times of flood peak exceedances in a river, and let Y_i be the height above a base level of the exceedance occurring at time T_i. (An "exceedance" is a local maximum in the flow, but between successive exceedances the flow must drop below the base level.) For each A, let $M(A) = \sup\{Y_i : T_i \in A\}$ be the maximum flood peak occurring in the time set A. Assume that the hypotheses of Exercise 1.11 are fulfilled.
a) Prove that for each A and u, $P\{M(A) > u\} = \exp[-\int_A \mu(dt)K(t, (u, \infty))]$.
b) Prove that the stochastic process $M_t = M([0, t])$ is a (nonhomogeneous) strong Markov process.

1.13 Let N be a Cox process directed by the random measure M. Prove that if M is infinitely divisible as a random measure, then N is infinitely divisible as a point process.

1.14 Verify (1.25)–(1.27).

1.15 Let $M = \sum U_i \varepsilon_{X_i}$ be the classical ρ-compound of a point process $N = \sum \varepsilon_{X_i}$ (Definition 1.37).
a) Calculate the Laplace functional and mean measure of M.
b) Prove that $\mathcal{L}_M$ depends continuously on $\mathcal{L}_N$ and ρ.
c) Calculate the Laplace functional, mean measure and zero-probability functional of the marked point process $\bar{N} = \sum \varepsilon_{(X_i, U_i)}$.

1.16 a) Let N be a Poisson process on E with mean measure μ, and let N' be the p-thinning of N, where p is a function from E to $[0,1]$. Prove that N' and $N'' = N - N'$ are independent Poisson processes, and determine their mean measures.

b) Let N be Poisson and let N' be a point process satisfying $N'(A) \leq N(A)$ for each A. Prove that if N' and $N'' = N - N'$ are independent, then N' is a p-thinning of N, where $p(x) = (d\mu_{N'}/d\mu_N)(x)$.

1.17 Let N_1 and N_2 be independent point processes on E. Prove that $N_1 + N_2$ is simple if and only if N_1 and N_2 are simple and $A_1 \cap A_2 = \emptyset$, where A_i is the set of fixed atoms of N_i [i.e., points x such that $P\{N_i(\{x\}) > 0\} > 0$].

1.18 a) Let N_1 and N_2 be independent point processes on E whose superposition $N_1 + N_2$ is a Poisson process. Prove that N_1 and N_2 are Poisson. [*Hint:* Use Raikov's theorem and Exercise 1.6.]

b) Construct an example of point processes N_1 and N_2 that are *not* independent but for which $N_1 + N_2$ is Poisson notwithstanding.

1.19 This exercise describes a Poisson process model of light traffic on an infinite one-dimensional road. Let X_i be the position in $\mathbf{R}$ at $t = 0$ of the ith vehicle, and let V_i be its velocity, which is random but does not vary over time. Assume that $N = \sum \varepsilon_{X_i}$ is a homogeneous Poisson process with rate 1 and that the V_i are i.i.d. with distribution G satisfying $G(\{0\}) = 0$. ("Light traffic" means that the assumption of independent velocities is credible.)

a) Prove that $\bar{N} = \sum \varepsilon_{(X_i,V_i)}$ is Poisson on $\mathbf{R}^2$ and determine its mean measure.

b) Prove that for each $t \in \mathbf{R}$, the point process $N^t = \sum \varepsilon_{X_i + tV_i}$, which gives vehicle positions at time t, is Poisson with rate 1.

c) For $y \in \mathbf{R}$, let $T_i(y) = (y - X_i)/V_i$ be the time, possibly negative, at which vehicle i passes the point y. Then $N^+(y) = \sum \mathbf{1}(V_i > 0)\varepsilon_{T_i(y)}$ gives times at which vehicles moving to the right pass y, and $N^-(y) = \sum \mathbf{1}(V_i < 0)\varepsilon_{T_i(y)}$ describes left-moving vehicles. Prove that for each y, $N^+(y)$ and $N^-(y)$ are independent Poisson processes and calculate their mean measures.

1.20 This exercise presents some properties of the queueing system $M/G/\infty$. Let $E = \mathbf{R}$ and let $N = \sum_{i=-\infty}^{\infty} \varepsilon_{T_i}$ be a homogeneous Poisson process with rate 1, whose points are arrival times of customers to a queueing system with infinitely many servers. (Taking $E = \mathbf{R}$ removes transient effects.) Let Y_i be the service time of the customer arriving at T_i. There is no wait for service, so this customer departs at time $T_i + Y_i$. Assume that the Y_i are independent of N and i.i.d. with distribution G.

a) Prove that the departure process $N^* = \sum_{i=-\infty}^{\infty} \varepsilon_{T_i + Y_i}$ is Poisson with rate 1. (That this is true regardless of G implies that observation of N^* alone yields no information about G.)

b) Suppose that N is an arbitrary point process but that the Y_i and N^* are as above, and that N^* is Poisson with rate 1. Prove or refute the assertion that N must also be Poisson.

1.21 Provide the details necessary to verify (1.38).

1.22 Given a point process N, let $p = P\{N \neq 0\}$ and

$$F(A) = P\{N(A) = 1 | N(E) = 1\}.$$

Let N' be Poisson with mean measure $\mu = [-\log(1-p)]F$. Prove that there exists a coupling (N, N') satisfying

$$P\{N \neq N'\} = \max\{P\{N(E) \geq 2\}, P\{N'(E) \geq 2\}\}.$$

1.23 Let N be the n-sample empirical process engendered by a probability ρ on E. Calculate the reduced Palm distributions of N.

1.24 Let M be the random measure $M(dt) = Y_t dt$ on $\mathbf{R}_+$, where Y is a positive random process. Show that $M^k(dt_1, \ldots, dt_k) = Y(t_1) \cdots Y(t_k) dt_1 \cdots dt_k$ for each k, and hence that

$$Q^*_M \left(\textstyle\sum_{i=1}^k \varepsilon_{t_i}, \Gamma \right) = E[Y(t_1) \cdots Y(t_k) \mathbf{1}(M \in \Gamma)]/E[Y(t_1) \cdots Y(t_k)].$$

1.25 Let N be a simple point process on E with Palm processes N_μ, $\mu \in \mathbf{M}_s$. Prove that for each μ and $\nu \in \mathbf{M}_s$, $N_{\mu+\nu} = (N_\mu)_\nu$.

1.26 Establish the assertions in Example 1.58.

1.27 Prove (1.47).

1.28 Let N be a stationary point process on $\mathbf{R}$. Prove that

$$P\{N = 0\} + P\{N((-\infty, 0)) = N((0, \infty)) = \infty\} = 1.$$

1.29 Show that if N is a homogeneous Poisson process then the Palm-Khintchine equations, in the differential form $(d/dt)P\{N(0,t) > k\} = p^*_k(t)$, are simply the Kolmogorov forward equations for the Markov process $N_t = N(0,t]$.

1.30 Prove Proposition 1.68.

1.31 This exercise concerns stationary renewal processes. Given a probability distribution F on $(0, \infty)$ with finite mean m, define the ("excess lifetime") distribution $G(t) = (1/m) \int_0^t [1 - F(u)] du$. Let $N = \sum_{i=-\infty}^\infty \varepsilon_{T_i}$ be a point process constructed as follows. (Recall that $U_i = T_i - T_{i-1}$ are the interarrival times.)
i) $\cdots < T_0 \leq 0 < T_1 < \cdots$;
ii) $(-T_0, U_1)$ and the U_i, $i \neq 1$, are independent;
iii) the U_i, $i \neq 1$, have distribution F;
iv) $(-T_0, U_1)$ has distribution $(1/m)\mathbf{1}(u < v)du F(dv)$.
Then N is a *stationary renewal process* with interarrival distribution F.
a) Prove that N is stationary.
b) Prove that $-T_0$ and T_1 both have distribution G.
c) Prove that U_1 has distribution $H(t) = (1/m) \int_0^t v F(dv)$.
d) Prove that $P^*(\Omega) = 1/m$, where P^* is the Palm measure of N.
e) Prove that under Palm distribution $P^*(\cdot)/P^*(\Omega)$ the U_i are i.i.d. with distribution F.

1.32 Let $N = \sum_{n=-\infty}^{\infty} \varepsilon_{T_n}$ be a simple, stationary point process with finite intensity ν and Palm measure P^*.
   a) Prove that $P^*\{T_0 \neq 0\} = 0$.
   b) Prove that N is simple if and only if $P^*\{N(\{0\}) \neq 1\} = 0$.
   c) Let $F^*(\cdot) = P^*\{T_1 \in (\cdot)\}$. Prove that $\int v F^*(dv) = 1$.
   d) Prove that $P\{-T_0 \in du, U_1 \in dv\} = 1(0 < u < v)F^*(dv)$.
   e) Prove that for each positive random variable Z,

$$E[Z] = E^* \left[\int_0^{T_1} Z \circ \theta_s \, ds \right].$$

(In particular, $E^*[T_1] = 1$.)
   f) Prove that under the Palm distribution $P_0 = P^*/\nu$ the interarrival sequence (U_i) is strictly stationary.

1.33 Let N be a Cox process on $\mathbf{R}^d$ with directing measure $M(dx) = Y(x)\lambda(dx)$, where $(Y(x))$ is a positive, measurable stochastic process parametrized by E and λ is Lebesgue measure. (In Chapter 7, Y is termed the directing intensity of N.) Prove that N is stationary if and only if Y is strictly stationary.

1.34 Verify (1.59).

1.35 Let (K_α) be linearly ordered, bounded convex sets in $\mathbf{R}^d$ such that $\delta(K_\alpha) \to \infty$. Prove that $\lambda(\{x : d(x, \partial K_\alpha) \leq 1\})/\lambda(K_\alpha) \to 0$.

1.36 Let N be a point process on E with mean measure satisfying $\mu_N(E) = \infty$. Prove that N is a mixed Poisson process if and only if $\mathcal{L}_{N_\mu}$ does not depend on μ (a.e. with respect to K_N).

1.37 Let $X_1, X_2, \ldots$ be i.i.d., positive random variables with distribution function F. Let $M_n = \max\{X_1, \ldots, X_n\}$ the nth partial maximum, and assume that there are constants $a_n > 0$ and b_n and a distribution function H such that $P\{(M_n - b_n)/a_n \leq x\} \to H(x)$ for each x. Then, H is one of the classical "extreme value distributions," and has the form

$$H(x) = \exp[-(1 - k(x - \eta)/\sigma)^{1/k}],$$

k and η are real and $\sigma > 0$. For each n, let N_n be the point process on $(0, \infty) \times (0, \infty)$ given by $N_n = \sum_{i=1}^{n} \varepsilon_{(i/(n+1),(X_i - b_n)/a_n)}$. Prove that N_n converges in distribution to a Poisson process N whose mean measure μ satisfies

$$\mu((t_1, t_2) \times (x, \infty)) = (t_2 - t_1)[1 - k(x - \eta)/\sigma]^{1/k}.$$

NOTES

General sources on the distributional — as opposed to martingale — theory of point processes are Cox and Isham (1980), at a relatively elementary level; Daley and Vere-Jones (1988), detailed yet comprehensible; Kallenberg (1983), elegant, measure-theoretic and complete, albeit austere and terse; Matthes *et al.* (1978), voluminous but very difficult reading; and Snyder (1975), which, although restricted to point processes on **R**, contains extensive discussion of engineering applications, especially to communication.

Section 1.1

The material here is standard. A proof of (1.1) is given in Kallenberg (1983). Poisson processes are, of course, the most fundamental of point processes, and their history on the real line is lengthy. That of Poisson processes on general spaces is briefer; two early sources are Doob (1953) and Dobrushin (1956). Cox process are treated first in Lundberg (1940) and Cox (1955); Grandell (1976) and Snyder (1975) are more modern expositions focusing on state estimation, which we treat in detail in Sections 7.2 and 7.3. Empirical processes, whose role — especially in statistics — is likewise fundamental, and the subject of Gaenssler (1984) and Pollard (1984); see also Kallenberg (1983). Feller (1971) remains the definitive reference on renewal theory, but does not include recent developments, especially those relating to regenerative processes. See Chapter 8 for additional sources as well as a summary of key components of the theory.

Section 1.2

Uniqueness results are not given here in the sharpest form; for example the class of sets for which the equalities in Theorem 1.12 must obtain need only be, in the terminology of Kallenberg (1983), a "DC-semiring." The added generality is not purely formal, but we do not need it here. A characteristic functional for random measures, with definition and properties analogous to those of the Laplace functional, is analyzed in von Waldenfels (1968).

Section 1.3

Kallenberg (1983) and Matthes *et al.* (1978) contain more detailed and complete presentations of convergence in distribution for point processes and (in the former) random measures. Lemma 1.20 is exceedingly important: it renders tightness (typically the most difficult step in proving convergence in distribution) quite routine. Proposition 1.23 is proved in Kallenberg (1983). An alternative methodology for demonstrating convergence in distribution of point processes on $\mathbf{R}_+$ is discussed in Theorems 2.25 and 2.26.

Section 1.4

Marked point processes are introduced in Matthes (1963a); see Matthes *et al.* (1978) and Franken *et al.* (1981) for expository treatments. The former also contains detailed discussion of clustered point processes. Matthes (1963a) first examined infinitely divisible point processes; P. M. Lee (1967) is another early reference. Despite its being difficult, Matthes *et al.* (1978) is *the* reference on infinitely divisible point processes. The Poisson cluster representation in Theorem 1.32 is due in stages to P. M. Lee (1967), Jagers (1972) and Matthes *et al.* (1978). Random measures with independent increments were first investigated, under the rubric "completely random measures," by Kingman (1967).

Section 1.5

Thinned and compound Poisson processes have been studied at least since the 1940s. Jagers and Lindvall (1974) and, especially, Kallenberg (1975a) formalized the terminology and established characterizations as limits of rarefied processes (an analogue of Theorem 1.41 holds for compound point processes). The two characterizations — Theorems 1.40 and 1.41 — are due to Mecke (1968) and Kallenberg (1975b).

Our definition of a null array is more restrictive and less elegant than that of Kallenberg (1983), whose assumes, in effect, that $k_n = \infty$ for each n, with the proviso that in most case all but finitely many of the N_{ni} are zero. Theorem 1.43 comes originally from Kerstan and Matthes (1964a). The principal results admit random measure analogues as well.

Section 1.6

The coupling in Theorem 1.44 is based on techniques for Bernoulli and Poisson random variables introduced by Serfling (1978). Theorem 1.46 appears initially here. For variations and refinements, see Deheuvels *et al.* (1988).

Section 1.7

Conditioning ideas identifiable as pertaining to what are now termed Palm distributions are indeed found in Palm (1943), but Khintchine (1960) is the first systematic investigation. Ryll-Nardzewski (1961) is another principal early work. Extensions to more general spaces and processes were achieved by Jagers (1973) and Papangelou (1974b); aspects of these are discussed in Section 2.6. A complete, up-to-date development is presented in Kallenberg (1983), where Palm distributions are linked with concepts from statistical physics such as Gibbs distributions (see Section 2.6 for additional discussion). As mentioned in the text, except in the context of stationary point processes, unreduced Palm distributions for point processes are of little serious interest. Proposition 1.55 is given for $\mu = \varepsilon_x$ in the unreduced case in Grandell (1976) and for general μ by Karr (1985a).

Section 1.8

The earliest general treatment of stationary point processes is Khintchine (1960); Daley and Vere-Jones (1988), Matthes *et al.* (1978) and Neveu (1977) are more recent expository presentations. Much of the general theory for spaces other than **R** was given first in Matthes (1963a), Kerstan and Matthes (1964a) and Franken *et al.* (1965). A proof of Proposition 1.60 can be found, for example, in Krickeberg (1974). Proposition 1.64 is due to Ryll-Nardzewski (1964) and Slivnjak (1962).

Orderliness and regularity are examined in Leadbetter (1968, 1972a), Daley and Vere-Jones (1972, 1988) and Daley (1974). The result now known as Dobrushin's lemma — that a stationary, orderly point process [with $(1/t)P\{N(0,t) > 2\} \to 0$ as $t \to 0$, a weakened form of (1.52)] is simple — is attributed to Dobrushin by Volkonskii (1960). Koroljuk's theorem, which identifies the limit in (1.53), is so-named in Khintchine (1960) but seems never to have appeared under Koroljuk's name. Khintchine's contribution to Corollary 1.67 is existence of the limit. The Palm-Khintchine equations of Theorem 1.69 are due to Khintchine (1960), and the elegant proof to Neveu (1977).

Elaboration of ergodicity for stationary point processes can be found in Daley and Vere-Jones (1988), Matthes *et al.* (1978) and Neveu (1977). Theorem 1.72 is due to Nguyen and Zessin (1979a).

Some omissions merit mention. Stationary Poisson cluster processes are understood in much detail; Matthes *et al.* (1978) is the most complete reference. Applications as diverse as cosmology (Neyman and Scott, 1972), earthquakes (Vere-Jones, 1970), queueing systems (Franken *et al.*, 1981, which treats in addition marked point processes that are stationary in only one coordinate), and stereology (Mecke and Stoyan, 1983) have been examined.

Exercises

1.1 A systematic analysis occupies several sections of Matthes *et al.* (1978). Conditional uniformity characterizes mixed Poisson processes among Cox processes.

1.3 Kallenberg (1975b) shows that this interchangeability property also characterizes mixed Poisson processes and mixed empirical processes. His definition of "symmetrically distributed" is different from ours, but equivalent.

1.4 See Serfozo (1977) for general discussion of random time changes for random measures.

1.6 This result is due to Rényi (1967).

1.7 Kallenberg (1983) employs this device systematically; see also the discussion following Proposition 1.23.

1.8 For the converse (i.e., with minor restrictions a completely monotone mapping Ψ with $\Psi(\emptyset) = 1$ is the zero-probability functional of some point process), see Kurtz (1974).

1.12 The model and results are due to Karr (1976a).

1.13 The converse is false even when E is a singleton.

1.16 See Karr (1985b) for an extension to arbitrary point processes.

1.17 Although intuitive, this property is nontrivial; see Franken *et al.* (1981) for a proof.

1.18 Çinlar (1972) presents this and related results. In connection with b), see Jacod (1975a).

1.19 The model is examined in Rényi (1964a) and Solomon and Wang (1972). Limit versions of these invariance properties are proved in Breiman (1963) and Stone (1968a).

1.20 This queue is, of course, classical. M. Brown (1970) and Thédéen (1964, 1967a) consider related properties and the extent to which they characterize Poisson processes (see also Exercise 6.24). More on departure processes, which have been an important motivation for study of non-Poisson point processes, is given in Brémaud (1981) and Franken *et al.* (1981).

1.21 See Dudley (1972), Gaenssler (1984) and Whitt (1973).

1.22 This result can be applied in conjunction with results in Section 6 to provide Poisson approximations to non-Bernoulli point processes.

1.28 Like Exercise 1.17, this property is intuitive but not trivial; it is due to Ryll-Nardzewski (1960). More generally, if N is stationary and A is a subset of E^k invariant under the transformation $(y_1, \ldots, y_k) \to (y_1 - x, \ldots, y_k - x)$ for each $x \in E$, then $P\{N^k(A) = 0\} + P\{N^k(A) = \infty\} = 1$. See Proposition 9.1.

1.31 See Neveu (1977) for more on stationary renewal processes. Further aspects, including inference, are treated in Chapter 9.

1.32 Note the analogy to Exercise 1.31.

1.36 This and several similar characterizations based on Palm distributions appear in Kallenberg (1983).

1.37 See Pickands (1971).

2
Point Processes: Intensity Theory

The compensator of a point process N on $\mathbf{R}_+$ is a random measure A interpretable as the local conditional mean of N given its strict past: informally, $A(dt) = E[N(dt)|\mathcal{F}^N([0,t))]$. The compensator is ordinarily smoother than N and in nice cases $A_t = \int_0^t \lambda_u du$ for some random process λ, the *stochastic intensity* of N. In this chapter we develop the theory of compensators and stochastic intensities for point processes on $\mathbf{R}_+$, which is linked intimately to that of martingales, because the difference between a point process and its compensator is a (mean zero) martingale, to which various interpretations can be assigned. In information terms, it is an innovation process conveying all information about N not derivable from the strict past. In statistical settings, by contrast, it constitutes noise (akin, by virtue of its being orthogonal to the past, to the residuals in a regression problem), representing aspects that cannot be estimated. A rather general theory of inference for point processes in which the statistical model is specified via the stochastic intensity is described in Chapter 5. In this setting martingales serve a fundamental role in computations of means, variances, and covariances, and as the basis for important central limit theorems. Applications to specific classes of point processes appear in Chapters 7, 8 and 9.

For point processes on general spaces the topic of conditioning is much more difficult, principally because of the absence of a linear order defining a natural way of observation. Nevertheless, useful results do exist; the central objects are the Palm distributions introduced in Section 1.7. In addition, we treat the theory of exvisible projections, developed along lines parallel to compensators.

The latter topic constitutes Section 6; the remainder of the chapter is

53

organized as follows. Section 1, after setting notation and terminology, portrays the structure of the internal history of a marked point process on $\mathbf{R}_+$. Stochastic Stieltjes integrals are treated briefly in Section 2. In Section 3 we define compensators of point processes and marked point processes and establish important properties. These include not only characterizations and explicit constructions but also applications to convergence in distribution and approximation of point processes. When the compensator is absolutely continuous, the derivative is a random process called the stochastic intensity; this special, especially important, and (fortunately) rather frequent case is the topic of Section 4. Only one result appears in Section 5: every $\mathcal{F}^N$-martingale is the integral of a predictable process with respect to the innovation martingale $M = N - A$; this theorem confirms rather convincingly the innovation interpretation of M and is a key result in the theory.

2.1   The History of a Marked Point Process

Let $N = \sum_{n=1}^{\infty} \varepsilon_{(T_n, Z_n)}$ be a marked point process on $\mathbf{R}_+$ with mark space $(E, \mathcal{E})$. Recall the interpretations: the nth event occurs at time T_n and has mark $Z_n \in E$. The interarrival times are denoted by $U_n = T_n - T_{n-1}$, with $U_1 = T_1$. Throughout we assume that the point process $\sum_{n=1}^{\infty} \varepsilon_{T_n}$ on $\mathbf{R}_+$ is simple. For $B \in \mathcal{E}$, let $N_t(B) = \sum_{n=1}^{\infty} \mathbf{1}(T_n \leq t, Z_n \in B)$, with $N_t = N_t(E)$. Here and elsewhere, when convenient we switch freely between measure and distribution function notation. As in Section 1.1, define

$$\mathcal{F}_t^N = \sigma(N_s(B) : 0 \leq s \leq t, B \in \mathcal{E}),$$

and further define $\tilde{\mathcal{F}}_t^N = \mathcal{F}_t^N \bigvee \mathcal{N}$, where $\mathcal{N}$ is the family of all P-null sets engendered by $\mathcal{F}_\infty^N$. Our principal goal in this section is to describe the structure of the filtration $\mathcal{F}^N = (\mathcal{F}_t^N)$, virtually all of which is inherited by $(\tilde{\mathcal{F}}_t^N)$; the former is the real object of interest, but it is sometimes necessary for technical reasons to work with the latter. We term each the *internal history* of N.

We digress briefly to recall some ideas from the "*théorie générale des processus.*" Although phrased in terms of $\mathcal{F}^N$ for concreteness, they apply to any filtration. There are two key concepts: stopping times and various forms of measurability for stochastic processes.

A random variable T with values in $[0, \infty]$ is a *stopping time* if $\{T \leq t\} \in \mathcal{F}_t^N$ for every t; that is, whether T has occurred before t can be determined by observation of N over $[0, t]$. Associated with the stopping time T are the σ-algebra

$$\mathcal{F}_T^N = \{\Lambda \in \mathcal{F}_\infty^N : \Lambda \cap \{T \leq t\} \in \mathcal{F}_t^N \text{ for all } t\},$$

which represents evolution of N until the random time T (see Proposition 2.3 for a more precise statement) and the σ-algebra

$$\mathcal{F}_{T-}^N = \sigma(\Lambda \cap \{T > t\} : t \geq 0, \Lambda \in \mathcal{F}_t^N) \vee \mathcal{F}_0^N.$$

The latter comprises the *strict past* of N at T in much the same way that $\mathcal{F}_{t-}^N = \mathcal{F}^N([0,t))$ is the strict past at the deterministic time t. We have $\mathcal{F}_{T-}^N \subseteq \mathcal{F}_T^N$ and $T \in \mathcal{F}_{T-}^N$. (Here and throughout we write that a random variable belongs to a σ-algebra to mean that it is measurable.)

Suppose now that $X = (X_t)$ is a random process. The least restrictive measurability relationship between X and $\mathcal{F}^N$ is that as a mapping on $\mathbf{R}_+ \times \Omega$, X be *measurable* with respect to $\mathcal{B}(\mathbf{R}_+) \times \mathcal{F}_\infty^N$ and that X be *adapted* to $\mathcal{F}^N$ in the sense that $X_t \in \mathcal{F}_t^N$ for each t. These minimal assumptions are too weak in many instances; they fail to imply, for example, that the process $Y_t = \int_0^t X_u du$ even be adapted. A stronger condition resolves this difficulty: X is *progressive* (sometimes, *progressively measurable*) if for every t the mapping $(u, \omega) \to X_u(\omega)$ on $[0, t] \times \Omega$ is measurable with respect to $\mathcal{B}([0, t]) \times \mathcal{F}_t^N$. Evidently, if X is progressive, then so is the process $\int_0^t X_u du$. Every left- or right-continuous process is progressive.

Predictability is a fundamental concept. In applications such as state estimation and martingale inference one must distinguish processes whose value for every t is determined by $\mathcal{F}_{t-}^N$ from those for which error-free prediction of the current state from the strict past is impossible. After some terminology this leads to the class of predictable processes.

The *predictable σ-algebra* $\mathcal{P}$ generated by N is the σ-algebra on $\mathbf{R}_+ \times \Omega$ generated by the π-system of sets $(s, t] \times \Lambda$, where $s < t$ and $\Lambda \in \mathcal{F}_s^N$. Other characterizations include:

a) $\mathcal{P} = \sigma((s, \infty) \times \Lambda : s \geq 0, \Lambda \in \mathcal{F}_s^N)$;

b) $\mathcal{P} = \sigma(\{(t, \omega) : t \leq T(\omega)\} : T$ is a bounded stopping time);

c) $\mathcal{P} = \sigma(X : X$ is adapted and left-continuous);

d) $\mathcal{P} = \sigma(X : X$ is adapted and continuous). A process X, then, is *predictable* if $X \in \mathcal{P}$ as a function on $\mathbf{R}_+ \times \Omega$.

Justification of the term resides in the property that if X is predictable and T is a finite stopping time, then X_T is $\mathcal{F}_{T-}^N$ measurable. By contrast, if X is merely progressive, then $X_T \in \mathcal{F}_T^N$ but no stronger statement holds in general. Every predictable process is progressive. For practical purposes predictable may be taken to be synonymous with *left-continuous* and adapted.

We now describe the structure of the internal history $\mathcal{F}^N$. The heuristic description is simple: the accumulating information changes only at the arrival times T_n, and whereas $T_n \in \mathcal{F}_{T_n-}^N$, the mark Z_n belongs to $\mathcal{F}_{T_n}^N$ but not $\mathcal{F}_{T_n-}^N$. Finally, Z_n is the only information in $\mathcal{F}_{T_n}^N$ not contained in $\mathcal{F}_{T_n-}^N$.

Here are the most elementary properties.

Proposition 2.1. a) For each t

$$\mathcal{F}_t^N = \sigma\big(1(T_n \leq s, Z_n \in B) : n \geq 1, 0 \leq s \leq t, B \in \mathcal{E}\big); \qquad (2.1)$$

b) The filtration $\mathcal{F}^N$ is right-continuous: for each t, $\mathcal{F}_t^N = \bigcap_{h>0} \mathcal{F}_{t+h}^N$. □

We next note properties of $\mathcal{F}^N$-stopping times.

Proposition 2.2. a) For each n, T_n is a stopping time of $\mathcal{F}^N$;
b) For each n,

$$\mathcal{F}_{T_n}^N = \sigma((T_1, Z_1), \ldots, (T_n, Z_n)), \qquad (2.2)$$

$$\mathcal{F}_{T_n-}^N = \sigma((T_1, Z_1), \ldots, (T_{n-1}, Z_{n-1}), T_n). □ \qquad (2.3)$$

The proof is deferred until after the next result, which generalizes (2.1) and is a key property.

Proposition 2.3. If T is a stopping time of $\mathcal{F}^N$, then

$$\mathcal{F}_T^N = \sigma(N_{T \wedge s}(B) : s \geq 0, B \in \mathcal{E}). \qquad (2.4)$$

Proof: That $\supseteq$ holds in (2.4) is clear, and we may prove the opposite inclusion under the restriction that T assumes only countably many values $0 \leq t_0 < t_1 < \cdots \leq \infty$. Let t be fixed and suppose that $\Lambda \in \mathcal{F}_t^N$; then since $\Lambda_k = \Lambda \cap \{T = t_k\}$ belongs to $\mathcal{F}_{t_k}^N$ there exist a countable dense subset V of $\mathbf{R}_+$ and a function f_k such that

$$1_{\Lambda_k} = f_k(N_t(B) : t \in V \cap [0, t_k], B \in \mathcal{E}) \qquad (2.5)$$

But $T \wedge t = t$ on Λ_k whenever $t \leq t_k$, and this converts (2.5) to $1_{\Lambda_k} = f_k(N_{T \wedge t}(B) : t \in V \cap [0, t_k], B \in \mathcal{E})$. ∎

The proof of (2.2) then proceeds as follows. To prove $\subseteq$, it suffices by (2.4) to observe that $N_{T_n \wedge t}(B) \in \sigma((T_1, Z_1), \ldots, (T_n, Z_n))$ for all n, t and B, which is obvious, as is $\supseteq$ in (2.2). For (2.3), $\supseteq$ holds because $T_n \in \mathcal{F}_{T_n}^N$, while for $\subseteq$ by the monotone class theorem we need only show that generators of $\mathcal{F}_{T_n}^N$ of the form $\{N_s(B) = k, t < T_n\}$ belong to $\sigma((T_1, Z_1), \ldots, (T_{n-1}, Z_{n-1}), T_n)$, which follows immediately from

$$\{N_s(B) = k, t < T_n\} = \{t < T_n\} \bigcap \{[\textstyle\sum_{i=1}^n 1(T_i \leq s, Z_i \in B)] = k\}.$$

Two additional results elucidate further properties of stopping times.

Proposition 2.4. Given a finite stopping time T of $\mathcal{F}^N$, for each n,

$$\mathcal{F}_T^N \cap \{T_n \le T < T_{n+1}\} = \mathcal{F}_{T_n}^N \cap \{T_n \le T < T_{n+1}\}. \tag{2.6}$$

Proof: For each t, B and k,

$$\{N_{T \wedge t}(B) = k\} \cap \{T_n \le T < T_{n+1}\} = \{N_{T_n \wedge t}(B) = k\} \cap \{T_n \le T < T_{n+1}\},$$

and (2.6) follows by two applications of (2.4) and the monotone class theorem. ∎

Proposition 2.5. Let T be a stopping time of $\mathcal{F}^N$. For each n there is a random variable $U_n \in \mathcal{F}_{T_n}^N$ such that on $\{T_n \le T\}$,

$$T \wedge T_{n+1} = (T_n + U_n) \wedge T_{n+1}. \tag{2.7}$$

Proof: Propositions 2.2 and 2.4 entail existence of functions g_n satisfying $T\mathbf{1}(T_n \le T < T_{n+1}) = g_n(T_1, Z_1, \ldots, T_n, Z_n)\mathbf{1}(T_n \le T < T_{n+1})$, so we may take $U_n = [g_n(T_1, Z_1, \ldots, T_n, Z_n) - T_n]^+$. ∎

As will be seen in the proofs of Theorems 2.18 and 2.34, the point of (2.7) is that $U_n \in \mathcal{F}_{T_n}^N$.

We conclude the section with two constructions of predictable processes with particular properties.

Proposition 2.6. For each n let Y_n be a $\mathcal{B}(\mathbf{R}_+) \times \mathcal{F}_{T_n}^N$-measurable mapping on $\mathbf{R}_+ \times \Omega$. Then the process $X_t = Y_n(t)$, $t \in (T_n, T_{n+1}]$, is predictable.

Proof: The processes $\mathbf{1}(T_n \le T < T_{n+1})$ are predictable, so X may be treated "piecewise." By the monotone class theorem we may suppose that $Y_n(t, \omega) = \mathbf{1}(t \in V)\mathbf{1}(\omega \in \Lambda)$ for $V \in \mathcal{B}(\mathbf{R}_+)$ and $\Lambda \in \mathcal{F}_{T_n}^N$, in which case

$$Y_n(t)\mathbf{1}(T_n \le T < T_{n+1}) = \mathbf{1}(t \in V)\mathbf{1}(\tilde{T}_n \le T < \tilde{T}_{n+1}),$$

where $\tilde{T}_n = T_n$ on Λ and is infinite on Λ^c. The right-hand side of this expression is evidently predictable. ∎

Proposition 2.7. Given $\Lambda \in \mathcal{F}_{T_n}^N$, there is a predictable process $X \ge 0$ such that $X_t = 0$ unless $t \in (T_{n-1}, T_n]$ and such that $X_{T_n} = \mathbf{1}_\Lambda$. ☐

2.2   Stochastic Stieltjes Integration

Here we summarize some ideas associated with stochastic integration that are needed in the book, but with no pretense of completeness. In particular, all integrals are simply stochastic Stieltjes integrals, because the integrators are either increasing processes (e.g., point processes) or differences of increasing processes (with, therefore, paths of bounded variation over compact intervals). Specifically, the latter are usually innovation martingales. A central idea is the intimate relationship among martingales, predictable processes, and integrals with respect to martingales.

Suppose that $(\Omega, \mathcal{F}, P)$ is a probability space, that Y is a measurable stochastic process with sample paths of locally bounded variation, and that X is a measurable process whose sample paths are locally bounded. Then the process $X * Y$ defined by

$$(X * Y)_t = \int_0^t X_s dY_s \tag{2.8}$$

is called the *stochastic Stieltjes integral* of X with respect to Y. In (2.8) and elsewhere the interval of integration is $(0, t]$; that the right-hand endpoint is included is crucial. For each t and ω the integral in (2.8) exists as a Lebesgue-Stieltjes integral and for each t, $(X * Y)_t$ is a random variable. However, as hinted in Section 1, further measurability typically is needed; ordinarily, we assume that X is predictable and Y is progressive, in which case $X * Y$ is progressive.

The definition (2.8) remains meaningful if Y is an increasing process $[t \to Y_t(\omega)$ is increasing for each $\omega]$ and X is nonnegative. In particular, if $N = \sum \epsilon_{T_n}$ is a point process on $\mathbf{R}_+$, then for X measurable and either locally bounded or nonnegative, the stochastic integral $X * N$ exists, and for each t,

$$(X * N)_t = \sum_{n=1}^{\infty} \mathbf{1}(T_n \leq t) X_{T_n}.$$

In view of (2.8), properties of $X * Y$ mirror those of Lebesgue-Stieltjes integrals and do not require special mention. For the sake of completeness, however, we do include two formulas for integration by parts, which are used later in the chapter.

Proposition 2.8. Let a and b be functions of locally bounded variation on $\mathbf{R}_+$, each right-continuous with left-hand limits. Then for each t,

$$a(t)b(t) - a(0)b(0) = \int_0^t a(s-)db(s) + \int_0^t b(s)da(s) \tag{2.9}$$

$$= \int_0^t a(s-)db(s) + \int_0^t b(s-)da(s)$$

$$+ \sum_{s \leq t} \Delta a(s) \Delta b(s), \tag{2.10}$$

where $a(s-)$ is the left-hand limit of a at s and $\Delta a(s) = a(s) - a(s-)$. $\square$

Suppose now that $\mathcal{H}$ is a filtration on $(\Omega, \mathcal{F}, P)$ and that M is a martingale with locally integrable variation, in the sense that for each $t > 0$,

$$E\left[\int_0^t d|M|_s\right] < \infty, \tag{2.11}$$

where $|M|$ is the total variation process associated with M: $|M|_t$ is the variation of the path $s \to M_s$ over $[0, t]$. One of the truly important results in stochastic integration is that the integral of a predictable process C with respect to M is another martingale.

Heuristically, the argument is simple but revealing. In differential form, the martingale M is characterized by the property that $E[dM_t|\mathcal{H}_t] = 0$ for each t, so that

$$E[d(C * M)_t|\mathcal{H}_t] = E[C_{t+}dM_t|\mathcal{H}_t] = C_{t+}E[dM_t|\mathcal{H}_t] = 0.$$

Note the crucial role of predictability in enabling one to bring C_{t+} outside the conditional expectation. Here are a more formal statement and more rigorous proof.

Theorem 2.9. Let M be a martingale fulfilling (2.11) and let C be a predictable process satisfying $E[\int_0^t |C_s| \, d|M|_s] < \infty$ for each t. Then the stochastic integral $C * M$ is a martingale.

Proof: Suppose that C is a "primitive" predictable process of the form $C(u, \omega) = \mathbf{1}(s < u \le t)\mathbf{1}(\omega \in \Lambda)$, where $s < t$ and $\Lambda \in \mathcal{H}_s$. Then $C * M$ exists, is adapted, and evidently satisfies $E[|(C * M)_t|] < \infty$ for each t. Moreover, for $u < v$ and $\Gamma \in \mathcal{H}_u$,

$$E[\{(C * M)_v - (C * M)_u\}; \Gamma] = E[(M_{v \wedge t} - M_{u \wedge t}); \Lambda \cap \Gamma] = 0,$$

which confirms that $C * M$ is a martingale. An application of the monotone class theorem completes the proof. ∎

2.3 Compensators

We begin with a point process $N = \sum_{n=1}^{\infty} \varepsilon_{T_n}$ on $\mathbf{R}_+$, with counting process $N_t = \sum_{n=1}^{\infty} \mathbf{1}(T_n \le t)$, and a filtration $\mathcal{H} = (\mathcal{H}_t)$ such that the following hold.

Assumptions 2.10. a) $\mathcal{H}$ satisfies "*les conditions habituelles de la théorie générale du processus*" (the "usual conditions" — see Dellacherie, 1972).

b) $\mathcal{F}_t^N \subseteq \mathcal{H}_t$ for each t (N is adapted to $\mathcal{H}$).

c) $E[N_t] < \infty$ for each t.

The first condition is required for technical reasons, and the second is obviously needed. The third is for simplicity: it permits us to avoid local martingales; moreover, it is fulfilled in many, if not most, important applications. When it fails one "localizes" the results given below (see Appendix B); the standard choice of localizing times is the arrival times T_n.

The notion of the compensator of a point process is the fundamental new idea of the section.

Definition 2.11. Let N be a point process on $\mathbf{R}_+$ satisfying Assumptions 2.10. The *compensator* of N with respect to $\mathcal{H}$ is the unique random measure A on $\mathbf{R}_+$ such that

    a) The process (A_t) is $\mathcal{H}$-predictable;

    b) For every nonnegative $\mathcal{H}$-predictable process C,

$$E\left[\int_0^\infty C\, dN\right] = E\left[\int_0^\infty C\, dA\right]. \tag{2.12}$$

Existence and uniqueness of the compensator can be proved, as in Dellacherie (1972), using projection methods. One first defines the *predictable projection* $(\tilde{X}_t)$ of an adapted, positive process (X_t) to satisfy $E[\tilde{X}_T] = E[X_T]$ for every finite predictable stopping time T (see Exercise 2.2). Roughly speaking, this means that $\tilde{X}_t = E[X_t|\mathcal{H}_{t-}]$ for suitably chosen versions of these conditional expectations. This definition is equivalent to

$$E\left[\int_0^\infty \tilde{X}\, dB\right] = E\left[\int_0^\infty X\, dB\right] \tag{2.13}$$

for every predictable random measure B [in the sense of Definition 2.11a)]. In this context, (2.12) becomes the adjoint of the operator (2.13).

Alternatively, the counting process (N_t) is a submartingale with respect to $\mathcal{H}$ and the compensator can be realized as the predictable increasing process in the Doob-Meyer decomposition of N (see Theorems 2.14 and B.6 for further details).

The compensator depends not only on $\mathcal{H}$ but also — which is crucial in martingale inference — on the probability P. However, we suppress both dependences whenever possible.

Here is the simplest example.

Example 2.12. (Poisson process). Let N be a Poisson process on $\mathbf{R}_+$ with diffuse mean measure μ and let $\mathcal{H} = \mathcal{F}^N$. Then the compensator A is deterministic and equal to μ. Indeed, the process $t \to \mu_t$ is trivially predictable, and for $C(u,\omega) = \mathbf{1}(s < u \leq t)\mathbf{1}(\omega \in \Lambda)$ [here $s < t$ and $\Lambda \in \mathcal{F}_s^N$ — it suffices by the monotone class theorem to consider only such processes], the independent increments character of N implies that

$$E\left[\int_0^t C\, dN\right] = E[(N_t - N_s); \Lambda] = (\mu_t - \mu_s)P\{\Lambda\} = E\left[\int_0^t C\, d\mu\right]. \; \square$$

The next example is more complicated, but shows clearly the role of Cox processes.

Example 2.13. (Cox process). Let N be a Cox process directed by the diffuse random measure M and let $\mathcal{H}_t = \mathcal{F}_\infty^M \vee \mathcal{F}_t^N$. Then the compensator of N is the directing measure M (see Exercise 2.10). However, for $\mathcal{H}_t = \mathcal{F}_t^N$, the compensator is in general different. $\square$

The following characterization of the compensator is essential to the dynamic viewpoint of point processes that accompanies the martingale approach, as well as in statistical inference. It states that the difference between a point process and its compensator is a martingale. This martingale, though, is schizophrenic: in some contexts it is a noise process for which estimation is impossible, while in others it contains all relevant information about the development of N.

Theorem 2.14. Given a point process N satisfying Assumptions 2.10 and a predictable random measure A, the following are equivalent:
 a) A is the compensator of N;
 b) The process $M_t = N_t - A_t$ is a (mean zero) martingale.

Proof: For $C(u,\omega) = \mathbf{1}(s < u \le t)\mathbf{1}(\omega \in \Lambda)$ a generator of the predictable σ-algebra $\mathcal{P}$, Definition 2.11 yields

$$E[(N_t - N_s); \Lambda] = E\left[\int_0^\infty C\,dN\right] = E\left[\int_0^\infty C\,dA\right] = E[(A_t - A_s); \Lambda], \quad (2.14)$$

which by Assumption 2.10c) becomes $E[(N_t - A_t); \Lambda] = E[(N_s - A_s); \Lambda]$, and hence b) holds if a) does. Conversely, if $N - A$ is a martingale, then A is integrable, so (2.14) obtains, and (2.12) follows by the monotone class theorem. $\blacksquare$

Together with Theorem 2.21 and Proposition 2.23, Theorem 2.14 constitutes the basis of variance and covariance computations, as well as important central limit theorems.

In infinitesimal form (2.14) becomes the heuristic expression

$$dA_t = E[dN_t | \mathcal{H}_{t-}]. \quad (2.15)$$

Increments in processes must be thought of as *forward* in time from t. Thus $dM_t = dN_t - dA_t$ is that portion of dN_t that cannot be foreseen from observation of N over $[0, t)$ and whatever other information is part of the strict past $\mathcal{H}_{t-}$. This motivates the following terminology.

Definition 2.15. The martingale $M = N - A$ is the $(P, \mathcal{H})$-*innovation martingale* of N.

The innovation interpretation is consistent with previous examples. Thus, since a Poisson process has independent increments, for $\mathcal{H} = \mathcal{F}^N$, the innovation martingale contains all information other than mean values, which could have been predicted even in total absence of observations. At another extreme is the "degenerate" (but good for counterexamples) case that $\mathcal{H}_t = \mathcal{F}^N_\infty$ for all t: there is no innovation at all — everything about N is revealed at $t = 0$ — and $A \equiv N$. At neither of these extremes lie processes for which all innovation occurs at a single random time, as illustrated by the following example.

Example 2.16. (Lattice process). Let $N = \sum_{n=0}^\infty \varepsilon_{X_n}$ be (the restriction to $\mathbf{R}_+$ of) the stationary point process of Example 1.58 and suppose that $\mathcal{H}_t = \mathcal{F}^N_t$. Then once $T_1 = X$, which is uniformly distributed on $[0, 1]$, has been observed, it is known exactly where all remaining points lie. Using Theorem 2.18,

$$A_t = \int_0^t (1 - x)^{-1} dx, \qquad t \in [0, T_1],$$

while for $n \geq 1$,

$$A_t = \int_0^{T_1} (1 - x)^{-1} dx + (n - 1) + \mathbf{1}(t \geq T_{n+1}), \qquad t \in (T_n, T_{n+1}].\ \square$$

Explicit calculation of compensators is possible (in principle, at least) using either of the next two results. The first is more general (and related to the construction of conditional intensities in Section 6) and correspondingly less useful in applications where there is special structure. The second posits a specific form for $\mathcal{H}$, but it is the "typical" form, so the result is broadly applicable after all.

Theorem 2.17. Let N be a point process on $\mathbf{R}_+$ satisfying Assumptions 2.10 and let A be its compensator. Then for each t, almost surely,

$$A_t = \lim_{n \to \infty} \sum_{k=0}^{2^n - 1} E\left[\{N((k+1)t/2^n) - N(kt/2^n)\} | \mathcal{H}(kt/2^n)\right].\ \square \qquad (2.16)$$

Theorem 2.18. Let N be a point process on $\mathbf{R}_+$ satisfying Assumptions 2.10 with respect to the filtration

$$\mathcal{H}_t = \mathcal{H}_0 \vee \mathcal{F}^N_t \qquad\qquad (2.17)$$

For each n, let $F_n(\omega, du)$ be a regular conditional distribution of $U_{n+1}(= T_{n+1} - T_n)$ given $\mathcal{H}_{T_n}$: $F_n(\omega, du) = P\{U_{n+1} \in du | \mathcal{H}_{T_n}\}(\omega)$. Then

$$A_t = A_{T_n} + \int_0^{t - T_n} \frac{F_n(dx)}{F_n[x, \infty)}, \qquad t \in (T_n, T_{n+1}]. \qquad (2.18)$$

Proof: We show that for each n and finite stopping time T, $E[N_{T \wedge T_n}] = E[A_{T \wedge T_n}]$, with A given by (2.18). By (2.17) and Proposition 2.5 there is $V_n \in \mathcal{H}_{T_n}$ (this is crucial!) satisfying (2.7). The rest of the proof is computational. On the one hand,

$$E \left[\sum_{j=0}^{n-1} \left[\int_0^{V_j \wedge U_{j+1}} F_j[x, \infty)^{-1} F_j(dx) \right] \mathbf{1}(T \geq T_j) \right]$$
$$= \sum_{j=0}^{n-1} E \left[E \left[\int_0^{V_j \wedge U_{j+1}} F_j[x, \infty)^{-1} F_j(dx) \Big| \mathcal{H}_{T_j} \right] \mathbf{1}(T \geq T_j) \right]$$
$$= \sum_{j=0}^{n-1} E \left[\int_0^{V_j} F_j(dx) \mathbf{1}(T \geq T_j) \right]$$

but on the other hand,

$$E[N_{T \wedge T_n}] = E \left[\sum_{j=0}^{n-1} (N_{T \wedge T_{j+1}} - N_{T \wedge T_j}) \mathbf{1}(T \geq T_j) \right]$$
$$= E \left[\sum_{j=0}^{n-1} \mathbf{1}(V_j \geq U_{j+1}) \mathbf{1}(T \geq T_j) \right]$$
$$= \sum_{j=0}^{n-1} E \left[\int_0^{V_j} F_j(dx) \mathbf{1}(T \geq T_j) \right]. \quad \blacksquare$$

The assumption (2.17) limits information not initially present at $t = 0$ to arise from the point process alone. In applications, of course, this may not be reasonable physically.

A consequence of Theorem 2.18 is the following uniqueness property.

Theorem 2.19. Let N be a point process defined over $(\Omega, \mathcal{F})$, let P and P' be probabilities with respect to which N has $\mathcal{H}$-compensators A and A', and assume that (2.17) is fulfilled. Then provided that $A = A'$ and $P = P'$ on $\mathcal{H}_0$, $P = P'$ on $\mathcal{H}_\infty$.

Proof: Theorem 2.18 shows that the conditional distributions $F_n(du) = P\{U_{n+1} \in du | \mathcal{H}_{T_n}\}$ and $F_n'(du) = P'\{U_{n+1} \in du | \mathcal{H}_{T_n}\}$ coincide for every n and from this the conclusion follows by Propositions 2.2 and 2.3. $\quad \blacksquare$

We elaborate on one interpretation of (2.18). Let T be a positive random variable whose distribution function F admits density $f = F'$. Then the function

$$h(t) = f(t)/[1 - F(t)] = P\{T \leq t + dt | T > t\}/dt \qquad (2.19)$$

is known in reliability and survival analysis as the hazard function of F. Comparison of (2.18) and (2.19) identifies the compensator as a cumulative *conditional hazard function*. See also Theorem 2.31 and associated discussion.

Here is one further example.

Example 2.20. (Renewal process). Let $N = \sum_{n=1}^{\infty} \varepsilon_{T_n}$ be a renewal process with interarrival distribution F. Then $F_n(\omega, \cdot) = F(\cdot)$ for all n and ω, and (2.18) becomes

$$A_t = -\sum_{n=1}^{\infty} \log[1 - F(t \wedge T_n)]. \tag{2.20}$$

See Section 8.1 for additional discussion and interpretations. □

The innovation martingale M has associated with it a predictable variation process $\langle M \rangle$; see Appendix B for fundamental properties. The predictable variation and, more generally, predictable covariation processes arising from marked point processes, not only play a fundamental computational role but also are the basis for central limit theorems for martingale estimators. We now calculate $\langle M \rangle$.

Theorem 2.21. Let N be a point process with compensator A and assume that the innovation martingale $M = N - A$ is square integrable. Let $\Delta A_t = A_t - A_{t-}$. Then

$$\langle M \rangle_t = \int_0^t (1 - \Delta A_s) dA_s. \tag{2.21}$$

Proof: The integration-by-parts formula (2.10) implies that

$$M_t^2 = 2\int_0^t M_{s-} dM_s + \sum_{s \leq t} (\Delta M_s)^2.$$

Expanding the summation and using the identity $(\Delta N)^2 = \Delta N$ yields

$$M_t^2 = \int_0^t (2M_{s-} + 1 - 2\Delta A_s) dM_s + \int_0^t (1 - \Delta A_s) dA_s,$$

where we have also used the relationship $M = N - A$. The preceding formula exhibits the submartingale M^2 as the sum of a martingale (Theorem 2.9) and a predictable increasing process, and thus constitutes a Doob-Meyer decomposition. However, that decomposition is unique (Theorem B.6), and hence (2.21) holds. ∎

In particular, if A is continuous in t (especially, if A is absolutely continuous, as in Section 4), then $\Delta A = 0$ and $\langle M \rangle = A$. The differential interpretation $d\langle M \rangle = dA$, together with (2.15), then yields

$$E[(dN_t - E[dN_t|\mathcal{H}_{t-}])^2|\mathcal{H}_{t-}] = dA_t = E[dN_t|\mathcal{H}_{t-}].$$

That is, a point process with a continuous compensator is locally and conditionally a Poisson process in the sense that its mean and variance are equal.

Now let $N = \sum \varepsilon_{(T_n, Z_n)}$ be a marked point process with mark space $(E, \mathcal{E})$. For each $B \in \mathcal{E}$ the counting process $N_t(B) = \sum \mathbf{1}(T_n \leq t, Z_n \in B)$

admits compensator $A_t(B)$. But one can also regard $N_t(\cdot)$ as a stochastic process parametrized by $\mathbf{R}_+$ and taking values in $\mathbf{M}_p$, and it is desirable, then, that for each t, $A_t(B)$ be a random measure in B, which can be accomplished using a construction identical to that used for regular conditional probabilities, and yields the following result.

Theorem 2.22. Let N be a marked point process such that the point process $\sum \epsilon_{T_n}$ satisfies Assumptions 2.10. Then there exists a unique random measure A on $\mathbf{R}_+ \times E$ such that

a) For each B the process $A_t(B) = A([0,t] \times B)$ is predictable;

b) For each nonnegative predictable process C (i.e., defined on $\mathbf{R}_+ \times E$ and measurable with respect to $\mathcal{P} \times \mathcal{E}$), $E[\int C\,dN] = E[\int C\,dA]$, where the integrals are over $\mathbf{R}_+ \times E$. $\square$

Predictably, the random measure A is called the $(P, \mathcal{H})-compensator$ of N. For each B, $M_t(B) = N_t(B) - A_t(B)$ is a martingale; moreover, $M_t(\cdot) = N_t(\cdot) - A_t(\cdot)$ is a measure-valued martingale (but the values need not be positive measures), which we term the *innovation martingale*.

For the case that the innovation martingales $M(B_1)$ and $M(B_2)$ associated with disjoint subsets B_1 and B_2 of E are square integrable, the following result calculates the predictable covariation process. The most useful form arises when there exists a stochastic intensity (Proposition 2.32) or, more generally, if either $A_t(B_1)$ or $A_t(B_2)$ is merely continuous in t, for then these martingales are orthogonal.

Proposition 2.23. Let N be a marked point process with compensator A and let B_1 and B_2 be disjoint sets for which the martingales $M(B_1)$ and $M(B_2)$ are square integrable. Then

$$\langle M(B_1), M(B_2) \rangle_t = -\int_0^t \Delta A_s(B_1)\Delta A_s(B_2)\,ds. \qquad (2.22)$$

Proof: On the one hand, application of Theorem 2.21 to the point process $N_t(B_1 \cup B_2)$, which has compensator $A_t(B_1 \cup B_2) = A_t(B_1) + A_t(B_2)$, gives

$$\langle M(B_1 \cup B_2) \rangle_t = \int_0^t (1 - \Delta A_s(B_1) - \Delta A_s(B_2))[dA_s(B_1) + dA_s(B_2)],$$

while

$$\begin{aligned}
\langle M(B_1 \cup B_2) \rangle_t &= \langle M(B_1) \rangle_t + \langle M(B_2) \rangle_t + 2\langle M(B_1), M(B_2) \rangle_t \\
&= \int_0^t (1 - \Delta A_s(B_1))dA_s(B_1) + \int_0^t (1 - \Delta A_s(B_1))dA_s(B_1) \\
&\quad + 2\langle M(B_1), M(B_2) \rangle_t
\end{aligned}$$

by another appeal to Theorem 2.21. The remainder of the proof is computational. ∎

The "negative correlation" represented by (2.22) has an obvious physical
basis: the processes $N(B_1)$ and $N(B_2)$ cannot jump simultaneously. Hence
to the extent that a jump is deemed relatively more likely to be in one of
these processes, it must be relatively less so in the other.

By combining (2.21) and (2.22) and using the inner product property of
predictable covariation, for general B_1 and B_2,

$$
\begin{aligned}
\langle M(B_1), M(B_2)\rangle_t \;=\; & \int_0^t [1 - \Delta A(B_1 \cap B_2)]dA(B_1 \cap B_2) \\
& - \int_0^t \Delta A(B_1 \cap B_2)\Delta A(B_1 \triangle B_2)ds \\
& - \int_0^t \Delta A(B_1 \setminus B_2)\Delta A(B_2 \setminus B_2)ds. \qquad (2.23)
\end{aligned}
$$

We illustrate with a simple special case.

Example 2.24. (Position-dependent marking). Let $\tilde{N}$ be a point process on $\mathbf{R}_+$ from which N is constructed by position-dependent marking
(Example 1.28), using a kernel $K : \mathbf{R}_+ \to E$. Then the $\mathcal{F}^N$-compensator of
N is $A(dt, dx) = \tilde{A}(dt)K(t, dx)$, where A is the $\mathcal{F}^{\tilde{N}}$-compensator of N. □

When (2.17) holds, Theorem 2.18 admits a rather obvious extension to
marked point processes. We do not state it formally, but see Theorem 2.30
for the corresponding result when there exists a stochastic intensity.

Another application of compensators is to convergence in distribution for
point processes, as exemplified by the following Poisson limit theorem.

Theorem 2.25. Let $N^1, N^2, \ldots$ be simple point processes on $\mathbf{R}_+$ having
compensators $A^1, A^2, \ldots$ for their internal histories, and let N be a Poisson
process with diffuse mean measure μ. If for each t,

$$
A_t^n \xrightarrow{d} \mu_t, \qquad (2.24)
$$

then $N^n \xrightarrow{d} N$.

Proof: We prove only the special case that there exists a constant C
such that

$$
A_\infty^n = \lim_{t \to \infty} A_t^n \le C \qquad (2.25)
$$

almost surely for each n. Details on removal of (2.25) are given in Kabanov
et al. (1980a).

By Theorem 1.21 it suffices to show that for $f \ge 0$ a simple function on
$\mathbf{R}_+$ with compact support,

$$
\lim_{t \to \infty} E\left[\exp\{-N^n(f) + \int_0^\infty (1 - e^{-f})d\mu\}\right] = 1. \qquad (2.26)
$$

The key point is that each process

$$Z_t^n = \exp[-\int_0^t f dN^n + \int_0^t (1 - e^{-f})dA^n]$$

is, under (2.25), a square integrable martingale, in consequence of which

$$E\left[\exp\{-\int_0^t f dN^n + \int_0^t (1 - e^{-f})dA^n\}\right] = 1. \qquad (2.27)$$

An heuristic argument is the following. For small values of Δt,

$$E\left[\exp\{-\int_t^{t+\Delta t} f dN^n + \int_t^{t+\Delta t}(1 - e^{-f})dA^n\}\Big|\mathcal{F}_t^{N_n}\right]$$
$$\cong E\left[(1 + [e^{-f(t)} - 1]\Delta N_t^n)(1 + [e^{f(t)} - 1]\Delta A_t^n)\Big|\mathcal{F}_t^{N_n}\right]$$
$$= 1 + (e^{-f(t)} - 1)E[\Delta N_t^n - dA_t^n|\mathcal{F}_t^{N_n}]$$
$$\quad - (e^{-f(t)} - 1)^2 E[\Delta N_t^n dA_t^n|\mathcal{F}_t^{N_n}]$$
$$= 1 - (e^{-f(t)} - 1)^2 (dA_t^n)^2 \cong 1,$$

where (2.15) is used at the next-to-last step. Hence by (2.27)

$$\left|E\left[\exp\{-N^n(f) + \int_0^\infty(1 - e^{-f})d\mu\}\right] - 1\right|$$
$$= \left|E\left[\exp\{-N^n(f) + \int_0^\infty(1 - e^{-f})d\mu\}\right.\right.$$
$$\quad \left.\left. - \exp\{-N^n(f) + \int_0^\infty(1 - e^{-f})dA^n\}\right]\right|$$
$$\leq E\left[\left|\exp\{\int_0^\infty(1 - e^{-f})d\mu\} - \exp\{\int_0^\infty(1 - e^{-f})dA^n\}\right|\right]$$

and this last quantity converges to zero by (2.24) and the assumption that f is simple; therefore, (2.26) holds. ∎

A variant yields an inequality for the total variation distance between a point process and a Poisson process. (See Section 1.6 for definitions and details.)

Theorem 2.26. Let N be a point process on $[0, t]$ with $\mathcal{F}^N$-compensator A and let N^* be a Poisson process with mean measure μ. Then

$$\alpha(N, N^*) \leq E\left[\|A - \mu\|_t\right] + E\left[\sum_{s \leq t}(\Delta A_s)^2\right], \qquad (2.28)$$

where $\|A - \mu\|_t$ is the total variation of the signed measure $A - \mu$ over $[0, t]$.

Proof: We prove (2.28) under the stipulation that the compensator A be continuous, so that the second term on the right-hand side is zero, and we further assume that $A_\infty = \infty$ a.s. It follows (Exercise 2.11) that if

$\tilde{A}_s = \inf\{t : A_t \geq s\}$, then $\tilde{N}_s = N(\tilde{A}_s)$ is a stationary Poisson process with rate 1. Hence for $t \geq 0$ and X_λ having a Poisson distribution with mean λ,

$$
\begin{aligned}
\alpha(N_t, X_\lambda) &\leq P\{N_t \neq X_\lambda\} \\
&\leq E[|N_t - \tilde{N}_\lambda|] \\
&= E\left[\int_0^\infty [\mathbf{1}(\tilde{A}_\lambda < s \leq t) + \mathbf{1}(t < s < \tilde{A}_\lambda)] dN_s\right] \\
&= E\left[\int_0^\infty [\mathbf{1}(\tilde{A}_\lambda < s \leq t) + \mathbf{1}(t < s < \tilde{A}_\lambda)] dA_s\right] \\
&= E[|A(\tilde{A}_\lambda) - A_t|] = E[|A_t - \lambda|],
\end{aligned}
$$

where predictability of the integrand is used in the fourth step and the last step is by continuity of A. Suppose now that $0 = t_0 < t_1 < \cdots < t_k$ and that $X_{\mu(t_j) - \mu(t_{j-1})}$ are independent, Poisson distributed random variables with the indicated means. The preceding computation implies that

$$
\begin{aligned}
&\alpha(\{N_{t_j} - N_{t_{j-1}} : 1 \leq j \leq k\}, \{X_{\mu(t_j) - \mu(t_{j-1})} : 1 \leq j \leq k\}) \\
&\leq E\left[\sum_{j=1}^k \left|(A_{t_j} - A_{t_{j-1}}) - [\mu(t_j) - \mu(t_{j-1})]\right|\right] = E[\|A - \mu\|],
\end{aligned}
$$

and this gives (2.28). ∎

We conclude the section by comparing the bound (2.28) with that implied by Theorem 1.44.

Example 2.27. (Bernoulli process). Let $N = \sum_{i=1}^k N_i$, where the N_i are i.i.d. Bernoulli processes on $[0, 1]$ with parameters (p, F), and suppose that F admits density f. Then the compensator of N satisfies

$$
dA_u = (k - N_{u-}) \frac{pf(u)}{1 - pF(u)} du. \tag{2.29}
$$

Suppose that N^* is Poisson with mean $\mu = kpF$. Then (2.28) yields

$$
\begin{aligned}
\alpha(N, N^*) &\leq \int_0^1 E\left[\left|(k - N_{u-}) \frac{pf(u)}{1 - pF(u)} - kpf(u)\right|\right] du \\
&= p \int_0^1 E\left[\left|\frac{k - N_{u-}}{1 - pF(u)} - 1\right|\right] f(u) du \\
&= p \int_0^1 E[|N_u - kpf(u)|] \frac{f(u)}{1 - pF(u)} du \\
&\leq 2kp^2 \int_0^1 f(u) dF(u) = kp^2.
\end{aligned}
$$

By contrast, for N^{**} Poisson with mean $\mu = k[-\log(1-p)]F$, Theorem 1.44 yields $\alpha(N, N^{**}) \leq kp^2/2$. Application of (2.28) to estimate $\alpha(N, N^{**})$ is very difficult computationally. □

2.4   Stochastic Intensities

Let N be a point process on $\mathbf{R}_+$ with compensator A. In many important applications (and all of Chapter 5) A admits a (jointly measurable) density with respect to Lebesgue measure, which is, of course, a random process in general, and results in Section 3 simplify, especially Theorem 2.21 and, for marked point processes, Proposition 2.23. Moreover, the interpretation of the compensator as an integrated conditional hazard function becomes manifestly clear. In the following discussion the filtration $\mathcal{H}$ is fixed and we require that Assumptions 2.10 be satisfied.

Definition 2.28. Let N be a simple point process on $\mathbf{R}_+$ with compensator A. A positive, predictable process $\lambda = (\lambda_t)$ satisfying $A_t = \int_0^t \lambda_s ds$ is the $(P, \mathcal{H})$-*stochastic intensity* of N.

It follows from Theorem 2.19 that if it exists, the stochastic intensity is unique. The heuristic interpretation [compare (2.15)] is that for each t,

$$\lambda_t dt = E[\Delta N_t | \mathcal{H}_{t-}] = P\{\Delta N_t = 1 | \mathcal{H}_{t-}\} = P\{\Delta N_t > 0 | \mathcal{H}_{t-}\}. \qquad (2.30)$$

(Recall that N is simple.) Further, there is a pervasive analogy between stochastic intensities and hazard functions of ordinary random variables, one aspect of which is given in Theorem 2.30 and another in Theorem 2.31; see also Exercise 2.13.

For a Poisson process N with mean measure $\mu(dt) = \alpha_t dt$ (α is the *intensity function* of N) the $\mathcal{F}^N$-stochastic intensity exists and is identically equal to α. If N is a Cox process directed by the absolutely continuous random measure $M(dt) = X_t\, dt$ and if $\mathcal{H}_t = \mathcal{F}_t^N \vee \mathcal{F}_\infty^X$, then (X_t) is the $\mathcal{H}$-stochastic intensity of N and is called the *directing intensity*.

Not every point process admits a stochastic intensity. By making $\mathcal{H}$ large enough that $\mathcal{F}_\infty^N \subseteq \mathcal{H}_0$, one forces N to be its own compensator. A somewhat less degenerate example is a renewal process with discrete interarrival distribution, which by (2.20) does not possess a stochastic intensity even with respect to its internal history.

Indeed, as intimated above, existence of a stochastic intensity depends as much on the filtration as on the point process; the smaller $\mathcal{H}$, the more likely a stochastic intensity is to exist. In particular, if N admits $\mathcal{H}$-stochastic intensity λ and if $\mathcal{F}_t^N \subseteq \mathcal{G}_t \subseteq \mathcal{H}_t$, for each t, then N admits $\mathcal{G}$-stochastic intensity $\tilde{\lambda}$ satisfying $\tilde{\lambda}_t = E[\lambda_t | \mathcal{G}_t]$ almost surely for each t (Exercise 2.16). Thus for a Cox process, the $\mathcal{F}^N$-stochastic intensity $\tilde{X}$ satisfies $\tilde{X}_t = E[X_t | \mathcal{F}_t^N]$. In Chapters 5 and 7 we examine state estimation problems that amount to calculations of stochastic intensities.

Existence and nature of a stochastic intensity depend as well on the probability P, but not in an entirely arbitrary manner (see Theorem 2.31).

In view of the preceding discussion, one should not expect "simple" sufficient conditions for existence of a stochastic intensity, and there do not seem to be any, even for the internal history. Theorem 2.30 does provide one condition; we give it after extending the concept of stochastic intensity to marked point processes.

Suppose now that $N = \sum \varepsilon_{(T_n, Z_n)}$ is a marked point process with mark space E.

Definition 2.29. A stochastic process $(\lambda_t(B) : t \geq 0, B \in \mathcal{E})$ is the $(P, \mathcal{H})$-*stochastic intensity* of N provided that
 a) For each t, $B \to \lambda_t(B)$ is a random measure on E;
 b) For each B, the process $\lambda_t(B)$ is the stochastic intensity of the counting process $N_t(B) = \sum_n \mathbf{1}(T_n \leq t, Z_n \in B)$.

The never-explicitly-stated extension of Theorem 2.18 to marked point processes has the following analogue.

Theorem 2.30. Let N be a marked point process and suppose that Assumptions 2.10 are satisfied for the filtration $\mathcal{H}_t = \mathcal{H}_0 \vee \mathcal{F}_t^N$. Assume that for each n there is a measure-valued process $f_n(u, dx)$ such that

$$P\{U_{n+1} \in du, Z_{n+1} \in dx | \mathcal{H}_{T_n}\} = f_n(u, dx)du.$$

Then the stochastic intensity of N is given by

$$\lambda_t(B) = \frac{f_n(t - T_n, B)}{\int_{t-T_n}^{\infty} f_n(u, E)du}, \qquad t \in (T_n, T_{n+1}]. \ \square \qquad (2.31)$$

In particular, if N is an ordinary point process and (2.17) holds, then provided that $P\{U_{n+1} \in du | \mathcal{H}_{T_n}\} = f_n(u)du$, N admits stochastic intensity

$$\lambda_t = \frac{f_n(t - T_n)}{\int_{t-T_n}^{\infty} f_n(u)du}, \qquad t \in (T_n, T_{n+1}]. \qquad (2.32)$$

See Exercise 2.14 for the special form of (2.32) when N is a renewal process.

Comparison of (2.32) and (2.19) makes manifest the interpretation of the stochastic intensity as a conditional hazard function. In further pursuit of this analogy, consider a random variable T with distribution function F and hazard function $h = f/(1 - F)$. Viewing T as a sample of size 1 from the distribution function F, we write the likelihood function $\ell(T, F) = f(T)$ in terms of h:

$$\ell(T, F) = \exp[-\int_0^t h(s)ds + \log h(T)]. \qquad (2.33)$$

A similar representation holds for point processes, with the stochastic intensity playing the role of the hazard function in (2.33). For simplicity we formulate it for point processes without marks.

Theorem 2.31. Let $(\Omega, \mathcal{F}, P)$ be a probability space over which is defined a simple point process N having $\mathcal{F}^N$-stochastic intensity λ. Suppose that P_0 is a probability measure on $(\Omega, \mathcal{F})$ with respect to which N is a stationary Poisson process with rate 1. Then $P \ll P_0$ and for each t,

$$dP/P_0\big|_{\mathcal{F}_t^N} = \exp\left[\int_0^t (1 - \lambda_s)ds + \int_0^t \log \lambda_s \, dN_s\right]. \qquad (2.34)$$

Conversely, if P_0 is as above and P is a probability measure on $(\Omega, \mathcal{F})$ satisfying $P \ll P_0$, then there exists a predictable process λ such that N has stochastic intensity λ with respect to P.

Proof: For $u \leq t$ define

$$Z_u = \exp\left[\int_0^u (1 - \lambda_s)ds + \int_0^u \log \lambda_s \, dN_s\right];$$

then it is easily verified using the independent increments property of Poisson processes that $(Z_u)_{u \leq t}$ is a martingale over $(\Omega, \mathcal{F}, P_0)$. By (2.12) and Theorem 2.19, for the direct part of the theorem it suffices to show that if P' is the probability defined by $dP' = Z_t dP_0$, then for each bounded, predictable process C,

$$E'\left[\int_0^t C_s dN_s\right] = E'\left[\int_0^t C_s \lambda_s ds\right]. \qquad (2.35)$$

But

$$
\begin{aligned}
E'\left[\int_0^t C_s dN_s\right] &= E_0\left[Z_t \int_0^t C_s dN_s\right] \\
&= E_0\left[\int_0^t Z_s C_s dN_s\right] \\
&= E_0\left[\int_0^t Z_{s-} \lambda_s C_s dN_s\right] \\
&= E_0\left[\int_0^t Z_{s-} C_s \lambda_s ds\right] \\
&= E_0\left[\int_0^t Z_s C_s \lambda_s ds\right] \\
&= E_0\left[Z_t \int_0^t C_s \lambda_s ds\right] \\
&= E'\left[\int_0^t C_s \lambda_s ds\right],
\end{aligned}
$$

and therefore (2.35) holds. The justifications for the steps above are as follows:

2) and 6) Z is a P_0-martingale;

3) By direct computation, $Z_{T_n} = Z_{T_n-} \lambda_{T_n}$ for each arrival time T_n;

4) The P_0-stochastic intensity of N is identically one;

5) Almost surely, $Z_{s-} = Z_s$ for almost every $s \in [0, t]$;

The proof of the converse is deferred to Section 5. ∎

Note that the not-yet-proved part of Theorem 2.31 provides — albeit indirectly — a sufficient condition for existence of a stochastic intensity.

With the aid of (2.34) we can interpret a point process as a dynamic, uncountable set of independent Bernoulli trials, one for each time t. Taking λ_t as the success probability for the trial at t [see (2.30)] we conclude that for observation over the time interval $[0, t]$ the probability of successes at times $T_1, \ldots, T_{N_t}$ is

$$\prod_{k=1}^{N_t} \lambda_{T_k} = \exp[-\int_0^t (-\log \lambda) dN], \qquad (2.36)$$

while by (2.6) and (2.32) the probability of no successes in the interval (T_{j-1}, T_j) is

$$1 - F_{j-1}(T_j - T_{j-1}) = \exp[-\int_{T_{j-1}}^{T_j} \lambda_s ds]. \qquad (2.37)$$

Multiplying the terms in (2.36) and (2.37), along with a term (similar to the latter) corresponding to the probability of no successes in (T_{N_t}, t), gives that the probability of successes at and only at the points $T_1, \ldots, T_{N_t}$ is

$$\exp[-\int_0^t \lambda_s ds + \int_0^t \log \lambda_s dN_s]. \qquad (2.38)$$

For the Poisson process, $\lambda \equiv 1$ and (2.38) reduces to e^{-t}; taking the quotient of these two probabilities yields the likelihood ratio (2.34) and also a heuristic proof of Theorem 2.31. Note the resemblance between (2.33) and (2.34), confirming that the stochastic intensity is a conditional hazard function.

We continue by noting explicitly the simplified forms of Theorem 2.21 and Proposition 2.23 that obtain for point processes with stochastic intensities.

Proposition 2.32. a) Let N be a point process on $\mathbf{R}_+$ with stochastic intensity λ and assume that the innovation martingale $M_t = N_t - \int_0^t \lambda_s ds$ is square integrable. Then for each t $\langle M \rangle_t = \int_0^t \lambda_s ds$;

b) Let N be a marked point process on $\mathbf{R}_+$ with mark space E and stochastic intensity $\lambda_t(B)$, and let B_1 and B_2 be disjoint sets for which the innovation martingales $(M_t(B_i))$, $i = 1, 2$, are square integrable. Then $M(B_1)$ and $M(B_2)$ are orthogonal: $\langle M(B_1), M(B_2) \rangle \equiv 0$. ☐

In particular, the process $(M_t(B_1)M_t(B_2))$ is a martingale. This property will be crucial in Chapter 5 for calculation of covariances.

We conclude the section with an example whose inference will be examined in Chapter 5.

Example 2.33. (System of Markov processes). Let S be a finite set and let $(X_t(1)), \ldots, (X_t(K))$ be independent Markov processes, each with state space S and generator A. Suppose that we can observe only "macro data," that is, how many processes are in each state at any given time. We may represent the observations as a marked point process in the following manner. Let $E = \{(i,j) : i, j \in S; i \neq j\}$ and for $(i,j) \in E$ define the counting process

$$N_t(i,j) = \sum_{k=1}^{K} \sum_{s \leq t} \mathbf{1}(X_{s-}(k) = i, X_s(k) = j),$$

so that $N_t(i,j)$ is the observed number of jumps from i to j among the K processes during the time interval $[0, t]$.

For each i and t, let $Y_t(i) = \sum_{k=1}^{K} \mathbf{1}(X_{t-}(k) = i)$ be the number of processes in state i at time $t-$, and let $\mathcal{H}_t = \bigvee_{k=1}^{k} \mathcal{F}_t^{X(k)}$. Then Assumptions 2.10 are fulfilled and N has $\mathcal{H}$-stochastic intensity

$$\lambda_t(i,j) = A(i,j)Y_t(i). \tag{2.39}$$

Intuitively, this is clear from the interpretation of off-diagonal elements of the generator as transition intensities. Furthermore, each process

$$M_t(i,j) = N_t(i,j) - A(i,j)\int_0^t Y_s(i)ds \tag{2.40}$$

is a square integrable martingale over every bounded time interval [for $K = 1$, (2.40) reduces to Dynkin's formula] and for $(i,j) \neq (i',j')$, these martingales are orthogonal. □

2.5 Representation of Point Process Martingales

Let N be a point process on $\mathbf{R}_+$ having compensator A with respect to the completed internal history $\tilde{\mathcal{F}}^N$. According to Theorem 2.9, if H is a bounded predictable process and $M = N_A$ is the innovation martingale, then the process $\tilde{M}_t = \int_0^t H_s dM_s$ is also a martingale. This section is devoted to the converse assertion that every (suitably regular) $\tilde{\mathcal{F}}^N$-martingale has this form. One consequent interpretation is that the innovation martingale M does indeed contain all information concerning the evolution of N. The sole theorem in the section is formulated for marked point processes but, for simplicity and clarity, proved only for point processes without marks.

Theorem 2.34. Let $N = \sum \epsilon_{(T_n, X_n)}$ be a marked point process on $\mathbf{R}_+$ with mark space E and let M be the innovation martingale with respect to the internal history $\tilde{\mathcal{F}}^N$. Let Assumptions 2.10 be satisfied and let $(\tilde{M}_t)$ be a uniformly integrable $\tilde{\mathcal{F}}^N$-martingale. Then there exists a process $\{H(t,x): t \geq 0, x \in E\}$, jointly measurable in (ω, t, x) and predictable in (ω, t) for each fixed x, such that

$$\tilde{M}_t = \tilde{M}_0 + \int_{(0,t] \times E} H(s,x) dM_s(dx) \tag{2.41}$$

for every t.

Proof: For a point process without marks, (2.41) reduces to

$$\tilde{M}_t = \tilde{M}_0 + \int_0^t H_s dM_s, \tag{2.42}$$

where (H_t) is predictable. For the remainder of the proof we assume that the compensator (A_t) is continuous.

Given this assumption, construction of the process H is effected as follows. First, for each n let $F_n(du) = P\{U_{n+1} \in du | \tilde{\mathcal{F}}_t^N\}$; we assume that these are regular conditional distributions (see Theorem 2.18). Second, by Proposition 2.2 for each n there exists a function $V_n(u, \omega)$ that is $\mathcal{B}(\mathbf{R}_+) \times \tilde{\mathcal{F}}_t^N$-measurable and such that

$$\tilde{M}_{T_{n+1}}(\omega) = V_n(U_{n+1}(\omega), \omega). \tag{2.43}$$

Finally, as a consequence of Proposition 2.4, for each n and t there is $W_n(t) \in \tilde{\mathcal{F}}_{T_n}^N$ satisfying

$$\tilde{M}_t \mathbf{1}(T_n < t < T_{n+1}) = W_n(t) \mathbf{1}(T_n < t < T_{n+1}). \tag{2.44}$$

We will show that

$$H_t = V_n(t - T_n) - \int_{t-T_n}^{\infty} \frac{V_n(u)}{F_n(t - T_n, \infty)} F_n(du), \quad t \in (T_n, T_{n+1}]. \tag{2.45}$$

[It follows from Theorem 2.18 and the continuity assumption that the distributions $F_n(\cdot)$ are continuous.]

To confirm (2.42) we first observe that since M is a uniformly integrable martingale, the optional sampling theorem implies that for $\Lambda \in \tilde{\mathcal{F}}_{T_n}^N$,

$$E[\tilde{M}_t \mathbf{1}(T_n \leq t < T_{n+1}); \Lambda] = E[\tilde{M}_{T_{n+1}} \mathbf{1}(T_n \leq t < T_{n+1}); \Lambda]; \tag{2.46}$$

we next evaluate each side of (2.46). Since $W_n(t) \in \tilde{\mathcal{F}}_{T_n}^N$, (2.44) implies that

$$
\begin{aligned}
E[\tilde{M}_t \mathbf{1}(T_n \leq t < T_{n+1}); \Lambda] &= E\left[\mathbf{1}(U_{n+1} > t - T_n)W_n(t)\mathbf{1}(T_n \leq t); \Lambda\right] \\
&= E\left[F_n(t - T_n, \infty)W_n(t)\mathbf{1}(T_n \leq t); \Lambda\right].
\end{aligned}
$$

Similarly, because $\{T_n \le t\} \cap \Lambda \in \tilde{\mathcal{F}}_{T_n}^N$,

$$
\begin{aligned}
E[\tilde{M}_{T_n}\mathbf{1}(T_n \le t < T_{n+1}); \Lambda] &= E\left[V_n(U_{n+1})\mathbf{1}(U_{n+1} > t - T_n)\mathbf{1}(T_n \le t); \Lambda\right] \\
&= E\left[\{\textstyle\int_{t-T_n}^\infty V_n(u)F_n(du)\}\mathbf{1}(T_n \le t); \Lambda\right],
\end{aligned}
$$

where we have also used (2.43).

It follows that for fixed n and t,

$$
\tilde{M}_t\mathbf{1}(T_n \le t > T_{n+1}) = \mathbf{1}(T_n \le t)\int_{t-T_n}^\infty \frac{V_n(u)}{F_n(t - T_n, \infty)}F_n(du) \qquad (2.47)
$$

almost surely, and the P-null set can be made independent of t (and thence of n) by the continuity assumption. Brute force differentiation of the right-hand side of (2.47) produces the following, valid for $t \in (T_n, T_{n+1})$:

$$
\begin{aligned}
dM_t &= \frac{-V_n(t - T_n)F_n(dt - T_n)F_n(t - T_n, \infty)}{F_n(t - T_n, \infty)^2} \\
&\quad + \int_{t-T_n}^\infty \frac{V_n(u)F_n(dt - T_n)}{F_n(t - T_n, \infty)^2}F_n(du) \\
&= -H_t dA_t,
\end{aligned}
$$

where the last equality holds by Theorem 2.18. Consequently, (2.42) holds on (T_n, T_{n+1}). Taking limits in (2.47) as $t \uparrow T_{n+1}$ yields

$$
\tilde{M}_{T_{n+1}} - \tilde{M}_{T_{n+1}-} = V_n(U_{n+1}) - \int_{U_{n+1}}^\infty \frac{V_n(u)}{F_n(U_{n+1}, \infty)}F_n(du),
$$

and therefore (2.42) holds at T_{n+1}.

It remains only to observe that the process H is predictable by virtue of Proposition 2.6. ∎

We can now complete the proof of Theorem 2.31.

Proof: (Of the converse of Theorem 2.31). We need to show that if $P \ll P_0$ then there is a predictable process λ that is the $(P, \mathcal{F}^N)$-stochastic intensity of N. Define the P_0-martingale $Z_t = E_0[dP/dP_0|\tilde{\mathcal{F}}_t^N]$. By Theorem 2.34 — applied to Z over the probability space $(\Omega, \mathcal{F}, P_0)$, over which the stochastic intensity of N is identically 1 — there is a predictable process H such that

$$
Z_t = 1 + \textstyle\int_0^t H_s(dN_s - ds). \qquad (2.48)
$$

Given a bounded, predictable process C vanishing outside $[0, t]$,

$$
E\left[\textstyle\int_0^t C_s dN_s\right] = E_0\left[Z_t\textstyle\int_0^t C_s dN_s\right]
$$

$$= E_0\left[\int_0^t Z_s C_s dN_s\right]$$

$$= E_0\left[\int_0^t (Z_{s-} + H_s)C_s dN_s\right]$$

$$= E_0\left[\int_0^t (Z_{s-} + H_s)C_s ds\right]$$

$$= E_0\left[\int_0^t (1 + H_s/Z_{s-})\mathbf{1}(Z_{s-} > 0)Z_{s-}C_s ds\right]$$

$$= E_0\left[Z_t\int_0^t (1 + H_s/Z_{s-})\mathbf{1}(Z_{s-} > 0)C_s ds\right]$$

$$= E\left[\int_0^t (1 + H_s/Z_{s-})\mathbf{1}(Z_{s-} > 0)C_s ds\right],$$

where the third equality is valid because $Z_{T_n} = Z_{T_n-} + H_{T_n}$ at each arrival time T_n of N and since if Z, a positive martingale is trapped at zero if it ever reaches zero, in which case, by (2.48), H becomes zero as well. Hence we may take $\lambda_t = (1 + H_t/Z_{t-})\mathbf{1}(Z_{t-} > 0)$. ∎

Among other consequences of Theorem 2.34 is that it defines the class of allowable error processes in martingale estimation (Chapter 5).

2.6   Conditioning in General Spaces

The foregoing makes clear the fundamental role of the linear order structure of $\mathbf{R}_+$ in the theory of conditioning for point processes given partial observations. Such ideas as the past of a point process, the strict past, stopping time, predictable process, and innovation seem linked inextricably to one's observing point processes on $\mathbf{R}_+$ in the natural order of increasing time. Even in higher-dimensional Euclidean spaces (where there are at least intrinsic partial orders), let alone in general LCCB spaces, there seems little prospect for a full-fledged analogue of the compensator/stochastic intensity approach. The central difficulty, of course, is that the theory of martingales is in many senses inherently one-dimensional.

Nevertheless, the situation is not hopeless: there do exist analogues of some concepts (notably, predictability) discussed in preceding sections. In this section we describe two approaches to conditioning for point processes on general spaces. The first, based on a concept termed exvisibility (a spatial version of predictability), mimics rather closely some ideas involving compensators presented in Section 3.

By contrast, the second approach is a general theory of conditional distributions $P\{N_B \in (\cdot)|\mathcal{F}^N(B^c)\}$, where B is a (bounded) subset of the underlying space E and N_B is the restriction of N to B: $N_B(A) = N(A \cap B)$, broad enough to range from analogues of compensators at one extreme to Palm distributions at the other. The heart of this theory is a suitably formulated extension of the statement in Proposition 2.2 that for a point process on $\mathbf{R}_+$ the known information changes only at the arrival times. For a point process N on a general space this becomes: if $B \subseteq A$ and $N(A \setminus B) = 0$, then $\mathcal{F}^N(A^c)$ and $\mathcal{F}^N(B^c)$ contain the same information pertaining to N_B. A precise formulation is embodied in the following consistency property.

Proposition 2.35. Let N be a point process and let A and B be bounded sets with $B \subseteq A$. Then on $\{N(A \setminus B) = 0\}$,

$$P\{N_B \in (\cdot)|\mathcal{F}^N(B^c)\} = \frac{P\{N_B \in (\cdot), N(A \setminus B) = 0|\mathcal{F}^N(A^c)\}}{P\{N(A \setminus B) = 0|\mathcal{F}^N(A^c)\}}. \qquad (2.49)$$

Proof: For $\Lambda \in \mathcal{F}^N(B^c)$ of the form $\Lambda = \{N_{A \setminus B} \in \Gamma_1, N_{A^c} \in \Gamma_2\}$ and $\Delta \in \mathcal{F}^N(B)$,

$$\begin{aligned}
&P\{N_B \in \Delta, N(A \setminus B) = 0, \Lambda\} \\
&= \mathbf{1}(0 \in \Gamma_1)E\left[P\{N_B \in \Delta, N(A \setminus B) = 0|\mathcal{F}^N(A^c)\}\mathbf{1}(N_{A^c} \in \Gamma_2)\right] \\
&= \mathbf{1}(0 \in \Gamma_1)E\left[\frac{P\{N_B \in \Delta, N(A \setminus B) = 0|\mathcal{F}^N(A^c)\}}{P\{N(A \setminus B) = 0|\mathcal{F}^N(A^c)\}}\right. \\
&\qquad \left. \times \mathbf{1}(N(A \setminus B) = 0)\mathbf{1}(N_{A^c} \in \Gamma_2)\right] \\
&= E\left[\frac{P\{N_B \in \Delta, N(A \setminus B) = 0|\mathcal{F}^N(A^c)\}}{P\{N(A \setminus B) = 0|\mathcal{F}^N(A^c)\}}\mathbf{1}(N(A \setminus B) = 0); \Lambda\right]
\end{aligned}$$

and (2.49) then holds by the monotone class theorem. ∎

The content of this result is that conditional probabilities given the larger σ-algebra $\mathcal{F}^N(B^c)$ — provided N has no points in $A \setminus B$ — are simply rescalings of conditional probabilities given $\mathcal{F}^N(A^c)$. That is, observation over a larger set (B^c in this case), in the absence of further points of N, yields no new information. That all conditional distributions might then be expressible via a single kernel is at least plausible (and will be confirmed in Theorem 2.44). If so, then that kernel should be obtainable from the reduced Palm distributions: intuitively, suppose that B were fixed in (2.49) and that we could take $A = B \cup (\text{supp } N)^c$. Then $N(A \setminus B) = 0$, and (heuristically) (2.49) becomes

$$\begin{aligned}
P\{N_B \in (\cdot)|\mathcal{F}^N(B^c)\} &= P\{N_B \in (\cdot)|\mathcal{F}^N(B^c \cap \text{supp } N)\} \\
&= Q_N(N_{B^c}, \cdot). \qquad (2.50)
\end{aligned}$$

However, we are jumping ahead. Before providing details of this general theory of conditioning we will explicate the theory of exvisibility, along with that of the conditional intensity of a point process; this theory parallels that in Sections 3–5.

Let N be a simple point process on E with mean measure belonging to **M**. Exvisibility ("visibility from the outside") is analogous to predictability but is not a generalization since the two do not coincide for point processes on $\mathbf{R}_+$.

Definition 2.36. a) The *exvisible σ-algebra* engendered by N is the σ-algebra $\mathcal{V}$ on $E \times \Omega$ generated by sets $B \times \Lambda$ where $B \in \mathcal{B}$ (i.e., B is a bounded Borel set) and $\Lambda \in \mathcal{F}^N(B^c)$;

b) A stochastic process X parametrized by E is *exvisible* with respect to N if the mapping $(y, \omega) \to X_y(\omega)$ is $\mathcal{V}$-measurable;

c) A random measure M on E is *exvisible* with respect to N if M is $\mathcal{F}^N$-measurable and if the process $X_y = M(\{y\})$ is exvisible.

Roughly speaking, a process is exvisible if for each y its value at y is determined by the restriction of N to $\{y\}^c$, and a random measure M is exvisible if for each y, knowledge of N on $\{y\}^c$ determines whether there is an atom of M at y and, if so, its size. In particular, every diffuse random measure is exvisible — diffuseness is the spatial analogue of continuity. The point process N may be exvisible with respect to itself: on $\mathbf{R}_+$ the lattice process Example 1.58 is exvisible. Or N may fail to be exvisible: a Poisson process (because of independent increments) is not exvisible.

Exvisibility, however, differs from predictability in at least two respects. First, Definition 2.36 is formulated only for the internal history of N, and this restriction is crucial to some arguments below. Second and more important, whereas predictability is one-sided, dealing with prediction of dN_t from $\mathcal{F}^N_{t-}$, exvisibility on $\mathbf{R}_+$ is *two-sided*, and concerns estimation of dN_t from $\mathcal{F}^N(\{t\}^c)$. Thus a point process may be exvisible without being predictable and the exvisible projection and compensator need not be the same.

Nevertheless, the two theories are similar in many senses. For example, as for compensators, a projection theorem is one basic result.

Theorem 2.37. Let M be a random measure measurable with respect to $\mathcal{F}^N$. Then there is a unique exvisible random measure A satisfying

$$E[\int X \, dM] = E[\int X \, dA] \qquad (2.51)$$

for every positive exvisible process X. $\square$

We term A the *exvisible projection* of M. The proof of Theorem 2.37 is analogous to that of existence of compensators: one first shows that a process $X \geq 0$ admits an exvisible projection $\tilde{X} \geq 0$ such that $E[X_Y] = E[\tilde{X}_Y]$ for every exvisible random element Y of E (whose graph $\{(Y(\omega), \omega) : \omega \in \Omega\}$ belongs to $\mathcal{V}$) and then (2.51) corresponds to the adjoint of this operator.

Parallelism between the two theories extends to a "martingale" characterization of the exvisible projection.

Theorem 2.38. Let M be an $\mathcal{F}^N$-measurable random measure with $\mu_M \in \mathbf{M}$ and let A be an exvisible random measure. Then the following are equivalent:

a) A is the exvisible projection of M;

b) For each $B \in \mathcal{B}$,

$$E[M(B)|\mathcal{F}^N(B^c)] = E[A(B)|\mathcal{F}^N(B^c)]. \tag{2.52}$$

Proof: For an elementary exvisible process $X_y(\omega) = 1(y \in B)1(\omega \in \Lambda)$, where $B \in \mathcal{B}$ and $\Lambda \in \mathcal{F}^N(B^c)$, $E[\int X\,dM] = E[M(B); \Lambda]$ and $E[\int X\,dA] = E[A(B); \Lambda]$. By (2.51), a) implies equality of the left-hand sides of these two expressions, yielding b). Conversely, b) implies equality of the right-hand sides, which gives a) via the monotone class theorem. ∎

By iterated conditioning we obtain the "martingale" version.

Corollary 2.39. The exvisible projection A of M satisfies

$$E[M(C)|\mathcal{F}^N(B^c)] = E[A(C)|\mathcal{F}^N(B^c)]$$

whenever $C \subseteq B \in \mathcal{B}$. ☐

In particular, Theorem 2.38 applies to N itself, and (2.52) admits an interpretation analogous to (2.15): if A is the exvisible projection of N, then for each y, $A(dy) = E[N(dy)|\mathcal{F}^N(\{y\}^c)]$. Observe once more, however, that when $E = \mathbf{R}_+$, this expression is two-sided, whereas (2.15) is one-sided.

Historical development of the theory originated in the direction represented for compensators by Theorem 2.17, whose analogue for general spaces is the following.

Theorem 2.40. Let N be a simple, integrable point process on E and let $\mathcal{D} = (D_{nj})$ be a null array of $\mathcal{B}$-partitions of E. Then there exists a random measure A such that for each $D \in \mathcal{D}$,

$$\begin{aligned} A(D) &= \lim_{n\to\infty} \sum_{D_{nj} \subseteq D} P\{N(D_{nj}) = 1|\mathcal{F}^N(D_{nj}^c)\} \\ &= \lim_{n\to\infty} \sum_{D_{nj} \subseteq D} E[N(D_{nj})|\mathcal{F}^N(D_{nj}^c)] \end{aligned} \tag{2.53}$$

almost surely and in L^1.

Proof: We may presume to work with regular versions of the conditional probabilities $P\{(\cdot)|\mathcal{F}^N(D^c)\}$, $D \in \mathcal{D}$, which further satisfy the regularity conditions that for $C \subseteq D$ in $\mathcal{D}$

$$P\{N(D \setminus C) = 0|\mathcal{F}^N(D^c)\} > 0 \tag{2.54}$$

and that on $\{N(D \setminus C) = 0\}$,

$$P\{(\cdot)|\mathcal{F}^N(C^c)\} = \frac{P\{(\cdot), N(D \setminus C) = 0|\mathcal{F}^N(D^c)\}}{P\{N(D \setminus C) = 0|\mathcal{F}^N(D^c)\}}. \tag{2.55}$$

We note that (2.54) is equivalent to the condition (Σ) of Papangelou (1974b) (see the chapter notes), while (2.55) holds by Proposition 2.35.
  Hence for $y \in C \subseteq D$,

$$P\{N(C) > 0|\mathcal{F}^N(C^c)\} = 1 - \frac{P\{N(D) = 0|\mathcal{F}^N(D^c)\}}{P\{N(D \setminus C) = 0|\mathcal{F}^N(D^c)\}},$$

and the right-hand side of this expression converges as $C \downarrow \{y\}$; consequently for each y there exists the limit

$$U_y = \lim_{D \downarrow \{y\}} P\{N(D) > 0|\mathcal{F}^N(D^c)\}.$$

Moreover, as will be justified momentarily, $A^1 = \sum_{y \in E} U_y \epsilon_y$ is a random measure on E.
  The key step is to show existence of a random measure A^2 satisfying

$$A^2(\{y\})N(\{y\}) \equiv 0, \tag{2.56}$$

such that for each D,

$$A^2(\cdot \cap D) = \frac{E[N_D(\cdot)\mathbf{1}(N(D) = 1)|\mathcal{F}^N(D^c)]}{P\{N(D) = 0|\mathcal{F}^N(D^c)\}} \tag{2.57}$$

on $\{N(D) = 0\}$, and finally, with A_d^2 denoting the diffuse component of A^2, such that

$$A^2 - A_d^2 = \sum_y [1 - N(\{y\})][U_y/(1 - U_y)]\epsilon_y. \tag{2.58}$$

In fact, one may take A^2 to be defined by (2.56) and (2.57), using (2.55) to ensure that the definition is consistent. Then for $y \in D$, (2.57) implies that on $\{N(D) = 0\}$,

$$A^2(\{y\}) = \frac{P\{N(\{y\}) = N(D) = 1|\mathcal{F}^N(D^c)\}}{P\{N(D) = 0|\mathcal{F}^N(D^c)\}}$$

$$= \frac{P\{N(\{y\}) = N(D) = 1 | \mathcal{F}^N(D^c)\}}{P\{N(D \setminus \{y\}) = 0 | \mathcal{F}^N(D^c)\}}$$

$$\times \frac{P\{N(D \setminus \{y\}) = 0 | \mathcal{F}^N(D^c)\}}{P\{N(D \setminus \{y\}) = 0 | \mathcal{F}^N(D^c)\} - P\{N(\{y\}) = N(D) = 1 | \mathcal{F}^N(D^c)\}}$$

$$= \frac{U_y}{1 - U_y},$$

so that (2.58) holds, and hence A^2 is a well-defined (i.e., locally finite) random measure. For details of the argument needed to check that A^2 is finite *near* points of N [it vanishes *at* points of N by (2.56)], see Kallenberg (1980).

Now let

$$A = A^1 + A_d^2; \tag{2.59}$$

we will show that A satisfies (2.53). For $D \in \mathcal{D}$, (2.55) implies that on $\{N(D) = 0\}$,

$$\sum_{D_{nj} \subseteq D} P\{N(D_{nj}) = 1 | \mathcal{F}^N(D^c)\}$$

$$= \sum_{D_{nj} \subseteq D} \frac{P\{N(D_{nj}) = N(D) = 1 | \mathcal{F}^N(D^c)\}}{P\{N(D \setminus D_{nj}) = 0 | \mathcal{F}^N(D^c)\}}$$

$$\rightarrow \int_D \frac{P\{N(dy) = N(D) = 1 | \mathcal{F}^N(D^c)\}}{P\{N(D \setminus \{y\}) = 0 | \mathcal{F}^N(D^c)\}} = A(D).$$

In general (by the monotone class theorem),

$$\sum_{D_{nj} \subseteq D} P\{N(D_{nj}) = 1 | \mathcal{F}^N(D_{nj}^c)\} \mathbf{1}(N(D_{nj}) = 0)$$

$$\rightarrow \int_D [1 - N(\{y\})] A(dy) = A(D) - \sum_{y \in D} U_y N(\{y\}). \tag{2.60}$$

But by the construction and argument in the first part of the proof,

$$\sum_{D_{nj} \subseteq D} P\{N(D_{nj}) = 1 | \mathcal{F}^N(D_{nj}^c)\} \mathbf{1}(N(D_{nj}) > 0) \rightarrow \sum_{y \in D} U_y N(\{y\}). \tag{2.61}$$

The first part of (2.53) follows at once from (2.60)–(2.61) and the second holds because N is simple and integrable. ∎

The random measure A of Theorem 2.40 is the *conditional intensity* of N. Under the assumptions of the theorem it is independent of $\mathcal{D}$ (see Papangelou, 1974b, and Kallenberg, 1980). Exvisible projections and conditional intensities are connected by the following result.

Proposition 2.41. For a point process N satisfying the hypotheses of Theorem 2.40 the conditional intensity and the exvisible projection coincide.

Proof: We show that the conditional intensity A satisfies the conditions of Theorem 2.38. For (2.52), if $D \in \mathcal{D}$, then by (2.53) and L^1 convergence

$$
\begin{aligned}
E[A(D)|\mathcal{F}^N(D^c)] &= \lim E\left[\textstyle\sum_{D_{nj} \subseteq D} E[N(D_{nj})|\mathcal{F}^N(D^c_{nj})] \,\middle|\, \mathcal{F}^N(D^c)\right] \\
&= E[N(D)|\mathcal{F}^N(D^c)],
\end{aligned}
$$

which yields (2.52) by standard approximations. To show that the conditional intensity is exvisible, we use (2.59). The diffuse component A_d^2 is trivially exvisible, while A^1 is exvisible by construction. ∎

Thus the approach of "conditioning on the complement of a small set," which on $\mathbf{R}_+$ is represented in part by the compensator, admits an analogue in general spaces. By contrast, the Palm distributions represent the opposite extreme of "conditioning on a subset of the support" of a point process. One can link these extremes. The key idea is expressed by (2.49): all relevant information is contained in the observed atom positions. Observation over a larger set containing no additional points of the process does not — except by scaling — change the conditional distributions of the unobserved part. As noted previously, this raises the hope that conditional probabilities of the form (2.62) are all expressible from a single kernel. The heuristic expression (2.50) would then also be valid and would imply that on $\{N(B) = 0\} \cap \{\nu(B^c) = 0\}$,

$$
\frac{P\{N_B \in d\nu | \mathcal{F}^N(B^c)\}}{P\{N(B) = 0 | \mathcal{F}^N(B^c)\}} = \frac{Q_N(N, d\nu)}{Q_N(N, \{0\})}. \tag{2.62}
$$

We now make these ideas precise.

Definition 2.42. The *Gibbs kernel* of N is the kernel G on $\mathbf{M} \times \mathbf{M}_p$ given by

$$
G(\mu, d\nu) = \frac{Q_N(\mu, d\nu)}{Q_N(\mu, \{0\})}.
$$

Definition 2.43. The *Gibbs measure* of N is the random measure M on $\mathbf{M}_p$ given by $M(\Gamma) = G(N, \Gamma)$.

Note the interpretation $G(\mu, d\nu) = P\{N - \mu \in d\nu\}/P\{N - \mu = 0\}$.

The relationship of Palm distributions and Gibbs measures is given by the next result.

Theorem 2.44. Let N be a simple, integrable point process with Gibbs measure M. Then for each bounded set B,

$$
\frac{P\{N_B \in d\nu | \mathcal{F}^N(B^c)\}}{P\{N(B) = 0 | \mathcal{F}^N(B^c)\}} = M(d\nu) \tag{2.63}
$$

on $\{N(B) = 0\} \cap \{\nu(B^c) = 0\}$.

Proof: Let C be the compound Campbell measure of N and suppose that $\Lambda \subseteq \{\mu : \mu(B) = 0\}$ and that $\Gamma \subseteq \{\mu : \mu(B^c) = 0\}$. Then

$$E\left[\frac{P\{N_B \in \Gamma | \mathcal{F}^N(B^c)\}}{P\{N(B) = 0 | \mathcal{F}^N(B^c)\}} \mathbf{1}(N(B) = 0, N_{B^c} \in \Lambda)\right] = P\{N_B \in \Gamma, N_{B^c} \in \Lambda\},$$

while

$$E\left[\frac{Q(N, \Gamma)}{Q(N, \{0\})} \mathbf{1}(N(B) = 0)\mathbf{1}(N_{B^c} \in \Lambda)\right]$$

$$= \int C(\{0\} \times d\nu)\mathbf{1}(\nu(B) = 0)\mathbf{1}(\nu_{B^c} \in \Lambda)\frac{Q(\nu, \Gamma)}{Q(\nu, \{0\})}$$

$$= \int C(d\mu \times \{0\})\mathbf{1}(\mu(B) = 0)\mathbf{1}(\mu_{B^c} \in \Lambda)\frac{Q(\nu, \Gamma)}{Q(\nu, \{0\})}$$

$$= C(\{\mu : \mu(B) = 0, \mu_{B^c} \in \Lambda\} \times \Gamma)$$

By symmetry of C,

$$C(\{\mu : \mu(B) = 0, \mu_{B^c} \in \Lambda\} \times \Gamma) = C(\Gamma \times \{\nu : \nu(B) = 0, \nu_{B^c} \in \Lambda\})$$
$$= P\{N_B \in \Gamma, N_{B^c} \in \Lambda\},$$

and this completes the proof. ∎

If N is a Poisson process on E with finite mean measure μ, then $Q(\nu, \cdot) = \mathcal{L}_N(\cdot)$ for all ν (Section 1.7) and hence $G(\nu, d\eta) = e^{\mu(E)}\mathcal{L}_N(d\eta)$.

More generally, let N be Poisson and let $\tilde{N}$ be a point process such that the likelihood ratio $f = d\mathcal{L}_{\tilde{N}}/d\mathcal{L}_N$ exists and is strictly positive. Then by the independent increments and conditional uniformity properties of N,

$$\frac{P\{\tilde{N}_B \in \Gamma | \mathcal{F}^{\tilde{N}}(B^c)\}}{P\{\tilde{N}(B) = 0 | \mathcal{F}^{\tilde{N}}(B^c)\}}$$

$$= \frac{E[f(N)\mathbf{1}(N_B \in \Gamma) | \mathcal{F}^N(B^c)]}{E[f(N)\mathbf{1}(N(B) = 0) | \mathcal{F}^N(B^c)]}$$

$$= \frac{1}{f(\tilde{N}_{B^c})} \sum_{k=0}^{\infty} \frac{1}{k!} \int_{E^k} f(\tilde{N}_{B^c} + \sum_1^k \varepsilon_{x_i})\mathbf{1}(\sum_1^k \varepsilon_{x_i} \in \Gamma)\mu^k(dx).$$

By introducing the "energy" (log-likelihood) function $U(\nu, \nu + \eta) = -\log[f(\nu + \eta)/f(\nu)]$, one obtains an explicit representation for the Gibbs kernel of $\tilde{N}$:

$$G_{\tilde{N}}(\nu, \Gamma) = \sum_{k=0}^{\infty} \int_{E^k} \exp[-U(\nu, \nu + \sum_1^k \varepsilon_{x_i})]\mathbf{1}(\sum_1^k \varepsilon_{x_i} \in \Gamma)\mu^k(dx)/k!.$$

It remains to connect the Gibbs kernel explicitly with the "conditioning on the complement of a point" represented by exvisible projections and conditional intensities. Suppose that D is a set belonging to the null array of Theorem 2.40 with $N(D) = 0$. Then, with other notation as in that theorem, by (2.53) and (2.63),

$$
\begin{aligned}
A(D) &= \lim_{n \to \infty} \sum_{D_{nj} \subseteq D} E[N(D_{nj})|\mathcal{F}^N(D_{nj}^c)] \\
&= \frac{1}{Q(N, \{0\})} \lim_{n \to \infty} \sum_{D_{nj} \subseteq D} P\{N(D_{nj}) = 0|\mathcal{F}^N(D_{nj}^c)\} \\
&\quad \times \int Q(N, d\nu) \mathbf{1}(\nu(D^c) = 0)\nu(D_{nj}) \\
&= \frac{1}{Q(N, \{0\})} \int Q(N, d\nu) \mathbf{1}(\nu(D^c) = 0) \int_D \nu(d\mathbf{x})(1 - U_{\mathbf{x}}).
\end{aligned}
$$

For $x \in D_{nj}$ [keep in mind that $N(D) = 0$]

$$
\begin{aligned}
1 - U_x &= \lim_{n \to \infty} P\{N(D_{nj}) = 0|\mathcal{F}^N(D_{nj}(\mathbf{x})^c)\} \\
&= \frac{Q(N, \{0\})}{Q(N, \{0\}) + Q(N, \{\varepsilon_x\})}.
\end{aligned}
$$

Consequently, on $\{N(D) = 0\}$,

$$
A(D) = \int_{\{\nu(D^c)=0\}} Q(N, d\nu) \int_D \frac{\nu(d\mathbf{x})}{Q(N, \{0\}) + Q(N, \{\varepsilon_x\})}. \tag{2.64}
$$

Atoms of A can arise only from the random measure $A^1 = \sum U_y \varepsilon_y$, and hence the component of A not embodied in (2.64) is simply the purely atomic random measure $\tilde{A}(B) = \int_B U dN$.

EXERCISES

2.1 Let $(\mathcal{H}_t)$ be a filtration on $(\Omega, \mathcal{F})$. Prove that every right- or left-continuous adapted process X is progressive.

2.2 Define a stopping time T to be *predictable* if there is a sequence of stopping times T_n such that $T_n \uparrow T$ but $T_n < T$ on $\{T > 0\}$ for each n, and *totally inaccessible* if $P\{T = S\} = 0$ for every predictable stopping time S. Prove that the arrival times $T_1, T_2, \ldots$ in a stationary Poisson process N are totally inaccessible. (The filtration is $\mathcal{F}^N$.)

2.3 Let $\mathcal{H}$ be a filtration with respect to which M is a martingale and A is an adapted increasing process. Prove that for each finite stopping time T such that $E[|M_T A_T|] < \infty$, $E[\int_0^T M_s dA_s] = E[M_T A_T]$. [*Hint:* Use the property that $M_s = E[M_T|\mathcal{H}_s]$ on $\{s < T\}$.]

2.4 Show by construction of an explicit example that the predictability hypothesis in Theorem 2.9 cannot be suppressed.

2.5 Let $\mathcal{F}^N$ be the internal history of a point process N on $\mathbf{R}_+$ and let M be a process that is simultaneously a martingale and predictable. Prove that almost surely $M_t = M_0$ for all $t > 0$ (i.e., M is constant over time).

2.6 Let N be a Poisson process on $\mathbf{R}_+$ with diffuse mean measure μ. Prove directly that $M_t = N_t - \mu_t$, is a martingale.

2.7 Let N be a Poisson process on $\mathbf{R}_+$ with diffuse mean measure μ and let $M = N - \mu$ be the $\mathcal{F}^N$-innovation martingale.
a) Compute each of the stochastic integrals $\int_0^t N_s dN_s$, $\int_0^t N_{s-} dN_s$, $\int_0^t N_s dM_s$ and $\int_0^t N_{s-} dM_s$.
b) Determine which of these processes are martingales and interpret in terms of the discussion of predictability in Section 2.

2.8 Verify (2.29).

2.9 Prove that a simple point process N with continuous, deterministic compensator is necessarily a Poisson process. [*Hint:* It suffices (why?) to show that N has independent increments.]

2.10 a) Let N be a Cox process on $\mathbf{R}_+$ directed by the diffuse random measure M and let $\mathcal{H}_t = \mathcal{F}_t^N \vee \mathcal{F}_\infty^M$. Prove that the $\mathcal{H}$-compensator of N is M.
b) Conversely, let N be a point process on $\mathbf{R}_+$ adapted to $\mathcal{H}$ and suppose that there exists an increasing process A, with $A_t \in \mathcal{H}_t$ for all t, such that A is the $\mathcal{H}$-compensator of N. Prove that N is a Cox process directed by A.

2.11 Let N be a point process on $\mathbf{R}_+$ having $\mathcal{F}^N$-compensator A and assume that $P\{\lim_{t\to\infty} A_t = \infty\} = 1$. Define $\tilde{A}_s = \inf\{t : A_t > s\}$ and let $\tilde{N}_s = N(A_s)$. Prove that $\tilde{N}$ is a stationary Poisson process with rate 1.

2.12 This exercise demonstrates how the history can affect the compensator in a nontrivial manner. Let (N_t) be a *Yule process* on $\mathbf{R}_+$, that is, a pure birth process with birth rate in state i equal to i [N is a Markov process with generator $A(i, i+1) = -A(i, i) = i$].
a) Prove that there is a random variable V such that $e^{-t} N_t \to V$ almost surely. (It is clear intuitively that the population grows exponentially.)
b) Prove that the $\mathcal{F}^N$-compensator of N is $A_t = \int_0^t N_{s-} ds$. Thus, N_{s-} is the $\mathcal{F}^N$-stochastic intensity of N.
c) Show that the compensator of N with respect to $\mathcal{H}_t = \mathcal{F}_t^N \vee \sigma(V)$, where V is as in a), is $B_t = (e^t - 1)V$, and conclude with the aid of Exercise 2.10 that N is a Cox process.

2.13 Let T be a positive random variable with distribution function F admitting density f. The function $h(t) = f(t)/[1 - F(t)]$ is the *hazard function* or *failure rate function* of F. (The terms come from survival analysis and reliability.) Verify that:
a) For each t, $h(t) = \lim_{s\to 0} P\{T \le t + s | T > t\}/s$.

b) For each t, $F(t) = 1 - \exp[-\int_0^t h(u)du]$, so that F is determined uniquely by h.

c) F is an exponential distribution if and only if h is constant.

d) The expression (2.33) holds.

2.14 Let N be a renewal process with interarrival distribution F admitting hazard function h. Prove that the $\mathcal{F}^N$-stochastic intensity of N is $\lambda_t = h(V_{t-})$, where $V_t = t - T_{N_t}$, the *backward recurrence time* at t, is the time since the most recent arrival (see Chapter 8).

2.15 (Continuation of Exercise 2.14). Explain why the process $h(V_t)$ is *not* the stochastic intensity of N.

2.16 Let N be a point process with $\mathcal{H}$-stochastic intensity (λ_t) and let $\mathcal{G}$ be another filtration with $\mathcal{F}_t^N \subseteq \mathcal{G}_t \subseteq \mathcal{H}_t$ for each t. Prove that there exists a $\mathcal{G}$-stochastic intensity $(\tilde{\lambda}_t)$ satisfying $\tilde{\lambda}_t = E[\lambda_t | \mathcal{G}_t]$ almost surely for each t.

2.17 Let $N(1)$ and $N(2)$ be independent point processes on $\mathbf{R}_+$ and suppose that $N(i)$ has $\mathcal{F}^{N(i)}$-stochastic intensity $\lambda_t(i)$. Prove that with respect to $\mathcal{H}_t = \mathcal{F}_t^{N(1)} \vee \mathcal{F}_t^{N(2)}$, the superposition $N = N(1) + N(2)$ has stochastic intensity $\lambda_t = \lambda_t(1) + \lambda_t(2)$.

2.18 This exercise describes an explicit construction of a point process with a prescribed stochastic intensity. Let (λ_t) be a nonnegative, predictable process and let $\tilde{N} = \sum \varepsilon_{(X_i, Y_i)}$ be a stationary Poisson process on $\mathbf{R}_+^2$ with intensity 1, such that $\tilde{N}$ and λ are independent. Define a point process $N = \sum \varepsilon_{T_i}$ by the following inductive procedure:

i) $T_1 = \inf\{X_i : Y_i \leq \lambda_0\}$ is the x-coordinate of the leftmost point of $\tilde{N}$ falling below the line $y = \lambda_0$.

ii) With $T_1, \ldots, T_n$ defined previously, $T_{n+1} = \inf\{X_i : Y_i \leq \lambda_{T_n}\}$.

Prove that N has stochastic intensity λ.

2.19 Verify (2.39).

WARNING! In Exercises 2.20–2.23 it may be difficult or impossible at this point to proceed beyond conditional probabilities or expectations yielded by Exercise 2.16. Techniques will be presented in Chapters 3, 5 and 7 for completing the calculations.

2.20 a) Let $\tilde{N}$ be a point process on $\mathbf{R}_+$ with stochastic intensity (λ_t), and let T be a positive random variable independent of N and with hazard function h. Let N be the point process $N_t = \tilde{N}_{T \wedge t}$. Calculate the stochastic intensity of N with respect to the filtration $\mathcal{H}_t = \mathcal{F}_t^{\tilde{N}} \vee \sigma(T)$ and with respect to the internal history $\mathcal{F}_t^N$.

b) Let $N(1)$ and $N(2)$ be independent Poisson processes on $\mathbf{R}_+$ with rates ν_1 and ν_2, and let T be a random variable independent of $N(l)$ and $N(2)$ and having an exponential distribution with parameter θ. Let N be the point process defined by $N_t = N_{T \wedge t}(1) + N_{(T-t)^+}(2)$; at the random time T, N switches from being Poisson with rate ν_1 to being Poisson with rate ν_2.

Prove that N is a Cox process (see Exercise 2.10) and calculate the stochastic intensity with respect to the internal history. (This is the *disruption problem*.)

2.21 Let $\tilde{N}$ be a Poisson process on $\mathbf{R}_+$ with rate c and let (X_t) be a Markov process with state space $\{0, 1\}$ and positive transition rates a $(0 \rightarrow 1)$ and b $(1 \rightarrow 0)$. Assume that $\tilde{N}$ and X are independent. Calculate the stochastic intensity of the partially observed Poisson process $N_t = \int_0^t X_s d\tilde{N}_s$ with respect to its internal history.

2.22 Let $\tilde{N}$ be a Poisson process with rate ν, representing arrival times of particles at a Geiger counter. Each arrival locks the counter for an interval of deterministic length τ, during which other arrivals that may occur have no effect whatever.
a) Calculate the stochastic intensity of the point process N of recorded arrivals. [*Hint:* Use Theorem 2.30.]
b) Generalize to the case that the locked times are i.i.d. random variables independent of $\tilde{N}$.

2.23 This exercise examines the classical model of *censored survival data*. Let $X_1, \ldots, X_n$ be i.i.d. nonnegative random variables with absolutely continuous distribution F; these are the survival times, say for patients undergoing some medical treatment. However, not all the X_i are observed: contact with some patients may be lost, or the study may terminate, for example. This phenomenon is called *censoring*, and may be modeled (in the simplest case) as follows. Let $Y_1, \ldots, Y_n$ i.i.d. random variables with absolutely continuous distribution G, such that (X_i) and (Y_i) are independent; the Y_i are censoring times. The observed data for patient i are then $Z_i = X_i \wedge Y_i$ and $D_i = 1(X_i \leq Y_i)$, so that it is known whether Z_i is a survival time or a censoring time. These data are summarized in the marked point process $N = \sum_{i=1}^n \varepsilon_{(Z_i, D_i)}$. Calculate the stochastic intensity of N with respect to the internal history.

2.24 Consider a waiting facility at which objects arrive according to a point process N with $\mathcal{F}^N$-stochastic intensity λ. There they are detained and dispatched in groups. Specifically, at a fixed time τ all objects present are dispatched, but there may also be one intermediate dispatching at a random time $T \leq \tau$ chosen to minimize the total waiting cost over the interval $[0, \tau]$. One can think of τ as a cycle length, so that the long-run average cost per unit time is $1/\tau$ times the cost over $[0, \tau]$. Let $c(s)$ be the cost of storing one object for s time units; we assume c to be strictly increasing and differentiable. With no intermediate dispatch the expected total cost (per cycle) is $E[\int_0^\tau c(\tau - s)dN_s]$. The cost *saving* from a dispatch at the stopping time T is $C(T) = E[N_T c(\tau - T)]$, which should be maximized in order that the total cost be minimized.
a) Show that $E\left[\int_0^\tau c(\tau - s)dN_s\right] = E\left[\int_0^\tau c(\tau - s)\lambda_s ds\right]$.
b) Show that $C(T) = E\left[\int_0^\tau \{c(\tau - s)\lambda_s - N_s c'(\tau - s)\}ds\right]$. [*Hint:* Use integration by parts — see Proposition 2.8.]
c) Verify that if N is a Poisson process with rate ν and $c(s) \equiv s$, then $C(T) = E\left[\int_0^\tau \{\nu(t - s) - N_s\}ds\right]$.

d) Show that for the special case in c), the optimal stopping time is $T^* = \inf\{s : N_s \geq \nu(\tau - s)\}$.

2.25 Prove the Poisson limit theorem associated with Exercise 1.21, using compensators and the following plan. Let $X_1, X_2, \ldots$ be i.i.d. positive random variables with density function f, let $\lambda = f(0)$, and for each n let $N^{(n)} = \sum_{i=1}^{n} \varepsilon_{nX_i}$. Use (2.29) to calculate the stochastic intensity $(\lambda_t^{(n)})$, and then show that for each t, $\lambda_t^{(n)} \to \lambda_t$ in probability. Deduce from this that the hypotheses of Theorem 2.25 are fulfilled and hence that $N^{(n)}$ converges in distribution to a Poisson process with rate λ.

2.26 Let N be a point process with stochastic intensity (λ_t) for which the innovation martingale M is square integrable. Calculate explicitly the representation (2.41) for the martingale $M_t^2 - \int_0^t \lambda_s ds$.

2.27 Prove that there exists no point process for which the innovation martingale is a Wiener process. [*Hint:* See Exercise 2.5.]

2.28 Let N be a Poisson process on E with diffuse mean measure μ. Calculate the exvisible projection of N directly, calculate the conditional intensity using (2.53), and verify that the two are identical.

2.29 Let X be a random variable uniformly distributed on $[0, 1]$ and let N be the point process $\sum_{n=1}^{\infty} \varepsilon_{X+n+1/n}$. Compute the compensator of N and the exvisible projection of N and show that the two differ.

2.30 Let N be a point process on a general space E with reduced Palm distributions $Q_N(\mu, d\nu)$. Show that $Q_N(0, \cdot) = \mathcal{L}_N$.

2.31 Let N be a point process with compound Campbell measure C_N. Prove that if $P\{N(E) < \infty\} = 1$, then C is symmetric: $C_N(\Lambda \times \Gamma) = C_N(\Gamma \times \Lambda)$ for all $\Lambda, \Gamma \in \mathcal{M}_p$.

2.32 Let $N = \sum_{n=1}^{\infty} \varepsilon_{T_n}$ be a point process on $\mathbf{R}_+$, and let $\mathcal{H}$ be a filtration. Prove that N is adapted to $\mathcal{H}$ if and only if T_n is a stopping time of $\mathcal{H}$ for each n.

2.33 Prove that (2.34) extends to that case that t there is replaced by a stopping time T such that $P_0\{T < \infty\} = 1$.

NOTES

Section 2.1

There are numerous references on the measure-theoretic foundations of the theory of stochastic processes. Important earlier works are Chung and Doob (1965) and Meyer (1969); Dellacherie (1972) is the first complete treatment, of which Dellacherie and Meyer (1980) is an expanded version. More recently, Brémaud (1981), Liptser and Shiryaev (1978), Jacod and Shiryaev (1987) and Kallianpur (1980) have given expository presentations. The first of these is oriented especially to point processes and contains several of the propositions in the section.

Section 2.2

The theory of stochastic integrals with respect to martingales and semimartingales, which is founded on that of Wiener and Poisson integrals, has been developed largely by the French school of probabilists, especially Jacod, the Japanese, especially Itô, the founder of the theory, and the Soviets. Their descriptions of it, even though no one would call them easy, are the best. These include Jacod (1979), Jacod and Shiryaev (1987) Metivier and Pellaumail (1980), and Dellacherie and Meyer (1980), along with numerous papers from the Strasbourg Séminaire des Probabilités. Gill (1980b) and Liptser and Shiryaev (1978) contain developments of stochastic Stieltjes integrals specifically related to point processes and innovation martingales. The integration-by-parts formulas of Proposition 2.8, which are derived via Fubini's theorem, are standard.

Section 2.3

Important general references are Brémaud and Jacod (1977), Dellacherie (1972), Gill (1980b), Jacod and Shiryaev (1987), Liptser and Shiryaev (1978), and Shiryaev (1981). All but the second and fourth are devoted entirely or in significant part to point processes. In Assumption 2.10 appear some of the "*conditions habituelles*" of the French school. Although compensators can be traced at least to the "*projection dual prévisible d'un processus croissant*" of Dellacherie (1972), as well as to work of Papangelou, the key paper is Jacod (1975b), which gives in definitive form the martingale characterization of compensators (Theorems 2.14 and 2.22), the uniqueness theorem (Theorem 2.19), the form of likelihood ratios [Theorem 2.31 is a special case — see also Theorem 5.2], and the representation for point process martingales (generalizing Theorem 2.34).

The role of compensators in statistical inference began to emerge in Aalen (1975, 1978) and Gill (1980b) (see Chapter 5 for details).

There is another important line of contributions to the theory in its formative stages. Motivated by questions of smoothing, filtering, and prediction in communications, electrical engineers developed results that were crucial in directing the course of later research. (Nearly all of these have by now been supplanted by more general versions.) In particular, Snyder (1972a, 1972b, 1975) concerning state estimation for Cox processes, Boel *et al.* (1975) on representation of point process martingales, and van Schuppen (1977) on state estimation with point process observations are key references. The concept, interpretation, and application of innovation are due to these and other engineers, and ultimately to Kalman (1960) (see also Bucy and Kalman, 1961, and Kailath, 1970).

Theorem 2.17 appears in Gill (1980b) but is more or less standard and was probably known beforehand; a shortcoming is that application of (2.16) for each t need not produce a process that is predictable. Theorem 2.18 is due for ordinary point processes to Dellacherie (1972). Hazard functions are treated in numerous references, especially on reliability and analysis of survival data (see, e.g., Barlow and Proschan, 1975; Kalbfleisch and Prentice, 1980; Lawless, 1980; and Nelson, 1982). A good source of martingale results such as Theorem 2.21 and Proposition 2.23, along with predictable variation and predictable covariation, is Jacod (1979);

some ideas and results date to earlier works such as Kunita and Watanabe (1967). A sketch of the theory of continuous-time martingales appears in Appendix B.

The Poisson convergence theorem (Theorem 2.25) is taken from Kabanov et al. (1980a), and Theorem 2.26 is due to T. C. Brown (1983). Jacod and Shiryaev (1987) is a book-length treatment of the paradigm of proving convergence in distribution for semimartingales by proving convergence of triplets of "predictable characteristics."

Section 2.4

Of the general sources noted for Section 3, only Brémaud (1981) and Gill (1980b) contain extended discussions of stochastic intensities; some material can be found in Brémaud and Jacod (1977) and Jacobsen (1982). Two more elementary (but sometimes oblique) treatments are Cox and Isham (1980) and Snyder (1975). Motivation for study of point processes admitting stochastic intensities stems from statistical inference (Aalen, 1975, 1978; Gill, 1980b), from queueing theory, particularly attempts to describe output processes and flows within networks of queues (see Brémaud, 1981, for discussion and references) and from communication theory. An early version of Theorem 2.31, the representation theorem for likelihood ratios, was given by Rubin (1972), but results of Jacod (1975b), valid for marked point processes with very general mark spaces, are definitive. Kailath and Segall (1975) is a related paper. In the engineering literature (see Snyder, 1975) and elsewhere, point processes with nondeterministic stochastic intensities have been termed *self-exciting*, but in the probability literature "self-exciting" usually connotes a stochastic intensity that is a deterministic function of the past (see, e.g., Proposition 7.32 and Exercise 9.19).

Section 2.5

In addition to Jacod (1975b), important papers on structure of point process martingales are Boel et al. (1975), M. H. A. Davis (1976), van Schuppen (1977), and (for multidimensional analogues) Yor (1976). Brémaud (1981) and Liptser and Shiryaev (1978) are good expository treatments.

Section 2.6

Of the conditioning concepts presented here, exvisibility and exvisible projections are due to van der Hoeven (1982), parallel most clearly the dual projection approach of Dellacherie (1972) (see also van der Hoeven, 1983). Conditional intensities were introduced by Papangelou (1974b), the source as well of two regularity conditions that are important to the theory. In the notation of the section, Papangelou's condition (Σ) states that for each bounded set B, $P\{N(B) = 0|\mathcal{F}^N(B^c)\} > 0$ almost surely on the set $P\{N(B) = 1|\mathcal{F}^N(B^c)\} > 0$ while condition (Σ^*) states that for each $B \in \mathcal{B}$, almost surely the conditional expectation $E[N(\cdot)|\mathcal{F}^N(B^c)]$ has no atoms in B. Papangelou (1974b) shows that a point process N satisfying (Σ) and (Σ^*) fulfills a version of (2.63). Results in Kallenberg (1980) generalize those of Papangelou (1974b) by removing a square integrability hypothesis. Both Kallenberg (1980) and van der Hoeven (1982) shed further light on (Σ) and (Σ^*).

The remaining ideas and results presented in the section are nearly all due to

Kallenberg (1983, Chaps. 12–14), a strikingly original work that connects conditioning for point process with ideas in statistical physics. In particular, Proposition 2.35, the importance of the consequences thereof, Definition 2.43, and Theorem 2.44 appear there. A less arduous version is Kallenberg (1984).

Exercises

2.2 This is yet another manifestation of the memorylessness property of exponential distributions. See Dellacherie (1972) concerning predictability, accessibility, and total inaccessibility of stopping times.

2.5 The assumption that the history be generated by a point process is crucial; the property is not true, for example, for the internal history of a Wiener process.

2.7 Predictability, or not, of the integrands is the key to differences among these integrals.

2.9 The characterization of Poisson processes is due to Watanabe (1960), while the Cox process analogue was first given by Brémaud (1975a) (see also T. C. Brown, 1981). Cox processes are characterized by $\mathcal{H}_0$-measurability of the compensator.

2.10 See Exercise 2.9.

2.11 See Jacod (1975b) and Brémaud and Jacod (1977). In addition, this property is related to the construction in Exercise 2.18.

2.21 Basawa and Prakasa Rao (1980) and Karlin and Taylor (1975) have more on Yule processes.

2.13 See also Section 8.2 and accompanying references.

2.14 See Proposition 8.10, Theorem 8.17 and associated discussion.

2.15 See Exercise 2.14.

2.16 See Brémaud and Jacod (1977).

2.18 This intuitive and elegant construction was gleaned from a lecture by Brémaud.

2.19 The model appears in Aalen (1978) (see also Jacobsen, 1982); Billingsley (1961a)is the key classical reference. In van der Plas (1983) a related discrete time model is analyzed using least squares methods.

2.20 See Liptser and Shiryaev (1978). The disruption problem is analyzed further in Chapters 3, 5 (especially Example 5.31) and 7.

2.21 For additional intensity computations, see Section 6.3 and Karr (1982)

2.22 This and other counter models are treated in Feller (1971) (see also Çinlar, 1975a).

2.23 The seminal reference is Kaplan and Meier (1958), which treats estimation of F from the censored data $(Z_i, D_i,)$, $i = 1, \ldots, n$. The ensuing literature is voluminous; see Chapter 5 for additional references.

2.24 Ross (1969) established the Poisson version. For the general case, see Brémaud and Jacod (1977).

2.25 See references for Exercise 1.21.

2.31 See Kallenberg (1983).

3
Inference for Point Processes: An Introduction

In this chapter we describe a general framework for statistical inference and state estimation for point processes. Not all of it is explored in detail in later chapters, but examples presented here illustrate many aspects. One crucial point to remember is that inference is a tripartite subject, whose components are

Statistical Inference in the sense of estimation and hypothesis testing (usually in a nonparametric setting)

State Estimation, the optimal reconstruction, realization by realization, of unobserved portions of a point process or of associated random variables or processes

Combined Statistical Inference and State Estimation, in which attributes of the probability law of a point process required to calculate state estimators must be replaced by statistical estimators.

We pause to recall the structure of a *statistical model*, which, formulated in generality, is comprised of

- The *sample space* $(\Omega, \mathcal{G})$, representing the set of all conceivable outcomes of a "random experiment" and a σ-algebra of events (sets of outcomes) of interest.

- A collection of *random mechanisms* that might govern the experiment, represented as an indexed (as distinct from "parametric") family $\mathcal{P} = \{P_\alpha : \alpha \in I\}$ of probability measures on $(\Omega, \mathcal{G})$. Here I is an arbitrary index set containing as few as two elements or so large that $\mathcal{P}$ is the

93

set of all probability measures on $(\Omega, \mathcal{G})$. One element of $\mathcal{P}$, the "true" law, actually does govern the experiment.

- *Observations* or *data* represented by a sub-σ-algebra $\mathcal{F}$ of $\mathcal{G}$. In many cases $\mathcal{F} \subseteq \sigma(X)$ for some stochastic process X. The form of the data may be dictated exogenously by the "physics" of a particular application or may be partly or entirely under control of the statistician.

Within this setting the central problem of statistical inference is to construct meaningful statements concerning the probability measure actually governing the experiment, based on the observed data. Such statements necessarily involve uncertainty, which must be described, and preferably quantified.

The goal of *statistical estimation* is to determine the operative probability, or functionals of it, as accurately as possible and to characterize the errors that arise. *Hypothesis testing* concerns choosing which of two (or several) values of the index α or disjoint subsets of the index set I is more compatible with the data.

Roughly speaking, "finite sample" theory concerns data associated with observations over a compact set. Since finiteness is the same as compactness in $\mathbf{N}$, this includes the case $\mathcal{F} = \sigma(X_1, \dots, X_n)$ forming the basis of classical finite sample theory. "Asymptotic" theory pertains to the case $\mathcal{F} \uparrow \mathcal{F}_\infty$, with $\mathcal{F}_\infty$ representing observation over a noncompact set and $\mathcal{F}$ observations over compact sets that increase to it.

Customary usage designates as *parametric* the case that the index set I is a subset of a finite-dimensional Euclidean space and as *nonparametric* all others, although often with a connotation that $\mathcal{P}$ is very large. The intermediate case that the "parameter" is infinite-dimensional yet $\mathcal{P}$ is severely restricted notwithstanding appears infrequently in classical statistics but is central for point processes (see, e.g., Chapter 6).

One must also collect data suited to methods available for its analysis and make effective use of limited sampling resources. For example, a point process on $\mathbf{R}$ may be observed for a fixed length of time, until a fixed number of arrivals occurs, or until sufficient data are collected to make statements of prescribed accuracy.

We do not treat in detail fundamental statistical concepts such as consistency, asymptotic normality, efficiency, sufficiency, completeness, and contiguity. However, it does seem useful to present some basics at this point. For simplicity we consider a parametric model $\mathcal{P} = \{P_\theta : \theta \in \Theta\}$, where Θ is an open subset of $\mathbf{R}^k$ for some k, and i.i.d. data $X_1, X_2, \dots$. Expectation with respect to P_θ is denoted by E_θ. A *statistic* is any random variable T that is a function of the observed data. An *estimator* is a statistic whose value is meant to estimate the true value of the unknown parameter θ, or of some function $q(\theta)$.

Regularity The model $\mathcal{P}$ is *regular* if there is a σ-finite dominating measure Q such that $P_\theta \ll Q$ for each θ, if the derivatives $p(\omega, \theta) = dP_\theta/dQ(\omega)$ are continuous in θ, and if the *Fisher information*

$$I(\theta) = \int \left(\frac{|\partial p(\omega, \theta)/\partial \theta|}{p(\omega, \theta)} \right)^2 Q(d\omega) \qquad (3.1)$$

is finite for each θ.

Unbiasedness An estimator T of $q(\theta)$ is *unbiased* if $E_\theta[T] = q(\theta)$ for all θ.

Minimum variance An unbiased estimator T of $q(\theta)$ has minimum variance within a class $\mathcal{T}$ of unbiased estimators if $E_\theta[(T - q(\theta))^2] \leq E_\theta[(S - q(\theta))^2]$ for all $S \in \mathcal{T}$. The *Cramér-Rao lower bound* on the variance of unbiased estimators is $1/I(\theta)$, where $I(\theta)$ is the Fisher information of (3.1).

Sufficiency A statistic T is *sufficient* given the observations $\mathcal{F}$ if the conditional distribution of the data given T is functionally independent of θ: for $\Gamma \in \mathcal{F}$, $P_\theta\{\Gamma|T\}$ is the same for θ. Heuristically, a sufficient statistic captures all information about θ contained in the data, and inference may and should be based on it alone. Conditioning on a sufficient statistic reduces variance: if T is sufficient then for any statistic S and any function q,

$$E_\theta[(E[S|T] - q(\theta))^2] \leq E_\theta[(S - q(\theta))^2]$$

for all θ. Sufficiency ensures that the conditional expectation $E[S|T]$ does not depend on θ, and hence may be computed from knowledge of the model but not the value of θ.

Completeness A statistic T is *complete* with respect to $\mathcal{P}$ if the only function g for which $E_\theta[g(T)] = 0$ identically in θ is $g \equiv 0$. If T is complete and sufficient with $\mathrm{Var}_\theta(T) < \infty$ for all θ and if S is an unbiased estimator of $q(\theta)$, then $T^* = E[S|T]$ is a *uniformly minimum variance unbiased* (UMVU) estimator of $q(\theta)$:

$$E_\theta[(T^* - q(\theta))^2] \leq E_\theta[(U - q(\theta))^2]$$

for all θ and every unbiased estimator U of $q(\theta)$

Consistency For each n let T_n be an estimator of θ based on the observations $X_1, \ldots, X_n$. The sequence (T_n) is *strongly consistent* if under P_θ, $T_n \to \theta$ almost surely as $n \to \infty$, and *weakly consistent* if $T_n \to \theta$ in probability. The latter ordinarily suffices, and is proved by showing that both the squared bias $E_\theta[T_n - \theta)^2]$ and the variance $E_\theta[(T_n - E_\theta[T_n])^2]$ converge to zero. Consistency means, obviously, that the true value of the parameter is recoverable in the limit.

Asymptotic normality Consistent estimators T_n are *asymptotically normal* if under each P_θ,

$$\sqrt{n}[T_n - \theta] \xrightarrow{d} N(0, \sigma^2(\theta)) \qquad (3.2)$$

as $n \to \infty$. Asymptotic normality is crucial to analysis of the estimation error $T_n - \theta$, whose exact distribution is typically intractable, and in particular to construction of confidence bounds. Of course, estimation of the asymptotic variance $\sigma^2(\theta)$, which depends on (T_n) as well, is necessary in order to construct confidence regions.

Efficiency An asymptotically normal sequence (T_n) is *efficient* if the asymptotic variance $\sigma^2(\theta)$ of (3.2) is minimal, uniformly in θ, among those for all asymptotically normal sequences. In particular, any sequence for which the asymptotic variance attains the Cramér-Rao lower bound $1/I(\theta)$ is efficient.

By the term *classical statistics* we mean the statistics of regular parametric families, especially exponential families, given i.i.d. observations, some aspects of which have just been noted. The great principles of estimation, *maximum likelihood*, the *method of moments* (or *substitution*) and *least squares*, and the principle of *likelihood ratios* for hypothesis testing, rule classical statistics. They will be almost as predominant for us, but the complicated nature of point processes forces greater reliance on the principle of *ad hoc methods exploiting special structure*.

3.1   Forms of Observation

We begin by describing different ways in which point processes may be observed. There are three principal dichotomies, not independent in theory or practice, pertinent to the form of data arising from observation of a point process N on a space E. These yield the basic statistical models associated with point processes; the main of these are given explicitly below, while fundamental issues of statistical inference are presented in Section 2.

The three principal dichotomies are:

- Whether the underlying space E is *compact* or *noncompact*

- Whether the observations are a *single realization* of N or i.i.d. *multiple copies*

- Whether the observation is *complete* or *partial.*

Particular kinds of partial observation will be introduced momentarily and others appear in later chapters, so with one proviso we concentrate on the first two dichotomies. The proviso is that in reality and in statements of our results, point processes can only be observed over compact sets regardless of whether there is one realization or multiple copies. This does not, however, preclude one's studying asymptotic properties as the set of observation increases to a noncompact set.

3.1.1 Complete Observation

Hence there are four cases of complete observation, two of which are important theoretically and in applications and understood for at least some classes of point processes. The third is a straightforward generalization of either or both of these, while the fourth is of evident practical interest but underdeveloped theoretically, albeit not without reason: it is just very difficult, especially if one seeks results of reasonable generality.

The important cases are

- Observation of *multiple copies* of a point process over a *compact* set

- Observation or a *single realization* of a point process on a *noncompact* space. (Recall that actual observations are over compact subsets.)

Their importance derives primarily from their admitting — given suitable assumptions — interesting, useful, and broadly valid asymptotic results, in the former as the number of observed copies increases to infinity and in the latter as the process is observed over compact sets increasing to E.

The third case, multiple copies of a point process on a noncompact space, represents a surfeit of data. Methods from either of the two principal cases are applicable and no new difficulties arise.

The remaining case, observation of a single realization of a point process on a compact set, *is* of theoretical and practical interest, but can be very difficult, as are "finite sample" problems in other statistical contexts. Those results that do exist deal with models in which special structure is a dominant feature.

We now give explicitly the statistical models for the two principal cases listed above, along with some variants.

Model 3.1. (i.i.d. copies of a point process on a compact space). Let N be a point process on a compact space E. The statistical model of complete observation of a sequence of i.i.d. copies of N is given as follows. The sample space is $(\Omega, \mathcal{G}) = (\mathbf{M}_p, \mathcal{M}_p)^{\mathbf{N}}$, on which are defined coordinate mappings $N_i(\omega) = \omega_i$. Let $\{\mathcal{L}_\alpha\}$ be an indexed collection of probability measures on $(\mathbf{M}_p, \mathcal{M}_p)$; these are the candidates the probability law of N. Then $\mathcal{P} = \{P_\alpha : \alpha \in I\}$, where for each α, $P_\alpha = \mathcal{L}_\alpha^{\mathbf{N}}$. Under P_α, the N_i are i.i.d. with law $\mathcal{L}_\alpha$. Complete observation of $N_1, \ldots, N_n$ is represented by the σ-algebra $\mathcal{F}_n = \bigotimes_{i=1}^n \mathcal{F}^{N_i}$, where $\mathcal{F}^N = \mathcal{F}^N(E) = \sigma(N(A) : A \in \mathcal{E})$. □

Principal variations of this model are to incorporate an underlying random mechanism associated with each point process or to allow only partial observation. The former arises in connection with Cox processes (Chapter 7) and the latter in ways described below in this section.

Model 3.2. (Single realization of a point process on a noncompact space). Let N be a point process on a noncompact space E. The statistical model for complete observation of N is given as follows. The sample space is $(\Omega, \mathcal{G}) = (\mathbf{M}_p, \mathcal{M}_p)$, with $N(\omega) = \omega$. Let $\{\mathcal{L}_\alpha : \alpha \in I\}$ be an indexed family of candidate probability laws for N; then simply $\mathcal{P} = \{\mathcal{L}_\alpha : \alpha \in I\}$. For A a compact subset of E, the σ-algebra representing complete observation of N over A is $\mathcal{F}^N(A) = \sigma(N(B) : B \subseteq A)$. □

Both of these are "one-sample" models in the sense that there is a single operative probability measure. Except for isolated instances, we do not treat two-sample problems, so we have not included them as general formulations.

3.1.2 Partial Observation

We next list some forms of partial observation of point processes, but with no pretense of completeness. Five forms play central roles in this book.

Restriction to a subset This occurs N is observed only over a subset B of E, i.e., the observations are $N_B(\cdot) = N(\cdot \cap B)$. In the situation of Model 3.1 it may be that each process N_i is observed over a subset B_i of E, so that the data σ-algebra becomes $\mathcal{F}_n = \bigotimes_{i=1}^n \mathcal{F}^{N_i}(B_i)$.

Mapping In this case instead of the point process $N = \sum \varepsilon_{X_i}$ on E, one observes the mapping (Definition 1.36) $Nf^{-1} = \sum \varepsilon_{f(X_i)}$ on a space E', where $f : E \to E'$ is measurable and possibly unknown. Of course, if f is known and one-to-one, there is no partial observation because N can be reconstructed without error from Nf^{-1}. However, Nf^{-1} might be only partially observed. Also, in many examples, f is not one-to-one. An important instance, to illustrate, is that E is a product space and f a coordinate projection, so that a point process $N = \sum \epsilon_{(Y_i, X_i)}$ is observable only in the form $N f^{-1} = \sum \epsilon_{X_i}$. This model includes such diverse effects as being able to observe only the marks in a marked point process, observation of secondary points but not cluster centers in a Poisson cluster process (Sections 1.4 and 6.5), observations of Poisson processes arising in positron emission tomography (Section 6.3), and line or transect sampling methods used in ecology and forestry.

Integral data Here one observes only the integral data $\mathcal{F}^N(g_1, \ldots, g_k) = \sigma(N(g_1), \ldots, N(g_k))$, where $g_1, \ldots, g_k$ are prescribed functions. One motivation is that observation devices inevitably introduce smoothing, which can be modeled by integral data. Similarly, for spatial point processes it may be possible to perform only a finite number of measurements corresponding to the g_j.

The three foregoing forms of partial observation affect only the form of data in the statistical model; the final two, by introduction of exogenous randomness, alter all components of the model.

Thinning In the heuristic sense we take this to mean that a random subset of the points of N is deleted and only the remaining points, which constitute a "thinning" of N, are observable. The main thinning mechanism we examine is p-thinning in the sense of Definition 1.38. In problems of statistical inference the thinning function p may be unknown as well as the law of N, and might need to be estimated from observations of N'. For state estimation, both $\mathcal{L}_N$ and p are known and one attempts to reconstruct N from complete or partial observations of N'. By way of illustration we give explicitly the following variant of Model 3.1, which is analyzed in Section 4.5.

Model 3.3. (i.i.d. copies of a p-thinned point process on a compact space). Let N be a point process on a compact space E. The statistical model for complete observation of a sequence of i.i.d. copies of a p-thinning of N is as follows. The sample space is $(\Omega, \mathcal{G}) = (\Omega_0, \mathcal{G}_0)^{\mathbf{N}}$, where $(\Omega_0, \mathcal{G}_0) = (\mathbf{M}_p, \mathcal{M}_p)^2$, with coordinates (N_i, N_i'). Let $\{\mathcal{L}_\alpha : \alpha \in I_0\}$ be an indexed family of candidate probability laws for N and let H denote a set of (measurable) functions $p : E \to [0, 1]$, the possible thinning functions. For $\alpha \in I_0$ and $p \in H$, let $Q_{\alpha, p}$ be the joint law on $(\Omega_0, \mathcal{G}_0)$ of a point process N with law $\mathcal{L}_\alpha$ and the p-thinning N' of N. Then $\mathcal{P} = \{Q_{\alpha, p}^{\mathbf{N}} : \alpha \in I_0, p \in H\}$. The data corresponding to complete observation of $N_1', \ldots, N_n'$ is $\mathcal{F}_n = \bigotimes_{i=1}^n \mathcal{F}^{N_i'}$. □

Stochastic integral Here the observations may constitute a purely atomic random measure rather than a point process. Let X be a measurable, nonnegative stochastic process on E (possibly but not necessarily independent of N) and let the observations be the stochastic integral $X * N$ defined by

$$X * N(A) = \int_A X \, dN. \qquad (3.3)$$

When X assumes only the values 0 and 1, $X * N$ is a point process and, indeed, a thinning of N, although not in general a p-thinning. In several places in the book, N is a Poisson process on $\mathbf{R}_+$ and X is a binary Markov process independent of N. In this case the stochastic integral $X * N$ is a Cox process and a renewal process, and has received widespread application (see also Exercises 3.15–3.16).

The reader will have noticed that the forms of partial observation are not unambiguously distinct. For example, thinning and stochastic integrals overlap each other and also restriction to a subset — the thinning function or the integrand X could be an indicator function — and the integral data theme enters stochastic integrals as well.

For point processes $N = \sum \varepsilon_{T_i}$ on $\mathbf{R}_+$ there is one additional distinction — of synchronous or asynchronous data.

Synchronous data Synchronous sampling is observation until the time T_n of the nth arrival, and yields as data σ-algebras

$$\mathcal{F}^N_{T_n} = \sigma(T_1, \ldots, T_n) = \sigma(U_1, \ldots, U_n), \qquad (3.4)$$

where the $U_i = T_i - T_{i-1}$ are the interarrival times. The interval $[0, T_n]$ over which N is observed is thus random.

Asynchronous data Asynchronous sampling corresponds to observing N over fixed time intervals $[0, t]$, and gives data

$$\mathcal{F}^N_t = \sigma(N_u : u \le t). \qquad (3.5)$$

3.2   Statistical Inference

Our goal in this section is to outline a conceptual framework for statistical inference for point processes and to elucidate it with "sample" problems meant to convey the breadth and scope of point process inference, but not as a complete description. The structure is broad enough to include most extant theory and applications even though only part of the theory and but a small fraction of the applications are treated in detail in the remainder of the book.

The overall structure consists of five parts that will be explained in detail below:

- Parametric estimation

- Nonparametric estimation

- Parametric tests of quantitative hypotheses

- Nonparametric tests of quantitative hypotheses

- Tests of qualitative hypotheses.

Each subdivides based on the forms of observation discussed in Section 1, and in particular admits two principal and rather different subareas corresponding to observation of a single realization of a point process and to observation of i.i.d. multiple realizations. The complete observation/partial observation classification influences methodology rather than conceptualization and so does not figure prominently in the current discussion.

One can formulate further subdivisions into "finite sample" results associated with observation over compact sets and "asymptotic" results as the set of observation increases to a noncompact set, either by increase of the set over which one point process is observed or in the number of observed copies.

The parametric/nonparametric distinction for estimation and quantitative hypothesis testing is more or less as in classical statistics. *Parametric* problems are those in which the law of the point process is specified by model assumptions except for finitely many (typically, one or two) real parameters. *Nonparametric estimation* includes all other estimation problems, no matter how restricted the family $\mathcal{P}$ of candidate probability laws. For example, estimation of a completely unknown mean measure is a nonparametric problem, as is estimation of the probability law $\mathcal{L}$ itself when $\mathcal{P}$ is, say, restricted to the Poisson processes on $\mathbf{R}_+$. Even estimation of a single real parameter is a nonparametric problem if sufficiently little is assumed about the law of the point process. An example is estimation of the intensity of a stationary point process whose law is not further restricted (see Example 3.7).

In classical statistics, estimation consists of *point estimation* and *interval estimation* (more generally, construction of *confidence regions* for multidimensional parameters), and is linked through the latter to hypothesis testing. For point processes most estimation procedures, especially in nonparametric settings, yield point estimators. In spaces of measures confidence regions are difficult even to specify, and the traditional requirement of compactness may be impossible to fulfill.

Nonparametric testing of quantitative hypotheses pertains to infinite-dimensional characteristics such as mean measures, Laplace functionals, zero-probability functionals, and the like, as well as to finite-dimensional characteristics when $\mathcal{P}$ is infinite-dimensional. For example, a test concerning the intensity of a point process assumed only to be stationary is a nonparametric test of a quantitative hypothesis.

By contrast, qualitative hypotheses concern the *structure* of a point process, for example whether it has independent increments or is stationary or is a Cox process. These are sometimes referred to as questions of "model fit." Tests of qualitative hypotheses are nearly always nonparametric because specification of the law of a point process except for finitely many parameters nearly always specifies structural characteristics completely. Obviously, qualitative hypotheses and hypotheses concerning quantitative characteristics must be tested using rather different methods, but underlying principles such as likelihood ratio tests are common to both.

Especially for qualitative hypotheses but also for quantitative hypotheses, one encounters difficulties with specification and analysis of hypotheses. Reasonable hypotheses for point processes are often too broad. When the null hypothesis is very broad, a test constructed to have acceptably low power (probability of rejection) when the null is true may lack power under the alternative hypothesis as well, and be unlikely to reject either. Even if the null hypothesis is not hopelessly broad, a broad alternative may contain processes that are too much like the null hypothesis process, again causing the test to lack power. Severely restricting the structural assumptions, which would have the desirable effect of narrowing both null and alternative hypotheses, may be contradicted by the "physics" of the problem.

As an example, it is less difficult to test whether a renewal process is a Poisson process (Example 3.14) than to test whether a point process assumed only to be stationary is Poisson (Example 3.15). Unfortunately, though, the renewal assumption may be unsupported on physical grounds, if not actually contradicted. Similarly, it is harder to test the hypothesis that a stationary point process is a renewal process than to test the hypothesis that it is Poisson, because the former is broader.

A related problem is computation of power, especially under alternative hypotheses. Even for rather simple tests, an approximation of power, let alone an exact computation, may be impossible. The technique of identifying a worst case representative of the alternative hypothesis, computing power for it, and then using the result as a lower bound for power elsewhere under the alternative, is often ineffectual. A practical consequence is that one cannot determine, for a given test, to which kinds of deviations from the null hypothesis it is especially sensitive or insensitive.

In the examples that follow and the remainder of the book, two methodologies for statistical inference for point processes dominate:

- Empirical inference (Chapter 4)

- Intensity-based inference (Chapter 5).

Each has a facet reflecting the principle of maximum likelihood and another reflecting the method of moments, or substitution. We elaborate briefly.

Empirical inference is based on the notion that objects should be estimated by corresponding empirical averages and is most important in the multiple-copy case. In terms of likelihood, the method is based on the role of the empirical distribution as the (unrestricted) maximum likelihood estimator of the true distribution and on the property that the maximum likelihood estimator of a (smooth) functional is the same functional of the maximum likelihood estimator.

Given i.i.d. copies $N_1, \ldots, N_n$ of a point process with unknown law $\mathcal{L}$, the maximum likelihood estimator of $\mathcal{L}$ is the *empirical measure*

$$\hat{\mathcal{L}} = (1/n)\sum_{i=1}^{n}\varepsilon_{N_i},$$

a random, purely atomic probability distribution on $(\mathbf{M}_p, \mathcal{M}_p)$. Since the Laplace functional is a functional of $\mathcal{L}$ — $L(f) = \int e^{-\mu(f)}\mathcal{L}(d\mu)$ — it is estimated by the same functional of $\hat{\mathcal{L}}$, namely the empirical Laplace functional

$$\hat{L}(f) = \int e^{-\mu(f)}\hat{\mathcal{L}}(d\mu) = (1/n)\sum_{i=1}^{n}e^{-N_i(f)} \tag{3.6}$$

analyzed in Section 4.2. The substitution justification of (3.6) is simply that $L(f) = E[e^{-N(f)}]$. A compelling virtue of empirical methods is generality in regard to the underlying space.

Intensity-based methods apply only to point processes on $\mathbf{R}_+$ and only if integrability assumptions are fulfilled, but within this setting they are extraordinarily powerful. They are justified in likelihood terms by Theorem 2.31, which expresses the likelihood function in terms of the stochastic intensity. Thus when unknown parameters or functions appear explicitly in the stochastic intensity (Chapter 5) and if the latter is computable or of postulated form, then maximum likelihood estimation and likelihood ratio tests are feasible.

The substitution basis for intensity methods resides in Theorem 2.14, which for a point process N with stochastic intensity λ asserts that the innovation process $M_t = N_t - \int_0^t \lambda_s ds$ is a martingale, and in interpretation of increments in M as noise. For example, for the multiplicative intensity model of Model 5.6, where the stochastic intensity is $\lambda_t(\alpha) = \alpha_t \lambda_t$ with λ an observable, predictable process and α an unknown, deterministic function, the difference $dM_t = dN_t - \alpha_t \lambda_t dt$ constitutes martingale noise, so that the indefinite integral $B_t(\alpha) = \int_0^t \alpha_s ds$ is estimated by the stochastic process

$$\hat{B}_t = \int_0^t [1/\lambda_s] dN_s. \tag{3.7}$$

The *estimation error* $B_t(\alpha) - \hat{B}_t$ is then a martingale (for details see Chapter 5).

The principle of "ad hoc methods exploiting special structure" cannot be overlooked. Indeed, it is fundamental in Chapters 6–9.

Now, rather than prolong the verbal explanations, we present a series of examples intended to illustrate the range of statistical inference problems and methods associated with point processes, and also to elucidate the general structure that we have been attempting to describe.

3.2.1   Parametric Estimation

As noted previously, this case entails severely restrictive structural assumptions.

Example 3.4. (Renewal process). Let $N = \sum \varepsilon_{T_n}$ be a renewal process whose interarrival times U_i have gamma density

$$f_{\alpha,\beta}(x) = (1/\Gamma(\alpha))\beta^\alpha x^{\alpha-1} e^{-\beta x}, \tag{3.8}$$

with both the shape parameter α and scale parameter β unknown.

Given the synchronous data $\mathcal{F}^N_{T_n}$ the problem is classical: $U_1, \ldots, U_n$ are i.i.d. with density (3.8) and standard estimation procedures for gamma distributions apply. Let $L(\alpha, \beta) = \sum_{i=1}^n \log[f_{\alpha,\beta}(U_i)]$ be the log-likelihood function. When α and β are both unknown, the likelihood equations

$$\partial L/\partial \alpha = \partial L/\partial \beta = 0$$

cannot be solved in closed form (although numerical solution is always an option). The conventional procedure is to use the method of moments: $f_{\alpha,\beta}$ has mean α/β and variance α/β^2. If either α or β is known, then (3.8) defines a one-parameter exponential family and a multitude of results is applicable.

For asynchronous data $\mathcal{F}^N_t$, one approach is to take $\hat{\alpha}(t) = \hat{\alpha}_{N_t}$ as estimator of α, and similarly for $\hat{\beta}(t)$, where $\hat{\alpha}_n$ is the $\mathcal{F}^N_{T_n}$-estimator of α. In view of Proposition 2.4 this causes relatively little loss of information for finite t, while asymptotic properties of the two estimators are related in straightforward, systematic manner (see Chapter 8 for details).

An alternative approach uses Theorem 2.31. If $h_{\alpha,\beta}$ denotes the hazard function for $f_{\alpha,\beta}$, then under $P_{\alpha,\beta}$, N has stochastic intensity $\lambda_t(\alpha,\beta) = h_{\alpha,\beta}(V_{t-})$, where V_t is the backward recurrence time of N at t. Using Theorem 2.31, the log-likelihood function, up to a constant term, is

$$L_t(\alpha,\beta) = H_{\alpha,\beta}(V_t) + \sum_{i=1}^{N_t} \left[H_{\alpha,\beta}(U_i) + \log h_{\alpha,\beta}(U_i) \right], \qquad (3.9)$$

where $H_{\alpha,\beta}(t) = \int_0^t h_{\alpha,\beta}(u)du$. Although not amenable to closed-form maximization, (3.9) can be handled numerically in specific applications. $\square$

Example 3.5. (Partially observed Poisson process). Let N be a Poisson process on $\mathbf{R}_+$ with unknown rate λ and let (X_t) be a Markov process independent of N, with state space $\{0,1\}$ and known transition intensities $a\ (0 \to 1)$ and $b\ (1 \to 0)$ (see Exercise 2.21). The observations are the point process $N'(A) = \int_A X\,dN$ of (3.3), which is both a Cox process and a renewal process. Techniques from either of these classes are available, especially for synchronous data (see Exercises 3.15–3.16), but direct methods exist as well. Suppose that X has as initial distribution the invariant distribution $(b/(a+b), a/(a+b))$; then for each t, $E[N_t'] = \lambda at/(a+b)$. The asynchronous estimators

$$\hat{\lambda} = \frac{a+b}{a}\,\frac{N_t'}{t} \qquad (3.10)$$

are strongly consistent and asymptotically normal. The latter property may be used to construct approximate confidence intervals and hypothesis tests for λ. Note that in (3.10) dependence of the estimator on the sample size t is suppressed; for notational economy we do this whenever possible.

If X were observable in addition to N, then (3.10) would be replaced by

$$\hat{\lambda}^* = N_t'/\int_0^t X_s ds = \int_0^t X\,dN/\int_0^t X\,ds, \qquad (3.11)$$

for which strong consistency and asymptotic normality again hold. $\square$

Here is a multiple realization problem.

Example 3.6. (Pólya process). Let N be a *Pólya process* on $[0,1]$: N is a mixed Poisson process with directing measure $M(dx) = Y\,dx$, where Y is a random variable with gamma density (3.8). The problem is to estimate α and β from complete observation of i.i.d. copies $N_1, N_2, \ldots$ of N. By conditional uniformity the only data from N_i that need be retained is $Z_i = N_i([0,1])$. The Z_i are i.i.d. with mixed Poisson distribution

$$P_{\alpha,\beta}\{Z_i = k\} = \int_0^\infty f_{\alpha,\beta}(x)e^{-x}x^k dx/k!,$$

so the problem reduces to a standard problem of estimating the parameters of a gamma mixing distribution from mixed Poisson data (see Simar, 1976, and Section 7.1 for details). ▯

3.2.2   Nonparametric Estimation

We present first a single realization problem; in such cases the quantitative characteristic being estimated is often a real parameter.

Example 3.7. (Stationary point process). Let N be a stationary point process on $\mathbf{R}$ with unknown, finite intensity ν: the mean measure of N is known up to the parameter ν, but the set $\mathcal{P}$ of candidate laws of N is very large indeed. The model is complete observation of a single realization of N. The obvious estimator of ν given the asynchronous data $\mathcal{F}_t^N$ is $\hat{\nu} = N_t/t$. Whether these estimators are even weakly consistent depends on $\mathcal{L}_N$. If N is ergodic (Definition 1.70), then by Theorem 1.71 there is even strong consistency. If not, which is the case for the Pólya process of Example 3.6 when it is defined on $\mathbf{R}_+$, recovery of ν from single realizations is impossible. In the ergodic case asymptotic normality does not hold without further hypotheses, typically to the effect that $N(A)$ and $N(B)$ are nearly independent if A and B are widely separated; these are termed mixing conditions. ▯

Our second example involves multiple realizations.

Example 3.8. (Superposed Bernoulli processes). Let E be a compact space and let $N_1, N_2, \ldots$ be i.i.d. copies of the superposition of Bernoulli processes given by $N = \sum_{j=1}^k U_j \varepsilon_{X_j}$, where k is a known, positive integer, the X_j are i.i.d. random elements of E with unknown distribution F, and the U_j are i.i.d. Bernoulli random variables with $p = P\{U_j = 1\}$ also unknown; further, the X_j and U_j are independent (see Section 1.6 for details). The aim is to estimate p and F from data representing complete observation of $N_1, \ldots, N_n$. From the empirical perspective one chooses the estimators

$$\hat{p} = (1/nk)\sum_{i=1}^n N_i(E) \tag{3.12}$$

and

$$\hat{F}(\cdot) = \frac{1}{nk\hat{p}} \sum_{i=1}^n \frac{N_i(\cdot)}{N_i(E)} \mathbf{1}(N_i(E) > 0),$$

which are maximum likelihood estimators for the statistical model $\mathcal{P} = \{P_{p,F} : p \in (0,1], F \text{ is a probability distribution on } E\}$. It is straightforward to demonstrate strong consistency and joint asymptotic normality.

An alternative is to estimate the Laplace functional

$$L_{p,F}(f) = E_{p,F}[e^{-N(f)}] = (1 - p + p\int e^{-f} dF)^k$$

by the empirical Laplace functionals $\hat{L}$ introduced in (3.6), which have asymptotic properties described in Section 4.2. Then, since

$$p = 1 - P\{N(E) = 0\}^{1/k}, \tag{3.13}$$

and with the notation $\ell_F(f) = \int e^{-f} dF$, one arrives at the estimators

$$\hat{p} = 1 - [(1/n)\sum_{i=1}^{n} \mathbf{1}(N_i(E) = 0)]^{1/k} \tag{3.14}$$
$$\hat{\ell}(f) = [\hat{L}(f)^{1/k} + \hat{p} - 1]/\hat{p} \tag{3.15}$$

by substitution. Note that (3.14) is an empirical zero-probability functional corresponding to (3.13). Consistency and asymptotic normality of the estimators (3.14)–(3.15) follow from results in Sections 4.2 and 4.3. ☐

This example shows a common property of point process inference problems: rarely is there only one way to approach a problem (although different approaches may yield the same final result).

3.2.3 Parametric Testing of Quantitative Hypotheses

We pursue some of the preceding examples.

Example 3.9. (Renewal process). To continue Example 3.4, tests involving α and β are carried out using established methodology for gamma distributions. For synchronous data the problem is one for i.i.d. gamma distributed random variables. If α and β are both unknown, the situation is cumbersome if exact results are required, whereas if either (say α) is known, then there exist (from the theory of exponential families) uniformly most powerful tests for one-sided hypotheses $H_0 : \beta \geq \beta_0$. ☐

Example 3.10. (Partially observed Poisson process). Let N' be the point process of Example 3.5 and assume that X is not observable. To test the hypotheses $H_0 : \lambda \geq \lambda_0$, $H_1 : \lambda < \lambda_0$ given, say, asynchronous data $\mathcal{F}_t^{N'}$, it is evident intuitively that the critical region should have the form $\Omega_c = \{N_t' \leq \delta\}$ for some critical value δ depending on λ_0, t and the power bound under H_0. The issues are calculation of δ and of power under the alternative hypothesis. This requires calculation or approximation of the distribution of N_t' under each probability P_λ, a problem treated in Exercise 3.16. ☐

Example 3.11. (p-thinned Poisson process). Let N be a Poisson process on a compact space E with known mean measure μ, let $p \in (0, 1]$ be unknown, and let the observations be a single realization of the p-thinned process N'. To test $H_0 : p \geq p_0$, $H_1 : p < p_0$ given the data $\mathcal{F}^N(A)$, we use conditional uniformity and the fact that the μ is known to conclude that $N'(A)$ is a sufficient statistic. Clearly, the critical region should have the form $\Omega_c = \{N'(A) \leq \delta\}$ for appropriately chosen δ. But under P_p and H_0, $N'(A)$ has a Poisson distribution with mean $p\mu(A)$ and hence we are really testing $H'_0 : E[N'(A)] \geq p_0\mu(A)$, $H'_1 : E[N'(A)] < p_0\mu(A)$ given a sample of size 1 of $N'(A)$, which is a standard problem. $\square$

3.2.4 Nonparametric Tests of Quantitative Hypotheses

In this context, the tests concern finite-dimensional numerical characteristics, but the model is infinite-dimensional.

Example 3.12. (Stationary point process). In the context of Example 3.7, consider testing the simple hypotheses $H_0 : \nu = \nu_0$, $H_1 : \nu = \nu_1 < \nu_0$ given asynchronous data $\mathcal{F}_t^N$. It seems clear that one should consider critical regions $\Omega_c = \{\sup_{s \leq t}(N_s/s) \leq \delta\}$, but calculation of δ and of the power of the test, even for simpler critical regions $\Omega_c = \{N_t/t \leq \delta\}$ are formidable and perhaps impossible. $\square$

Example 3.13. (Superposed Bernoulli processes). For the processes of Example 3.8, tests involving p are relatively straightforward because both of the estimators of p constructed there have distributions not depending on F. Indeed, such tests in either case are fundamentally tests for binomial distributions, for which there exists substantial theory. Tests for F are more difficult, but can be effected using methods applicable to empirical processes on general spaces (see Section 4.1). $\square$

3.2.5 Tests of Qualitative Hypotheses

Here problems of computation of power and lack of power under the alternative may become acute. We mention two examples related to Section 3.5.

Example 3.14. (Renewal process). For a renewal process N on $\mathbf{R}_+$, consider testing the composite null hypothesis that N is a Poisson process against various alternatives, given a single realization. Alternatives of interest range from rather specific, for example that N has a gamma but not exponential interarrival distribution, for which the problem is parametric, to broad, for example that the interarrival distribution has a hazard function that is increasing, to very broad — N is simply not Poisson. To the extent that one views the interarrival distribution as an infinite-dimensional quantitative characteristic of N, the problem belongs in the category of nonparametric testing of quantitative hypotheses; however, the procedures below test qualitative structure that characterizes Poisson processes.

For synchronous data $\mathcal{F}_{T_n}^N$ the problem is to test the null hypothesis that the interarrival times $U_1, \ldots, U_n$ have an exponential distribution. To test this hypothesis directly, one could one a χ^2 goodness-of-fit test or tests based on the Kolmogorov-Smirnov statistic, the Cramér-von Mises statistic or the Anderson-Darling statistic (see notes for Chapter 4), but for any of these the unknown intensity λ of N must be estimated and incorporated into the test statistic, which complicates matters.

The *uniform conditional test* removes λ from consideration, at the expense of introducing a conditional test. (In a conditional test, all statements, including those about power, are valid conditionally on the value of some statistic.) The test is based on the qualitative property of conditional uniformity (Exercise 1.1): conditional on the value of T_n, the ratios $V_i = T_i/T_n$, $i = 1, \ldots, n - 1$, under the null hypothesis but regardless of the value of λ, are order statistics from the uniform distribution on $[0, 1]$. This property can be tested using previously mentioned statistics; no unknown parameters need be estimated. The power of the test under alternatives is not well understood, however.

Arguably a better test is the *gap test*, which is based on the spacings of the order statistics $U_{(j)}$ associated with the interarrival times U_j. Under the null hypothesis the normalized spacings $U_j^n = (n - j + 1)[U_{(j)} - U_{(j-1)}]$, $j = 1, \ldots, n$, are independent and exponentially distributed, which one can test using the uniform conditional test. Tests based on transformed data have been reported on empirical evidence to have power superior to those based on the untransformed data, the null distribution theory is known (Pyke, 1965), and there is some information available concerning power under renewal alternatives.

The uniform conditional test for asynchronous data $\mathcal{F}_t^N$ is constructed as follows. Under the Poisson null hypothesis and regardless of λ, conditional on N_t the values $V_i = T_i/t$, $i = 1, \ldots, N_t$, are order statistics from the uniform distribution on $[0, 1]$, which may be tested as described previously. Testing for conditional uniformity using a χ^2 goodness-of-fit test is equivalent to the *dispersion test for homogeneity*, which is based on a form of integral data. The interval $[0, t]$ is partitioned into subintervals $I_1, \ldots, I_k$ of equal length and the test statistic is the dispersion $D = (1/\bar{N}) \sum_{j=1}^{k} [N(I_j) - \bar{N}]^2$, where $\bar{N} = (1/k) \sum_{j=1}^{k} N(I_j)$. $\square$

Things are decidedly worse with less stringent underlying assumptions.

Example 3.15. (Stationary point process). Let N be a stationary point process on $\mathbf{R}$ with known intensity $\nu = 1$, and consider testing the null hypothesis that N is a Poisson process given a single realization. Here, even though the null hypothesis is simple, it is not clear how to proceed. The tests discussed in Example 3.14 could be used, but except for a few alternatives, nothing is known about their power. Presumably one must either narrow the structural assumptions or consider restricted alternatives other than renewal alternatives, or both, or more. Some specific examples appear in later chapters. $\square$

As we have stated before, this problem "sampler" is in no sense exhaustive. Other specific examples appear later in this chapter, in other chapters, in exercises, and in the literature.

3.3   State Estimation

Let N be a point process on E with *known* probability law $\mathcal{L}$. Suppose that the observations are the restriction of N to a subset A of E: $\mathcal{F}^N(A) = \sigma(N(B) : B \subseteq A)$.

In a state estimation problem the objective is to *reconstruct in optimal fashion* from the observations either the unobserved portions of N or an associated but unobservable random process. This is done realization by realization (ω-wise), and the term *state estimation* emphasizes that unobserved random variables, rather than statistical parameters, are being estimated. Indeed, the law of N must be assumed to be known completely. In Section 4 and later chapters, though, we do address simultaneous statistical inference and state estimation.

A *state estimator* of an unobservable random variable is another random variable, measurable with respect to the observations, that approximates it. We seek state estimators that are optimal with respect to the virtually universal criterion of *mean squared error* (MSE). To illustrate, suppose that we seek to estimate the unobserved portion of N [i.e., $N_{A^c}(\cdot) = N(\cdot \cap A^c)$]. To do this, for each function f on A^c we need a state estimator $\hat{N}(f)$ satisfying

- $\hat{N}(f)$ is measurable with respect to $\mathcal{F}^N(A)$ [physically, $\hat{N}(f)$ is determined by the observations]

- For every $Z \in \mathcal{F}^N(A)$, $E[(\hat{N}(f) - N(f))^2] \leq E[(Z - N(f))^2]$.

Furthermore, $f \rightarrow N(f)$ is an integral having linearity and continuity properties that should be retained by the optimal state estimators.

For fixed f the optimal solution $\hat{N}(f)$ is termed the *minimum mean squared error* (MMSE) state estimator of $N(f)$, and is, of course, the conditional expectation $\hat{N}(f) = E[N(f)|\mathcal{F}^N(A)]$. Even the linearity and continuity properties hold nearly as desired. It is still necessary to ensure that $\hat{N}(\cdot) = E[N(\cdot)|\mathcal{F}^N(A)]$ is a measure but in specific problems we are always able to do so.

If the story ended here there would be no story. In state estimation problems the central issues are computation of state estimators and description of how state estimators evolve as the point process is observed over larger sets. We illustrate in symbolic form and more explicitly in examples.

As a conditional expectation, the MMSE state estimator $\hat{N}(f)$ is a function of A, f, and, especially, the observations $N_A(\cdot)$. Symbolically,

$$\hat{N}(f) = H(A, N_A, f) \tag{3.16}$$

for some functional H. The first computational issue is to calculate H, which also depends on $\mathcal{L}$, *explicitly*. This may or may not be possible. According to Section 1.6, the Palm distributions of N are the key tools in such calculations, but unfortunately, Palm distributions themselves often cannot be calculated. Even in those cases that H can be computed "explicitly" the resulting expression may be too complicated to use, especially if the set A changes frequently. Thus one is always led to the issue of recursive computation, which we take up momentarily.

Example 3.16. (Empirical process). Let $N = \sum_{i=1}^{k} \varepsilon_{X_i}$ be an empirical process on E, with both k and the distribution F of the X_i known. Given a subset of A of E and a function f on A^c, and provided that $F(A) < 1$,

$$E[N(f)|\mathcal{F}^N(A)] = [k - N(A)](\int_{A^c} f dF)/F(A^c). \tag{3.17}$$

The state estimator is so simple that recursive computation hardly seems necessary; nevertheless, recursive methods may be easier, and are discussed in Example 3.17. □

In broad terms, a *recursive* method of computation enables one to revise state estimators as additional observations are obtained, rather than compute *da capo* (from the beginning) the state estimator that incorporates both previous and newly obtained observations. Recursive computation possesses both a "macro-level" component relevant to point processes on general spaces and a "micro-level" component restricted to point processes on $\mathbf{R}_+$. We illustrate first in stylized terms.

The macro component deals with discrete increases of the set A. Suppose that observations $\mathcal{F}^N(A)$ have been obtained previously and that the MMSE state estimator $E[N(\cdot)|\mathcal{F}^N(A)]$ has been calculated, by whatever means. At a later time N is observed over an additional set A' disjoint from A, so that the cumulative observations are $\mathcal{F}^N(A + A')$. To compute the revised state estimator $E[N(\cdot)|\mathcal{F}^N(A + A')]$ one could start from scratch, but unless there is a very simple method for explicit computation, starting over may be prohibitively difficult and time consuming. At a minimum, one should compute only the increment $E[N(\cdot)|\mathcal{F}^N(A + A')] - E[N(\cdot)|\mathcal{F}^N(A)]$, and in the process this increment should be decomposed into two parts, one representing only the effect of having observed the point process over a larger set and not a function of the new observations $N_{A'}$ (but certainly a function of the previous observations), and the second incorporating the both sets of observations but in such a manner that contributions of old and new are resolved. In symbolic form, these ideas imply an expression

$$E[N|\mathcal{F}^N(A + A')] - E[N|\mathcal{F}^N(A)]$$
$$= J(A, A', N_A, f) + K(A, A', N_A, N_{A'}, f), \qquad (3.18)$$

with K vanishing if $N_{A'} = 0$.

The micro aspect of recursive computation centers around making sense of the idea of the new observations being over an infinitesimal set dx and of an infinitesimal analogue of (3.18):

$$E[N|\mathcal{F}^N(A + dx)] - E[N|\mathcal{F}^N(A)]$$
$$= J(A, dx, N_A, f) + K(A, dx, N_A, N(dx), f), \qquad (3.19)$$

where $x \notin A$. This expression should not be taken too literally; it is intended mainly to be suggestive. If the terms on the right-hand side were measures in dx, then it would be possible to construct, by integration, any state estimator from infinitesimal building blocks of the form (3.19), provided that there were an agreed-upon starting point and an agreed-upon order in which observations were obtained, "one point at a time." This occurs only for point processes on $\mathbf{R}_+$, in which case (3.19) becomes a stochastic differential equation

$$dE[N(f)|\mathcal{F}_t^N] = X_t dt + Y_t dM_t, \qquad (3.20)$$

where X and Y are predictable random processes depending on f and the past history of N, and where M is the innovation martingale (Theorem 2.14) associated with N. The processes X and Y are termed the *predictable change* and *filter gain*. When there is no innovation, the second term on the right-hand side of (3.20) vanishes and there is only predictable change in the state estimator, resulting solely from observing the point process over a larger set. See also the discussion relating to Proposition 2.35.

We illustrate by expanding Example 3.16.

Example 3.17. (Empirical process). Let N be the empirical process of Example 3.16 and suppose that A and A' are disjoint with $F(A + A') < 1$. Suppose that f is supported outside A, and for notational simplicity let $J(B) = (\int_{B^c} f dF)/F(B^c)$. Then

$$E[N(f)|\mathcal{F}^N(A + A')] - E[N(f)|\mathcal{F}^N(A)]$$
$$= [k - N(A)][J(A + A') - J(A)] - N(A')J(A + A'),$$

which is evidently of the form of (3.18).

Suppose now that $E = \mathbf{R}_+$ and F has hazard function h; then N has $\mathcal{F}^N$-stochastic intensity $\lambda_t = h(t)(k - N_{t-})$. Letting $J(t) = \int_t^\infty f dF/[1 - F(t)]$, we obtain a specific example of (3.20):

$$dE[N(f)|\mathcal{F}_t^N] = h(t)[k - N_{t-}]f(t)dt - J(t)dM_t. \quad \square \qquad (3.21)$$

Expressions such as (3.20)–(3.21), as well as the discussion of conditioning in Section 2.6, identify the fundamental role of stochastic intensities and Palm distributions in state estimation for point processes. Further confirmation appears in later chapters; a prominent example is state estimation for Cox processes, treated in Section 7.2. Yet another link is Theorem 2.34, the representation theorem for point process martingales. Evidently, a process $(E[N(f)|\mathcal{F}_t^N])$ is a martingale; the representation of it as the integral of a predictable process with respect to the innovation martingale is closely related to the stochastic differential equation (3.20). See also Section 5.4, where we solve a very general state estimation problem using the martingale representation theorem.

The state estimation problem for Cox processes concerns unobservable directing measures, which are not functionals of N, but are of physical significance notwithstanding. The issues outlined in this section are equally pertinent here and sometimes more difficult. We illustrate with an example treated at length in Sections 7.2 and 7.4.

Example 3.18. (Mixed Poisson process). Let N be a mixed Poisson process on a compact space E with directing measure $Y\nu$, where ν is a known, finite, diffuse measure on E and Y is a positive random variable with known distribution F. The observations are $\mathcal{F}^N = \mathcal{F}^N(E)$; the multiplier Y, an underlying random mechanism, is not observable. General theory developed in Chapter 7 implies that

$$E[Y|N] = \frac{\int F(dx)e^{-x\nu(E)}x^{N(E)+1}}{\int F(dx)e^{-x\nu(E)}x^{N(E)}}. \qquad (3.22)$$

The MMSE estimator of the directing measure M is thus the diffuse random measure $E[M|N] = E[Y|N]\nu$. Recursive versions appear in Chapter 7. $\quad \square$

3.4   Combined Inference and State Estimation

Without knowledge of the probability law of the point process N, calculation of MMSE state estimators is not possible. Nonetheless, there are many situations in which the law of N is unknown but state estimation is the paramount goal. In such cases statistical inference and state estimation must be effected simultaneously, which we call the problem of *combined inference and state estimation.*

In principle this problem is solvable if there exist consistent estimators of those attributes of the law of N needed to compute MMSE state estimators. It is approached using the *principle of separation*:

- First, derive the relevant state estimation formula under the assumption that $\mathcal{L}_N$ is known.

- Then, examine this formula to determine which characteristics are required to calculate the state estimator.

- Next, use the observations, together with data from other copies of the process, if available, to estimate these characteristics.

- Then, construct a *pseudo-state estimator* by substituting estimated for exact attributes of $\mathcal{L}_N$, at the same time using the observations in the same way they would be used in the "true" state estimator.

- Finally, analyze properties of the pseudo-state estimators in comparison to the true state estimators, especially asymptotically.

The combined problem therefore incorporates all the questions, classifications, and difficulties introduced already, along with some new wrinkles because the observations of N play a dual role, as data for statistical inference and as partial observations on which state estimation is based. Rather than describe fully the resultant classification, we mention and illustrate principal cases. The important distinction is whether there is but a single realization of N or there are i.i.d. copies.

For several reasons the single realization case is the less interesting and important. As discussed in Section 2, single realization data are most immediately useful for estimating real parameters (even though the setting may be nonparametric), but most state estimation formulas involve more about the law of N than just a few real parameters. Furthermore, the roles of the observations as data for statistical inference and for state estimation are difficult to unravel, whereas in the multiple-copy case there is a convenient way to sidestep this issue. Finally, asymptotics are less interesting than in the multiple-copy case in terms of both the kinds of results that can be proved and the assumptions required to prove them. Even so, the single realization case is devoid of neither theoretical substance nor practical applications, so we illustrate it in symbolic form.

Let N be a point process on a noncompact space E, with law $\mathcal{L}$ known only incompletely. Assume that the observations are $\mathcal{F}^N(A)$, with A compact, and that we wish to perform state estimation for the unobserved portion of N. According to (3.16), there is a functional H such that

$$E_{\mathcal{L}}[N(f)|\mathcal{F}^N(A)] = H(\mathcal{L}^0, N_A, f),$$

where $\mathcal{L}^0$ denotes characteristics of $\mathcal{L}$ required to compute the MMSE state estimator (the notation emphasizes dependence of the state estimator itself on $\mathcal{L}$). The form of H — this is a crucial point! — is fully determined by the structural assumptions of the statistical model and can be computed without knowledge of $\mathcal{L}$. Let $\hat{\mathcal{L}}^0$ be as estimator of $\mathcal{L}^0$ based on the observations $\mathcal{F}^N(A)$; then the pseudo-state estimator is

$$\hat{E}[N(f)|\mathcal{F}^N(A)] = H(\hat{\mathcal{L}}^0, N_A, f). \tag{3.23}$$

The notation employs the caret to distinguish an estimator. There is evident dual dependence of the pseudo-state estimator (3.23) on N, once as data in construction of $\hat{\mathcal{L}}^0$ and again as observations serving as "input" for state estimation. If the true state estimator can be computed, and if the estimator $\hat{\mathcal{L}}^0$ can, then so can the pseudo-state estimator.

One then wishes to analyze the difference between the pseudo-state estimator and the true state estimator. Finite sample results are few and often rather uninformative. Hence one is led to seek asymptotic results as $A \uparrow E$. The goals, given that $\hat{\mathcal{L}}^0 \to \mathcal{L}^0$ in an appropriate sense, are to show that $\hat{E}[N(f)|\mathcal{F}^N(A)] - E_{\mathcal{L}}[N(f)|\mathcal{F}^N(A)]$ converges to zero and to relate the rates of convergence. But then f cannot remain fixed, for if it does, then once A is large enough to contains the support of f, then $E_{\mathcal{L}}[N(f)|\mathcal{F}^N(A)]$ — and consequently also $\hat{E}[N(f)|\mathcal{F}^N(A)]$ — degenerates to $N(f)$. For f to vary with A requires that the variation be systematic, which entails special structure on E (e.g., that of a group), and so the restrictions mount.

The picture is less gloomy if instead of unobserved portions of N one is estimating an unobservable random mechanism associated with N. Here is an illustration.

Example 3.19. (Yule process). Let N be a Yule process with unknown birth rate $c > 0$: N_t is the size at t of a population that, when its size is i, reproduces at rate ci. By Exercise 2.12 there is a random variable V

satisfying $e^{-ct}N_t \to V$ almost surely under P_c. Also, N is a mixed Poisson process directed by the random measure $M = V\nu$, where $\nu_t = e^{ct} - 1$. Suppose that we wish to perform state estimation for V having observed one realization of N over $[0, t]$; keep in mind that c is unknown. The $\mathcal{F}^N$-stochastic intensity of N is $\lambda_t = cN_{t-}$, and together with the substitution principle embodied in (3.7), this leads to the (martingale) estimators

$$\hat{c} = N_t / \int_0^t N_s ds. \tag{3.24}$$

This is also essentially the maximum likelihood estimator obtained using the property that interarrival times in N are independent and exponentially (albeit not identically) distributed. Let F_c denote the distribution of V under P_c; then by (3.22),

$$E_c[V|\mathcal{F}_t^N] = \frac{\int F_c(dx)e^{-x\nu_t}x^{N_t+1}}{\int F_c(dx)e^{-x\nu_t}x^{N_t}},$$

and it is straightforward to combine this expression with (3.24) to obtain a pseudo-state estimator. $\square$

For combined inference and state estimation in the case of i.i.d. copies $N_1, N_2, \ldots$ of N (here we take E to be compact) the outlook is brighter still. Suppose that $N_1, \ldots, N_n$ have been observed previously over E and that N_{n+1}, is now observed over the subset A. Let $\mathcal{L}^0$ again signify aspects of $\mathcal{L} = \mathcal{L}_N$ not specified by model assumptions. For example, it may be assumed only that N is a Cox process with diffuse directing measure. The true state estimator

$$E_\mathcal{L}[N_{n+1}(f)|\mathcal{F}^{N_{n+1}}(A)] = H(\mathcal{L}^0, (N_{n+1})_A, f)$$

is now superseded by the pseudo-state estimator

$$\hat{E}[N_{n+1}(f)|\mathcal{F}^{N_{n+1}}(A)] = H(\hat{\mathcal{L}}^0, (N_{n+1})_A, f), \tag{3.25}$$

where $\hat{\mathcal{L}}^0$ is an estimator of $\mathcal{L}$ based on $N_1, \ldots, N_n$, for example, a functional of the empirical measure $\hat{\mathcal{L}} = \frac{1}{n}\sum_{i=1}^n \varepsilon_{N_i}$. We have the option whether to include the partial observations $(N_{n+1})_A$ when constructing $\hat{\mathcal{L}}^0$. In practice one never deliberately ignores data, so $(N_{n+1})_A$ should be included, but for proving results, however, basing $\hat{\mathcal{L}}^0$ only on $N_1, \ldots, N_n$ renders the two stochastic arguments in (3.25) independent, and proofs become less complicated. Asymptotically, there is no difference provided that the estimators are consistent, as they are in all specific cases we consider.

The limit theorems are more clearly delineated and valid under fewer structural assumptions than in the single-realization model. Typically, it will hold that $\hat{\mathcal{L}}^0 \to \mathcal{L}^0$ at rate $n^{-1/2}$, in the sense that $n^{1/2}[\hat{\mathcal{L}}^0 - \mathcal{L}^0]$ has a nondegenerate limit distribution. Then, the main goal is to show that the error processes $n^{1/2}\{\hat{E}[N_{n+1}|\mathcal{F}^{N_{n+1}}(A)] - E_{\mathcal{L}}[N_{n+1}|\mathcal{F}^{N_{n+1}}(A)]\}$ also have a nondegenerate limit distribution (often a mixture of Gaussian distributions). Such a rate is evidently the best possible, because the effectiveness of pseudo-state estimators as approximations to the true state estimators cannot exceed that of the estimation procedure itself. In some cases, however, one must settle for slower rates of convergence. Section 7.4 exhibits both situations for Cox processes.

In "highly nonparametric" contexts the only functional form H that can be employed might involve the Palm distributions of N, giving rise to the problem of estimating Palm distributions from the N_i; this problem and implications for combined inference and state estimation are treated in Sections 4.4 and 7.4.

We conclude the section with a simple example.

Example 3.20. (Empirical process). Let $N_1, N_2, \ldots$ be i.i.d. copies of the empirical process N of Example 3.16. Assume that k is known but that the distribution F is unknown. With $N_1, \ldots, N_n$ observed previously, the maximum likelihood estimator of F is the empirical measure $\hat{F} = \frac{1}{nk}\sum_{i=1}^{k} N_i$, so that if N_{n+1}, is now observed over A, then the pseudo-state estimator associated with the state estimator (3.17) is

$$\hat{E}[N_{n+1}(f)|\mathcal{F}^{N_{n+1}}(A)] = [k - N_{n+1}(A)](\int_{A^c} f d\hat{F})/\hat{F}(A^c),$$

whose asymptotic properties are readily derivable. ☐

3.5 Example: Ordinary Poisson Processes

The sample problems in preceding sections have been chosen deliberately to depict a variety of point processes, but lack the coherence of a narrower context. In this section, by contrast, we consider only the most fundamental class of point processes, the ordinary (stationary; homogeneous) Poisson processes on $\mathbf{R}_+$, and address a number of the problems posed in Sections 1–4. For brevity many proofs are relegated to the exercises.

The statistical model is that $N = \sum_{n=1}^{\infty} \varepsilon_{T_n}$ is a Poisson process on $\mathbf{R}_+$ with unknown rate λ. The family of candidate probability laws is $\mathcal{P} = \{P_\lambda : \lambda \in (0, \infty)\}$; both synchronous and asynchronous data are examined. In terms of the Sections 1 and 2, the underlying space is noncompact and the statistical setting is parametric; since much of the rest of the book focuses on multiple realizations, we restrict attention here to statistical inference, state estimation, and combined inference and state estimation based on single realizations of N.

3.5.1   Estimation of λ

Estimation of λ from the synchronous data $\mathcal{F}_{T_n}^N$ of (3.4) is a special case of Example 3.4: the interarrival times $U_1, \ldots, U_n$ are independent and identically exponentially distributed with mean $\theta = 1/\lambda$. [The conventional statistical parameterization for exponential densities is via the mean, *viz.*, $f(x, \theta) = (1/\theta)e^{-x/\theta}$, $\theta \in (0, \infty)$.] The following properties are well known.

Proposition 3.21. a) The maximum likelihood estimator of λ given $\mathcal{F}_{T_n}^N$ is $\hat{\lambda} = 1/\hat{\theta}$, where $\hat{\theta} = (1/n) \sum_{i=1}^{n} U_i$;

b) For each λ, $\hat{\lambda} \to \lambda$ almost surely under P_λ;

c) Under P_λ, $\sqrt{n}[\hat{\lambda} - \lambda] \xrightarrow{d} N(0, \lambda^2)$. $\square$

For each n, $n\hat{\theta}$ has an Erlang$(n, 1/\theta)$ distribution, so one can construct exact confidence intervals for λ. Alternatively, the asymptotic normality in Proposition 3.21c) can be used to construct approximate large-sample confidence intervals.

It seems evident that for asynchronous data $\mathcal{F}_t^N$ the appropriate estimator of λ is

$$\hat{\lambda} = N_t/t. \tag{3.26}$$

We will discuss shortly four justifications for (3.26) — all principles of estimation lead to it — but first here is a preliminary result.

Lemma 3.22. Given the data $\mathcal{F}_t^N$, N_t is a sufficient statistic for λ. $\square$

The empirical method of inference appeals first to Lemma 3.22, in order to base inference solely on N_t, which has a Poisson distribution with mean λt. For a Poisson distribution with unknown mean μ and, as we have, a sample of size 1, the maximum likelihood estimator and method of moments estimator are both simply the single sample value. Thus $\widehat{\lambda t} = N_t$, which gives (3.26).

For the intensity method, under P_λ, N has deterministic, time-independent stochastic intensity $\lambda_t \equiv \lambda$. Theorem 2.31 implies that for each λ we have $P_\lambda \ll P_1$ on $\mathcal{F}_t^N$ (and vice versa) with

$$\log dP_\lambda/dP_1 \big|_{\mathcal{F}_t^N} = t(1 - \lambda) + N_t \log \lambda.$$

Maximizing this likelihood function with respect to λ again yields $\hat{\lambda}$ of (3.26).

Finally, the martingale estimation principle (3.7) asserts that for each s, dN_s estimates λds within innovation noise, and therefore $N_t = \int_0^t dN_s$ estimates $\lambda t = \int_0^t \lambda ds$.

The following properties are elementary.

Proposition 3.23. Let $\hat{\lambda}$ be given by (3.26). Then under each probability P_λ,

a) $\hat{\lambda} \to \lambda$ almost surely;

b) $\sqrt{t}[\hat{\lambda} - \lambda] \overset{d}{\to} N(0, \lambda)$. $\square$

Example 3.7 shows that the estimators $\hat{\lambda}$ of (3.26) have a certain robustness property. If the estimators (3.26) were used to estimate the intensity of a point process that is not Poisson but is stationary and ergodic, they would remain strongly consistent. Similarly, $1/\hat{\lambda}$ is a robust estimator of the mean interarrival time if N is a renewal process rather than a Poisson process.

3.5.2 Hypothesis Tests for λ

The estimation procedures just described yield corresponding tests of hypotheses involving λ; we shall not explain them in detail. Instead, we present some possibly less familiar ideas associated with asymptotics of tests based on likelihood ratios. A sequential test will also be discussed.

To begin, consider testing the simple hypotheses

$$
\begin{aligned}
H_0: &\quad \lambda = \lambda_0 \\
H_1: &\quad \lambda = \lambda_1
\end{aligned}
\tag{3.27}
$$

given the data $\mathcal{F}_t^N$. By Theorem 2.31 the log-likelihood ratio is

$$
\log dP_{\lambda_1}/dP_{\lambda_2}\Big|_{\mathcal{F}_t^N} = t(\lambda_0 - \lambda_1) + N_t \log(\lambda_1/\lambda_0),
\tag{3.28}
$$

and likelihood ratio tests are easy to devise, except that N_t has a discrete distribution, so that exact specification of power under H_0 may be not feasible.

In the limit, perfect discrimination between possible values of λ is possible.

Proposition 3.24. (Dichotomy theorem). For $\lambda_0 \neq \lambda_1$, $P_{\lambda_0} \perp P_{\lambda_1}$ on $\mathcal{F}_\infty^N$.

Proof: It suffices to observe that the probabilities P_λ are concentrated on disjoint sets, which follows from the strong consistency statement in Proposition 3.23. $\blacksquare$

Note that singularity occurs on $\mathcal{F}_\infty^N$ even though P_{λ_0} and P_{λ_1} are mutually absolutely continuous on $\mathcal{F}_t^N$ for every $< \infty$. Any two of the probabilities P_λ are identical or orthogonal on $\mathcal{F}_\infty^N$ and, given enough data (i.e., for t sufficiently large), any reasonable approach can decide the test (3.27) without error.

More interesting asymptotics arise if one considers instead local behavior of likelihood ratios, by permitting, in effect, alternative hypotheses that vary with t and converge to the null hypothesis as $t \to \infty$. Such methods and results fall under the rubrics *local asymptotic normality* and *contiguity*, and, among other things, delineate precisely the asymptotic capabilities of likelihood ratio tests. It is not difficult to guess that alternative hypotheses should converge to the null at rate $1/\sqrt{t}$, so we replace (3.27) by

$$H_0: \quad \lambda = \lambda_0$$
$$H_1: \quad \lambda = \lambda_0 + h\sqrt{\lambda_0/t},$$

where $h \in \mathbf{R}$ is fixed and the data is $\mathcal{F}_t^N$. These alternatives are called *contiguous*. Let L_t be the logarithm of the corresponding likelihood ratio.

Proposition 3.25. Under the hypotheses of the preceding paragraph, with respect to P_{λ_0}, $L_t \xrightarrow{d} N(-h^2/2, h^2)$.

Proof: From (3.28),

$$
\begin{aligned}
L_t &= N_t \log(1 + h/\sqrt{\lambda_0 t}) - h\sqrt{\lambda_0 t} \\
&= N_t[h/\sqrt{\lambda_0 t} - h^2/2\lambda_0 t] - h\sqrt{\lambda_0 t} + o_P(1) \\
&= \frac{h}{\sqrt{\lambda_0}}\sqrt{t}[\hat{\lambda} - \lambda_0] - \frac{\hat{\lambda}}{\lambda_0}\frac{h^2}{2} \\
&\xrightarrow{d} (h/\sqrt{\lambda_0})N(0, \lambda_0) - h^2/2,
\end{aligned}
$$

where the term $o_P(1)$ converges in P_{λ_0}-probability to zero and is omitted from subsequent expressions, and where the convergence in distribution is by Proposition 3.23. ∎

The testing implication of Proposition 3.25 is that likelihood ratio tests can discriminate between alternatives that are contiguous at rate slower than $1/\sqrt{t}$ but not at faster rates.

Likelihood ratios can also be used to construct sequential tests of hypotheses such as (3.27). Point processes are so often observed in real time that one is obliged to consider sequential methods, especially since an assured amount of information can be obtained thereby. By contrast, observation for a fixed length of time, because of the resultant random sample size N_t, yields statements whose precision is known only after the data are analyzed; this is one conditional aspect of the uniform conditional test discussed in Example 3.14. We now construct a sequential probability ratio test (SPRT) for the simple hypotheses $H_0 : \lambda = \lambda_0$, $H_1 : \lambda = \lambda_0 + \Delta$, where $\Delta < 0$. The probabilities α of rejecting H_0 when it holds and β of rejecting H_1 when it holds are both specified in advance. (For fixed-sample-size tests, by contrast, the Neyman-Pearson theory dictates that α be prescribed and minimizes β; the value that results may or may not be acceptably low.) Construction of the SPRT involves specifying boundaries ℓ and u, and if ever the likelihood ratio (3.28), viewed as a stochastic process, falls below ℓ, the test is terminated and H_0 is accepted; if it exceeds u, then H_1 is accepted. The termination time is a stopping time T, finite almost surely under either hypothesis, but not bounded above. It can further be shown that (approximately) $\ell = \log[\beta/(1 - \alpha)]$ and $u = \log[(1 - \beta)/\alpha]$.

3.5.3 Other Hypothesis Tests

Examples 3.14 and 3.15 discuss testing whether a point process is Poisson; here we mention some related problems and techniques that have been applied to them.

1) *Tests for Trend.* It is often of interest to test whether a point process is stationary against the alternative that there is some kind of trend. In full generality, even if it were well posed, the question is intractable, but sufficiently specific cases can be handled, one of which is the following.

Example 3.26. (Poisson process with trend). Let N be a Poisson process with intensity function $\lambda_t = e^{\alpha + \beta t}$, where α and β are both unknown. We wish to test the null hypothesis $H_0 : \beta = 0$ that there is no trend against alternatives under which $\beta \neq 0$. This can be done using a likelihood ratio test. Of course, one must estimate α under H_0, which is easy: for in this case, N is an ordinary Poisson process with rate $\lambda = e^{\alpha}$. Both α and β must be estimated under H_1 which is hardly more difficult: the log-likelihood function is

$$L = e^{\alpha}(1 - e^{\beta t})/\beta + \alpha N_t + \beta \sum_{i=1}^{N_t} T_i, \tag{3.29}$$

and one then substitutes into the appropriate log-likelihood ratio. Only large-sample distribution theory is available, however. □

2) *Comparison of independent Poisson processes.* This two-sample problem can be approached in a variety of ways, some leading to conditional tests. For independent Poisson processes $N(1)$ and $N(2)$ with rates $\lambda(1)$ and $\lambda(2)$, observed over time intervals $[0, t_1]$ and $[0, t_2]$, conditional on $N_{t_1}(1) +$

$N_{t_2}(2) = n$, $N_{t_1}(1)$ has a binomial distribution with parameters n and $p = t_1\lambda_1/(t_1\lambda_1 + t_2\lambda_2)$. The hypothesis $\lambda(1) = \lambda(2)$ is equivalent to $p = t_1/(t_1 + t_2)$, a known value, and can be tested with techniques available for binomial distributions. Similarly, if $N(i)$ is observed until n_i arrivals occur, then the "variance ratio" $R = (n_1/n_2)/[T_{n_1}(1)/T_{n_2}(2)]$ has an F-distribution; note that use of R allows unconditional tests. Nonparametric tests such as the sign test or the Wilcoxon signed rank test have the advantage of robustness if, for example, the $N(i)$ are not Poisson but are renewal processes.

3.5.4   State Estimation

Independence of increments in a Poisson process renders the state estimation problem for observations $\mathcal{F}_t^N$ entirely trivial: for f vanishing on $[0, t]$, $E[N(f)|\mathcal{F}_t^N] = \lambda \int f(x)dx$.

3.5.5   Combined Inference and State Estimation

There is sufficient structure in a Poisson process to produce nondegenerate asymptotics for the single-realization problem of combined inference and state estimation. Let f be a function vanishing outside $[0, 1]$, let $I(f) = \int f(x)dx$, and for each t, let $f_t(x) = f(x - t)$. The behavior as $t \to \infty$ of the difference between the true state estimator $E_\lambda[N(f_t)|\mathcal{F}_t^N] = \lambda I(f)$ and the pseudo-state estimator $\hat{E}[N(f_t)|\mathcal{F}_t^N] = \hat{\lambda}I(f)$, where $\hat{\lambda}$ is given by (3.26), is described by the final result of the chapter.

Proposition 3.27. Under each P_λ,

$$\sqrt{t}\left\{\hat{E}[N(f_t)|\mathcal{F}_t^N] - E_\lambda[N(f_t)|\mathcal{F}_t^N]\right\} \xrightarrow{d} N(0, \lambda I(f)). \square$$

A few more Poisson process inference problems appear in the exercises. The theme common to all is "exploitation of special structure."

EXERCISES

3.1 Let $(\Omega, \mathcal{F}, P)$ be a probability space, let $\mathcal{G}$ be a sub-σ-algebra of $\mathcal{F}$, and let Z be a random variable with $E[Z^2] < \infty$. Prove that the MMSE state estimator of Z given the observations $\mathcal{G}$ is the conditional expectation $E[Z|\mathcal{G}]$.

3.2 Let $X, Y_1, \ldots, Y_n$ be random variables with a joint normal distribution. Prove that the conditional expectation $E[X|Y_1, \ldots, Y_n]$ is the orthogonal projection of X onto the vector space $V = \text{sp}\{Y_1, \ldots, Y_n\}$. Discuss explicit computation of $E[X|Y_1, \ldots, Y_n]$.

3.3 Let N be the Pólya process of Example 3.6, defined on $\mathbf{R}$ rather than $\mathbf{R}_+$.
a) Prove that N is stationary.
b) Prove that N is not ergodic and that, in fact, $N_t/t \to V$ almost surely as $t \to \infty$.
c) Prove that the invariant σ-algebra $\mathcal{I}$ (Definition 1.70) satisfies $\mathcal{I} = \sigma(V)$.

3.4 Consider the Bernoulli process $N = U\varepsilon_X$ with known parameters p and F.
a) For each t calculate the MMSE state estimators $P\{X \in A, U = 1|\mathcal{F}_t^N\}$, where $A \in \mathcal{B}(\mathbf{R}_+)$.
b) Describe asymptotic behavior of these estimators as $t \to \infty$.

3.5 Establish strong consistency and asymptotic normality of the estimators $\hat{p}$ given by (3.12).

3.6 Let $N_1, N_2, \ldots$ be superposed Bernoulli processes as in Example 3.8, with k there known but F and p unknown. Suppose that $N_1, \ldots, N_n$ have been observed over all of E.
a) Show that $\sum_{i=1}^n N_i(E)$ is a sufficient statistic for p.
b) Construct and analyze a test of the hypotheses $H_0 : p = p_0$, $H_1 : |p - p_0| > \Delta$, where $\Delta > 0$ is prescribed.

3.7 Verify (3.17).

3.8 Verify (3.21).

3.9 Let $\tilde{N}$ be a Poisson process on $\mathbf{R}_+$ with unknown rate λ and let T be a positive random variable independent of N and with known distribution G. The observations are the point process $N_t = \tilde{N}_{t \wedge T}$.
a) Devise estimators $\hat{\lambda}$ given the data $\mathcal{F}_t^N$.
b) Show that no estimators of λ can be even weakly consistent.

3.10 (Continuation of Exercise 3.9) Suppose that λ is known. Calculate the MMSE state estimators $E[\tilde{N}_t|\mathcal{F}_t^N]$ and obtain a recursive representation for them.

3.11 For the Yule process of Example 3.19, show that the estimators $\hat{c}$ of (3.24) are strongly consistent and asymptotically normal.

3.12 Consider the counter model of Exercise 2.22 with deterministic deadtime τ. Suppose that the rate λ of the arrival process $\tilde{N}$ is unknown and that only the point process $N = \sum_{n=1}^{\infty} \varepsilon_{T_n}$ of recorded arrivals is observable.
a) Show that N is a renewal process and calculate the interarrival distribution.
b) Using a), or otherwise, develop estimators $\hat{\lambda}$ of λ given both synchronous data $\mathcal{F}_{T_n}^N$ and asynchronous data $\mathcal{F}_t^N$, and establish their asymptotic properties.

3.13 (Continuation of Exercise 3.12) Suppose now that λ is known but the deadtime τ is unknown.
a) For each n calculate the maximum likelihood estimator $\hat{\tau}$ of τ given the data $\mathcal{F}_{T_n}^N$.
b) Describe asymptotic properties of these estimators.

c) Develop a likelihood ratio test of the hypotheses $H_0 : \tau = \tau_0$, $H_1 : \tau = \tau_1 < \tau_0$ given the data $\mathcal{F}_t^N$.

3.14 (Continuation of Exercise 3.12) Assume that λ and τ are both known and consider state estimation for the arrival process $\tilde{N}$ given observation of the record process N. For each t calculate the MMSE state estimator $E[\tilde{N}_t | \mathcal{F}_t^N]$, and devise a recursive method (it need not be a stochastic differential equation) for computation.

3.15 Let N' be the partially observed Poisson process of Example 3.5 and Exercise 2.21.

a) Prove that N' is a renewal process, calculate the Laplace transform

$$\ell(\alpha) = \int_0^\infty e^{-\alpha t}[1 - G(t)]dt,$$

where G is the interarrival distribution, and compute the mean interarrival time.

b) Suppose that the transition rates a and b are known but that the rate λ of the underlying Poisson process is not. Use a) to construct estimators of λ given the data $\mathcal{F}_{T_n}^{N'}$.

c) Characterize the asymptotic behavior of the state estimators in b).

3.16 (Continuation of Exercise 3.15) Assume that a, b and λ are all known.

a) Prove that N' is a Cox process.

b) Use a), or another method, to calculate the distribution of N_t' for each t.

3.17 (Continuation of Exercise 3.15) Assume once more that λ is unknown (but a and b are known)

a) Prove that the estimators $\hat{\lambda}$ of (3.10), based on the data $\mathcal{F}_t^{N'}$, are strongly consistent and asymptotically normal.

b) Prove that the same is true of the estimators $\hat{\lambda}^*$ of (3.11), which correspond to the data $\mathcal{F}_t^N \bigvee \mathcal{F}_t^X$.

3.18 Let N be a Poisson process with unknown rate λ and let X be a binary Markov process independent of N with known transition rates a ($0 \to 1$) and b ($1 \to 0$); so far the setup is as in Exercises 2.21 and 3.15–3.17. Suppose, however, that one observes the process $Z_t = X_t N_t$, which of course is not a point process. Devise estimators $\hat{\lambda}$ of λ given the observations $\mathcal{F}_t^Z$ and discuss their asymptotic behavior as $t \to \infty$.

3.19 (Continuation of Exercise 3.18) Suppose now that the rate λ of N is known.

a) For each t calculate the MMSE state estimators $E[N_u | \mathcal{F}_t^Z]$ for all $u \geq 0$.

b) Describe their asymptotic properties as $t \to \infty$.

c) Use a) and results from Exercise 3.18 to construct pseudo-state estimators $\hat{E}[N_u | \mathcal{F}_t^Z]$ each t and u.

d) Characterize asymptotic behavior of the difference between the true and pseudo-state estimators as $t \to \infty$.

3.20 Let (X_t) be the Markov process of Exercise 3.18; assume that a is known but b is not. Suppose that $X_0 = 1$ and let $T_1, T_2, \ldots$ be the successive times at which

X returns to state 1. The observations are the point process $N = \sum_{n=1}^{\infty} \varepsilon_{T_n}$.

a) Prove that N is a renewal process and calculate the interarrival distribution.

b) Use a) to devise estimators $\hat{b}$ of b given the synchronous data $\mathcal{F}_{T_n}^N$, and discuss their large sample properties.

c) Suppose now that a is unknown as well. Construct and analyze a test of the hypothesis $H_0 : a = b$ given the synchronous data $\mathcal{F}_{T_n}^N$.

3.21 Let N be a stationary point process on $\mathbf{R}$ with unknown intensity ν (see Example 3.7). Let P^* be the Palm measure. In Exercise 1.32 it is shown that under the Palm distribution $P_0 = P^*/\nu$, the interarrival sequence (U_i) is strictly stationary.

a) Show that if N is P-ergodic (Definition 1.70), then (U_i) is P_0-ergodic.

b) Construct estimators of ν given the data $\sigma(U_1, \ldots, U_n)$ and describe their behavior as $n \to \infty$ under the assumption that N is ergodic.

3.22 Let N be the point process on $\mathbf{R}$ given by $N = \sum_{n=-\infty}^{\infty} \varepsilon_{X+nY\theta}$, where X and Y are independent random variables uniformly distributed on $[0, 1]$ and $\theta > 0$ is unknown.

a) Prove that N is stationary.

b) Calculate the intensity $\nu(\theta)$ of N.

c) Discuss estimation of θ given a single realization of N over $[0, t]$.

3.23 (Continuation of Exercise 3.22) Suppose that $\theta = 1$ and that the observations are $\mathcal{H}_t = \sigma(N((0, u]) : 0 \leq u \leq t)$. For each t calculate the MMSE state estimators $E[X|\mathcal{H}_t]$ and $E[Y|\mathcal{H}_t]$.

3.24 Consider the following variant of the hypothesis-testing problem of Example 3.12. The observations are a single realization over $[0, t]$ of a stationary point process N with unknown intensity ν, and the hypotheses are $H_0 : \nu = \nu_0$, $H_1 : \nu = \nu_1 > \nu_0$. Consider tests based on critical regions of the form $\Omega_c = \{N_t/t \geq \delta\}$.

a) Show that given α it is possible to choose δ in such a manner that $P\{\Omega_c\} < \alpha$ for all stationary point processes satisfying the null hypothesis.

b) Show that, however, regardless of the value of δ there are stationary point processes [*Hint:* Consider mixed Poisson processes.] satisfying the alternative hypothesis for which $P\{\Omega_c\}$ is arbitrarily small. What does this mean concerning the power of the test?

c) Show that b) holds for critical regions $\Omega_c = \{N_s/s \geq \delta$ for some $s \leq t\}$.

3.25 Prove the following results concerning estimation of the rate of an ordinary Poisson process.

a) Proposition 3.21.

b) Lemma 3.22.

c) Proposition 3.23.

3.26 Let N be a Poisson process on $\mathbf{R}_+$ with unknown rate λ and suppose that we have the asynchronous data $\mathcal{F}_k^N$ corresponding to observation of N over

$[0, k]$. Here we examine alternatives to the estimators $\hat{\lambda}$ of (3.26). Since $P_\lambda\{N(i, i+1] = 0\} = e^{-\lambda}$ for each i, one could estimate λ using

$$\hat{\lambda}^* = -\log[(1/k)\textstyle\sum_{i=1}^k \mathbf{1}(N(i, i+1] = 0)].$$

a) Show that the estimators $\hat{\lambda}^*$ are strongly consistent and asymptotically normal as $k \to \infty$.

b) Compare the limiting variances of $\sqrt{n}[\hat{\lambda} - \lambda]$ (see Proposition 3.23) and $\sqrt{k}[\hat{\lambda}^* - \lambda]$ and argue when and why one of the estimators should be preferred to the other.

3.27 From the likelihood function L of (3.29) associated with observation of a non-homogeneous Poisson process N with intensity function $\lambda_t = e^{\alpha + \beta t}$ determine the maximum likelihood estimators $\hat{\alpha}$ and $\hat{\beta}$.

3.28 Let N be a Poisson process on $\mathbf{R}_+$ with rate λ and let A be a bounded set. Calculate the MMSE state estimators $E[N(B)|\mathcal{F}^N(A)]$.

3.29 Let $N(1)$ and $N(2)$ be independent Poisson processes on $\mathbf{R}_+$ with rates $\lambda(1)$, $\lambda(2)$, and let $N = N(1)+N(2)$. For each t calculate $E[N_t(1)|\mathcal{F}_t^N]$ and develop a recursive method of computation.

3.30 Let $N(1), N(2), \ldots$ be i.i.d. Poisson processes on $\mathbf{R}_+$ with known rate λ and suppose that one can observe only the superposition $N = \sum_{i=1}^m N_i$ where $m > 0$ is an unknown integer.

a) Show that for each t, N_t is a sufficient statistic for m given the data $\mathcal{F}_t^N$.

b) Calculate the maximum likelihood estimators $\hat{m}$ and discuss their properties as $t \to \infty$.

3.31 Let $N(1), N(2), \ldots$ be i.i.d. Poisson processes on $\mathbf{R}_+$ with known rate λ and suppose that one observes only the superposition $N = \sum_{i=1}^M N(i)$, where M is a nonnegative, integer-valued random variable independent of the $N(i)$. For each t compute the MMSE state estimator $E[M|\mathcal{F}_t^N]$.

3.32 Let $\tilde{N} = \sum_{n=-\infty}^\infty \varepsilon_{T_n}$ be a stationary Poisson process on $\mathbf{R}$ with unknown rate λ and let Z_i, $i \in \mathbf{Z}$, be i.i.d. random variables, independent of N, with unknown distribution G.

a) Discuss estimation of λ from the point process $N = \sum_{n=-\infty}^\infty \varepsilon_{T_n+Z_n}$. What about estimation of G?

b) Suppose now that λ and G are known. Compute $E[\tilde{N}(f)|\mathcal{F}_t^N]$ for each f and t.

3.33 Formulate and prove a theorem on local asymptotic normality, as $n \to \infty$, of the likelihood ratios associated with Proposition 3.21.

3.34 Let $U_1, \ldots, U_n$ be independent, exponentially distributed random variables with parameter $\lambda = 1$, let $U_{(1)}, \ldots, U_{(n)}$ be their order statistics, and define the normalized spacings $U_j' = (n - j + 1)[U_{(j)} - U_{(j-1)}]$ $j = 1, \ldots, n$. Prove that $U_1', \ldots, U_n'$ are independent and identically exponentially distributed.

NOTES

As general references on mathematical statistics we mention Bickel and Doksum (1977), Cox and Hinckley (1974), Ferguson (1967), LeCam (1986), Lehmann (1959, 1983), Wilks (1962), and Zacks (1971); there are many others. Statistical inference for point processes is treated in some generality in Brillinger (1978) (which draws extensive but not always convincing parallels between inference for point processes and inference for time series), Cox and Lewis (1966), and Krickeberg (1982), although only the latter includes general spaces. The literature on martingale inference for point processes on $\mathbf{R}_+$ is discussed in notes for Chapter 5.

Maximum likelihood estimators are, in most senses, the most important. Since several versions of the generic arguments below appear in the book, here is a brief summary of their theory. Let $X_1, X_2, \ldots$ be i.i.d. random variables with density function $f(x, \theta)$, where $\theta \in \mathbf{R}$ is unknown. Crucial objects are the n-sample log-likelihood functions $L_n(\theta) = \sum_{i=1}^{n} \log f(X_i, \theta)$ and their first two derivatives with respect to θ:

$$L_n'(\theta) = \sum_{i=1}^{n} \frac{f'(X_i, \theta)}{f(X_i, \theta)}, \tag{3.30}$$

known as the *score function*, and $L_n''(\theta)$. The *maximum likelihood estimator* (MLE) $\hat{\theta}$ satisfies the *likelihood equation*

$$L_n'(\hat{\theta}) = 0. \tag{3.31}$$

Consistency of the MLEs in a regular model is shown by showing that the random function $Z_n(u) = \prod_{i=1}^{n}[f(X_i, \theta + u)/f(X_i, \theta)]$ satisfies $\sup_{|u|>\gamma} Z_n(u) \to 0$, in the sense of convergence in probability, for each $\gamma > 0$.

Asymptotic normality, under the assumption that the Fisher information $I(\theta)$ given by (3.1) is finite, is then demonstrated by a Taylor expansion:

$$L_n'(\theta) = L_n'(\hat{\theta}) + [\hat{\theta} - \theta]L_n''(\theta^*) = [\hat{\theta} - \theta]L_n''(\theta^*),$$

by (3.31), where θ^* lies between $\hat{\theta}$ and θ. That is,

$$\frac{1}{\sqrt{n}} \sum_{i=1}^{n} \frac{f'(X_i, \theta)}{f(X_i, \theta)} = \sqrt{n}[\hat{\theta} - \theta] \frac{1}{n} \sum_{i=1}^{n} \frac{d^2}{d\theta^2} \log f(X_i, \theta). \tag{3.32}$$

By the law of large numbers and consistency, almost surely with respect to P_θ,

$$\frac{1}{n} \sum_{i=1}^{n} \frac{d^2}{d\theta^2} \log f(X_i, \theta^*) \to E_\theta \left[\frac{d^2}{d\theta^2} \log f(X_1, \theta) \right] = -I(\theta).$$

The random variables $f'(X_i, \theta)/f(X_i, \theta)$ are i.i.d. with mean zero and finite P_θ-variance $I(\theta)$, so by the central limit theorem

$$\frac{1}{\sqrt{n}} \sum_{i=1}^{n} \frac{f'(X_i, \theta)}{f(X_i, \theta)} \xrightarrow{d} N(0, I(\theta)),$$

and consequently

$$\sqrt{n}[\hat{\theta} - \theta] \xrightarrow{d} N(0, 1/I(\theta)).\tag{3.33}$$

When (3.33) obtains, MLEs attain the Cramér-Rao lower bound, and hence are efficient.

Hypothesis testing is dominated in a similar manner by the Neyman-Pearson approach and likelihood ratio tests. In this framework, *power*, the probability of rejecting the null hypothesis H_0, is required to fall below a specified value whenever the null hypothesis holds and is to be maximized, subject to this constraint, when the alternative hypothesis holds. A test is defined by a *test statistic* constructed from the data and a *critical region*, such that if the test statistic falls in the critical region, H_0 is rejected.

A test is *uniformly most powerful* if its power is maximal for all parameter values associated with the alternative. As a general rule, uniformly most powerful tests exist only for tests of one-sided hypotheses concerning real parameters.

The Neyman-Pearson lemma implies that one should use *likelihood ratio tests*, whose test statistics are of the form

$$T_n = \frac{\sup\{L_n(\theta) : \theta \text{ satisfies } H_1\}}{\sup\{L_n(\theta) : \theta \text{ satisfies } H_0\}} = \frac{L_n(\hat{\theta}_1)}{L_n(\hat{\theta}_0)},$$

where $\hat{\theta}_q$ is the MLE under H_q. Rejection occurs if T exceeds a critical value. Under broad assumptions, $2 \log T_n$ has a limiting χ^2-distribution under H_0.

Section 3.1

Model 3.1 provides the setting for all of Chapter 4, with Model 3.3 occupying Section 4.5. As examples in this chapter suggest, only in the presence of rather stringent structural assumptions is single-realization inference (Model 3.2) feasible. Cox and Lewis (1966) contains more on the synchronous/asynchronous differentiation.

Section 3.2

General sources on nonparametric statistics include Hollander and Wolfe (1973) as well as the broader books of Bickel and Doksum (1977) and Wilks (1962); each tries at least implicitly to explain a parametric-nonparametric distinction.

Two facets of classical statistics are largely absent from this book. One is Bayesian *statistical* inference, which we have omitted with the knowledge that relatively work exists concerning point processes. However, of course, state estimation is the Bayesian method *par excellence*, but not with an arbitrary prior distribution. Second, there is neither a body of problems nor a set of methods corresponding to linear models and regression in classical statistics, primarily because of the discrete, nonnegative character of point processes, In Chapter 5, though we do describe one model with a regression-like structure, but it is analyzed using martingale techniques.

Foundation issues and concepts for estimation and hypothesis testing are treated in the general sources mentioned above; there one can find general principles such as maximum likelihood estimation, the method of moments and likelihood ratio tests, along with more specific notions. For estimation problems we concentrate nearly exclusively on point estimation (as opposed to construction of confidence regions) and usually ignore optimality issues. For parametric problems these can often be resolved via classical theory (treated very mathematically in Ibragimov and Has'minskii, 1981, or LeCam, 1986), while for nonparametric problems they are often just too difficult. Ordinarily we are satisfied with estimators that are consistent and asymptotically normal, the latter in the sense that an error process converges in distribution to a Gaussian process. Similarly, for tests of hypotheses, computation of power under the null hypothesis and analysis of asymptotics may be all that one can achieve.

Now for the specific examples. Inference procedures for gamma distributions (Examples 3.4 and 3.9) are presented in Barlow and Proschan (1975), Bickel and Doksum (1977), and Wilks (1962). Here as well as in Chapter 8, the reliability literature is a fertile source of details; gamma distributions, for example, often appear in statistics books only as an incompletely presented example of a two-parameter exponential family. Serfozo (1975) contains a powerful and general method for inferring asymptotic properties of continuous time stochastic processes from those of discrete parameter processes embedded at random times satisfying minimal stability conditions. The stochastic integral of Examples 3.5 and 3.10, viewed as a partially Poisson process, is treated by Grandell (1976) and Karr (1982); see also Karr (1984a), Exercises 2.21 and 3.15–3.17, and Chapters 6, 7, and 10. Smith and Karr (1985) includes numerical calculations of maximum likelihood estimators of all three parameters from precipitation data. In Example 3.6 the issue is estimation of an unknown mixing distribution from i.i.d. samples when the distributions being mixed are known; Albrecht (1982b), Lindsay (1983a, 1983b), Simar (1976) and Tucker (1963) address mixed Poisson distributions specifically. Estimation of the intensity of a stationary point process on **R** is treated in Brillinger (1978), Cox (1965), and Cox and Lewis (1966), and for general spaces by Krickeberg (1982) and in Chapter 9.

The problem of testing whether a renewal process is Poisson (Example 3.14) has an extensive literature; see Cox and Lewis (1966) and Lewis (1965, 1972a) for exposition and further references. Epstein (1960) is an early source. The uniform conditional test and dispersion test are described in Lewis (1965) and the gap test in Lewis (1972a) and Pyke (1965); these tests seem to work best against some fairly specific but sometimes not very explicitly described alternatives. Tests for uniformity, and for other specific distributions, are discussed in the notes for Chapter 4; to use them to test directly whether interarrival times are exponential entails the complication of estimating the unknown mean. Less has been written about the more difficult problem of testing whether a stationary point process is Poisson (Example 3.15). Davies (1977) and Ripley (1976d) use moment and cumulative measures (see Chapter 9) to analyze narrowly defined alternative hypotheses. Sundt (1982) shows that the easily posed problem of testing whether a mixed Poisson process is Poisson is intractable.

Section 3.3

Brémaud (1981), Liptser and Shiryaev (1978), and Snyder (1975) are key expository sources on state estimation for point processes, although each treats only point processes on $\mathbf{R}_+$. All recursive methods in detail; the third is particularly rich in intuition, interpretation, and illustrative examples. In connection with state estimation for mixed Poisson processes (Example 3.18), see Grandell (1976), Karr (1983; 1984b), Krickeberg (1982), Snyder, (1975), and Section 7.2. Section 5.4 contains the solution of a very general state estimation problem on $\mathbf{R}_+$. Additional insight can be gained from the voluminous literature on state estimation for diffusion processes.

Section 3.4

The problem of combined inference and state estimation is not new, but seems not to have previously been analyzed systematically. Results for specific models appear in Sections 4.4, 6.3, 7.4, 8.3, 10.1 and 10.3. Inference for Yule processes (Example 3.19) is discussed in Basawa and Prakasa Rao (1980) and Feigin (1976).

Section 3.5

Birnbaum (1954) is one of the earliest substantive references to inference for Poisson processes *per se*, although the methods of Propositions 3.21 and 3.23 predate it. Proposition 3.24 is a special case of Theorem 6.12, references for which are given in the notes for Section 6.2. The concept of local asymptotic normality (Proposition 3.25) is due to LeCam (1960a); Greenwood and Shiryaev (1985), LeCam (1986) and Roussas (1972) are oriented in part to inference for stochastic processes. Sequential estimation and test procedures for Poisson processes are described in Birnbaum (1954), Dvoretsky *et al.* (1953a, 1953b) and Vardi (1979). (Wald (1947) is the seminal reference on sequential methods in general, with Govindarajulu (1975) a more modern treatment.) Bartholomew (1956), Boswell (1966), Cox and Lewis (1966), and Saw (1975) address tests of Poisson processes for trend. Comparison of Poisson processes is considered by Birnbuam (1954), Durbin (1961), Lewis (1965), and Qureishi (1964); the methods are based on conditional tests involving binomial distributions (for asynchronous data) or techniques for comparing χ^2-distributions (for synchronous data). Section 5.3 presents different methods that have desirable robustness properties.

Exercises

3.11 Maximum likelihood estimation of c can be based on the property that inter-arrival times for a Yule process are independent and exponentially distributed (see Basawa and Prakasa Rao, 1980).

3.12 See Feller (1966) for fundamental properties of this counter model.

3.14 See Karr (1982) and Kingman (1963); the latter characterizes processes that are simultaneously renewal processes and Cox processes (see also Exercise 3.16 and Theorem 8.24).

3.16 See Grandell (1976) and Karr (1982); the latter treats the estimators of (3.10) and (3.11).

3.19 The "linear interpolation" property here is yet another manifestation of conditional uniformity.

3.20 Karr (1984a) considers additional aspects, including state estimation.

3.21 See Krickeberg (1982), Neveu (1977), and Chapter 9.

3.24 This is a fairly graphic demonstration that some kinds of structural restrictions are not stringent enough to permit construction of tests with reasonable power.

3.26 Empirical zero-probability functionals are treated in generality in Section 4.3.

3.27 Cox and Lewis (1966) treat this model as a particular alternative hypothesis representing trend; for small t the trend is approximately linear — the most elementary sort of trend — without the intensity function's becoming negative.

3.30 See Cox and Lewis (1966). This exercise shows that seemingly different problems can be identical.

3.31 See also Exercises 1.20 and 6.24. M. Brown (1970) examines estimation of G when both $\tilde{N}$ and N are observable but points cannot be paired between them.

3.34 See Pyke (1965).

4
Empirical Inference
for Point Processes

Let $N_1, N_2, \ldots$ be i.i.d. copies of a point process N on a *compact* space E, and let $\mathcal{L}$ denote their common distribution. This chapter is devoted to analysis of the *empirical measures*

$$\hat{\mathcal{L}} = (1/n)\sum_{i=1}^{n}\varepsilon_{N_i} \tag{4.1}$$

as estimators of $\mathcal{L}$, especially their asymptotic properties, and to estimation of functionals $H(\mathcal{L})$ using corresponding functionals $H(\hat{\mathcal{L}})$. That the N_i are themselves point processes, although obviously central to our development, is less crucial to some tools and ideas used in this chapter. There is an extensive literature dealing with empirical processes engendered by i.i.d. random elements of general measurable spaces, which we apply in conjunction with the special structure of $\mathbf{M}_p$ to derive characteristics of the $\hat{\mathcal{L}}$. But the order of deduction can also be reversed: a random element X_i of E is the "mean" of the point process $N = \varepsilon_{X_i}$, and so point process techniques can be used to establish properties of empirical processes. Both points of view are pursued.

We are in the setting of Model 3.1, and in most cases assume no restriction on $\mathcal{L}$. Consequently, $\hat{\mathcal{L}}$ is maximum likelihood estimator of $\mathcal{L}$ and if the functional H is smooth, then $H(\hat{\mathcal{L}})$ is likewise the maximum likelihood estimator of $H(\mathcal{L})$. The method of moments basis for the empirical measures is that $E_{\mathcal{L}}[\hat{\mathcal{L}}(\Gamma)] = \mathcal{L}(\Gamma)$ for each $\Gamma \in \mathcal{M}_p$. Although it does not follow in general that $H(\hat{\mathcal{L}})$ is also the method of moments estimator of $H(\mathcal{L})$, this is the case if H is an integral functional, i.e., $H(\mathcal{L}) = \int_{\mathbf{M}_p} \tilde{H}(\eta)\mathcal{L}(d\eta)$. Laplace functionals and zero-probability functionals are perhaps the most important of this kind.

In Section 1 we examine set-indexed empirical measures $\{\hat{\mathcal{L}}(\Gamma) : \Gamma \in \mathcal{M}_p\}$

and function-indexed analogues $\{\hat{\mathcal{L}}(h) : h \in C_K(\mathbf{M}_p)\}$, which are more tractable in many ways. In the former case this leads to strong uniform consistency results such as Theorems 4.2, 4.3 and 4.5, while for the latter we also derive "infinite-dimensional" central limit theorems, of which Theorems 4.6 and 4.7 are examples. Sections 2–4 concern estimation of three key functionals of $\mathcal{L}$: the Laplace functional, zero-probability functional, and reduced Palm distributions. The first two are integral functionals of $\mathcal{L}$; estimators are easy to formulate and their properties are straightforward. Palm distributions, which are derivatives, present a more difficult problem. Our approach is to view them as limits of ratios of integral functionals of $\mathcal{L}$, which are estimated by corresponding limits involving $\hat{\mathcal{L}}$. Section 5 applies material from Section 2 to the case that the observed point processes are i.i.d. p thinnings of the N_i, where both $\mathcal{L}_N$ and the thinning function p are unknown. In Section 6 we explore some implications of point process results concerning ordinary empirical processes and apply conditional uniformity to produce strong approximations of Poisson processes by Wiener processes.

4.1 Empirical Measures

This section surveys fundamental asymptotic properties of empirical measures. We merge methodology exploiting special structure of the space $\mathbf{M}_p$ in which the point processes $\hat{\mathcal{L}}$ take their values with the theory of empirical processes associated with i.i.d. random elements of general measurable spaces. Interpretation may be eased if it is borne in mind that empirical measures are analogous to empirical distribution functions, that strong uniform consistency results generalize the Glivenko-Cantelli theorem, and that for statistical purposes the central limit theorems are directed to obtaining, among other things, analogues of the Kolmogorov-Smirnov limit distribution.

The setting is this. Let E be a compact metric space; then under the vague topology, $\mathbf{M}_p$ is a complete, separable metric space and is even locally compact (Exercise 4.1). Let $N_1, N_2, \ldots$ be i.i.d. copies of a point process N on E with unknown law $\mathcal{L}$. Our model is fully nonparametric: $\mathcal{L}$ is presumed to have no particular structure or form.

We begin by noting "finite-dimensional" asymptotic properties of the empirical measures $\hat{\mathcal{L}}$ of (4.1). These are three: strong consistency, asymptotic normality and a law of the iterated logarithm.

Proposition 4.1. a) For each $\mathcal{L}$ and each $\Gamma \in \mathcal{M}_p$, $\hat{\mathcal{L}}(\Gamma) \to \mathcal{L}(\Gamma)$ almost surely with respect to $P_{\mathcal{L}}$. Furthermore,

$$\hat{\mathcal{L}} \xrightarrow{w} \mathcal{L} \tag{4.2}$$

almost surely, where $\xrightarrow{w}$ denotes weak convergence of probability measures on $\mathbf{M}_p$;

b) For each $\mathcal{L}$ and for $\Gamma_1, \ldots, \Gamma_k \in \mathcal{M}_p$,

$$\sqrt{n} \left[(\hat{\mathcal{L}}(\Gamma_1), \ldots, \hat{\mathcal{L}}(\Gamma_k)) - (\mathcal{L}(\Gamma_1), \ldots, \mathcal{L}(\Gamma_k)) \right] \xrightarrow{d} N(0, R),$$

where $R(i, j) = \mathcal{L}(\Gamma_i \cap \Gamma_j) - \mathcal{L}(\Gamma_i)\mathcal{L}(\Gamma_j)$;

c) For each $\mathcal{L}$ and each Γ, almost surely

$$\limsup_{n \to \infty} \frac{\sqrt{n}[\hat{\mathcal{L}}(\Gamma) - \mathcal{L}(\Gamma)]}{\sqrt{2 \log \log n}} = \mathcal{L}(\Gamma)[1 - \mathcal{L}(\Gamma)]. \square$$

The asymptotic properties in Proposition 4.1 are weak and uninformative esthetically and for purposes of statistical inference. Given a subset $\mathcal{C}$ of $\mathcal{M}_p$, we define for each n the *discrepancy*

$$D_n(\mathcal{C}, \mathcal{L}) = \sup_{\Gamma \in \mathcal{C}} |\hat{\mathcal{L}}(\Gamma) - \mathcal{L}(\Gamma)|.$$

For estimation of $\mathcal{L}$, much more information is contained in uniform convergence statements of the form $D_n(\mathcal{C}, \mathcal{L}) \to 0$, at least if $\mathcal{C}$ is reasonably large. Similarly, for testing the hypothesis $H_0 : \mathcal{L} = \mathcal{L}_0$, $D_n(\mathcal{C}, \mathcal{L}_0)$ is more sensitive than a statistic $|\hat{\mathcal{L}}(\Gamma) - \mathcal{L}_0(\Gamma)|$ with Γ fixed. Use of D_n as a test statistic necessitates knowledge at least of its distribution under H_0, which is one motivation for the central limit theorems presented in this section. Similar issues arise for function-indexed empirical measures.

Uniformity for functions will be applied in Section 2 to convergence of empirical Laplace functionals; uniformity for sets is applied in Section 3 to empirical zero-probability functionals. To obtain uniform convergence we pursue a direct approach that yields a strong uniform consistency theorem for function-indexed processes; perform a more detailed analysis of the weak convergence $\hat{\mathcal{L}} \xrightarrow{w} \mathcal{L}$ in Proposition 4.1a), which provides information concerning setwise convergence and uniform convergence for both sets and functions; and apply the general theory of empirical processes to derive uniformity results for sets.

We begin with the direct approach.

Theorem 4.2. Let $\mathcal{K}$ be a subset of $C(\mathbf{M}_p)$ that is uniformly bounded in the topology of uniform convergence on compact subsets. Then almost surely under $P_{\mathcal{L}}$, $\sup_{h \in \mathcal{K}} |\hat{\mathcal{L}}(h) - \mathcal{L}(h)| \to 0$.

Proof: For each k let $\mathbf{M}_p(k) = \{\eta \in \mathbf{M}_p : \eta(E) \leq k\}$, a compact set. Given $\varepsilon > 0$, choose and fix k such that $\mathcal{L}(\mathbf{M}_p \setminus \mathbf{M}_p(k)) < \varepsilon$. Then by (4.2) almost surely $\hat{\mathcal{L}}(\mathbf{M}_p \setminus \mathbf{M}_p(k)) < 2\varepsilon$ for all n sufficiently large. Letting $\check{h}(\eta) = h(\eta)\mathbf{1}(\eta \in \mathbf{M}_p(k))$, we have that for large n,

$$\sup_{\mathcal{K}} |\hat{\mathcal{L}}(h) - \mathcal{L}(h)| \leq \sup_{\mathcal{K}} |\hat{\mathcal{L}}(\check{h}) - \mathcal{L}(\check{h})| + \sup_{\mathcal{K}} \hat{\mathcal{L}}(|h - \check{h}|) + \sup_{\mathcal{K}} \mathcal{L}(|h - \check{h}|)$$

$$\leq \sup_{\mathcal{K}} |\hat{\mathcal{L}}(\check{h}) - \mathcal{L}(\check{h})| + 6\varepsilon C,$$

where $C = \sup_{h \in \mathcal{K}} \|h\|_\infty$, which is finite. To show that the remaining supremum converges to zero almost surely, again let $\varepsilon > 0$ be fixed. Then by the Arzéla-Ascoli theorem there is $\delta > 0$ such that whenever $d(\eta, \nu) < \delta$ in $\mathbf{M}_p(k)$, [d denotes the Prohorov metric — see Section 1.6], then

$$\sup_{\mathcal{K}} |\check{h}(\eta) - \check{h}(\nu)| < \varepsilon.$$

By compactness, there is a partition $\mathbf{M}_p(k) = \sum_{j=1}^{M} \Gamma_j$ with diam $\Gamma_j < \varepsilon$ for each j. Also, let $\eta_j \in \Gamma_j$ be fixed. The multi-dimensional Glivenko-Cantelli theorem implies that almost surely

$$\sup_{|\alpha_j| \leq C} \left| \hat{\mathcal{L}}(\textstyle\sum_{j=1}^{M} \alpha_j \mathbf{1}_{\Gamma_j}) - \mathcal{L}(\textstyle\sum_{j=1}^{M} \alpha_j \mathbf{1}_{\Gamma_j}) \right| \to 0. \qquad (4.3)$$

(This could also be argued using the Vapnik-Chervonenkis criterion to be introduced presently.) For $h \in \mathcal{K}$,

$$\begin{aligned}
|\hat{\mathcal{L}}(\check{h}) - \mathcal{L}(\check{h})| \leq\ & \left| \hat{\mathcal{L}}(\check{h}) - \hat{\mathcal{L}}\left(\textstyle\sum_{j=1}^{M} \check{h}(\eta_j)\mathbf{1}_{\Gamma_j}\right) \right| \\
& + \left| \hat{\mathcal{L}}\left(\textstyle\sum_{j=1}^{M} \check{h}(\eta_j)\mathbf{1}_{\Gamma_j}\right) - \mathcal{L}\left(\textstyle\sum_{j=1}^{M} \check{h}(\eta_j)\mathbf{1}_{\Gamma_j}\right) \right| \\
& + \left| \mathcal{L}\left(\textstyle\sum_{j=1}^{M} \check{h}(\eta_j)\mathbf{1}_{\Gamma_j}\right) - \mathcal{L}(\check{h}) \right| \\
\leq\ & 2\varepsilon + \left| \hat{\mathcal{L}}\left(\textstyle\sum_{j=1}^{M} \check{h}(\eta_j)\mathbf{1}_{\Gamma_j}\right) - \mathcal{L}\left(\textstyle\sum_{j=1}^{M} \check{h}(\eta_j)\mathbf{1}_{\Gamma_j}\right) \right|,
\end{aligned}$$

and now only application of (4.3) is needed to complete the proof. ∎

We next consider what can be extracted from more detailed study of the conclusion $\hat{\mathcal{L}} \overset{w}{\to} \mathcal{L}$ in Proposition 4.1. As is well known, this implies that $\hat{\mathcal{L}}(\Gamma) \to \mathcal{L}(\Gamma)$ for every $\mathcal{L}$-*continuity set* Γ [satisfying $\mathcal{L}(\partial \Gamma) = 0$, where $\partial \Gamma$ is the boundary of Γ]. There are uniform versions that derive solely from (4.2). We list one, due to Billingsley and Tøpsøe (1967).

Theorem 4.3. Let $\mathcal{C}$ be a class of bounded Borel subsets of $\mathbf{M}_p$ such that $\mathcal{L}(\partial \Gamma) = 0$ for every $\Gamma \in \mathcal{C}$ and suppose that $\{\partial \Gamma : \Gamma \in \mathcal{C}\}$ is compact in the Hausdorff topology on the set of bounded, closed subsets of $\mathbf{M}_p$. Then almost surely $D_n(\mathcal{C}, \mathcal{L}) \to 0$. □

Theorem 4.3 is clearly cumbersome, yet Proposition 4.1 implies that for any $\Gamma \in \mathcal{M}_p$, not even an $\mathcal{L}$-continuity set, $\hat{\mathcal{L}}(\Gamma) \to \mathcal{L}(\Gamma)$ almost surely, so there *must* be uniform convergence results that are both usable and general.

There are, and they come from the theory of empirical processes. We will describe the setting and one result in some detail; application to empirical zero-probability functionals is given in Section 3.

We seek classes $\mathcal{C} \subseteq \mathcal{M}_p$ such that $D_n(\mathcal{C}, \mathcal{L}) \to 0$ almost surely under $P_{\mathcal{L}}$ for each $\mathcal{L}$. Obviously $\mathcal{C}$ must be restricted, for if $\mathcal{C}$ were $\mathcal{M}_p$, then $D_n(\mathcal{C}, \mathcal{L})$ would be the variation distance between $\hat{\mathcal{L}}$ and $\mathcal{L}$ (see Section 1.6) and if $\mathcal{L}$ were diffuse (the usual case in applications), no convergence would obtain. As it turns out — not surprisingly, since the most concrete way to visualize the information contained in a σ-algebra is as the capability to separate points — one way to limit the size of $\mathcal{C}$ is to restrict its power to "shatter" finite subsets of $\mathbf{M}_p$. The associated theory, due originally to Vapnik and Chervonenkis (1971), is very general. We outline its main features, but retain the point process context.

The keystone is a combinatorial way of limiting the size of $\mathcal{C}$. Given a finite subset Λ of $\mathbf{M}_p$, let $\Delta(\mathcal{C}, \Lambda) = |\{\Lambda \cap \Gamma : \Gamma \in \mathcal{C}\}|$ be the number of subsets of Λ that can be realized by intersecting Λ with some element of $\mathcal{C}$. Here $|A|$ is the cardinality of A. Let $v_r(\mathcal{C}) = \sup_{|\Lambda|=r} \Delta(\mathcal{C}, \Lambda)$; then evidently $v_r(\mathcal{C}) \le 2^r$ for all r. A set Λ with $|\Lambda| = r$ and $\Delta(\mathcal{C}, \Lambda) = 2^r$ is *shattered* by $\mathcal{C}$. The larger $\mathcal{C}$, the more sets it can shatter. Vapnik and Chervonenkis (1971) established the following dichotomy.

Proposition 4.4. For a given class $\mathcal{C}$, either $v_r(\mathcal{C}) = 2^r$ for all r or there is an integer s such that for all r,

$$v_r(\mathcal{C}) \le r^s + 1. \; \Box \tag{4.4}$$

The class $\mathcal{C}$ is termed a *Vapnik-Chervonenkis class* (VC class) if (4.4) holds for some s. The smallest such s is the VC *exponent* of $\mathcal{C}$. Clearly, a VC class is restricted in size; that this is the right condition in order to derive strong uniform consistency is not obvious, but true notwithstanding, modulo one measurability detail: there is no way other than assumption to ensure that the discrepancy $D_n(\mathcal{C}, \mathcal{L})$ is a random variable.

Here is the main result.

Theorem 4.5. If $\mathcal{C}$ is a VC subclass of $\mathcal{M}_p$ such that $D_n(\mathcal{C}, \mathcal{L})$ is a random variable for each n, then $D_n(\mathcal{C}, \mathcal{L}) \to 0$ almost surely with respect to $P_{\mathcal{L}}$. $\Box$

For the proof, we refer to Gaenssler (1984) or Pollard (1984).

The virtue of Theorem 4.5 is that the law of the N_i effectively does not enter the hypotheses: if C is a VC class, then strong uniform consistency holds. Consequently, $D_n(C, \mathcal{L}_0)$ is a suitable statistic for testing the hypothesis $H_0 : \mathcal{L} = \mathcal{L}_0$. However, as noted before, without at least a large-sample approximation to its distribution under $P_{\mathcal{L}_0}$ its usefulness is severely curtailed. This leads one, therefore, to seek central limit theorems complementing the consistency properties.

There is no hope for a central limit theorem entirely independent of $\mathcal{L}$: as Proposition 4.1 demonstrates, limiting covariances depend on $\mathcal{L}$. For set-indexed processes, additional issues arise that we address momentarily, but first here is one central limit theorem for function-indexed empirical measures. Discussion of the metric entropy condition imposed in it appears in the chapter notes.

Theorem 4.6. Let k be fixed and let $\mathbf{M}_p(k)$ be as in Theorem 4.2. Let $\mathcal{K}$ be a compact set of continuous functions on $\mathbf{M}_p(k)$ having finite metric entropy with respect to the uniform metric on $C(\mathbf{M}_p(k))$. Then

$$\{\sqrt{n}[\hat{\mathcal{L}}(h) - \mathcal{L}(h)] : h \in \mathcal{K}\} \xrightarrow{d} \{G(h) : h \in \mathcal{K}\}$$

as random elements of $C(\mathcal{K})$ (endowed with the uniform topology), where G is a continuous Gaussian process with mean zero and covariance function $R(h_1, h_2) = \mathcal{L}(h_1 h_2) - \mathcal{L}(h_1)\mathcal{L}(h_2)$. $\square$

The proof is substantially that of Theorem 4.10, which is proved in Section 2, so we omit it.

For set-indexed processes further complications arise. Given a class C of subsets of $\mathbf{M}_p$, define *empirical error processes*

$$E_n(\Gamma) = \sqrt{n}[\hat{\mathcal{L}}(\Gamma) - \mathcal{L}(\Gamma)], \quad \Gamma \in C.$$

The goal is to identify conditions on C and $\mathcal{L}$ under which the E_n converge in distribution to a continuous Gaussian process $G = \{G(\Gamma) : \Gamma \in C\}$, which by Proposition 4.1 has covariance function

$$R(\Gamma_1, \Gamma_2) = \mathcal{L}(\Gamma_1 \cap \Gamma_2) - \mathcal{L}(\Gamma_1)\mathcal{L}(\Gamma_2). \tag{4.5}$$

In order that this convergence make sense, several issues must be confronted:

- There must be a metric space V in which the E_n and G take their values, but usually V is not separable and G must then be confined to a separable subspace V_0.

- Continuity of G entails a metric on $\mathcal{C}$ itself.

- Finally, measurability of the E_n as random elements of V must also be arranged.

We do not describe fully here how these problems are resolved within the milieu of empirical processes, but we do sketch how the metric on $\mathcal{C}$ and the spaces V_0 and V, all depending on $\mathcal{L}$, are constructed. We then present one central limit theorem with some potential for inference.

For the metric on $\mathcal{C}$, one takes essentially the only choice: $\tilde{d}(\Lambda_1, \Lambda_2) = \mathcal{L}(\Lambda_1 \triangle \Lambda_2)$; then the covariance function R of (4.5) is uniformly continuous. The space V_0 is the linear space of functions $\varphi : \mathcal{C} \to \mathbf{R}$ that are bounded and $\tilde{d}$-uniformly continuous. Then, if $\mathcal{C}$ satisfies a strengthened form of compactness, existence can be shown of a Gaussian process G taking values in V_0, and hence continuous, with covariance function R. Next, V is the linear space of functions $\psi = \varphi + \sum_{i=1}^{k} a_i \varepsilon_{\mu_i}$, where $\varphi \in V_0$, $k \geq 0$, the a_i are real (not necessarily positive) and the μ_i are elements of $\mathbf{M}_p$. One endows V with the metric $d(\psi_1, \psi_2) = \sup_{\Gamma \in \mathcal{C}} |\psi_1(\Gamma) - \psi_2(\Gamma)|$. While the E_n take values in V, it is not true without further restrictions on $\mathcal{C}$ that they are measurable with respect to the Borel σ-algebra $\mathcal{V}$ on V, which is not countably generated. Nor is separability of (V_0, d) immediate; it must be assumed, either directly or indirectly.

Finally, it is necessary to deal with convergence in distribution of random elements of nonseparable metric spaces. Given a nonseparable metric space (V, d), let $\mathcal{V}_b$ denote the σ-algebra generated by the bounded d-balls; then $\mathcal{V}_b$ is contained strictly in the Borel σ-algebra $\mathcal{V}$. For random elements $X, X_1, X_2, \ldots$ measurable with respect to a σ-algebra $\mathcal{G}$ satisfying $\mathcal{V}_b \subseteq \mathcal{G} \subseteq \mathcal{V}$, we say that (X_n) converges in distribution to X (still written $X_n \overset{d}{\to} X$) if there is a separable subspace $\hat{V}$ of V such that $P\{X \in \hat{V}\} = 1$ and if $E[g(X_n)] \to E[g(X)]$ for every bounded, continuous and $\mathcal{V}_b$-measurable function g on V. See Gaenssler (1984) for details; in general, properties mimic those of convergence in distribution for random elements of separable metric spaces.

Finally, if the metric space $(\mathcal{C}, \tilde{d})$ is totally bounded (this is an important limitation on the size of $\mathcal{C}$), then (V_0, d) is separable and the error processes E_n are measurable random elements of $(V, \mathcal{V}_b)$.

The following result imposes restrictions stringent enough to overcome all the difficulties.

Theorem 4.7. Let $\ell \geq 1$ be fixed, let K be a compact subset of $\mathbf{R}^\ell$ and let $H : \mathbf{M}_p \times K \to \mathbf{R}$ be a function such that

 i) For each $z \in K$, $H(\cdot, z)$ is continuous on $\mathbf{M}_p$;

ii) H satisfies the Lipschitz condition

$$\sup\nolimits_{\eta \in M_p}\ \sup\nolimits_{z \neq z' \in K} \frac{|H(\eta, z) - H(\eta, z')|}{|z - z'|} < \infty;$$

iii) Uniformly in $z \in K$, $P_{\mathcal{L}}\{-\epsilon \leq H(N, z) < \epsilon\} = O(\epsilon)$ as $\epsilon \downarrow 0$.
Then for $\mathcal{C}$ the family of sets $\{\eta : H(\eta, z) \geq 0\}$, where z ranges over K,

a) The metric space (V_0, d) is separable;

b) The error processes E_n are measurable random elements of $(V, \mathcal{V}_b)$;

c) There exists a $\tilde{d}$-continuous Gaussian process G with covariance function given by (4.5);

d) $E_n \overset{d}{\to} G$ under $P_{\mathcal{L}}$. □

We move on to estimation of functionals of point processes. The simplest is the mean measure μ, with which we conclude the sections, and which we estimate by substitution from empirical measures:

$$\hat{\mu}(f) = \int_{M_p} \eta(f) \hat{\mathcal{L}}(d\eta) = (1/n)\sum_{i=1}^n N_i(f).$$

One then obtains the following analogue of Proposition 4.1

Proposition 4.8. a) If $\mu \in M$, then $\hat{\mu} \overset{w}{\to} \mu$ almost surely, where $\overset{w}{\to}$ denotes weak convergence in M_p;

b) If $\mu \in M$, then $\sup_{\Lambda} |\hat{\mu}(f) - \mu(f)| \to 0$ almost surely for each compact subset $\mathcal{K}$ of $C^+(E)$;

c) If $E[N(E)^2] < \infty$, then as random measures on E the processes $\sqrt{n}[\hat{\mu} - \mu]$ converge in distribution to a Gaussian random measure G with covariance function $R(f, g) = \rho(f \otimes g)$, where ρ is the covariance measure of the N_i;

d) Assume that there is $\delta > 0$ such that $E[N(E)^{2+\delta}] < \infty$. Then almost surely the set $\{\sqrt{n}[\hat{\mu} - \mu]/\sqrt{2 \log \log n} : n \geq 3\}$ is relatively compact in the space of signed Radon measures on E (see Appendix A), with limit set $\{\eta : |\eta(f)| \leq \sqrt{R(f, f)}$ for all $f \in C(E)\}$. □

4.2 Empirical Laplace Functionals

Given the i.i.d. copies $N_1, N_2, \ldots$ of N, the Laplace functional $L(f) = E_{\mathcal{L}}[e^{-N(f)}]$ is an integral functional of $\mathcal{L}$: $L(f) = \int_{M_p} e^{-\eta(f)} \mathcal{L}(d\eta)$. Since the empirical measure $\hat{\mathcal{L}}$ is both maximum likelihood estimator and of moments estimator of $\mathcal{L}$, the *empirical Laplace functional*

$$\hat{L}(f) = \int_{M_p} e^{-\eta(f)} \hat{\mathcal{L}}(d\eta) = (1/n)\sum_{i=1}^n e^{-N_i(f)}$$

is, similarly, maximum likelihood and method of moments estimator of L. In this section we develop asymptotic properties of empirical Laplace functionals, some of them consequences of results in Section 1, others derived directly. Our interpretation is that empirical Laplace functionals are stochastic processes parametrized by $C^+(E)$, and the goal is to derive limit theorems in this setting. As it turns out, this space is too large to yield precise results and one must work instead with (for consistency) compact subsets or (for asymptotic normality) compact subsets with finite metric entropy.

Empirical Laplace functionals can serve as tools for statistical estimation and hypothesis testing, in particular since — as Theorems 4.9 and 4.10 confirm — there is a rather satisfactory asymptotic theory for them. The reason for this is that the Laplace functional and the empirical Laplace functionals depend continuously on $f \in C^+(E)$. For inference (and, on esthetic grounds, in general) hypotheses of theorems should exhibit only minimal dependence on the unknown Laplace functional; in this respect as well, the results for empirical Laplace functionals are quite satisfactory: Theorem 4.9 entails no assumptions, while Theorem 4.10 imposes only a first moment assumption.

We now formulate and prove the main results of the section. By Theorem 1.12 we choose L, which equals L_{N_i} for each i, as the index for the statistical model, which remains unrestricted, and we denote by P_L the corresponding probability measure.

Here is the main strong uniform consistency property.

Theorem 4.9. For $\mathcal{K}$ be a compact subset of $C^+(E)$, almost surely with respect to P_L, $\sup_{\mathcal{K}} |\hat{L}(f) - L(f)| \to 0$.

Proof: Define a mapping $f \to h_f$ of $\mathcal{K}$ into $C(\mathbf{M}_p)$ by $h_f(\eta) = e^{-\eta(f)}$, and observe that $L(f) = \mathcal{L}(h_f) = \int h_f d\mathcal{L}$, and, similarly, $\hat{L}(f) = \hat{\mathcal{L}}(h_f)$. By Theorem 4.2 it suffices to show that $\mathcal{K}' = \{h_f : f \in \mathcal{K}\}$ is a uniformly bounded and compact subset of $C(\mathbf{M}_p)$. The former is evident since $0 \le h_f \le 1$ for each f while since the mapping $f \to h_f$ continuous, $\mathcal{K}'$ is the continuous image of a compact set and hence is compact. ∎

Theorem 4.9 is useful for statistical estimation of L as well as for hypothesis testing. In the latter context the hypothesis $H_0 : L = L_0$ can be tested using the discrepancy

$$D_n(\mathcal{K}, L_0) = \sup_{f \in \mathcal{K}} |\hat{L}(f) - L_0(f)|,$$

provided that some knowledge is available concerning its P_{L_0}-distribution. Among other things, Theorem 4.10 is a step in this direction.

Theorem 4.10. Let $\mathcal{K}$ be a compact subset of $C^+(E)$ having finite metric entropy with respect to the uniform metric (see the chapter notes for a brief discussion) and assume that $E_L[N(E)] < \infty$. Then under P_L,

$$\{\sqrt{n}[\hat{L}(f) - L(f)] : f \in \mathcal{K}\} \overset{d}{\to} \{G(f) : f \in \mathcal{K}\}$$

as random elements of the metric space $C(\mathcal{K})$, where G is a continuous Gaussian process with mean zero and covariance function

$$R(f, g) = L(f + g) - L(f)L(g). \tag{4.6}$$

Proof: Convergence in distribution is a direct consequence of Araujo and Giné (1980, Theorem 7.16 and Corollary 7.17), for which one views the functionals $e^{-N_i(\cdot)} - L(\cdot)$ as i.i.d. centered random elements of $C(\mathcal{K})$; the hypothesis $E_L[N(E)] < \infty$ is required to fulfill the assumptions there. These same results also imply the existence of a continuous Gaussian process G with covariance function R. ∎

The same argument proves Theorem 4.6 as well; there the compact space $\mathbf{M}_p(k)$ plays the role of E in Theorem 4.10. In both cases we are dealing with i.i.d. random elements of the set of continuous functions on a compact metric space.

Concerning the discrepancy $D_n(\mathcal{K}, L_0)$, Theorem 4.10 and the continuous mapping theorem have the following consequence.

Corollary 4.11. Under the assumptions of Theorem 4.10, we have $\sqrt{n}D_n(\mathcal{K}, L) \overset{d}{\to} \sup_{f \in \mathcal{K}} |G(f)|$. ☐

One stumbling block remains, however. To use $D_n(\mathcal{K}, L_0)$ as a test statistic for the hypothesis $H_0 : L = L_0$, one must calculate the distribution of the supremum $M = \sup_{f \in \mathcal{K}} |G(f)|$. In general, this is impossible, however.

On the bright side, empirical Laplace functionals sometimes exhibit surprisingly good behavior even in parametric situations.

Example 4.12. Let $E = [0, 1]$ and suppose that $N_1, N_2, \ldots$ are i.i.d. homogeneous Poisson processes with unknown rate λ. In the absence of this information one would use the empirical Laplace functional $\hat{L}$ given by (4.5). But under the Poisson model, the maximum likelihood estimator of λ is $\hat{\lambda} = (1/n) \sum_{i=1}^n N_i(E)$, which by substitution yields as restricted estimator of the Laplace functional

$$\hat{L}^*(f) = \exp[-\hat{\lambda} \int_0^1 (1 - e^{-f})].$$

Concerning asymptotic behavior of $\hat{L}^*(\cdot)$, one can show that for all λ,

$$\sup_{f \in C^+} |\hat{L}^*(f) - L_\lambda(f)| \leq |\hat{\lambda} - \lambda| \sup_{f \in C^+} \int_0^1 (1 - e^{-f}) \to 0 \qquad (4.7)$$

almost surely, which is stronger than can be concluded using Theorem 4.9. Moreover, if Y, with distribution $N(0, \lambda)$, is such that $\sqrt{n}[\hat{\lambda} - \lambda] \overset{d}{\to} Y$ under P_λ, then the error processes $\{\sqrt{n}[\hat{L}^*(f) - L_\lambda(f)] : f \in C^+\}$ converge in distribution to the process $G(f) = Y[\int_0^1 (1 - e^{-f})] \exp[-\lambda \int_0^1 (1 - e^{-f})]$.

On heuristic grounds $\hat{L}^*$ is preferred to $\hat{L}$ because it is tailored to the Poisson model, but this estimator is risky because it may be useless if the Poisson assumption fails. The empirical Laplace functional has the advantage of robustness, but perhaps at the price of sacrificing less-than-optimal behavior when the Poisson hypothesis holds. This happens sometimes but not always. The proper tool for comparison is *asymptotic relative efficiency*, in other words, comparison of asymptotic variances. For fixed f the asymptotic variance of $\hat{L}(f)$ is [by (4.6)] $L(2f) - L(f)^2 = \exp[-\lambda \int_0^1 (1 - e^{-2f})] \exp[-2\lambda \int_0^1 (1 - e^{-f})]$, while that of $\hat{L}^*(f)$ is $\lambda[\int_0^1 (1 - e^{-f})]^2 \exp[-2\lambda \int_0^1 (1 - e^{-f})]$. Hence the asymptotic efficiency of $\hat{L}^*(f)$ relative to $\hat{L}(f)$ is

$$e(\lambda) = \frac{[\lambda \int_0^1 (1 - e^{-2f})]^2 \exp[-\lambda \int_0^1 (1 - e^{-f})]}{1 - \exp[-\lambda \int_0^1 (1 - e^{-f})]}.$$

For small values of λ the asymptotic relative efficiency is less than 1, indicating that the estimator $\hat{L}^*(f)$ is superior; however, as $\lambda \to \infty$ the asymptotic relative efficiency converges to zero and hence for large λ the empirical Laplace functional $\hat{L}(f)$ is the better estimator. $\square$

4.3   Empirical Zero-Probability Functionals

From the integral representation

$$z(A) = P_{\mathcal{L}}\{N(A) = 0\} = \int_{M_p} 1(\eta(A) = 0)\mathcal{L}(d\eta)$$

of the zero-probability functional, we obtain by substitution the *empirical zero-probability functionals*

$$\hat{z}(A) = \int_{M_p} 1(\eta(A) = 0)\hat{\mathcal{L}}(d\eta) = (1/n)\sum_{i=1}^n 1(N_i(A) = 0),$$

where $A \in \mathcal{E}$. By contrast with the continuous integrand $\eta \to e^{-\eta(f)}$ appearing in empirical Laplace functionals, here there is the discontinuous integrand $\eta \to 1(\eta(A) = 0)$; the difficulty, discontinuity of $\eta \to \eta(A)$ rather than that of the indicator function, is so substantial that beyond finite-dimensional results we have only one fairly general strong uniform consistency theorem (Theorem 4.13) and a much less general central limit theorem (Theorem 4.14). The latter is, however, a bit more general than may first seem apparent.

We note the elementary results. Let $A_1, \ldots, A_k$ be sets in $\mathcal{E}$. Then for each $\mathcal{L}$, $(\hat{z}(A_1), \ldots, \hat{z}(A_k)) \to (z(A_1), \ldots, z(A_K))$ almost surely, and as random k-vectors,

$$\sqrt{n}\left[(\hat{z}(A_1), \ldots, \hat{z}(A_k)) - (z(A_1), \ldots, z(A_k))\right] \overset{d}{\to} Y,$$

where Y has a normal distribution with mean zero and covariance matrix $R(i, j) = z(A_i \cup A_j) - z(A_i)z(A_j)$.

Inspection of R reveals problems that must be addressed to extend these results to infinite subclasses $\mathcal{D}$ of $\mathcal{E}$. To begin, one requires a metric on $\mathcal{D}$ with respect to which the covariance function $R(A_1, A_2) = z(A_1 \cup A_2) - z(A_1)z(A_2)$ is continuous. (Without continuity of R there is no hope for continuity of a Gaussian process G with covariance function R.) Then the empirical zero-probability functionals and the Gaussian limit G must take values in a suitable metric space, ...(see discussion in Section 1).

All this can be overcome, but only (it seems) under severe restrictions on $\mathcal{D}$. Before doing so, however, we show how the Vapnik-Chervonenkis approach of Theorem 4.5 yields strong uniform consistency for empirical zero-probability functionals.

Theorem 4.13. Let $\mathcal{D}$ be a VC subclass of $\mathcal{E}$; then for each $\mathcal{L}$, almost surely with respect to $P_{\mathcal{L}}$, $\sup_{A \in \mathcal{D}} |\hat{z}(A) - z(A)| \to 0$.

Proof: The key point is to show that for each positive integer k, the family $\mathcal{C}_k$ of sets $\{\eta : \eta(E) = k, \eta(A) = 0\}$, $A \in \mathcal{D}$, is a VC class of subsets of $\mathbf{M}_p(k) = \{\eta \in \mathbf{M}_p : \eta(E) = k\}$ Given a finite subset $\{\eta_1, \ldots, \eta_\ell\}$ of $\mathbf{M}_p(k)$, with $\eta_i = \sum_{j=1}^k \varepsilon_{x_{ij}}$, then for each i and A, $\eta_i(A) = 0$ if and only if $(x_{i1}, \ldots, x_{ik})$ belongs to $(E \setminus A)^k$ and hence it suffices to show that $\{(E \setminus A)^k : A \in \mathcal{D}\}$ is a VC class in E^k. But $\{E \setminus A : A \in \mathcal{D}\}$ is a VC class in $\mathcal{E}$ so what needs to be shown is that if $\mathcal{H}$ is a VC class in E, the $\mathcal{H}^k = \{B^k : B \in \mathcal{H}\}$ is a VC class in E^k. But indeed, if r is such that $\mathcal{H}$ can shatter no subset of E with cardinality r or greater, then $\mathcal{H}^k$ cannot shatter any subsets of E^k of cardinality $(r - 1)^k + 1$. Hence $\mathcal{C}_k$ is a VC class in $\mathbf{M}_p(k)$.

Theorem 4.5, applied to the subprobability measures $\mathcal{L}_k(\Gamma) = \mathcal{L}(\Gamma \cap \mathbf{M}_p(k)) = P_{\mathcal{L}}\{N \in \Gamma, N(E) = k\}$, with associated empirical measures $\hat{\mathcal{L}}_k(\Gamma) = (1/n)\sum_{i=1}^n \mathbf{1}(N_i \in \Gamma, N_i(E) = k)$, implies that

$$\sup_{A \in \mathcal{D}} |\hat{\mathcal{L}}_k(\{\eta(A) = 0\}) - \mathcal{L}_k(\{\eta(A) = 0\})| = \sup_{\Gamma \in \mathcal{C}_k} |\hat{\mathcal{L}}_k(\Gamma) - \mathcal{L}_k(\Gamma)| \to 0 \quad (4.8)$$

almost surely for each k. Since $\mathcal{L}_k(\mathbf{M}_p) = P_{\mathcal{L}}\{N(E) = k\}$ and analogously for $\hat{\mathcal{L}}_k(\mathbf{M}_p)$, standard arguments imply that $\sum_{k=0}^{\infty} |\hat{\mathcal{L}}_k(\mathbf{M}_p) - \mathcal{L}_k(\mathbf{M}_p)| \to 0$. Thus given $\varepsilon > 0$ and k_0 large enough that $\sum_{i=k_0}^{\infty} \mathcal{L}_k(\mathbf{M}_p) < \varepsilon$, then almost surely $\sum_{i=k_0}^{\infty} \hat{\mathcal{L}}_k(\mathbf{M}_p) < 2\varepsilon$ for all n sufficiently large. For such n,

$$\sup_{A \in \mathcal{D}} |\hat{z}(A) - z(A)| \le 3\varepsilon + \sup_{\Gamma \in \mathcal{C}_k} |\hat{\mathcal{L}}_k(\Gamma) - \mathcal{L}_k(\Gamma)|, \qquad (4.9)$$

and the proof is concluded by combining (4.8) and (4.9). ∎

We present a central limit analogue of Theorem 4.13 in what seems at first to be a very special case; thereafter, we argue that the result is less restrictive than it appears initially.

Theorem 4.14. Let $E = [0, 1]$ and let $\mathcal{D} = \{D_t : t \in [0, 1]\}$, where $D_t = (t, 1]$, and assume that the mean measure μ of the N_i is diffuse. Then

$$\{\sqrt{n}[\hat{z}(D_t) - z(D_t)]\} \xrightarrow{d} \left\{ Y + [1 - z(D_0)]B\left(\frac{z(D_t) - z(D_0)}{1 - z(D_0)}\right) \right\}, \qquad (4.10)$$

where Y is a random variable with distribution $N(0, z(D_0)[1 - z(D_0)])$ and B is a Brownian bridge on $[0, 1]$ independent of Y.

Proof: For each i, let $X_i = \inf\{u : N_i([0, u]) = N_i(E)\}$ and note that $X_i = 0$ if $N_i(E) = 0$. The X_i are i.i.d. random variables taking values in $[0, 1]$ with distribution $F(t) = P\{X_i \le t\} = z(D_t)$; hence the $\hat{z}(D_t)$ are the empirical distribution functions $\hat{F}(t)$ associated with the sequence (X_i) and, moreover, $\sqrt{n}[\hat{z}(D_t) - z(D_t)] = \sqrt{n}[\hat{F}(t) - F(t)]$. By Proposition 4.1,

$$\sqrt{n}[\hat{z}(D_0) - z(D_0)] = \sqrt{n}[\hat{F}(0) - F(0)] \xrightarrow{d} N(0, F(0)[1 - F(0)]). \qquad (4.11)$$

On $(0, 1]$ the distribution function F is continuous and hence by standard theory of empirical processes on $\mathbf{R}$ (see Csörgö and Révész 1981, or Gaenssler, 1984), if $G_n(t) = [\hat{F}(t) - \hat{F}(0)]/[1 - \hat{F}(0)]$ and $G(t) = [F(t) - F(0)]/[1 - F(0)]$, then

$$\{\sqrt{n}[G_n(t) - G(t)]\} \xrightarrow{d} \{B(G(t))\}, \qquad (4.12)$$

where B is a Brownian bridge independent of the limit random variable in (4.11). Now, (4.10) follows from (4.11)–(4.12) by straightforward calculations. ∎

One can generalize Theorem 4.14 as follows. Let E be arbitrary and let $\mathcal{D} = \{D_t : 0 \le t \le 1\}$ be a linearly ordered, decreasing family of subsets of E with $E[N(D_1)] = 0$. Then one can interpret $\mathbf{1}(N_i(D_t) = 0)$ as $\mathbf{1}(X_i \le t)$ for some random variable X_i with distribution function $F(t) = z(D_t)$ and (4.10) holds even without notational changes.

4.4 Estimation of Reduced Palm Distributions

As described in Section 2.6, reduced Palm distributions play a fundamental role in calculation of conditional distributions for a point process N given observations $\mathcal{F}^N(B^c)$ and are, furthermore, central to state estimation for Cox processes on general spaces (Section 7.2, especially Theorem 7.6) and to combined inference and state estimation for Cox processes (Section 7.4). Thus it is important to estimate the reduced Palm distributions given observation of i.i.d. copies N_i. We maintain the approach of the preceding sections, expressing Palm distributions as functionals of the law of the N_i and taking as estimators of them the same functionals of the empirical measures. However, now the situation is more complicated because the reduced Palm distributions $Q_N(\mu, \cdot)$ are derivatives in μ. We represent them as limits of integral functionals of $\mathcal{L}$, which are estimated, once again, with corresponding functionals of $\hat{\mathcal{L}}$. As in the remainder of the chapter the setting is fully nonparametric; however, we do suppose that N is simple.

Rather than estimate reduced Palm distributions themselves we estimate their Laplace functionals $L(\mu, g) = E[e^{-N_\mu(g)}]$, where the N_μ are the *reduced Palm processes* of N. (See Section 1.7.) Our estimators not only are simple conceptually — they are essentially histogram estimators — and computable but also are strongly uniformly consistent and asymptotically normal.

With $L(\mu, g)$ written in the heuristic form

$$L_N(\mu, g) = E[e^{-\{N(g)-\mu(g)\}} | N - \mu \geq 0],$$

we identify the main issue: approximation of the null event $\{N - \mu \geq 0\}$ by events of positive probability. We so in the following manner. Let $\{A_{nj} : n \geq 1, 1 \leq j \leq \ell_n\}$ be a null array of partitions of E. Then we use the estimators

$$\hat{L}(\mu, g) = \frac{e^{\mu(g)} \sum_{i=1}^{n} e^{-N_i(g)} \prod_{j=1}^{\ell_n} \mathbf{1}(N_i(A_{nj}) \geq \mu(A_{nj}))}{\sum_{i=1}^{n} \prod_{j=1}^{\ell_n} \mathbf{1}(N_i(A_{nj}) \geq \mu(A_{nj}))}.$$

Because the partitions are successively finer, for given μ eventually all its points lie in distinct partition sets and since eventually each $N_i(A_{nj})$ is either zero or one as well, the product of indicator functions is indeed an approximation to $\mathbf{1}(N_i - \mu \geq 0)$.

Large sample properties of these estimators are given in the next two results.

Theorem 4.15. Assume that
a) $\mu \to L(\mu, g)$ is continuous for each g;

b) There exists $t > 0$ such that $E[e^{tN(E)}] < \infty$;

c) For each k, $\sum_{n=1}^{\infty} \ell_n^k/n^2 < \infty$.

Then for each compact subset $\mathcal{K}$ of $C^+(E)$ and compact subset $\mathcal{K}'$ of $\mathbf{M}_p$,

$$\lim_{n \to \infty} \sup_{\mu \in \mathcal{K}'} \sup_{g \in \mathcal{K}} |\hat{L}(\mu, g) - L(\mu, g)| = 0 \qquad (4.13)$$

almost surely.

Proof: For each k let $\mathbf{M}_p(k) = \{\mu \in \mathbf{M}_p : \mu(E) = k\}$; in proving (4.13) we may and do assume that $\mathcal{K}'$ is a subset of $\mathbf{M}_p(k)$ for some fixed k. We form the decomposition

$$
\hat{L}(\mu, g) = \frac{e^{\mu(g)} E\left[e^{-N(g)} \prod_{j=1}^{\ell_n} \mathbf{1}(N(A_{nj}) \geq \mu(A_{nj}))\right]}{E\left[\prod_{j=1}^{\ell_n} \mathbf{1}(N(A_{nj}) \geq \mu(A_{nj}))\right]}
$$

$$
\times \frac{(1/n) \sum_{i=1}^{n} e^{-N_i(g)} \prod_{j=1}^{\ell_n} \mathbf{1}(N_i(A_{nj}) \geq \mu(A_{nj}))}{E\left[\prod_{j=1}^{\ell_n} \mathbf{1}(N(A_{nj}) \geq \mu(A_{nj}))\right]}
$$

$$
\times \frac{E\left[\prod_{j=1}^{\ell_n} \mathbf{1}(N(A_{nj}) \geq \mu(A_{nj}))\right]}{(1/n) \sum_{i=1}^{n} \prod_{j=1}^{\ell_n} \mathbf{1}(N_i(A_{nj}) \geq \mu(A_{nj}))}
$$

$$
= \text{I} \times \text{II} \times \text{III}.
$$

We shall show that $\text{I} \to L(\mu, g)$, while $\text{II} \to 1$ and $\text{III} \to 1$, all in the almost sure sense.

Suppose now that μ lies in $\mathbf{M}_p(k)$. Then in view of b), for all sufficiently large n,

$$\prod_{j=1}^{\ell_n} \mathbf{1}(N_i(A_{nj}) \geq \mu(A_{nj})) = (1/k!) \int_{E^k} \prod_{j=1}^{\ell_n} \mathbf{1}(\sum_{i=1}^{k} \varepsilon_{x_i} \geq \mu(A_{nj})) N^{(k)}(dx),$$

so that with $\Gamma_n(\mu) = \{\nu : \prod_{j=1}^{\ell_n} \mathbf{1}(\nu(A_{nj}) \geq \mu(A_{nj})) = 1\} = \{\nu : \nu(A_{nj}) \geq \mu(A_{nj})$ for $j = 1, \ldots, \ell_n\}$,

$$
E\left[\prod_{j=1}^{\ell_n} \mathbf{1}(N_i(A_{nj}) \geq \mu(A_{nj}))\right]
$$

$$
\cong (1/k!) E\left[\int_{E^k} \mathbf{1}(\sum_{i=1}^{k} \varepsilon_{x_i} \in \Gamma_n(\mu)) N^{(k)}(dx)\right]
$$

$$
= K_N(\Gamma_n(\mu) \cap \mathbf{M}_p(k)).
$$

(See Section 1.7 for notations, definitions and key formulas.) Similarly,

$$
e^{\mu(g)} E\left[e^{-N(g)} \prod_{j=1}^{\ell_n} \mathbf{1}(N_i(A_{nj}) \geq \mu(A_{nj}))\right]
$$

$$
\cong \int_{\Gamma_n(\mu) \cap \mathbf{M}_p(k)} K_N(d\nu) L(\nu, g).
$$

and therefore

$$|\mathrm{I} - L(\mu, g)| \le \frac{1}{K_N(\Gamma_n(\mu))} \int_{\Gamma_n(\mu)} |L(\eta, g) - L(\mu, g)| \, K_N(d\eta) \to 0$$

by a) and b), since $\Gamma_n(\mu) \downarrow \{N - \mu \ge 0\}$.

To deal with the term III, we note that the sets $\Gamma_n(\mu)$ may be chosen so that there are at most ℓ_n^k of them, denoted by $\Gamma_{n\ell}$, so that given $\epsilon > 0$,

$$
\begin{aligned}
P\{\sup_\mu |\mathrm{III} - 1| > \epsilon\} &= P\left\{\max_\ell \left|\frac{(1/n)\sum_{i=1}^n \mathbf{1}(N_i \in \Gamma_{n\ell})}{P\{N \in \Gamma_{n\ell}\}} - 1\right| > \epsilon\right\} \\
&\le \sum_{\ell=1}^{\ell_n} P\left\{\left|\frac{(1/n)\sum_{i=1}^n \mathbf{1}(N_i \in \Gamma_{n\ell})}{P\{N \in \Gamma_{n\ell}\}} - 1\right| > \epsilon\right\} \\
&= \mathrm{O}(\ell_n^k/n^2)
\end{aligned}
$$

by Chebyshev's inequality; hence $\mathrm{III} \to 1$ almost surely by c) and the Borel-Cantelli lemma.

The argument that $\mathrm{II} \to 1$ almost surely is deduced from that for III by an absolutely continuous change of probability measure (see Karr, 1985a, for details).

Uniformity in $\mu \in \mathcal{K}'$ is established by standard approximation and equicontinuity reasoning, and hence (4.13) holds. ∎

Our central limit theorems is a compromise between generality and usefulness.

Theorem 4.16. Let g be a fixed element of $C_+(E)$, and assume that in addition to the conditions of Theorem 4.15,

$$\lim_{n\to\infty} \sqrt{n} \int_{\mathbf{M}_p} \left|\frac{E[e^{-\{N(g)-\mu(g)\}}\mathbf{1}(N \in \Gamma_n(\mu))]}{P\{N \in \Gamma_n(\mu)\}}\right| C(d\mu \times \mathbf{M}_p) = 0. \quad (4.14)$$

Then as signed random measures on $\mathbf{M}_p$,

$$\sqrt{n}[\hat{L}(\mu, g) - L(\mu, g)]C(d\mu \times \mathbf{M}_p) \overset{d}{\to} G_1(d\mu) - L(\mu, g)G_2(d\mu),$$

where $G = (G_1, G_2)$ is a bivariate Gaussian random measure on $\mathbf{M}_p$ with covariance function R given by (4.15) below.

Proof: Recall that g is fixed. Introduce the centered random measures

$$
\begin{aligned}
M_i(\Gamma) &= e^{-N_i(g)}\mathbf{1}(N_i \in \Gamma) - E[e^{-N(g)}\mathbf{1}(N \in \Gamma)] \\
\tilde{M}_i(\Gamma) &= \mathbf{1}(N_i \in \Gamma) - P\{N \in \Gamma\}
\end{aligned}
$$

on $\mathbf{M}_p$. The pairs $(M_i, \tilde{M}_i)$ are i.i.d. and hence by Karr (1979, Theorem 2.2), there is a mean-zero Gaussian process $G = (G_1, G_2)$, with each component a random measure, such that $(1/\sqrt{n})\sum_{i=1}^n (M_i, \tilde{M}_i) \overset{d}{\to} G$. (See also Proposition 4.8.) Derivation of the covariance function R of G is elementary, with the result that if $r(f, \Gamma) = E[e^{-N(f)}\mathbf{1}(N \in \Gamma)]$, then the 4-vector $(G(\Lambda_1, \Gamma_1), G(\Lambda_2, \Gamma_2))$ has covariance matrix

$$R((\Lambda_1, \Gamma_1)), G(\Lambda_2, \Gamma_2)) = \begin{bmatrix} R_{11} & R_{12} \\ R_{21} & R_{22} \end{bmatrix}, \tag{4.15}$$

where

$$R_{11} = \begin{bmatrix} r(2g, \Gamma_1) - r(g, \Gamma_1)^2 & r(g, \Gamma_1 \cap \Lambda_1) - r(g, \Gamma_1)r(0, \Lambda_1) \\ r(g, \Gamma_1 \cap \Lambda_1) - r(g, \Gamma_1)r(0, \Lambda_1) & r(0, \Lambda_1) - r(0, \Lambda_1)^2 \end{bmatrix}$$

where R_{22} is defined analogously with Γ_1, Λ_1 replaced by Γ_2 and Λ_2, and where

$$R_{12} = \begin{bmatrix} r(2g, \Gamma_1 \cap \Gamma_2) - r(g, \Gamma_1)r(g, \Gamma_2) & r(g, \Gamma_1 \cap \Lambda_2) - r(g, \Gamma_1)r(0, \Lambda_2) \\ r(g, \Lambda_1 \cap \Gamma_2) - r(0, \Lambda_1)r(g, \Gamma_2) & r(0, \Lambda_1 \cap \Lambda_2) - r(0, \Lambda_1)r(0, \Lambda_2) \end{bmatrix},$$

with R_{12} defined by obvious analogy. Note also that for fixed f, $r(f, \cdot)$) is a measure on $\mathbf{M}_p$ with $r(f, \cdot) \ll r(0, \cdot)$.

Suppose that H is a bounded, continuous function on $\mathbf{M}_p$; the argument that follows is a function space version of the "delta method." For each n,

$$\sqrt{n}\int H(\mu)[\hat{L}(\mu, g) - L(\mu, g)]K_N(d\mu)$$
$$= \sqrt{n}\int H(\mu)e^{\mu(g)}\left[\frac{\sum_{i=1}^n e^{-N_i(g)}\mathbf{1}(N_i \in \Gamma_n(\mu))}{\sum_{i=1}^n \mathbf{1}(N_i \in \Gamma_n(\mu))} - \frac{r(g, \Gamma_n(\mu))}{r(0, \Gamma_n(\mu))}\right]K_N(d\mu)$$
$$+ \sqrt{n}\int H(\mu)\left[\frac{E[e^{-\{N(g)-\mu(g)\}}\mathbf{1}(N \in \Gamma_n(\mu))]}{P\{N \in \Gamma_n(\mu)\}} - L(\mu, g)\right]K_N(d\mu),$$

and the second term, which represents the asymptotic bias of the estimators $\hat{L}(\mu, g)$, converges to zero by (4.14).

Given measures τ and σ on $\mathbf{M}_p$, let

$$h_n(\tau, \sigma) = \int H(\mu)e^{\mu(g)}[\tau(\Gamma_n(\mu))/\sigma(\Gamma_n(\mu))]K_N(d\mu).$$

Then concerning the remaining term in the expression above, we have

$$\sqrt{n}\int H(\mu)e^{\mu(g)}\left[\frac{\sum_{i=1}^n e^{-N_i(g)}\mathbf{1}(N_i \in \Gamma_n(\mu))}{\sum_{i=1}^n \mathbf{1}(N_i \in \Gamma_n(\mu))} - \frac{r(g, \Gamma_n(\mu))}{r(0, \Gamma_n(\mu))}\right]K_N(d\mu)$$

$$
= \sqrt{n}\left[h_n\left((1/n)\sum_{i=1}^n e^{-N_i(g)}\mathbf{1}(N_i \in (\cdot)), (1/n)\sum_{i=1}^n \mathbf{1}(N_i \in (\cdot))\right)\right.
$$
$$
\left. - h_n(r(g,\cdot), r(0,\cdot))\right]
$$
$$
= \sqrt{n}\, h_n'(r(g,\cdot), r(0,\cdot))\left[(1/n)\sum_{i=1}^n (M_i, \tilde{M}_i)\right] + o_P(1), \tag{4.16}
$$

where h_n' is the Fréchet derivative of h_n and the term $o_P(1)$ converges to zero in probability. It is straightforward to verify that for measures η, λ, σ and τ on $\mathbf{M}_p$, the latter three absolutely continuous with respect to the first,

$$
\lim_{n\to\infty} h_n'(\lambda, \eta)[\tau, \sigma] = \int H(\mu)e^{\mu(g)}\frac{d\tau}{d\eta}(\mu)K_N(d\mu)
$$
$$
- \int H(\mu)e^{\mu(g)}\frac{d\lambda}{d\eta}(\mu)\frac{d\sigma}{d\eta}(\mu)K_N(d\mu).
$$

Combining this with (4.16) produces

$$
\sqrt{n}\int H(\mu)[\hat{L}(\mu, g) - L(\mu, g)]K_N(d\mu)
$$
$$
\overset{d}{\to} \int H(\mu)G_1(d\mu) - \int H(\mu)L(\mu, g)G_2(d\mu),
$$

where G is the bivariate Gaussian random measure introduced above. Since H was arbitrary, the proof is complete. ∎

These estimators are applied in Section 7.4 to the problem of combined inference and state estimation for Cox processes on general spaces.

4.5 Inference for Thinned Point Processes

Our setting is now Model 3.3: there are underlying point processes N_i constituting i.i.d. copies of a point process N with unknown law $\mathcal{L}$ and Laplace functional L, but these processes are not observable. Instead, the observations are p-thinnings (Definition 1.38) $N_1', N_2', \ldots$ of the N_i, where $p : E \to [0, 1]$ is an unknown function. Hence the N_i' are i.i.d. copies of a p-thinning N' of N, which has law $\mathcal{L}'$, Laplace functional L', mean measure $\mu'.\ldots$ Principal problems — once more in a full nonparametric model — are to estimate from observation of the N_i' the Laplace functional L of the underlying processes N_i *and* the thinning function p. For reference we recall that [see (1.39)–(1.40)] $L'(f) = L(-\log[1 - p + pe^{-f}])$ which, provided that $p(x) > 0$ for all x, can be inverted as

$$
L(g) = L'(-\log[(p - 1 + e^{-g})]/p]), \tag{4.17}
$$

and that

$$
\mu'(dx) = p(x)\mu(dx), \tag{4.18}
$$

so that estimation of p is a density estimation problem. Indeed, in light of (4.17) and results in Section 2 concerning empirical Laplace functionals, estimation of p is the novel aspect of the section.

Not only at the level of generality of the preceding paragraph, but even in specific cases the model formulated there is not identifiable. (A statistical model with probability measures P_α, $\alpha \in I$ and data $\mathcal{F}$ is *identifiable* if $P_{\alpha_1} \neq P_{\alpha_2}$ on $\mathcal{F}$ whenever $\alpha_1 \neq \alpha_2$ in I. Identifiability is nearly indispensable for meaningful inference.) For the problem at hand the set of candidate laws for the N_i' is $\{\mathcal{L}_{L,p}\}$ with L varying over the set of Laplace functionals of point processes on E and p over the set of measurable functions from E to $[0, 1]$. For the data $\mathcal{F}$ representing the entire sequence (N_i') it is possible that $P_{L_1, p_1} = P_{L_2, p_2}$ on $\mathcal{F}$ even though $L_1 \neq L_2$ and $p_1 \neq p_2$. (See Exercise 4.15; in the extreme case that $p \equiv 0$, nothing about L can be discerned from the N_i'.) One can engender identifiability in a number of ways; we choose the following conditions, which are stipulated for the remainder of the section.

Assumptions 4.17. a) The mean measure μ of the N_i is known;

b) $\mu\{x : p(x) = 0\} = 0$.

Assumption 4.17b) is essentially unavoidable (we actually assume that p is bounded away from zero), but there are alternatives to Assumption 4.17a). We impose this particular choice because it entails only first moment knowledge and because together with (4.18) it reduces estimation of p to a "true" density estimation problem. Our approach, based on histogram estimators, remains "primitive." Finally, if the N_i were observable, then μ could be estimated using techniques in Proposition 4.8.

We now construct estimators of p and L. For the former, as in Section 4, let $\{A_{nj} : n \geq 1, 1 \leq j \leq \ell_n\}$ be a null array of partitions of E; since μ is known we may and do assume that $\mu(A_{nj}) > 0$ for each n and j. Motivated by (4.18) we take

$$\hat{p}(x) = (1/n\mu(A_{nj}))\sum_{i=1}^{n} N_i'(A_{nj}), \quad x \in A_{nj};$$

the motivation is (4.18). Let $\hat{L}'$ be the empirical Laplace functionals associated with the N_i': $\hat{L}'(f) = (1/n)\sum_{i=1}^{n} e^{-N_i'(f)}$. Then on the basis of (4.17) we take

$$\hat{L}(g) = \hat{L}'(-\log[(\hat{p} - 1 + e^{-g})/\hat{p}]), \tag{4.19}$$

as estimator of the Laplace functional L of the N_i.

Results in Section 2 for empirical Laplace functionals allow us to concentrate on the estimators $\hat{p}$. Both they and the $\hat{L}$ are strongly uniformly consistent under rather broad assumptions.

Theorem 4.18. Suppose that Assumptions 4.17 hold and that in addition

i) μ is diffuse;
ii) p is continuous and $p(x) > 0$ for all $x \in E$;
iii) $E[N(E)^4] < \infty$;
iv) There is $\delta < 1$ such that as $n \to \infty$,

$$\ell_n \max_j E[N(A_{nj})^3]/\mu(A_{nj})^3 = O(n^\delta). \qquad (4.20)$$

Then almost surely
a) $\sup_{x \in E} |\hat{p}(x) - p(x)| \to 0$;
b) For each compact subset $\mathcal{K}$ of $C^+(E)$, $\sup_{g \in \mathcal{K}} |\hat{L}(g) - L(g)| \to 0$.

Proof: We prove a) first, and it is then quite easy to deduce b) from a) and Theorem 4.10, although one does need the full strength of the uniform convergence there. Let $p_n(x) = \mu'(A_{nj})/\mu(A_{nj})$; then i) and the martingale convergence theorem imply that $\|p_n - p\|_\infty \to 0$. Thus it suffices to show that with

$$R_n(x) = \sum_{i=1}^n N_i'(A_{nj})/\mu'(A_{nj}), \qquad x \in A_{nj},$$

we have $\|R_n - 1\|_\infty \to 0$ almost surely.
  With $\varepsilon > 0$ fixed, for each n

$$
\begin{aligned}
&P\{\sup_x |R_n(x) - 1| > \varepsilon\} \\
&= P\{\max_j |\sum_{i=1}^n [N_i'(A_{nj}) - \mu'(A_{nj})]| /\mu'(A_{nj}) > n\varepsilon\} \\
&\leq \frac{\ell_n}{\varepsilon^4 n^4} \max_j E\left[(\sum_{i=1}^n [N'(A_{nj}) - \mu'(A_{nj})])^4 \right]/\mu'(A_{nj})^4 \\
&= O\left(\frac{\ell_n}{n^2} \max_j \mathrm{Var}(N'(A_{nj}))^2/\mu'(A_{nj})^4 \right) \\
&= O\left(\frac{\ell_n}{n^2} \max_j E[N'(A_{nj})^3]/\mu'(A_{nj})^3 \right)
\end{aligned}
$$

by the Cauchy-Schwarz inequality and since p is bounded away from zero. From (4.20) and the standard Borel-Cantelli argument we conclude that $\|R_n - 1\|_\infty \to 0$ almost surely.
  To prove b), observe that for each g,

$$
\begin{aligned}
&|\hat{L}(g) - L(g)| \\
&\leq \left| \hat{L}'(-\log[(\hat{p} - 1 + e^{-g})/\hat{p}]) - \hat{L}'(-\log[(p - 1 + e^{-g})]/p) \right| \\
&\quad + \left| \hat{L}'(-\log[(p - 1 + e^{-g})/p]) - L'(-\log[(p - 1 + e^{-g})/p]) \right|.
\end{aligned}
$$

By a), $\hat{p}(x) > 0$ for all x once n is sufficiently large [hence $\hat{L}$ is well defined, which is not obvious from (4.19)]. The set $\mathcal{K}' = \{-\log[(p - 1 + e^{-g})/p] : g \in$

$\mathcal{K}\}$ is a compact subset of $C^+(E)$ by compactness of $\mathcal{K}$, and hence the second term above converges to zero almost surely, uniformly in g, by Theorem 4.9. For each g,

$$\left| \hat{L}'(-\log[(\hat{p}-1+e^{-g})/\hat{p}]) - \hat{L}'(-\log[(p-1+e^{-g})/p]) \right|$$

$$\leq \ [(1/n)\textstyle\sum_{i=1}^n N_i'(E)] \left[\|1/\hat{p} - 1/p\|_\infty + \left\| \log \frac{p-1+e^{-g}}{\hat{p}-1+e^{-g}} \right\|_\infty \right].$$

In this expression the first factor converges to the finite constant $\mu'(E)$ and the second converges to zero, uniformly in $g \in \mathcal{K}$, by a) and the Arzéla-Ascoli theorem. ∎

The estimators $\hat{p}$ may exceed one for small values of n, but evidently Theorem 4.18 remains valid if $\hat{p}$ is replaced by $\min\{\hat{p}, 1\}$. Only the condition (4.20) merits comment: it adjudicates the rate at which $\ell_n \to \infty$. A faster rate is better for convergence of the bias term $\|p_n - p\|_\infty$ to zero, but a slower rate is necessary in order that the numerator of $\hat{p}$ be effective as an estimator of $\mu'(A_{nj})$, in other words that the variance converge to zero; as usual, these requirements conflict. The condition can be relaxed, of course, if one settles for weak uniform consistency.

We now consider central limit aspects.

Theorem 4.19. Assume that
i) $|p(x) - p(y)| \leq C d(x, y)$ for some constant C;
ii) As $n \to \infty$,

$$\sqrt{n} \, \max_j \, \sqrt{\mu(A_{nj})} \, \text{diam} \, A_{nj} \ \to \ 0 \tag{4.21}$$

$$\sqrt{n} \, \min_j \, \mu(A_{nj}) \ \to \ \infty; \tag{4.22}$$

iii) As $A \downarrow \emptyset$,

$$\text{Var}(N(A))/\mu(A) \to 1. \tag{4.23}$$

Let $x \in E$ be fixed and let $j = j(n, x)$ satisfy $x \in A_{nj}$. Then

$$\sqrt{n\mu(A_{nj})}[\hat{p}(x) - p(x)] \xrightarrow{d} N(0, p(x)). \tag{4.24}$$

Proof: In the decomposition

$$\sqrt{n\mu(A_{nj})}[\hat{p}(x) - p(x)] \ = \ (1/\sqrt{n\mu(A_{nj})})\textstyle\sum_{i=1}^n [N_i'(A_{nj}) - \mu'(A_{nj})]$$

$$+ \sqrt{n\mu(A_{nj})}[\mu'(A_{nj})/\mu(A_{nj}) - p(x)]$$

the second term converges to zero by i) and (4.21); see the comments following the proof. Since iii) implies that $\text{Var}(N'(A_{nj}))/\mu(A_{nj}) \to p(x)$, it suffices to show that the central limit theorem holds for the triangular array of random variables

$$X_{ni} = \frac{N_i'(A_{nj}) - \mu'(A_{nj})}{\sqrt{n \ \text{Var}(N'(A_{nj}))}}, \qquad i \leq n.$$

For each n, $E[X_{ni}] = 0$ for all i, $\sum_{i=1}^{n} \text{Var}(X_{ni}) = 1$ and $P\{|X_{ni}| > \epsilon\} = O(n\epsilon^2)$ uniformly in i, so it remains only to verify the Lindeberg condition. For $\eta > 0$,

$$\begin{aligned}
\sum_{i=1}^{n} E[X_{ni}^2 \mathbf{1}(|X_{ni}| > \eta)] &= nE[X_{n1}^2 \mathbf{1}(|X_{n1}| > \eta)] \\
&= E[Y_{n1}^2 \mathbf{1}(|Y_{n1}| > \sqrt{\eta})],
\end{aligned}$$

where $Y_{n1} = \sqrt{\text{Var}(N_1'(A_{nj}))}[N_1'(A_{nj}) - \mu'(A_{nj})]$, and this last quantity evidently converges to zero. ∎

Under a condition analogous to (4.23), the limits (4.24) are independent for distinct values of x.

Proposition 4.20. Suppose that in addition to the hypotheses of Theorem 4.19, for disjoint sets $A \downarrow \emptyset$ and $B \downarrow \emptyset$,

$$\text{Cov}(N(A), N(B)) = o\left(\sqrt{\mu(A)\mu(B)}\right). \qquad (4.25)$$

Then for $x \neq y$ the limit random variables in (4.24) associated with x and y are independent. ☐

Nevertheless, a limit distribution for $\|\hat{p} - p\|_\infty$, suitably normalized, exists.

Theorem 4.21. Assume that the hypotheses of Theorem 4.19 and Proposition 4.20 hold, and let $b(n) = \sqrt{2\log n - \log\log n - \log 2\pi}$. Then for each t,

$$\lim_{n \to \infty} P\left\{ b(\ell_n) \left[\sup_x \sqrt{n\mu(A_{nj})} \, |\hat{p}(x) - p(x)| - b(\ell_n) \right] \leq t \right\}$$

$$= \exp[-\textstyle\int_E \sqrt{p(x)} e^{-t/p(x)} \mu(dx)]. \ ☐$$

The proof is given in Karr (1985b).

Our discussion of statistical estimation concludes with an example.

Example 4.22. (Mixed Poisson process). Let N be a mixed Poisson process on E with directing measure $M = Y\mu$, where $E[Y] = 1$, so that μ is the mean measure of N. The thinned processes N_i' are mixed Poisson processes directed by the random measures $M_i'(dx) = Y_i p(x)\mu(dx)$. Although there are nonparametric techniques available for estimation of $\mu'(dx) = p(x)\mu(dx)$ and the distribution of the Y_i that are tailored to mixed Poisson processes (see Section 7.1), if the mixed Poisson assumption were suspect it would be desirable to use the more robust estimators from this section. Concerning the various conditions, we note first that

$$E[N(A)^3] \;=\; E[Y^3]\mu(A)^3 + 3E[Y^2]\mu(A)^2 + \mu(A) \quad (4.26)$$
$$\mathrm{Cov}(N(A), N(B)) \;=\; \mathrm{Var}(Y)\mu(A)p(B) + \mu(A \cap B). \quad (4.27)$$

Hence (4.20) is satisfied by the choices $\ell_n = n^{1/4}$, $\mu(A_{nj}) = \mu(E)n^{-1/4}$, in which case strong uniform consistency obtains. From (4.27), $\mathrm{Var}(N(A)) = \mathrm{Var}(Y)\mu(A)^2 + \mu(A)$ so that (4.23) is satisfied; similarly, for $A \cap B = \emptyset$, $\mathrm{Cov}(N(A), N(B)) = \mathrm{Var}(Y)\mu(A)\mu(B)$ and so (4.25) holds. We may satisfy (4.21) with $\ell_n = n^{2/5}$, $\mu(A_{nj}) = \mu(E)n^{-2/5}$, and diam $A_{nj} \cong n^{-2/5}$, but unfortunately, these choices fail to fulfill (4.20), although they do satisfy an analogous condition for weak uniform consistency. $\square$

The obvious state estimation problem for thinned point processes is this: given observation of a p-thinning N' of an unobservable point process N, with both the law of N and the thinning function p known, reconstruct N. Of course, since N' is observable we only need reconstruct $N - N'$.

A general solution is expressible in terms of the reduced Palm distributions (Definition 1.51) of N.

Theorem 4.23. Let N' be the p-thinning of N, and let $Q(\mu, d\nu)$ denote the reduced Palm distributions of N. Then

$$P\{N - N' \in d\nu | N'\} = \frac{\exp[-\int - \log[1 - p(x)]\nu(dx)]Q(N', d\nu)}{\int \exp[-\int - \log[1 - p(x)]\eta(dx)]Q(N', d\eta)}. \quad (4.28)$$

Proof: In fact, (4.28) is simply Bayes' theorem. The knowledge that $N' = \mu$, implies nothing except that N must have a point at each point of μ, and so by (1.62),

$$P\{N \in \mu + \Gamma | N' = \mu\} \;=\; P\{N - \mu \in \Gamma | N' = \mu\}$$
$$=\; P\{N - \mu \in \Gamma | N - \mu \geq 0\} = Q(\mu, \Gamma),$$

Put differently, the probability that the deleted component $N - N'$ is ν, given that $N' = \mu$, is $Q(\mu, d\nu)$. The conditional independence assumption implies that a component $\nu = \sum_j \varepsilon_{y_j}$ of N is deleted with probability

$$\prod_j [1 - p(y_j)] = e^{-\int [-\log(1-p)]d\nu}.$$

Then, (4.28) follows at once. ∎

Note that the denominator of (4.28) is the Laplace functional of the reduced Palm distribution $Q(N, \cdot)$ and could be estimated using the techniques presented in Section 4.

4.6 Strong Approximation of Poisson Processes

In this section, we use the powerful concept of strong approximation for uniform empirical processes to produce strong approximations of Poisson processes by Wiener and Kiefer processes.

We begin with some background. Suppose that $E = [0, 1]$ and that the X_i are independent and identically uniformly distributed. Introducing the uniform empirical distribution functions $\hat{F}_n(x) = (1/n) \sum_{i=1}^n \mathbf{1}(X_i \leq x)$ and the uniform empirical error processes $E_n(x) = \sqrt{n}[\hat{F}_n(x) - F(x)]$, we have that

$$\{E_n(x) : 0 \leq x \leq 1\} \xrightarrow{d} \{B(x) : 0 \leq x \leq 1\}, \tag{4.29}$$

where B is a Brownian bridge. That is, B has the representation $B(x) = W(x) - xW(1)$, where W is a standard Wiener process, and hence has covariance function

$$R(x, y) = x \wedge y - xy. \tag{4.30}$$

An important tool for analysis of limit properties such as (4.29) is *strong approximation*: a statement of convergence in distribution is replaced by a statement of almost sure convergence accompanied by a rate of convergence (this necessitates a sequence of Brownian bridges B_n). Provided that the rate of convergence is sufficiently rapid, results on path properties of the Brownian bridge (for example, concerning modulus of continuity and oscillation behavior) carry over to the E_n. To pursue these ramifications (among them the invariance principle) would lead us too far afield at this point; rather, we refer the reader to Csörgő and Révész (1981), especially Chapter 5, and the references there. The following result of Komlós *et al.* (1975, 1976) contains the best possible rate of convergence.

Theorem 4.24. There exists a probability space on which are defined i.i.d. uniform random variables X_i and Brownian bridges B_n such that almost surely

$$\sup_{0 \leq x \leq 1} |E_n(x) - B_n(x)| = \mathrm{O}((\log n)/\sqrt{n}), \quad n \to \infty. \;\square$$

We omit the proof; this is a deep and difficult result.

If Brownian bridges can be used to approximate empirical processes, it seems plausible that partial sum processes for i.i.d. random variables can be approximated by Wiener processes, and indeed this is so (see Komlós et al., 1975, 1976). But then conditional uniformity of Poisson processes should allow one to construct strong approximations of them. In the next result we do so.

Theorem 4.25. There exists a probability space on which are defined a homogeneous Poisson process $(N_t)_{t \geq 0}$ with rate 1 and Wiener process W such that almost surely as $t \to \infty$

$$\sup_{0 \leq x \leq 1} \left| \sqrt{t}(N_{xt}/N_t - x) - (1/\sqrt{t})[W(tx) - xW(t)] \right| = O((\log t)/\sqrt{t}).$$
(4.31)

Proof: Let $Z_1, Z_2, \ldots$ be i.i.d. random variables having a Poisson distribution with mean 1; by the strong approximation theorem of Komlós et al. (1975, 1976) for partial sums of i.i.d. random variables (see also Csörgő and Révész, 1981, Theorem 2.6.1), these are definable on a probability space that also supports a Wiener process W in such a way that for $N_n = \sum_{i=1}^n Z_i$,

$$|N_n - n - W(n)| = O(\log n)$$
(4.32)

almost surely. Extend the sequence (N_n) to become a Poisson process (N_t) by conditional uniformity: in each interval $(n - 1, n]$ distribute $Z_n = N_n - N_{n-1}$ points independently and uniformly; these are the points of N in this interval. With n and $k \leq n$ fixed and $u \in (k - 1, k]$,

$$|N_u - u - W(u)| \leq |N_u - N_k| + |u - k| + |W(u) - W(k)| + |N_k - k - W(k)|,$$

and consequently,

$$
\begin{aligned}
\sup_{u \leq n} |N_u - u - W(u)| &\leq \max_{k \leq n} |N_k - k - W(k)| + \max_{k \leq n} Z_k \\
&\quad + \max_{k \leq n} \sup_{x \leq 1} |W(k) - W(k - x)| + 1 \\
&= O(\log n) + O(\log n) + O(\sqrt{\log n}) + O(1)
\end{aligned}
$$

by (4.32), an elementary calculation and (Csörgő and Révész, 1981, Corollary 1.2.3). Therefore, $\sup_{u \leq t} |N_u - u - W(u)| = O(\log t)$, so that within error $N_t(\log t)/\sqrt{t}$, uniformly in x,

$$
\begin{aligned}
\sqrt{t}(N_{xt}/N_t - x) &= (\sqrt{t}/N_t)[xt + W(xt) - xN_t] \\
&= \frac{W(xt) - xW(t)}{\sqrt{t}} + \frac{N_t - t}{\sqrt{t}} \frac{W(xt) - xW(t)]}{t} \frac{t}{N_t}
\end{aligned}
$$

$$+ x \frac{\sqrt{t}}{N_t}[t - N_t + W(t)]$$

$$= \frac{W(xt) - xW(t)}{\sqrt{t}} + \frac{N_t - t}{\sqrt{t}} \frac{W(xt) - xW(t)}{t} \frac{t}{N_t}$$

$$+ O(\sqrt{t} \log t / N_t)$$

$$= (1/\sqrt{t})[W(xt) - xW(t)] + O((\log t)/\sqrt{t}),$$

since $N_t/t \to 1$, $\sqrt{t}[N_t - t] = O(\sqrt{\log \log t})$ and $(1/t)[W(xt) - xW(t)] = O(\sqrt{(\log \log t)/t})$ by the functional law of the iterated logarithm of Strassen (1964). ∎

Note that each process $B_t(x) = (1/\sqrt{t})[W(tx) - xW(t)]$ in Theorem 4.25 is a Brownian bridge; thus for large t the process $\sqrt{t}(N_{xt}/N_t - x)$ is approximately a Brownian bridge; (4.31) provides a bound on the approximation error.

There is further connection between Poisson processes and strong approximation. The earliest strong approximation theorem for uniform empirical processes (Brillinger, 1969) constructed the uniform random variables (a triangular array rather than a single sequence) from a Poisson process with arrival times T_n by letting the X_{ni} have order statistics T_i/T_n, $i = 1, \ldots, n$.

The final result of the chapter is an alternative strong approximation for Poisson processes on $[0, 1]$. The key idea is to invoke conditional uniformity by viewing a Poisson process N^n on $[0, 1]$ with mean measure $n \, dx$ as $N^n = \sum_{i=1}^{S_n} \epsilon_{X_i}$, where the X_i are independent and identically uniformly distributed and $S_n = \sum_{i=1}^{n} Y_i$, with the Y_i i.i.d., independent of the X_i, and Poisson distributed with mean 1. By approximating the uniform empirical error processes $E_k(\cdot)$ by a Kiefer process and the normalized partial sums $(1/\sqrt{n})[S_n - n]$ by a Wiener process, we obtain a strong approximation for the normalized Poisson processes $\{(1/\sqrt{n})[N^n(x) - nx] : 0 \le x \le 1\}$. Not only is the result of intrinsic interest, but also one can deduce from it a variety of consequences, including properties of empirical zero-probability functionals for Poisson processes (see Section 6.1).

A *Kiefer process* $\{K(x, k) : 0 \le x \le 1, k \in \mathbf{N}\}$, constructed from a two-dimensional Wiener process $\tilde{W}(x, y)$ via the transformation

$$K(x, k) = \tilde{W}(x, k) - x\tilde{W}(1, k), \qquad (4.33)$$

is a mean zero Gaussian process with covariance

$$E[K(x_1, k_1)K(x_2, k_2)] = (x_1 x_2 - x_1 \wedge x_2)(k_1 \wedge k_2). \qquad (4.34)$$

For fixed k, $B_k(x) = K(x, k)/\sqrt{k}$ is a Brownian bridge, and the processes $\tilde{B}_k(x) = K(x, k) - K(x, k - 1)$ are independent Brownian bridges.

Theorem 4.26. There exists a probability space on which are defined Poisson processes N^n on $[0,1]$, a Kiefer process K and a Wiener process W independent of K such that as $n \to \infty$

$$\sup_x (1/\sqrt{n})|N^n(x) - nx - K(x,n) - W(n)x| = O((\log n)^2/n^{1/4}) \quad (4.35)$$

almost surely.

Proof: By the Komlós *et al.* (1975, 1976) strong approximation theorem for uniform empirical processes, there is a probability space supporting the sequence (X_i) and a Kiefer process K such that almost surely

$$\sup_x |\sqrt{k} E_k(x) - K(x,k)| = O((\log k)^2). \quad (4.36)$$

By the same authors' strong approximation for partial sums, there is a probability space on which are defined Poisson-distributed summands Y_j and a Wiener process W such that

$$|S_n - n - W(n)| = O(\log n). \quad (4.37)$$

Independence can be effected by construction, in which case $N^n = \sum_{i=1}^{S_n} \epsilon_{X_i}$ is indeed a Poisson process with rate n. Now let $F_k(x) = \sqrt{k} E_k(x) + x$ denote the empirical distribution functions engendered by the X_i. Then, with error bounds uniform in x,

$$
\begin{aligned}
(1/\sqrt{n})(N^n(x) - nx) &= (1/\sqrt{n})(S_n F_{S_n}(x) - nx) \\
&= (1/\sqrt{n})(S_n F_{S_n}(x) - S_n x) - (1/\sqrt{n})[S_n - n]x \\
&\cong \frac{S_n[F_{S_n}(x) - x]}{\sqrt{n}} + \frac{x W(n)}{\sqrt{n}} \\
&= (1/\sqrt{n})\sqrt{S_n} E_{S_n}(x) + (1/\sqrt{n})x W(n) \\
&\cong (1/\sqrt{n})K(x, S_n) + (1/\sqrt{n})x W(n),
\end{aligned}
$$

where the error bound at the third step is $O((\log n)/\sqrt{n})$ by (4.37) and that at the fifth step is $(\log S_n)^2/\sqrt{S_n}$ by (4.36).

But $(\log S_n)^2/\sqrt{S_n} = (\log n)^2/\sqrt{n}$ within the error bound in (4.35), so it remains to analyze the term $(1/\sqrt{n})[K(x,n) - K(x, S_n)]$. We have

$$
\begin{aligned}
&\sup_x |K(x,n) - K(x, S_n)|/\sqrt{n} \\
&\leq \sup_x |\tilde{W}(x,n) - \tilde{W}(x, S_n)|/\sqrt{n} + \sup_x |x\tilde{W}(1,n) - x\tilde{W}(1, S_n)|/\sqrt{n} \\
&\leq 2\sup_x |\tilde{W}(x,n) - \tilde{W}(x, S_n)|/\sqrt{n} \\
&= O((\log n)^2/n^{1/4})
\end{aligned}
$$

by (4.33) and by a result of Csörgö and Révész (1981, Theorem 1.2.1) on the increments of a Wiener process. This confirms (4.35). ∎

EXERCISES

4.1 Prove that if E is a compact metric space, then the space $M_p(E)$ is locally compact.

4.2 Let $N_1, N_2, \ldots$ be i.i.d. point processes with unknown law $\mathcal{L}$.
a) Show that the empirical measures $\hat{\mathcal{L}}$ are the maximum likelihood estimators of $\mathcal{L}$.
b) Show that if H is a bijection of the set of probability measures on M_p into a topological space V, then $H(\hat{\mathcal{L}})$ is the maximum likelihood estimator of $K(\mathcal{L})$.

4.3 Prove Proposition 4.1.

4.4 Verify that (4.6) gives the proper covariance function for the limit process of Theorem 4.10.

4.5 In connection with Example 4.12,
a) Prove (4.7).
b) Prove the central limit theorem stated there for the empirical Laplace functional error processes $\{\sqrt{n}[\hat{L}^*(f) - L(f)] : f \in C^+(E)\}$.
c) Verify the computation of the asymptotic relative efficiency.

4.6 Let $N_1, N_2, \ldots$ be i.i.d. copies of a homogeneous Poisson process N on $[0, 1]$ with unknown rate λ. With $\hat{\lambda} = (1/n) \sum_{i=1}^{n} N_i([0, 1])$, the maximum likelihood estimator of λ, prove that the specialized empirical zero-probability functionals $\hat{z}(A) = e^{-\hat{\lambda}|A|}$, where $|A|$ is the Lebesgue measure of A, are strongly uniformly consistent and asymptotically normal. (Establish the latter for as large a class of sets as possible.)

4.7 Let N be a Poisson process on a compact metric space E with diffuse mean measure μ. Prove that if $f \geq 0$ is a function such that $\{x : f(x) \notin \mathbf{N}\} = 0$, then $N(f)$ has a discrete distribution.

4.8 Let $(D, \mathcal{D}, \rho)$ be a finite measure space. Prove that the function $d(A, B) = \rho(A \triangle B)$ is a pseudometric on $\mathcal{D}$.

4.9 Let N be a point process with mean measure μ and zero-probability functional z. Prove that z is μ-continuous: if $A, A_1, A_2, \ldots$ are sets with $\mu(A_k \triangle A) \to 0$, then $z(A_k) \to z(A)$.

4.10 Let $\hat{z}$ be the empirical zero-probability functional engendered by i.i.d. point processes $N_1, \ldots, N_n$.
a) Show that if the N_i are Poisson, then for any sets A and B, the correlation between $\hat{z}(A)$ and $\hat{z}(B)$ is nonnegative.
b) Construct an example in which $\hat{z}(A)$ and $\hat{z}(B)$ are negatively correlated whenever $A \cap B = \emptyset$.

4.11 Let $\mathcal{C}$ be the class of subintervals of $[0, 1]$.

a) Prove that $\mathcal{C}$ is a Vapnik-chervonenkis class and calculate its VC exponent.

b) Prove that if $\mathcal{H}$ is a VC class in the space E, then for each k, $\mathcal{H}^k = \{A^k : A \in \mathcal{H}\}$ is a VC class in E^k.

c) Prove that the collection of sets $\{\mu : \mu(A) = 0\}$, $A \in \mathcal{C}$, is *not* a VC class in $M_p([0, 1])$. Why does this not violate Theorem 4.13?

4.12 Prove that $[0, 1]$ has finite metric entropy with respect to the Euclidean metric.

4.13 Let $\mathcal{K}$ be the set of Lipschitz functions f on $[0, 1]$ ($|f(x) - f(y)| \le |x - y|$ for each x and y). Prove that $\mathcal{K}$ has finite metric entropy with respect to the uniform metric.

4.14 Let E be a compact metric space. Prove that there exists a subset $\mathcal{K}$ of $C^+(E)$ such that

i) $\mathcal{K}$ has finite metric entropy with respect to the uniform metric.

ii) If N_1 and N_2 are point processes on E with $N_1(f) \stackrel{d}{=} N_2(f)$ for all $f \in \mathcal{K}$, then $N_1 \stackrel{d}{=} N_2$.

4.15 Show that there exist Poisson processes N_1 and N_2 with mean measures $\mu_1 \ne \mu_2$ and functions $p_1 \ne p_2$ such that the p_1-thinning N_1' has the same distribution as the p_2-thinning N_2' of N_2.

4.16 For the mixed Poisson process model of Example 4.22,

a) Establish (4.26).

b) Prove (4.27) and from it derive the formula given for $\text{Var}(N(A))$.

4.17 Let $(N_1, N_1'), (N_2, N_2'), \ldots$ be i.i.d. copies of a pair (N, N') consisting of a Poisson process N on a compact metric space E with unknown, diffuse mean measure μ and a p-thinning N' of N, where $p : E \to [0, 1]$ is an unknown function. Suppose that both N_i and N_i' are observable for each i and that it is desired to estimate the thinning function p with μ treated as a nuisance parameter. Given a null array $\{A_{nj}\}$ partitions of E, reasonable estimators are given by

$$\hat{p}(x) = \sum_{i=1}^{n} N_i'(A_{nj}) / \sum_{i=1}^{n} N_i(A_{nj}), \qquad x \in A_{nj}.$$

Using Theorem 4.18 as a model, develop and verify conditions for strong uniform consistency of these estimators.

4.18 Let X be a random variable with distribution function F and let F^{-1} be the left-continuous inverse of F: $F^{-1}(x) = \inf\{t : F(t) \ge x\}$, $0 < x < 1$.

a) Prove that $X \stackrel{d}{=} F^{-1}(U)$, where U is uniformly distributed on $[0, 1]$.

b) Prove that if F is continuous, then $F(X) \stackrel{d}{=} U$.

4.19 Take as the definition of the Wiener process W on $[0, 1]$ that $x \to W_x$ is continuous almost surely; that W has independent and stationary increments; and that W_x has distribution $N(0, x)$ for each x.

a) Prove that $E[W_x W_y] = x \wedge y$ for each x and y.

b) Prove that the process $B_x = W_x - xW_1$ has covariance function R given by (4.30).

4.20 Let $X_1, \ldots, X_n$ be i.i.d. random variables uniformly distributed on $[0, 1]$, $\hat{F}(x) = (1/n) \sum_{i=1}^{n} 1(X_i \leq x)$ be the empirical distribution function, and let $E_n(x) = \sqrt{n}[\hat{F}(x) - x]$ be the associated uniform empirical error process. Prove that
a) $\hat{F}(\cdot)$ is a strong Markov process.
b) The process $\{E_n(x)/(1 - x)\}$ is a martingale with respect to the filtration generated by $\hat{F}$.

4.21 Let E_n be as in Exercise 4.20. Prove directly that the finite-dimensional distributions of (E_n) converge to those of a Brownian bridge.

4.22 Let N be a Poisson process on $\mathbf{R}$ with rate 1. Prove that the sequence (N_n) defined by $N_n(t) = (1/\sqrt{n})[N_{nt} - nt]$, $0 \leq t \leq 1$, converges in distribution to a Wiener process on $[0, 1]$.

4.23 Let W be a Wiener process on $[0, 1]$.
a) Show that for each a, $P\{\sup_{t \leq 1} |W_t| > a\} = 2P\{|W_1| > a\}$.
b) Discuss how to use a) and Exercise 4.22 to approximate the distribution of $\sup_{t \leq n} |N_t - t|$, where N is a Poisson process with rate 1.

4.24 Let $\{K(x, k) : 0 \leq x \leq 1, k \in \mathbf{N}\}$ be the Kiefer process defined by (4.33), where the two-dimensional Wiener process $\tilde{W}$ is, by definition, a mean zero Gaussian process with $E[\tilde{W}(x_1, y_1)\tilde{W}(x_2, y_2)] = (x_1 \wedge x_2)(y_1 \wedge y_2)$.
a) Prove that K is a mean zero Gaussian process with covariance function (4.34).
b) Prove that for each k, $B_k(x) = K(x, k)/\sqrt{k}$ is a Brownian bridge.
c) Prove that the processes $\tilde{B}_k(x) = K(x, k) - K(x, k - 1)$ are independent Brownian bridges.

NOTES

Section 4.1

None of the material on empirical processes is new, but its systematic application to empirical measures engendered by point processes does seem to be novel. Good expository treatments of empirical processes are Csörgő and Révész (1981), Gaenssler (1984), Gaenssler and Stute (1979), Pollard (1984) and Serfling (1980). Theorem 4.2 is based on ideas applied in Karr (1985b) to empirical Laplace functionals. Uniformity in weak convergence is examined in Billingsley and Tøpsøe (1967), from which Theorem 4.3 is taken, as well as in Gaenssler and Stute (1976) and Tøpsøe (1977). Proposition 4.4, Theorem 4.5, and the idea that combinatorial conditions could substitute for topological hypotheses are all due to Vapnik and Chervonenkis (1971). Theorem 4.5 and the highly similar Theorem 4.10 rest ultimately on Dudley (1974) and Jain and Marcus (1975); Araujo and Giné (1980) is a general presentation of central limit theorems in $C(S)$, where S is a compact metric space. Theorem 4.7 is taken from Gaenssler (1984). Proposition 4.8 is given for measure-valued Markov chains in Karr (1979).

This chapter contains almost no mention of hypothesis tests, which are, of course, are also significant. Suppose that $X_1, X_2, \ldots$ are i.i.d. random variables with unknown distribution function F and let $\hat{F}(x) = \frac{1}{n} \sum_{i=1}^{n} 1(X_i \leq x)$ be the sequence of empirical distribution functions. Tests of the null hypothesis $H_0 : F = F_0$ can be based on the *Kolmogorov-Smirnov statistic* $D_n = \sqrt{n}\|\hat{F} - F_0\|_\infty$, for which universal tail bounds and an exact asymptotic distribution theory under H_0 are known; on the *Cramér-von Mises statistic*

$$W_n^2 = n \int_{-\infty}^{\infty} [\hat{F}(x) - F_0(x)]^2 F_0(dx);$$

and on the *Anderson-Darling statistic*

$$A_n^2 = n \int_{-\infty}^{\infty} \frac{[\hat{F}(x) - F_0(x)]^2}{F_0(x)[1 - F_0(x)]} F_0(dx),$$

which can be extended to permit other weight functions. For details, including extension to composite hypotheses (see Durbin, 1973a, 1973b; or Serfling, 1980).

Section 4.2

The principal results, Theorems 4.9 and 4.10, are due to Karr (1985b), where they were applied in the context of thinned point processes. Dudley (1973, 1974, 1978) and Gaenssler (1984) analyze the role of metric entropy conditions in proving central limit theorems for i.i.d. random elements of "large" metric spaces such as function spaces.

The metric entropy of a compact subset A of a metric space (V, d) is a measure of the massiveness of A; a set with finite metric entropy is "nonmassive" in a particular manner. For each $\varepsilon > 0$, let $n(A, \varepsilon)$ be the smallest integer m for which there exist open sets $O_1, \ldots, O_m$ with diam $O_j < \varepsilon$ for each j, such that $A \subseteq \bigcup_{j=1}^{n} O_j$; roughly speaking, $n(A, \varepsilon)$ is the number of balls of radius ε needed to cover A. If

$$\int_0^1 \sqrt{\log n(A, \varepsilon^2)} \, d\varepsilon < \infty,$$

then A is said to have *finite metric entropy* with respect to d. Compactness of A implies that $n(A, \varepsilon^2) < \infty$. for each ε and evidently $n(A, \varepsilon^2)$ increases as ε decreases, and so only the lower limit of integration matters. The role of finite metric entropy in establishing central limit theorems is to ensure continuity of the Gaussian limit process G and then, by approximation, to provide tightness of the processes error processes such as $\sqrt{n}[\hat{L} - L]$. Convergence of finite-dimensional distributions is elementary, and convergence in distribution then ensues.

Metric entropy of sets of continuous functions is treated in Kolmogorov and Tihomirov (1961) and Lorentz (1966). For example, for $E = [0, 1]^d$ the set of functions with order of smoothness $p + \alpha$ (derivatives of orders $1, \ldots, p$ exist and are uniformly bounded, and those of order p are Lipschitz with exponent α) has finite metric entropy provided that $p + \alpha > d/2$. In particular, given $\alpha > 1/2$, the set of functions f on $[0, 1]$ with $\|f\|_\infty < 1$ and $|f(x) - f(y)| < |x - y|^\alpha$ for all x, y

has finite metric entropy. Less is known for a general compact space E, but it is always possible (see Exercise 4.14) to construct a compact set $\mathcal{K} \subseteq C^+(E)$ such that $\mathcal{K}$ has finite metric entropy and such that point processes N_1 and N_2 on E for which $N_1(f) \overset{d}{=} N_2(f)$ for all $f \in \mathcal{K}$ are identically distributed.

Section 4.3

Theorems 4.13 and 4.14 appear here for the first time. General discussion of the Brownian bridge, which also appears in Section 6, is given in Karlin and Taylor (1975) and at a more advanced level in Csörgő and Révész (1981), where there is extensive development of sample path properties not only of the Brownian bridge and the Wiener process but also of the two-dimensional Wiener process and the Kiefer process (see Section 6).

Section 4.4

The material here comes from Karr (1985a). See Kallenberg (1983), Karr (1988), Section 2.6, and Section 7.2 for discussion of reduced Palm distributions in conditioning for point processes, and Grenander (1981) and Tapia and Thompson (1978) concerning histogram density estimators, to which our estimators are analogous.

Section 4.5

The principal source is Karr (1985b), which contains Theorems 4.18, 4.19, and 4.21. New here is the general state estimation result for thinned point processes, Theorem 4.23; it illustrates once more the importance of reduced Palm distributions in conditioning. Underlying it is a very general Bayes' theorem of Kallianpur and Striebel (1968) (see also Liptser and Shiryaev, 1978). Brillinger (1979) analyzes related problems, while in Section 6.3 inference for thinned Poisson processes is investigated by specialized methods.

Section 4.6

As part of its systematic presentation of strong approximation — especially for partial sum processes and uniform empirical error processes — Csörgő and Révész (1981) contains Theorem 4.24 as well as the associated approximation for partial sums by a Wiener process; the definitive forms of these results are due to Komlós *et al.* (1975, 1976). It is interesting in terms of point processes to note that in an early version of Theorem 4.24 (Brillinger, 1969) uniformly distributed random variables are constructed from a Poisson process by means of conditional uniformity. Theorems 4.25 and 4.26 are new. Pyke (1968) contains an "in distribution" analogue of one of the consequences of the latter. Basic properties of the Kiefer process, first employed by Kiefer (1972), can be found in Csörgő and Révész (1981).

Exercises

4.5 Specialized empirical Laplace and zero-probability functionals for Poisson processes on general spaces are developed in Section 6.1.

4.6 See also Section 6.1.

4.11 See Gaenssler (1984) for this and related examples.

4.16 The mixed Poisson model is treated in detail in Karr (1984b).

4.17 See Karr (1985b).

4.18 The quantile transformation is a standard and important tool for reducing assertions concerning general empirical processes to corresponding properties of uniform empirical processes (see, e.g., Csörgő and Révész, 1981).

4.20 See Durbin (1973a) for these and related properties.

4.21 Convergence of finite-dimensional distributions was shown in Doob (1949), with weak convergence in $C[0, 1]$ established in Donsker (1952).

4.22 A martingale proof is contained in Exercise 5.9.

4.23 See Pyke (1959) for an early analysis.

4.27 These and deeper properties and applications of the Kiefer process are given in Csörgő and Révész (1981).

5
Martingale Inference for
Point Processes: General Theory

In the same way that Chapter 4 is associated with Chapter 1, this chapter is linked with Chapter 2. Our topic is the "martingale method" of inference for point processes on $\mathbf{R}_+$ that admit stochastic intensities; except for the nature of the dependence of the stochastic intensity on unknown parameters, we impose few structural restrictions, so that the level of generality is more comparable to Chapter 4 than to Chapters 6–9. As intimated in Chapters 2 and 3, the martingale method can be interpreted as a conditional form of the method of moments, in which martingales are construed as noise processes. Thus given an unobservable process, an observable process differing from it by a martingale is a martingale estimator. Although this may seem more a problem of state estimation than statistical inference, in models for which good results are available, the unobservable process being estimated is comprised mainly or entirely of an unknown function and is only minimally random. Moreover, in the state estimation context it is not the noise interpretation of a martingale that prevails but rather the innovation interpretation.

The principal advantage of the martingale method is generality: it provides coherent rules for construction of estimators that apply when other methods fail, effective, systematic methods for calculation of means, variances, and covariances (perhaps its main contribution to applications), and potent techniques for establishing asymptotic normality of estimators.

One consequence of our assumptions concerning stochastic intensities is that the family of candidate probability laws is dominated (by the law of a Poisson process), so that legitimate likelihood functions and ratios are available. Typically, however, the index set is so large that these likelihood functions are unbounded, rendering conventional maximum likelihood estimation impossible. In such situations the martingale method is appealing because it remains effective notwithstanding. As an alternative, one can construct consistent sequences of restricted maximum likelihood estimators, as we do in Theorem 5.18.

Similarly, likelihood ratio tests can be constructed, but small-sample distribution theory is difficult. By contrast, the martingale method yields test statistics as easily as it yields martingale estimators, and asymptotic properties can be derived readily, via martingale central limit theorems. Indeed, perhaps the most compelling argument in support of the martingale method is the ease with which estimators and test statistics can be calculated and analyzed.

Organization of the sections is as follows. Section 1 formulates the fundamental statistical model for point processes with stochastic intensities, outlines the basis of the martingale method of estimation, and introduces two important examples. Large-sample properties of martingale estimators — mean square consistency and asymptotic normality — are described in Section 2, which begins with two very general limit theorems for sequences of martingales. Thereafter, asymptotic behavior of martingale estimators is examined in a general setting. Section 3 is a detailed treatment of the multiplicative intensity model. We consider consistency and asymptotic normality of martingale and sieve maximum likelihood estimators, as well as hypothesis testing. Section 4 treats state estimation when the observations constitute a point process on $\mathbf{R}_+$ admitting a stochastic intensity. An unobservable or partially observable state process is to be reconstructed with minimal error from these observations. The central issue is recursive computation of MMSE state estimators; our principal tool is the martingale representation from Theorem 2.34. Section 5 is devoted to the Cox regression model and several extensions. This model has been a focus and inspiration of recent research.

5.1 Statistical Models Based on Stochastic Intensities

We consider marked point processes on $[0, 1]$ with finite mark space $E = \{1, \ldots, K\}$. Let $(\Omega, \mathcal{G}, P)$ be a probability space on which are defined

- A marked point process $N = \sum_n \varepsilon_{(T_n, X_n)}$ with mark space E

- A filtration $\mathcal{H}$ satisfying Assumptions 2.10a) and b), with $\mathcal{H}_1 \subseteq \mathcal{G}$

- A K-variate, positive, $\mathcal{H}$-predictable process $\lambda = (\lambda_t(1), \ldots, \lambda_t(K))$, assumed *bounded*.

As in Chapter 2, put $N_t(k) = \sum_n \mathbf{1}(T_n \leq t, X_n = k)$, $k = 1, \ldots, K$, and let $N_t = \sum_k N_t(k)$. We further stipulate that for each k the $(P, \mathcal{H})$-stochastic intensity of $N(k)$ is $\lambda(k)$. The boundedness assumption on λ remains in force throughout Sections 1–3.

Since in the statistical model to be formulated momentarily different mechanisms governing the random experiment correspond to different probabilities on $(\Omega, \mathcal{G})$, it is necessary to indicate dependence of the stochastic intensity on the probability as well as the filtration $\mathcal{H}$; whenever the latter is clearly understood, dependence on it is suppressed. There follow, we recall, the properties:

1. For each k, $M_t(k) = N_t(k) - \int_0^t \lambda_s(k)ds$ is a square integrable $(P, \mathcal{H})$-martingale;

2. The predictable variation of $M(k)$ is $\langle M(k)\rangle_t = \int_0^t \lambda_s(k)ds$;

3. For $k \neq j$ the martingales $M(k)$ and $M(j)$ are orthogonal.

The statistical model incorporates an arbitrary index set I, and with respect to the probability P_α, N has $\mathcal{H}$-stochastic intensity constructed by multiplying the observable "baseline" stochastic intensity λ by another predictable process $H(\alpha)$, which in important applications is deterministic, but depends on the unknown index α.

Model 5.1. (Stochastic intensity model). Let $(\Omega, \mathcal{G}, P)$, N, $\mathcal{H}$ and λ satisfy the conditions above. Let I be an arbitrary index set and for each $\alpha \in I$, let $H(\alpha)$ be a K-variate, nonnegative, $\mathcal{H}$-predictable process. The *stochastic intensity model* with *baseline intensity* λ is given as follows.

- The sample space is $(\Omega, \mathcal{G})$;

- For each $\alpha \in I$ let P_α be a probability on $(\Omega, \mathcal{G})$ such that $N(k)$ has $(P_\alpha, \mathcal{H})$-stochastic intensity

$$\lambda_t(\alpha, k) = H_t(\alpha, k)\lambda_t(k). \qquad (5.1)$$

The set $\mathcal{P}$ of candidate laws consists of those P_α for which

$$E_\alpha\left[\sum_{k=1}^K \int_0^1 H_t(\alpha, k)\lambda_t(k)dt\right] < \infty; \qquad (5.2)$$

- The observations are the filtration $\mathcal{F}_t = \mathcal{F}_t^N \vee \mathcal{F}_t^\lambda$ corresponding to complete observation of the marked point process N *and* the baseline stochastic intensity λ over $[0, t]$. $\square$

The general goal, obviously, is to make inferences concerning the unknown index α from the given observations.

Using (5.1), we see that α influences N only by multiplying the baseline intensity by the α-dependent factor $H(\alpha)$. As the examples below illustrate, the model is very general, but reasonably precise results obtain only when $H(\alpha)$ is not very — or not at all — random. However, before presenting examples we demonstrate that Model 5.1 can always be realized in such a manner that $\mathcal{P} = \{P_\alpha\}$ is a dominated family.

Theorem 5.2. Let $(\Omega, \mathcal{G}, P)$, N, $\mathcal{H}$ and λ be as above and suppose that for each $\alpha \in I$, $H(\alpha)$ is a K-variate, nonnegative, $\mathcal{H}$-predictable process satisfying

$$\sum_{k=1}^K \int_0^1 H_t(\alpha, k)\lambda_t(k)dt < \infty \qquad (5.3)$$

almost surely with respect to P. Then for each α, with respect to the probability P_α defined by

$$\frac{dP_\alpha}{dP} = \exp\left[\sum_{k=1}^K \{\int_0^1 \lambda(k)[1 - H(\alpha, k)]ds + \int_0^1 \log H(\alpha, k)dN(k)\}\right], \quad (5.4)$$

N has $\mathcal{H}$-stochastic intensity $\lambda_t(\alpha) = H_t(\alpha)\lambda_t$.

Proof: Commencing with (5.4) we often omit variables of integration. Moreover, dN-integrals always include the right-hand endpoint of the interval of integration. If the process $Y_t(\alpha)$ defined by (5.4) is a martingale over $(\Omega, \mathcal{H}_t, P)$, then by imitating the proof of Theorem 2.31 we can conclude that under P_α, N has a stochastic intensity $\lambda(\alpha)$. Hence we need to verify that for $s < t$,

$$E\left[\exp\left\{\sum_1^K \left[\int_s^t \lambda(k)[1 - H(\alpha, k)]du + \int_s^t \log H(\alpha, k)dN(k)\right]\right\} \middle| \mathcal{H}_s\right] = 1,$$

which we do in heuristic but illuminating fashion. By iteration properties of conditional expectations it suffices to establish this for values of $t - s$ small enough that $N_t(k) - N_s(k)$ is with high probability zero or 1 for each k. In this case

$$E\left[\exp\left\{\sum_1^K \int_s^t \lambda(k)[1 - H(\alpha, k)]du + \sum_1^K \int_s^t \log H(\alpha, k)dN(k)\right\} \middle| \mathcal{H}_s\right]$$

$$\cong E\left[\exp\left\{\sum_1^K \lambda_s(k)[1 - H(\alpha, k)](t - s) + \sum_1^K \log H(\alpha, k)\Delta N_s(k)\right\} \middle| \mathcal{H}_s\right]$$

$$\cong \; E\left[\left\{1 - (t-s)\sum_1^K \lambda_s(k)[1 - H_s(\alpha,k)]\right\}\prod_1^K H_s(\alpha,k)^{\Delta N_s(k)}\Big|\mathcal{H}_s\right]$$

$$= \; E\left[\left\{1 - (t-s)\sum_1^K \lambda_s(k)[1 - H_s(\alpha,k)]\right\}\right.$$
$$\left.\times \; \prod_1^K[1 - \Delta N_s(k) + H_s(\alpha,k)\Delta N_s(k)]\Big|\mathcal{H}_s\right]$$

$$\cong \; E\left[\left\{1 - (t-s)\sum_1^K \lambda_s(k)[1 - H_s(\alpha,k)]\right\} - \sum_1^K[1 - H_s(\alpha,k)]\Delta N_s(k)\Big|\mathcal{H}_s\right]$$

$$= \; 1 + \sum_1^K[1 - H_s(\alpha,k)]E[(t-s)\lambda_s(k) - \Delta N_s(k)|\mathcal{H}_s],$$

which equals 1 by predictability of H and since $N_t(k) - \int_0^t \lambda(k)ds$ is a $(P,\mathcal{H})$-martingale for each k. ∎

For likelihood-based inference, (5.4) is the starting point. Since $\mathcal{P}$ is a dominated family, we may employ the *log-likelihood function*

$$L(\alpha) = \sum_{k=1}^K \left\{\int_0^1 \lambda(k)[1 - H(\alpha,k)]ds + \int_0^1 \log H(\alpha,k)dN(k)\right\},$$

either to be maximized in order to estimate α or to form likelihood ratio statistics for hypothesis tests. However, at least two difficulties must be overcome. First, without further restrictions, L is not expressed solely in terms of the data (N,λ) and the unknown index α, because of the presence of the unobservable processes $H(\alpha,k)$. This difficulty vanishes if $\mathcal{H} = \mathcal{F}^N$ [at least in principle, although explicit computation of $H(\alpha,k)$ from the observations may not be feasible] but also in the more important case that either $H(\alpha)$ is deterministic, as in the multiplicative intensity model, or that the past is observable, as in Section 5.

A more ingrained problem is that except in rather degenerate cases the likelihood function is unbounded above; even so, it is possible, using the method of sieves, to construct consistent estimators by maximum likelihood techniques (see Section 3).

We now explain the rationale behind the martingale method. There are two interrelated aspects: *what* to estimate and *how*. The "how" is actually the more fundamental in that it comprises the general principle of the martingale method: *choose estimators for which the estimation error process is a martingale*. This, in turn, dictates what can be estimated. Theorem 2.34 identifies as fundamental $(P_\alpha, \mathcal{F}^N)$-martingales the innovations $M_t(\alpha,k) = N_t(k) - \int_0^t \lambda_s(\alpha,k)ds$, in terms of which every $(P_\alpha, \mathcal{F}^N)$-martingale M^* is expressible as the integral of a predictable process:

$$M_t^* = M_0^* + \sum_{k=1}^K \int_0^t Y_s(k)dM_s(\alpha,k), \tag{5.5}$$

where the $Y(k)$ are predictable. Therefore, (5.5) defines the class of allowable error processes. The integral there is the error associated with estimation of the K-variate process $\int_0^t Y(k)H(\alpha, k)\lambda(k)ds$ by the stochastic integrals $\int_0^t Y(k)dN(k)$.

This brings us nearly to the class of processes to be estimated. Since the baseline intensity is observable there is no sense in its being part of what is estimated, but however, one must account for its possibly being equal to zero. We introduce the following very convenient notation: for $x \geq 0$,

$$x^{-1} = (1/x)\mathbf{1}(x > 0).$$

All of this leads to the following formalization.

Definition 5.3. Let Y be a predictable process with

$$\sum_{k=1}^K \int_0^1 |Y_s(k)|H_s(\alpha, k)\lambda_s(k)ds < \infty \tag{5.6}$$

almost surely with respect to each P_α. The *martingale estimator* of the K-variate process

$$B_t(\alpha, k) = \int_0^t Y_s(k)H_s(\alpha, k)\mathbf{1}(\lambda_s(k) > 0)ds \tag{5.7}$$

is the process

$$\hat{B}_t = \int_0^t Y_s(k)\lambda_s(k)^{-1}dN_s(k). \tag{5.8}$$

There is also a differential interpretation: for each t, $Y_t(k)H_t(\alpha, k)\lambda_t(k)dt$ is equal to $Y_t(k)dN_t(k)$ up to the "martingale noise" $dM_t(k)$, which has conditional mean zero given past observations.

Note that whereas the estimated process B is continuous in t, the martingale estimator $\hat{B}$ is purely discontinuous; in some circumstances this may be undesirable.

Confirmation that the error processes are martingales is provided by the following result, along with additional properties.

Proposition 5.4. Let Y be a bounded, predictable process and let B and $\hat{B}$ be given by (5.7)–(5.8). Assume that there is $\delta > 0$ such that for each k and t, $\lambda_t(k) > \delta$ whenever $\lambda_t(k) > 0$. Then under each probability P_α the processes $\hat{B}(k) - B(\alpha, k)$ are orthogonal mean zero square integrable martingales with

$$\langle \hat{B}(k) - B(\alpha, k)\rangle_t = \int_0^t Y_s(k)^2 \lambda_s(k)^{-1}H_s(\alpha, k)ds. \tag{5.9}$$

Proof: By (5.5), $\hat{B}_t(k) - B_t(\alpha, k) = \int_0^t Y(k)\lambda(k)^{-1}dM(\alpha, k)$. The integrand is bounded and predictable, and $M(\alpha, k)$ is a square integrable martingale; consequently, the stochastic integral is a martingale (Theorem 2.9) and the boundedness assumptions ensure that it is square integrable. By Theorem B.12 and the property that $d\langle M(\alpha, k)\rangle_t = H_t(\alpha, k)\lambda_t(k)dt$ (Proposition 2.32),

$$
\begin{aligned}
\langle \hat{B}(k) - B(\alpha, k)\rangle_t &= \int_0^t [Y(k)^2/\lambda(k)]\lambda(k)^{-1}d\langle M(\alpha, k)\rangle \\
&= \int_0^t Y(k)^2 \lambda(k)^{-1} H(\alpha, k)ds. \;\blacksquare
\end{aligned}
$$

We note a simple but extremely useful by-product: means and variances of martingale estimators are straightforward to calculate — and hence also to estimate. This is arguably the most important strength of the martingale method in applications.

Proposition 5.5. For each α, t and k, $E_\alpha[\hat{B}_t(k)] = E_\alpha[B_t(\alpha, k)]$ and

$$
E_\alpha[(\hat{B}_t(k) - B_t(\alpha, k))^2] = \int_0^t E_\alpha[Y_s(k)^2 \lambda_s(k)^{-1} H_s(\alpha, k)]ds. \tag{5.10}
$$

For $k \neq j$ and all t,

$$
E_\alpha[\{\hat{B}_t(k) - B_t(\alpha, k)\}\{\hat{B}_t(j) - B_t(\alpha, j)\}] = 0. \;\square \tag{5.11}
$$

The mean squared error (effectively a variance) in (5.10) is needed in order to apply central limit theorems for martingale estimators, and therefore must be estimated. A beauty of the martingale method is that this is straightforward: the process

$$
C_t(\alpha, k) = \int_0^t Y_s(k)^2 \lambda_s(k)^{-1} H_s(\alpha, k)ds
$$

is of the form (5.7) for $\tilde{Y}_s(k) = Y_s(k)^2/\lambda_s(k)$, and hence has martingale estimator

$$
\hat{C}_t(k) = \int_0^t [Y_s(k)^2/\lambda_s(k)]\lambda_s(k)^{-1}dN_s(k),
$$

which estimates the mean squared error.

In a strict sense, Model 5.1 is completely general: by taking $\lambda \equiv 1$ and I to be the set of all predictable processes H satisfying (5.2), the point process can have an arbitrary stochastic intensity, but of course at this level of generality one can deduce only comparably uninformative conclusions. The interesting and useful cases, as intimated above, are those that the α-dependent factor H is primarily deterministic. We now introduce this case, then discuss two specific instances.

Model 5.6. (Multiplicative intensity model). Let I be a subset of the set of functions $\alpha : [0, 1] \to \mathbf{R}_+^K$, each component of which is left-continuous with right-hand limits and satisfies $\int_0^1 \alpha_s(k)ds < \infty$. The *multiplicative* *intensity model* is Model 5.1 with $H_t(\alpha, k) = \alpha_t(k)$ for each k, identically in ω, so that under P_α, $N(k)$ has stochastic intensity $\lambda_t(\alpha, k) = \alpha_t(k)\lambda_t(k)$. $\square$

The goal is to estimate the unknown function $\alpha = (\alpha(1), \ldots, \alpha(K))$. The martingale method — this is potentially a shortcoming — estimates integrals of the $\alpha(k)$ instead. Since N and λ are observable, the process

$$B_t(\alpha, k) = \int_0^t \alpha_s(k)\mathbf{1}(\lambda_s(k) > 0)ds$$

has martingale estimator

$$\hat{B}_t(k) = \int_0^t \lambda_s(k)^{-1}dN_s(k). \tag{5.12}$$

If $\lambda(k) > 0$ for all k almost surely with respect to P (and hence with respect to each P_α), the martingale estimator of $B_t(\alpha, k) = \int_0^t \alpha(k)ds$ is simply $\hat{B}_t(k) = \int_0^t [1/\lambda(k)]dN(k)$. Note that these estimators are very easy to calculate, a not insignificant advantage of the martingale method.

The multiplicative intensity model is the broadest setting in which there exists a good asymptotic theory given i.i.d. copies of an underlying process; this theory is described in Section 3.

Perhaps it is not evident even how to formulate limit theorems; roughly, they arise as stochastic intensities converge to infinity, with scaling effected by the Y-processes in (5.7)–(5.8). For i.i.d. processes one achieves this by a sufficiency reduction in Proposition 5.14 that permits attention to be restricted to superpositions. The i.i.d. assumption is used to verify the hypotheses of general limit theorems that, as explained in Section 2, hold more broadly.

Two specific instances of the multiplicative intensity model are pursued at some length in the text and exercises. The first yields in a special case an estimator identical to the empirical distribution function.

Example 5.7. (i.i.d. random variables). Let $X_1, \ldots, X_n$ be i.i.d. random variables with values in $[0, 1]$ and distribution function F admitting hazard function $h = F'/(1 - F)$. Let $\alpha = h$, take $K = 1$, and let P be such that the X_i are uniformly distributed on $[0, 1]$. With respect to P_h the point process $N = \sum_{i=1}^n \varepsilon_{X_i}$ has $\mathcal{F}^N$-stochastic intensity $\lambda_t(h) = h(t-)(n - N_{t-})$ (see Example 3.17). As estimator of the integrated hazard function $B_t(h) = \int_0^t h(s)\mathbf{1}(N_s < n)ds$, the martingale method gives

$$\hat{B}_t = \int_0^t (n - N_{s-})^{-1}dN_s = \sum_{j=1}^{N_t} 1/(n - j + 1). \tag{5.13}$$

Typically, though, it is more germane to estimate the distribution function $F(t) = 1 - e^{-\int_0^t h\,ds}$ rather than h. One way to accomplish this is the following. Let $\lambda_t = n - N_{t-}$ be the baseline stochastic intensity. Then if one treats $\hat{B}_t$ as estimator of $\int_0^t h\,ds$ (a good approximation for large n) and substitutes into (5.13) there results $\hat{F}(t) = 1 - e^{-\int_0^t [1/\lambda]dN}$, which formally satisfies the differential equation

$$d\hat{F}(t) = \frac{1 - \hat{F}(t)}{\lambda_t}dN_t. \tag{5.14}$$

The exact solution of (5.14), the Nelson-Aalen or *product limit estimator*

$$\hat{F}(t) = 1 - \prod_{s \leq t}[1 - \Delta N_s/(n - N_{s-})], \tag{5.15}$$

in this case reduces to the empirical distribution function.

One can also give a direct argument for (5.15). In survival analysis, the X_i represent lifetimes, for example of individuals receiving a particular medical treatment. Just before time s, $n - N_{s-}$ individuals remain alive (they are said to be still *at risk*), so $1 - \Delta N_s/(n - N_{s-})$ is an estimator of the probability of an individual's surviving past s given survival until s. Multiplication of these estimated conditional probabilities over $s \leq t$ and subtraction from 1 yield (5.15).

The product limit estimator, unlike the empirical distribution function, accommodates *censored survival data* (see Exercise 2.23). Suppose that, in addition to the lifetimes X_i, there are i.i.d. random variables $Y_1, \ldots, Y_n$, the censoring times, with (possibly unknown) distribution function G, such that $(X_1, \ldots, X_n)$ and $(Y_1, \ldots, Y_n)$ are independent. For each j one observes not X_j but rather $Z_j = \min\{X_j, Y_j\}$, and $D_j = \mathbf{1}(X_j \leq Y_j)$, so that it is known whether Z_j is a survival time $(D_j = 1)$ or a censoring time $(D_j = 0)$. The point process $N_t = \sum_{j=1}^n \mathbf{1}(Z_j \leq t, D_j = 1)$ of recorded deaths has stochastic intensity $\lambda_t(h) = h(t-)\lambda_t$, where now $\lambda_t = \sum_{j=1}^n \mathbf{1}(Z_j \geq t)$ is the number of individuals still at risk (neither dead nor censored) at time t. One may then repeat the steps above to obtain the product limit estimator

$$\hat{F}(t) = 1 - \prod_{s \leq t}[1 - \Delta N_s/\lambda_s], \tag{5.16}$$

which is no longer an empirical distribution function.

Thus not only can the martingale method accommodate censoring but also its estimator reduces to the classical estimator in the absence of censoring. For this model asymptotics arise as $n \to \infty$ and are discussed in Example 5.17. $\square$

The second special case is the Markov process model of Example 2.33.

Example 5.8. (Independent Markov processes). This model is introduced in Example 2.33. Let $(X_t(1)), \ldots, (X_t(K))$ be independent Markov processes with finite state space S and unknown generator A, let the mark space be $E = \{(i,j): i,j \in S, i \neq j\}$, and let $\mathcal{H}_t = \bigvee_{\ell=1}^{K} \sigma(X_s(\ell): 0 \leq s \leq t)$. For each $(i,j) \in E$ let $N_t(i,j) = \sum_{\ell=1}^{K} \sum_{s \leq t} \mathbf{1}(X_{s-}(\ell) = i, X_s(\ell) = j)$ be the number of observed $i \to j$ transitions among the n processes during the interval $[0,s]$, and let $\lambda_t(i) = \sum_{\ell=1}^{K} \mathbf{1}(X_{t-}(\ell) = i)$ be the number of processes in state i at time $t-$. Then the $(P_A, \mathcal{H})$-stochastic intensity of N is given by $\lambda_t(A, i, j) = A(i,j)\lambda_t(i)$. For the dominating measure P we take the probability corresponding to $A(i,j) = 1$ for all $i \neq j$. The martingale estimator of the process $B_t(A, i, j) = A(i,j) \int_0^t \mathbf{1}(\lambda(i) > 0)ds$ is $\hat{B}_t(i,j) = \int_0^t \lambda(i)^{-1}dN(i,j)$; thus one estimator of A is

$$\hat{A}(i,j) = \int_0^1 \lambda(i)^{-1} dN(i,j) / \int_0^1 \mathbf{1}(\lambda(i) > 0)ds. \tag{5.17}$$

Alternatively, the process $B_t(A, i, j) = A(i,j) \int_0^t \lambda(i)ds$ has martingale estimator $\hat{B}_t(i,j) = N_t(i,j)$, which engenders the more intuitive and classical estimator

$$\hat{A}^*(i,j) = N_1(i,j) / \int_0^1 \lambda(i)ds; \tag{5.18}$$

$A^*(i,j)$ is the number of $i \to j$ transitions divided by the total occupation time of i, so plausibly estimates the rate of transition, namely $A(i,j)$. However, the martingale method accommodates time-dependent generators (Jacobsen, 1982), whereas "classical" methods leading directly to (5.18), which is the nonparametric maximum likelihood estimator (see, e.g., Billingsley, 1961a), cannot.

Asymptotic properties as $n \to \infty$ are discussed in the exercises. $\square$

5.2   Asymptotics for Martingale Estimators

Initially, our setting is that of Model 5.1, and our concern with limit behavior of the martingale estimators of Definition 5.3. We begin with two fundamental martingale limit theorems. Next we develop a general asymptotic theory for martingale estimators associated with a sequence of versions of Model 5.1, with broad rather than specific stability conditions. In Section 3 we specialize to i.i.d. copies of a multiplicative intensity process and finally to the models of Examples 5.7 and 5.8.

To embark on this program, let $(\Omega, \mathcal{G}, P)$ be a probability space on which is defined a filtration $\mathcal{H}$ satisfying the "*conditions habituelles.*" Several forms of consistency for martingale estimators can deduced from the following inequality.

Proposition 5.9. (Lenglart inequality). Let X and Y be adapted, right-continuous, nonnegative processes with Y nondecreasing and predictable and

$Y_0 = 0$. Suppose that Y dominates X in the sense that for every finite stopping time T,

$$E[X_T] \le E[Y_T]. \tag{5.19}$$

Then for each $\varepsilon, \eta > 0$ and each finite stopping time T,

$$P\{\sup_{s \le T} X_s > \varepsilon\} \le \eta/\varepsilon + P\{Y_T > \eta\}.$$

Proof: Let $X_t^* = \sup_{s \le t} X_s$; then evidently,

$$P\{X_T^* > \varepsilon\} \le P\{X_T^* \ge \varepsilon Y_T < \eta\} + P\{Y_T \ge \eta\}.$$

With $S = \inf\{t : Y_t \ge \eta\}$, predictability of Y implies that $\mathbf{1}(Y_T < \eta)X_T^* \le X_{T \wedge S}^*$, so the proof is complete if we show that for every finite stopping time T, $\varepsilon P\{X_T^* > \varepsilon\} \le E[Y_T]$. With T fixed, let $R_n = T \wedge \inf\{s \le n : X_s \ge \varepsilon\} \wedge n$; then successively using monotonicity of Y, (5.19) and the definition of R_n,

$$E[Y_T] \ge E[Y(R_n)] \ge E[X(R_n)] \ge \varepsilon P\{X(R_n) \ge \varepsilon\} = \varepsilon P\{X_{T \wedge n}^* \ge \varepsilon\},$$

and the proof follows by letting $n \to \infty$. ∎

Despite its simplicity this inequality is central, because the square of a square integrable martingale is dominated by the predictable variation process.

Before presenting general central limit theorems for sequences of martingales, let us consider what form it should have. Suppose that $M^1, M^2, \ldots$ are mean zero square integrable martingales. The basic question is: Under what conditions does there exist a mean zero continuous Gaussian martingale M such that $M^n \overset{d}{\to} M$?

Since a martingale has orthogonal increments, the Gaussian limit has independent increments, so there is a continuous, nondecreasing function v, the *variance function*, which characterizes the law of M, satisfying $E[M_t^2] = E[\langle M \rangle_t] = v_t$ for each t. Minimally we must have $E[(M_t^n)^2] = E[\langle M^n \rangle_t] \to v_t$ for every t, but a stronger variance stabilization condition, namely that $\langle M^n \rangle_t \to v_t$ in probability (equivalently, in distribution) for each t, is needed for central limit behavior. One can think of this condition as simultaneously matching second moments and imposing the asymptotic determinism engendered by independence assumptions in the classical central limit theorem. Finally, since M is continuous, jumps in the martingales M^n must be asymptotically negligible. The hypothesis imposed below is stronger than needed to prove the theorem; variants are noted following the proof.

Theorem 5.10. Let $M^1, M^2, \ldots$ be mean zero square integrable martingales and let v be a continuous, nondecreasing function on $[0, 1]$ with $v_0 = 0$. Suppose that

a) For each t, $\langle M^n \rangle_t \xrightarrow{d} v_t$;

b) There are constants $c_n \downarrow 0$ such that

$$\lim_{n \to \infty} P\{\sup_{t \leq 1} |\Delta M_t^n| \leq c_n\} = 1. \tag{5.20}$$

Then there exists a continuous Gaussian martingale M, with $\langle M \rangle_t = v_t$ for all t, such that $M^n \xrightarrow{d} M$ on the function space $D[0, 1]$.

Proof: The proof divides into two main steps.

1) We first show that the sequence (M^n) is tight in $D[0, 1]$; then (5.20) implies that every subsequential limit in distribution is continuous. For each stopping time T and positive ε, η and δ, the processes $X_t = (M_{T+t}^n - M_T^n)^2$ and $Y_t = \langle M^n \rangle_{T+t} - \langle M^n \rangle_T$ satisfy the hypotheses of Proposition 5.9; consequently,

$$P\left\{ \sup_{T \leq s \leq T+t} |M_s^n - M_T^n| > \varepsilon \right\} \leq \eta/\varepsilon^2 + P\{\langle M^n \rangle_{T+\delta} - \langle M^n \rangle_T > \eta\}. \tag{5.21}$$

From (5.21) we conclude (see Aldous, 1978 or Billingsley, 1968) that tightness of (M^n) ensues from that of $(\langle M^n \rangle)$, which in turn follows from a).

2) It suffices now to show that for every subsequence $(M^{n'})$ converging in distribution to some $\tilde{M}$, $\tilde{M} \overset{d}{=} M$, for which it is enough to establish that $\tilde{M}$ and $\tilde{M} - v$ are martingales. Except for integrability details that we gloss over, the continuous mapping theorem implies that we only need demonstrate that each of these processes is the limit in distribution of a sequence of martingales, which for $\tilde{M}$ is true by definition. By a) the sequences $(M^{n'})^2 - \langle M^{n'} \rangle$ and $(M^{n'})^2 - v$ have the same limit in distribution, and the functional $J : C[0, 1] \to D[0, 1]$ defined by $J(f) = f^2 - v$ is continuous. Hence, again by the continuous mapping theorem,

$$(M^{n'})^2 - \langle M^{n'} \rangle = J(M^{n'}) + v - \langle M^{n'} \rangle \xrightarrow{d} J(\tilde{M}) = \tilde{M}^2 - v,$$

so that $\tilde{M}^2 - v$ is the limit in distribution of a sequence of martingales. $\blacksquare$

Conditions weaker than (5.20) suffice to yield the same conclusion; these include the Lindeberg condition that

$$\lim_{n \to \infty} E\left[\sum_{t \leq 1} (\Delta M_t^n)^2 1(|\Delta M_t^n| \geq \varepsilon) \right] = 0 \tag{5.22}$$

for every $\varepsilon > 0$ and the still weaker asymptotic rarefaction of jumps condition of Rebolledo (1980).

For completeness and later reference, we record the multivariate version of Theorem 5.10.

Theorem 5.11. Let K be fixed, for each n let $M^n(1), \ldots, M^n(K)$ be orthogonal mean zero square integrable martingales, and let $v : [0, 1] \to \mathbf{R}_+^K$ be continuous and nondecreasing in each component with $v_0 = 0$. Suppose that for each t and k, $\langle M^n(k) \rangle_t \overset{d}{\to} v_t(k)$ and that each sequence $M^n(k)$ satisfies (5.20) [or (5.22)]. Then there exists a K-variate continuous Gaussian process M, with each component a martingale and with $\langle M(k), M(j) \rangle = 1(i = j)v(k)$, such that $M^n \overset{d}{\to} M$ on $D[0, 1]^K$. $\square$

That is, the components of M are independent Gaussian martingales and $M(k)$ has variance function $v(k)$. Theorem 5.11 is proved by reduction to the one-dimensional case using the Cramér-Wold device (see Rebolledo, 1978).

We now study implications of Proposition 5.9 and Theorem 5.11 concerning asymptotic behavior of martingale estimators. First we present general results for a sequence of versions of Model 5.1 in which intensities are required merely to converge to infinity. In Section 3 we show that they apply to data constituting i.i.d. copies of a multiplicative intensity process.

The general formulation for limit theorems is a bit involved (but not difficult once one grasps it), and proceeds as follows. Fixed and *independent of the sequence index n* are

- The underlying probability space $(\Omega, \mathcal{G}, P)$

- The index set I

- The family $\mathcal{P} = \{P_\alpha : \alpha \in I\}$ of candidate probability laws

- The mark space $E = \{1 \ldots, K\}$.

For each n there are

- A marked point process N^n with mark space E

- A filtration $\mathcal{H}^n$ satisfying Assumptions 2.10a) and b)

- A bounded, nonnegative, K-variate, $\mathcal{H}^n$-predictable process λ^n, the *baseline intensity*, such that $N^n(k)$ has $(P, \mathcal{H}^n)$-stochastic intensity $\lambda^n(k)$

- For each $\alpha \in I$, a nonnegative, K-variate, $\mathcal{H}^n$-predictable process $H^n(\alpha)$ such that $N^n(k)$ has $(P_\alpha, \mathcal{H}^n)$-stochastic intensity

$$\lambda_t^n(\alpha, k) = H_t^n(\alpha, k)\lambda_t^n(k). \tag{5.23}$$

Thus the point processes, filtrations and stochastic intensities vary with n but the probabilities do not. We assume, without further mention, that

$$E_\alpha \left[\sum_{k=1}^K \int_0^1 H_t^n(\alpha, k) \lambda_t^n(k) dt \right] < \infty$$

for each n and α.

In addition to these objects we have for each n and k an $\mathcal{H}^n$-predictable process Y^n satisfying (5.6). The observations are the point processes N^n, the baseline stochastic intensities λ^n and the processes Y^n.

We estimate the processes

$$B_t^n(\alpha, k) = \int_0^t Y_s^n(k) H_s^n(\alpha, k) 1(\lambda_s^n(k) > 0) ds,$$

using martingale estimators

$$\hat{B}_t^n(k) = \int_0^t Y_s^n(k) \lambda_s^n(k)^{-1} dN_s^n(k). \tag{5.24}$$

Finally, we can pose the questions of interest:

- Are the martingale estimators $\hat{B}^n$ *consistent*, in the sense that $\hat{B}^n - B^n(\alpha) \to 0$ under P_α, for each α?

- Are the estimation error processes $\hat{B}^n - B^n(\alpha)$, suitably scaled, *asymptotically normal*, in the sense that under each P_α they satisfy a martingale central limit theorem?

We provide sufficient conditions for positive response to each.

The martingale method leads naturally to quadratic mean consistency, uniformly in t.

Theorem 5.12. Suppose that

$$E_\alpha[\int_0^1 Y^n(k) \lambda^n(k)^{-1} dN(k)] \to 0$$

for each α and k. Then the martingale estimators of (5.24) are uniformly consistent in mean square: for each α and k, as $n \to \infty$,

$$E_\alpha[\sup_{t \leq 1} (\hat{B}_t^n(k) - B_t^n(\alpha, k))^2] \to 0.$$

Proof: By Proposition 5.4, $\hat{B}^n(k) - B^n(\alpha, k)$ is a square integrable P_α-martingale. Theorem B.15 applied to the submartingale $(\hat{B}^n(k) - B^n(\alpha, k))^2$ gives

$$
\begin{aligned}
E_\alpha[\sup_t (\hat{B}_t^n(k) - B_t^n(\alpha, k))^2] &\leq 4E_\alpha[(\hat{B}_1^n(k) - B_1^n(\alpha, k))^2] \\
&= 4E_\alpha[\langle \hat{B}^n(k) - B^n(\alpha, k) \rangle_1] \\
&= 4E_\alpha[\int_0^1 Y^n(k)^2 \lambda^n(k)^{-1} H^n(\alpha, k) ds] \\
&= 4E_\alpha[\int_0^1 Y^n(k) \lambda^n(k)^{-1} dN^n(k)],
\end{aligned}
$$

the last equality holding because (5.23) gives the P_α-stochastic intensity of N^n. ∎

Various alternatives exist. Using the Burkholder-Davis-Gundy martingale inequalities, one can extend Theorem 5.12 to L^p-uniform consistency for $p \in [1, \infty)$, and even obtain necessary and sufficient conditions. In another direction, weak uniform consistency, i.e.,

$$\sup_{t \leq 1} |\hat{B}^n_t(k) - B^n_t(\alpha, k)| \to 0 \tag{5.25}$$

in P_α-probability, can be deduced directly, in sharper form, using (5.9) and Proposition 5.9, as follows. The martingale error process $(\hat{B}^n(k) - B^n(\alpha, k))$ is dominated in the sense of (5.19) by its predictable variation $\langle \hat{B}^n(k) - B^n(\alpha, k) \rangle$, and Proposition 5.9 shows that (5.25) obtains provided only that

$$\int_0^1 Y^n_s(k)^2 \lambda^n_s(k)^{-1} H^n_s(\alpha, k) ds \to 0 \tag{5.26}$$

in P_α-probability for each α.

The most general central limit theorem amounts only to a restatement of Theorem 5.11.

Theorem 5.13. Suppose that for each α,

$$\sum_{k=1}^K \int_0^1 Y^n_s(k)^2 \lambda^n_s(k) H^n_s(\alpha, k) ds < \infty$$

almost surely with respect to P_α, and let b_n be positive constants increasing to ∞. Assume that

a) For each α and k the P_α-martingales $\sqrt{b_n}[\hat{B}^n(k) - B^n(\alpha, k)]$ satisfy (5.20);

b) For each α there is a function $v(\alpha)$ fulfilling the properties in Theorem 5.11 such that

$$b_n \int_0^t [Y^n_s(k)/\lambda^n_s(k)]^2 H^n_s(\alpha, k) \lambda^n_s(k) ds \overset{d}{\to} v_t(\alpha, k) \tag{5.27}$$

for each t and k.

Then for each α there exists a continuous Gaussian martingale $M(\alpha)$ with $\langle M(\alpha, k), M(\alpha, j) \rangle = \mathbf{1}(k = j)v(\alpha, k)$ such that on $D[0, 1]^K$,

$$\sqrt{b_n}[\hat{B}^n - B^n(\alpha)] \overset{d}{\to} M(\alpha). \square$$

An intuitive basis, with $Y^n \equiv 1$, is that a point process with a stochastic intensity is locally and conditionally Poisson (see the discussion following Theorem 2.21), while a Poisson process, by Theorem 4.25, is approximately a Wiener process. Consequently, hence the more general point process, when scaled, is locally a Wiener process, i.e., approximately a Gaussian martingale.

5.3   The Multiplicative Intensity Model

In this section, we consider both martingale and maximum likelihood estimation for the multiplicative intensity model. Among other things, we describe one way to circumvent difficulties regarding likelihood functions noted below.

The setting is this: with respect to P_α we have i.i.d. copies $(N^{(i)}, \lambda^{(i)})$ of a pair (N, λ) such that the point process N has stochastic intensity $\alpha_t \lambda_t$. For simplicity, we assume that there are no marks. The index set I consists of all left-continuous, right-hand limited, nonnegative functions $\alpha \in L^1[0, 1]$.

The n-sample log-likelihood function is then

$$L_n(\alpha) = \int_0^1 \lambda^n(s)(1 - \alpha_s)ds + \int_0^1 \log \alpha_s \, dN_s^n, \qquad (5.28)$$

where $N^n = \sum_{i=1}^N N^{(i)}$ and $\lambda^n = \sum_{i=1}^n \lambda^{(i)}$. These functions are unbounded above in α, because they can be made arbitrarily large by taking each $\alpha(k)$ to be large near the points of N and nearly zero everywhere else.

5.3.1   Martingale Estimation

To analyze limit behavior of martingale estimators we apply results from Section 2 preceding to the sequence defined by the N^n, λ^n and $\mathcal{H}_t^n$. That this entails no loss of information is implied by the following result.

Proposition 5.14. a) For each n, (N^n, λ^n) is a sufficient statistic for α given the data $(N^{(1)}, \lambda^{(1)}), \ldots, (N^{(n)}, \lambda^{(n)})$;

b) For each k, $N^n(k)$ has $(P_\alpha, \mathcal{H}^n)$-stochastic intensity $\alpha(k)\lambda^n(k)$. $\square$

We estimate the deterministic integrals

$$B_t(\alpha, k) = \int_0^t \alpha_s(k)\mathbf{1}(E_\alpha[\lambda_s(k)] > 0)ds$$

with the (effectively) martingale estimators

$$\hat{B}_t^n(k) = \int_0^t \lambda_s^n(k)^{-1}dN_s(k). \qquad (5.29)$$

The stochastic intensities λ^n do converge to infinity, and so we utilize Theorems 5.12 and 5.13 to show that these estimators are consistent and asymptotically normal.

Theorem 5.15. For each α and k, as $n \to \infty$

$$E_\alpha[\sup_{t \leq 1}(\hat{B}_t^n(k) - B_t(\alpha, k))^2] \to 0.$$

Proof: The processes $\bar{B}_t^n(\alpha, k) = \int_0^t \alpha(k)\mathbf{1}(\lambda^n(k) > 0)ds$ have as martingale estimators the $\hat{B}^n$ of (5.29). According to Theorem 5.12,

$$E_\alpha[\sup_{t \leq 1}(\hat{B}_t^n(k) - \bar{B}_t^n(\alpha, k))^2] \to 0$$

provided that

$$E_\alpha[\int_0^1 \alpha(k)\lambda^n(k)^{-1}ds] \to 0$$

(we have omitted some easy calculations), which holds by the law of large numbers and the dominated convergence theorem. To complete the proof it is necessary to show that

$$E_\alpha[\sup_{t\leq 1}(\bar{B}_t^n(\alpha, k) - B_t(\alpha, k))^2] \to 0. \tag{5.30}$$

A stronger statement, (5.33), is established in the proof of Theorem 5.16. ∎

Theorem 5.16. Assume that for each α and k,

$$\int_0^1 E_\alpha[\lambda_s(k)]^{-1}ds < \infty. \tag{5.31}$$

Then the error processes $\sqrt{n}[\hat{B}^n \quad B(\alpha)]$ converge in Γ_α-distribution to a continuous K-variate Gaussian martingale $M(\alpha)$ satisfying

$$\langle M(\alpha, k), M(\alpha, j)\rangle_t = 1(k = j)\int_0^t E_\alpha[\lambda_s(k)]^{-1}\alpha_s(k)ds. \tag{5.32}$$

Proof: We verify the hypotheses of Theorem 5.13. Jumps in the martingale $\sqrt{n}[\hat{B}^n(k) - \bar{B}^n(\alpha, k)]$ arise only from $\hat{B}^n(k)$, whose jumps are of size $\lambda^n(k)^{-1} \cong n^{-1}$; hence (5.22) holds for $c_n = n^{-1/4}$, for example. To check (5.27) we note that

$$n\int_0^t [\lambda^n(\alpha, k)/\lambda^n(k)]ds = n\int_0^t [\alpha/\lambda^n(k)]ds \to \int_0^t E_\alpha[\lambda(k)]^{-1}\alpha\, ds$$

by the law of large numbers and (5.31). It remains to show that

$$\sqrt{n}\, \sup_{t\leq 1} |\bar{B}_t^n(\alpha, k) - B_t(\alpha, k)| \to 0 \tag{5.33}$$

in P_α-probability. [This also proves (5.30).] With $I_s = 1(E_\alpha[\lambda_s(k)] > 0)$,

$$\sqrt{n}\, \sup_{t\leq 1} |\bar{B}_t^n(\alpha, k) - B_t(\alpha, k)| = \sqrt{n}\sup_{t\leq 1} \left|\int_0^t \alpha_s(k)[1(\lambda_s^n(k) > 0) - I_s]ds\right|$$

$$= \sqrt{n}\int_0^1 \alpha_s(k)I_s 1(\lambda_s^n(k) = 0)ds;$$

taking expectations with respect to P_α gives

$$E_\alpha\left[\sqrt{n}\, \sup_t |\bar{B}_t^n(\alpha, k) - B_t(\alpha, k)|\right] = \int_0^1 \alpha_s(k)I_s\sqrt{n}P_\alpha\{\lambda_s(k) = 0\}^n ds,$$

which converges to zero by dominated convergence. ∎

To apply Theorem 5.16, it is necessary to estimate the variances in (5.32), but this can be done using the martingale method! The processes

$$C_t(\alpha, k) = n\int_0^t \lambda_s^n(k)^{-1}\alpha_s(k)ds,$$

which converge to the variance in (5.32), have martingale estimators

$$\hat{C}_t(\alpha, k) = n\int_0^t (1/\lambda_s^n(k))\lambda_s^n(k)^{-1}dN_s^n(k),$$

which are consistent by a repetition of the proof of Theorem 5.15.

We illustrate for the special cases of Examples 5.7 and 5.8.

Example 5.17. (i.i.d. random variables). Let $X_1, X_2, \ldots$ be i.i.d. random variables in $[0, 1]$ with unknown continuous hazard function h. Each point process $N^{(i)} = \epsilon_{X_i}$ has stochastic intensity $\lambda_t^{(i)} = h(t)\mathbf{1}(X_i \geq t)$; consequently, the superposition $N = \sum_{i=1}^n N^{(i)}$ has stochastic intensity $\lambda_t^n = \sum_{i=1}^n \lambda_t^{(i)} = h(t)(n - N_{t-}^n)$. As discussed in Example 5.7, the martingale estimator for $B_t^n(h) = \int_0^t h(s)\mathbf{1}(N_s^n < n)ds$ is

$$\hat{B}_t^n = \int_0^t (n - N_{s-}^n)^{-1}dN_s^n = \sum_{j=1}^{N_t^n} 1/(n - j + 1). \tag{5.34}$$

Theorem 5.15 does not apply directly, but Theorem 5.12 does, to yield $E_h[\sup(\hat{B}_t^n - B_t^n(h))^2] \to 0$ for each h, while Theorem 5.13 gives that under P_h, $\sqrt{n}[\hat{B}^n - B^n(h)] \overset{d}{\to} M(h)$, where $M(h)$ is a continuous Gaussian martingale with $\langle M(h) \rangle_t = F_h(t)/G_h(t)$, with F_h and $G_h = 1 - F_h$ the distribution and survivor functions associated with h. Verification of (5.27) is straightforward: with $b_n = n$ and $Y^n \equiv 1$,

$$n\int_0^t [\lambda_s^n(h)/(\lambda_s^n)^2]ds = n\int_0^t (\lambda_s^n)^{-1}h(s)ds \to \int_0^t [h(s)/G_h(s)]ds = F_h(t)/G_h(t).$$

In the absence of censoring, asymptotic behavior of the product limit estimator (5.15) is that of the empirical distribution function (see Section 4.6). For the censored case, see Gill (1980) or Jacobsen (1982). □

For the Markov process model of Example 5.8, while the martingale method is useful for formulating the estimators $\hat{A}$ of (5.17) and $\hat{A}^*$ of (5.18), it is less useful for their analysis, which can be effected directly. For example, since

$$\hat{A}^*(i, j) = \frac{(1/n)\sum_{\ell=1}^n N_1^{(\ell)}(i, j)}{(1/n)\sum_{\ell=1}^n \int_0^1 \mathbf{1}(X_s(\ell) = i)ds},$$

with ℓ referring to the ℓth process and $P_s(i, j)$ denoting the transition function, it follows by the strong law of large numbers that [assuming $X_0(\ell) = i_0$ for all ℓ]

$$\hat{A}^*(i, j) \to \frac{A(i, j)\int_0^1 P_s(i_0, i)ds}{\int_0^1 P_s(i_0, i)ds} = A(i, j)$$

almost surely; the corresponding central limit theorem is equally easy to derive. Limit behavior of the estimators $\hat{A}$ is treated in Exercise 5.14.

5.3.2 Maximum Likelihood Estimation

We next pursue an indirect approach to maximum likelihood estimation, the "method of sieves" of Grenander (1981, Section 8.1). The basic ideas are:

- For each $h > 0$ devise a subset I_h of I within which there exists for each sample size n a maximum likelihood estimator $\hat{\alpha} = \hat{\alpha}(n, h)$ satisfying $L_n(\hat{\alpha}) \geq L_n(\alpha)$ for all $\alpha \in I_h$.

- Choose the I_h to increase as h, the *sieve mesh*, decreases and such that $\bigcup_{h>0} I_h$ is a dense subset of I.

- Choose h to depend on n in such a manner that $\|\hat{\alpha}(n, h_n) - \alpha\|_1 \to 0$, so that the estimators are consistent.

The family (I_h) is termed a *sieve*.

Note the resemblance to kernel methods of density estimation (cf. Ramlau-Hansen, 1983) and, in particular, the simultaneous limit processes as the sample size increases and the sieve mesh decreases. Not only consistency but also computational tractability is an issue in regard to choice of the sieve; we examine this issue following Theorem 5.20.

We begin with a strong consistency theorem, giving conditions under which (5.39) holds almost surely. Thereafter we discuss c_n-consistency in the sense of convergence in probability. It is important to keep in mind that we estimate α itself rather than indefinite integrals; note also that the L_1-convergence is suited exactly to the role of α.

First, however, we must introduce a sieve. For each $h > 0$, let I_h denote the family of absolutely continuous functions α satisfying

$$h \leq \alpha \;\; \leq \;\; 1/h \tag{5.35}$$
$$|\alpha|'/\alpha \;\; \leq \;\; 1/h \tag{5.36}$$

everywhere on $[0, 1]$. As the sieve mesh h decreases to 0, the I_h increase to a dense subset of I.

We show that for each n and h, there exists a maximum likelihood estimator $\hat{\alpha}(n, h)$ relative to I_h, then we choose $h = h_n$ converging to zero sufficiently slowly that $\hat{\alpha}(n, h_n) \to \alpha$ in $L^1[0, 1]$.

Theorem 5.18. Assume that

i) The function $m_s(\alpha) = E_\alpha[\lambda_s]$ is bounded and bounded away from zero on $[0, 1]$;

ii) The "entropy" $H(\alpha) = -\int_0^1 [1 - \alpha_s + \alpha_s \log \alpha_s] m_s(\alpha) ds$ is finite;

iii) With $\mathrm{Var}_\alpha(\lambda_s)$ denoting the variance of λ_s under P_α,

$$\int_0^1 \mathrm{Var}_\alpha(\lambda_s) ds < \infty; \tag{5.37}$$

iv) $E_\alpha[N_1^2] < \infty$.

Then

a) For each n and h there exists a (not necessarily unique) maximum likelihood estimator $\hat{\alpha}(n, h) \in I_h$ satisfying

$$L_n(\hat{\alpha}(n, h)) \geq L_n(\alpha) \tag{5.38}$$

for all $\alpha \in I_h$;

b) For $h_n = n^{-1/4+\eta}$, with $0 < \eta < 1/4$ and $\hat{\alpha} = \hat{\alpha}(n, h_n)$,

$$\|\hat{\alpha}(n, h_n) - \alpha\|_1 \to 0 \tag{5.39}$$

almost surely with respect to P_α.

Proof: We follow the pattern of argument in Grenander (1981, Section 8.2, Theorem 1). To begin, it is shown there that

i) Each I_h is uniformly bounded and equicontinuous;

ii) $\{\alpha' : \alpha \in I_h\}$ is uniformly bounded for each h.

With n (and ω) fixed and $(\alpha(j))$ a sequence in I_h such that $L_n(\alpha(j)) \to \sup_{\alpha \in I_h} L_n(\alpha)$, it follows that there is a subsequence, which we continue to denote by $(\alpha(j))$, such that the derivatives $\alpha'(j)$ converge in $L^1[0, 1]$ and such that the real sequence $(\alpha_0(j))$ converges. Hence there is $\alpha(\infty) \in I_h$ such that $\alpha(j) \to \alpha(\infty)$ uniformly. Since $\alpha(\infty)$ is continuous and bounded away from zero, $L_n(\alpha(\infty)) = \lim_{j\to\infty} L_n(\alpha(j)) = \sup_{\alpha \in I_h} L_n(\alpha)$, proving a).

To establish b), we note first that by reasoning in Grenander (1981), given $\alpha \in I$ and $\delta > 0$, for sufficiently small $h > 0$ there exists a regularization $\tilde{\alpha} \in I_h$ — depending on δ and h, which we suppress — such that $\|\alpha - \tilde{\alpha}\|_\infty < \delta$ and

$$\left| \int m(\alpha)[1 - \tilde{\alpha} + \alpha(\log \tilde{\alpha})]ds - \int m(\alpha)[1 - \alpha + \alpha(\log \alpha)]ds \right| < \delta. \tag{5.40}$$

Recalling that (N, λ) is a generic copy, let

$$L(\beta) = \int \lambda(1 - \beta)ds + \int (\log \beta)dN$$

be the log-likelihood function. By Jensen's inequality, the expectation of a log-likelihood function is maximized at the "true" index value; consequently,

$$H(\alpha) = E_\alpha[L(\alpha)] \leq -E_\alpha[L(\beta)] = -\int m(\alpha)[1 - \beta + \alpha(\log \beta)]ds. \tag{5.41}$$

Suppose now that $\alpha \in I$ and $\delta > 0$ are fixed and let $\hat{\alpha} = \hat{\alpha}(n, h)$ be as in a), with n and h variable. Then for n sufficiently large and h sufficiently small,

$$
\left| -\int m(\alpha)[1 - \hat{\alpha} + \alpha(\log \hat{\alpha})]ds - H(\alpha) \right|
$$

$$
= -\int m(\alpha)[1 - \hat{\alpha} + \alpha(\log \hat{\alpha})]ds - H(\alpha)
$$

$$
= -\int m(\alpha)[1 - \hat{\alpha} + \alpha(\log \hat{\alpha})]ds
$$
$$
+ (1/n)\left[\int \lambda^n (1 - \hat{\alpha})ds + \int (\log \hat{\alpha})dN^n \right]
$$
$$
- (1/n)\left[\int \lambda^n (1 - \hat{\alpha})ds + \int (\log \hat{\alpha})dN^n \right]
$$
$$
+ (1/n)\left[\int \lambda^n (1 - \tilde{\alpha})ds + \int (\log \tilde{\alpha})dN^n \right]
$$
$$
- (1/n)\left[\int \lambda^n (1 - \tilde{\alpha})ds + \int (\log \tilde{\alpha})dN^n \right]
$$
$$
+ \int m(\alpha)[1 - \tilde{\alpha} + \alpha(\log \tilde{\alpha})]ds
$$
$$
- \int m(\alpha)[1 - \tilde{\alpha} + \alpha(\log \tilde{\alpha})]ds - H(\alpha)
$$

$$
\leq \left| \int m(\alpha)[1 - \hat{\alpha} + \alpha(\log \hat{\alpha})]ds \right.
$$
$$
\left. - (1/n)\left[\int \lambda^n (1 - \hat{\alpha})ds + \int (\log \hat{\alpha})dN^n \right] \right|
$$
$$
+ \left| \int m(\alpha)[1 - \tilde{\alpha} + \alpha(\log \tilde{\alpha})]ds \right.
$$
$$
\left. - (1/n)\left[\int \lambda^n (1 - \tilde{\alpha})ds + \int (\log \tilde{\alpha})dN^n \right] \right|
$$
$$
+ \delta,
$$

where we have used (5.38)–(5.41).

Assuming for the moment that by proper choice of large n and small h, the first two terms converge to zero almost surely, we will have demonstrated that

$$
-\int m(\alpha)[1 - \hat{\alpha} + \alpha(\log \hat{\alpha})]ds \to H(\alpha),
$$

almost surely, which is the same as

$$
\int m(\alpha)[\hat{\alpha} - \alpha(\log \hat{\alpha})]ds \to \int m(\alpha)[\alpha - \alpha(\log \alpha)]ds. \qquad (5.42)
$$

Following Grenander (1981) one final time, we infer from (5.42) that for $k(y) = y - \log(1 + y)$, $\int k(\hat{\alpha}/\alpha - 1)m(\alpha)\alpha ds \to 0$, almost surely, which shows that $\hat{\alpha} \to \alpha$ in $L^1(m_s(\alpha)ds)$ and, in view of i), suffices to confirm b).

Of the two remaining terms we analyze only the first, which is the more difficult; the second is handled analogously. Using integration by parts, (5.35) and (5.36), we see that

$$
|\int m(\alpha)[1 - \hat{\alpha} + \alpha(\log \hat{\alpha})]ds - (1/n)\left[\int \lambda^n (1 - \hat{\alpha})ds + \int (\log \hat{\alpha})dN^n \right]|
$$
$$
\leq \left| \int [1 - \hat{\alpha} + \alpha(\log \hat{\alpha})][\lambda^n /n - m(\alpha)]ds \right|
$$
$$
+ \left| (\log \hat{\alpha}_1)(1/n)[N_1^n - \int_0^1 \alpha_s \lambda_s^n ds] \right|
$$

$$+ \left| (1/n)\int_0^1 [\hat{\alpha}'_t/\alpha'_t][N^n_t - \int_0^t \alpha_s \lambda^n_s ds] dt \right|$$

$$\leq (1/h)\int_0^1 |\lambda^n_s/n - m_s(\alpha)| ds + (2/nh)\sup_t \left| N^n_t - \int_0^t \alpha_s \lambda^n_s ds \right|.$$

We consider the terms in order.

For fixed s, the random variables $\lambda^{(i)}_s - m_s(\alpha)$ are i.i.d. and by (5.37) have mean zero and finite second moment. Consequently, by the Marcinkiewicz-Zygmund strong law of large numbers (Chow and Teicher, 1978, page 122), for $h = h_n = n^{-1/4+\eta}$,

$$(1/h)|\lambda^n_s/n - m_s(\alpha)| \to 0 \qquad (5.43)$$

almost surely. From Fubini's theorem we infer that almost surely (5.43) holds almost everywhere on $[0, 1]$ with respect to Lebesgue measure. This last statement, (5.37) and the dominated convergence theorem give that almost surely

$$(1/h)\int |\lambda^n_s/n - m_s(\alpha)| ds \to 0.$$

Let $M^n_t = N^n_t - \int_0^t \alpha \lambda^n ds$ be the innovation martingale, whose quadratic variation process satisfies $[M^n]_t = N^n_t$. By a martingale inequality of Burkholder (1973), applied with $p = 4$ to M^n, for each $\varepsilon > 0$

$$P_\alpha\{(1/nh)\sup_t |M^n_t| > \varepsilon\} \leq E_\alpha[\sup_t|M^n_t|^4]/(nh\varepsilon)^4 = O(E_\alpha[[M^n]^2_1]/(nh\varepsilon)^4),$$

while by assumption iv),

$$E_\alpha[[M^n]^2_1] = E_\alpha[(N^n_1)^2] = n \operatorname{Var}(N_1) + n^2 E_\alpha[N_1]^2.$$

Thus, $P_\alpha\{(1/nh)\sup |M^n_t| > \varepsilon\} = O(1/n^2 h^4 \varepsilon^4)$ and for the prescribed choice $h_n = n^{-1/4+\eta}$ it follows that $(1/nh)\sup|N^n_t - \int_0^t \alpha \lambda^n ds| \to 0$ almost surely with respect to P_α, which completes the proof. ∎

The relatively slow rate of convergence in Theorem 5.18 is a disadvantage in applications since, roughly speaking, the sieve mesh h determines the accuracy of $\hat{\alpha}$ as an approximation to α. At the expense of weak rather than strong consistency, the rate of convergence can be improved significantly; moreover, one hypothesis from Theorem 5.18 can be eliminated and rates of convergence can be deduced.

Theorem 5.19. Suppose that assumptions i) – iii) of Theorem 5.18 are fulfilled and let (c_n) be a sequence with $c_n \to \infty$ and $c_n/\sqrt{n} \to 0$. Then for $h_n = \sqrt{c_n/\sqrt{n}}$ and $\hat{\alpha} = \hat{\alpha}(n, h_n)$

$$\lim_{d \to \infty} \limsup_{n \to \infty} \sup_{\|\alpha^* - \alpha\|_1 < \epsilon} P_{\alpha^*}\{c_n\|\hat{\alpha} - \alpha^*\|_1 > d\} = 0.$$

for every $\alpha \in I$ and $\varepsilon > 0$. ☐

That is, the estimators are c_n-consistent; see for example Millar (1983). For the proof, and that of Theorem 5.20 as well, we refer to Karr (1987a).

Note that one can come arbitrarily close to $\sqrt{n}$-consistency, but that because (h_n) depends on (c_n) to actually attain $\sqrt{n}$-consistency would require that $h_n \to 0$ arbitrarily slowly. There is one (important) case where this can be established. When α is known to belong to the sieve set I_h for some $h > 0$, then I_h is taken as the parameter space, and we obtain the following result.

Theorem 5.20. If $\alpha \in I_h$ and if hypotheses i) – iii) of Theorem 5.18 are satisfied, then the maximum likelihood estimators $\hat{\alpha} = \hat{\alpha}(n, h)$ are $\sqrt{n}$-consistent: for each $\varepsilon > 0$,

$$\lim_{d \to \infty} \limsup_{n \to \infty} \sup_{\alpha^* \in I_h, \|\alpha^* - \alpha\|_1 < \varepsilon} P_{\alpha^*}\{\sqrt{n}\|\hat{\alpha} - \alpha^*\|_1 > d\} = 0. \; \square \qquad (5.44)$$

We use this result momentarily to prove a central limit theorem for $\hat{\alpha}$.

Computation of the sieve maximum likelihood estimators is somewhat difficult. One must solve the function space optimization problem

$$\text{maximize} \; - \int_0^1 \alpha_s \lambda_s^n ds + \sum_i \log \alpha_{T_i}, \qquad (5.45)$$

where the T_i are the points of N^n, subject to the constraints (5.35)–(5.36). Similar problems arise in the calculation of nonparametric estimators of density functions using penalized likelihood functions (see Geman and Hwang, 1982, concerning the relationship to sieve methods), but the presence of λ_s^n in (5.45) and the form of the constraints preclude a closed form solution.

Alternative sieves include the *histogram sieve* defined via the sieve sets

$$I(m) = \{\alpha : \alpha \text{ is constant on } J_{mk} = ((k-1)/m, k/m], k = 1, \ldots, m\},$$

which is more convenient computationally. Indeed, in this case the maximum likelihood estimator in $I(m)$ satisfies

$$\hat{\alpha}_t(n, m) = N^n(J_{mk})/\int_{J_{mk}} \lambda_s^n ds, \qquad t \in J_{mk}.$$

If one integrates these estimators in order to estimate indefinite integrals of α, there result the estimators

$$\left(\int_0^t \alpha_s \mathbf{1}(\lambda_s^n > 0) ds \right)^\wedge = \int_0^t \hat{\alpha}_s \mathbf{1}(\lambda_s^n > 0) ds$$

$$\cong (1/m) \sum_{k=1}^{[mt]} N^n(J_{mk})/\int_{J_{mk}} \lambda_s^n ds \cong \int_0^t (\lambda_s^n)^{-1} dN_s^n,$$

where $[x]$ denotes the integer part of x. Thus, we obtain a maximum likelihood interpretation of the martingale estimator $\hat{B}$ of (5.12).

We now consider two aspects of asymptotic normality. The first is local asymptotic normality of likelihood ratios. We present such a theorem for the multiplicative intensity model, which not only is of independent interest but also leads to a central limit theorem for the estimators $\hat{\alpha}$.

We use martingale methods to examine the behavior of log-likelihood functions as random functions of t and, in particular, describe limit properties of differences $L_n(\alpha + \alpha^*/\sqrt{n}) - L_n(\alpha)$ as $n \to \infty$. The log-likelihood process, indexed by $t \in [0, 1]$, is

$$L_n(\alpha, t) = \int_0^t \lambda_s^n (1 - \alpha_s)ds + \int_0^t (\log \alpha_s)dN_s^n.$$

By Theorem 5.10, under the probability P_α,

$$\left((1/\sqrt{n})[N_t^n - \int_0^t \alpha_s \lambda_s^n ds] \right)_{0 \leq t \leq 1} \xrightarrow{d} (M_t(\alpha))_{0 \leq t \leq 1}, \qquad (5.46)$$

where $M(\alpha)$ is a Gaussian martingale with variance function given by $v_t(\alpha) = \int_0^t m_s(\alpha)\alpha_s ds$.

Theorem 5.21. Let α and α^* be fixed elements of I and suppose that

$$\int_0^1 [(\alpha_s^*)^{k+1}/\alpha_s^k]m_s(\alpha)ds < \infty, \qquad k = 1, 2.$$

Then under P_α the stochastic processes

$$\left(L_n(\alpha + \alpha^*\sqrt{n}, t) - L_n(\alpha, t) + \tfrac{1}{2}\int_0^t [(\alpha_s^*)^2/\alpha_s]m_s(\alpha)ds \right)_{0 \leq t \leq 1} \qquad (5.47)$$

converge in distribution to a Gaussian martingale with variance function

$$v_t(\alpha, \alpha^*) = \int_0^t [(\alpha_s^*)^2/\alpha_s]m_s(\alpha)ds.$$

Proof: Tightness of the sequence (5.47) is an immediate consequence of (5.46) and (5.48) below, so we need only identify the limit, in view of whose independent increments it suffices to consider one-dimensional distributions. For each n,

$$L_n(\alpha + \alpha^*/\sqrt{n}, t) - L_n(\alpha, t)$$

$$= -\frac{1}{\sqrt{n}}\int_0^t \lambda_s^n \alpha_s^* ds + \int_0^t \log\left(1 + \frac{1}{\sqrt{n}}\frac{\alpha_s^*}{\alpha_s}\right) dN_s^n$$

$$= -\frac{1}{\sqrt{n}}\int_0^t \lambda_s^n \alpha_s^* ds + \int_0^t \left[\frac{1}{\sqrt{n}}\frac{\alpha_s^*}{\alpha_s} - \frac{1}{2n}\left(\frac{\alpha_s^*}{\alpha_s}\right)^2\right] dN_s^n$$

$$+ \int_0^t O\left(\frac{1}{n^{3/2}}[\alpha_s^*/\alpha_s]^3\right) dN_s^n$$

$$= \frac{1}{\sqrt{n}} \int_0^t \left(\frac{\alpha_s^*}{\alpha_s}\right) [dN_s^n - \alpha_s \lambda_s^n ds] + \frac{1}{2n} \int_0^t \left(\frac{\alpha_s^*}{\alpha_s}\right)^2 dN_s^n$$

$$+ O\left(\frac{1}{\sqrt{n}} \int_0^1 \left(\frac{\alpha_s^*}{\alpha_s}\right)^3 \frac{dN_s^n}{n}\right) \tag{5.48}$$

uniformly in t. The error converges in probability to zero by the law of large numbers and Slutsky's theorem. That the error bound is uniform in t confirms the tightness asserted at the beginning of the proof. Again by the law of large numbers, the second term converges in probability to $-(1/2)\int_0^t [\alpha_s^*/\alpha_s]^2 m_s(\alpha)ds$, while by (5.46) and the continuous mapping theorem the first term converges in distribution to $\int_0^t (\alpha_s^*/\alpha_s)dM_s(\alpha)$. Denoted by $M(\alpha, \alpha^*)$, the limit is the integral of a predictable process with respect to a Gaussian martingale and is hence a Gaussian martingale with variance function given above. ∎

Albeit of interest in its own right, Theorem 5.21 leads as well to a central limit theorem for the maximum likelihood estimators $\hat{\alpha}$, provided that the assumptions of Theorem 5.20 be fulfilled. Interestingly, this theorem, which applies to indefinite integrals of the $\hat{\alpha}$, is the same central limit theorem satisfied by the martingale estimators of (5.12) (cf. Theorem 5.16).

Theorem 5.22. Suppose that $\alpha \in I_h$ for some fixed $h > 0$ and that the assumptions of Theorems 5.20 and 5.21 are satisfied, and let $\hat{\alpha}(n, h)$ be as in the former. Then the stochastic processes $Z_t^n = \sqrt{n} \int_0^t [\hat{\alpha}_s - \alpha_s]ds$ converge in P_α-distribution to a Gaussian martingale with variance function

$$w_t(\alpha) = \int_0^t [\alpha_s/m_s(\alpha)]ds.$$

Proof: With n fixed, define a transformation T of $L^1[0,1]$ into itself by

$$T(\alpha)_t = \int_0^t \lambda_s^n (1 - \alpha_s)ds + \int_0^t (\log \alpha_s)dN_s^n.$$

The first, second and third Fréchet derivatives of T are given, following straightforward computations, by

$$T'(\alpha)[\beta]_t = -\int_0^t \beta_s \lambda_s^n ds + \int_0^t [\beta_s/\alpha_s]dN_s^n \tag{5.49}$$

$$T''(\alpha)[\beta][\gamma]_t = -\int_0^t [\beta_s \gamma_s/\alpha_s]dN_s^n \tag{5.50}$$

$$T'''(\alpha)[\beta][\gamma][\delta]_t = 2\int_0^t [\beta_s \gamma_s \delta_s/\alpha_s]dN_s^n. \tag{5.51}$$

Then, with $1/\sqrt{n}$ denoting a constant function, by Luenberger (1969, Section 7.3)

$$
\begin{aligned}
L_n(\alpha + 1/\sqrt{n}) &- L_n(\alpha) \\
&= T'(\alpha)[1/\sqrt{n}] + \tfrac{1}{2}T''(\alpha)[1/\sqrt{n}][1/\sqrt{n}] \\
&\quad + O\left(T'''(\alpha)[1/\sqrt{n}][1/\sqrt{n}][1/\sqrt{n}]\right) \\
&= T'(\alpha)[1/\sqrt{n}] + \tfrac{1}{2}T''(\alpha)[1/\sqrt{n}][1/\sqrt{n}] \\
&\quad + O\left(T'''(\alpha)\frac{1}{\sqrt{n}}\int_0^1 \frac{1}{\alpha_s^3}\frac{dN_s^n}{n}\right) \\
&= T'(\alpha)[1/\sqrt{n}] + \tfrac{1}{2}T''(\alpha)[1/\sqrt{n}][1/\sqrt{n}] + O_P(1), \qquad (5.52)
\end{aligned}
$$

via the law of large numbers [this is essentially the same argument used to derive (5.48)]. The sense of the convergence in (5.52) is with respect to the norm on $L^1[0,1]$. Moreover,

$$
\tfrac{1}{2}T''(\alpha)[1/\sqrt{n}][1/\sqrt{n}]_t = -\frac{1}{2}\int_0^t \frac{1}{\alpha_s^2}\frac{dN_s^n}{n} \;\rightarrow\; -\frac{1}{2}\int_0^t \frac{m_s(\alpha)}{\alpha_s}ds, \qquad (5.53)
$$

in the sense of convergence in probability. Theorem 5.21, together with (5.52)–(5.53), implies that $T'(\alpha)[1/\sqrt{n}] \xrightarrow{d} M$, where M is a Gaussian martingale with variance function $v_t(\alpha,1) = \int_0^t [m_s(\alpha)/\alpha_s]ds$.

On the other hand, by the mean value theorem for functionals, for some $r \in (0,1)$,

$$
\begin{aligned}
T'(\alpha)[1/\sqrt{n}]_t &= T''(\hat{\alpha} + r(\alpha - \hat{\alpha}))[\alpha - \hat{\alpha}][1/\sqrt{n}] \\
&= -\int_0^t \left[\frac{\alpha - \hat{\alpha}}{\sqrt{n}(\hat{\alpha} + r(\alpha - \hat{\alpha}))^2}\right] dN^n \\
&= \sqrt{n}\int_0^t \frac{(\hat{\alpha} - \alpha)^2}{\alpha^2}\frac{dN^n}{n} + O_P\left(\frac{1}{\sqrt{n}}\int_0^1 \frac{(\hat{\alpha} - \alpha)^2}{\alpha^3}dN^n\right) \\
&= \sqrt{n}\int_0^t \frac{\hat{\alpha} - \alpha}{\alpha^2}\frac{dN^n}{n} + O_P(1) \\
&\cong \sqrt{n}\int_0^t (\hat{\alpha}_s - \alpha_s)[m_s(\alpha)/\alpha_s]ds,
\end{aligned}
$$

from which we infer that (as processes) $\sqrt{n}\int_0^t [\hat{\alpha}_s - \alpha_s][m_s(\alpha)/\alpha_s]ds \xrightarrow{d} M_t$. From this (also as processes)

$$
\sqrt{n}\int_0^t [\hat{\alpha}_s - \alpha_s]ds \xrightarrow{d} \int_0^t [\alpha_s/m_s(\alpha)]^2 dM_s,
$$

with the limit a Gaussian martingale with variance function $w_t(\alpha)$. ∎

5.3.3 Hypothesis Tests

We now consider testing for the multiplicative intensity model, which is one case where tests can be formulated specifically as well as analyzed. The tests concern whether a multiplicative intensity process has a prescribed multiplier and whether two multiplicative intensity processes (possibly with different baseline intensities) have the same — but unspecified — multiplier. As expected, results deal mainly with asymptotic distribution theory under the null hypothesis.

Let I be the set of functions $\alpha : [0, 1] \to \mathbf{R}_+^K$ with $\alpha(k) \in L^1[0, 1]$ for each k; recall that with respect to P_α the point process $N(k)$ has stochastic intensity $\lambda_t(\alpha, k) = \alpha_t(k)\lambda_t(k)$, where λ is predictable and functionally independent of N, but with law depending on α. By P we denote the probability corresponding to $\alpha \equiv 1$, yielding the log-likelihood function (5.28). When we deal with a sequence (N^n, λ^n) the log-likelihood functions are written $L_n(\alpha)$.

Our initial problem is to test the simple hypothesis $H_0 : \alpha = \alpha^*$, where α^* is a prescribed element of I, which we address for a sequence of superpositions of i.i.d. copies of a process (N, λ). For some results this much structure is not necessary, although it may be difficult to construct other specific cases in which the hypotheses are fulfilled.

Let $B_t^n(\alpha, k) = \int_0^t \alpha_s(k)\mathbf{1}(\lambda_s^n(k) > 0)ds$, whose martingale estimator is

$$\hat{B}_t^n(k) = \int_0^t \lambda_s^n(k)^{-1}dN_s^n(k).$$

Under H_0, the processes $\hat{B}^n(k) - B^n(\alpha^*, k)$ are orthogonal square integrable martingales. For the martingale method the principle of test construction is: *Use test statistics that are martingales.* Thus we can construct a whole family of martingale test statistics

$$Z_t^n(k) = \int_0^t Y_s^n(k)[d\hat{B}_s^n(k) - dB_s^n(\alpha^*, k)], \tag{5.54}$$

where Y^n is a bounded, $\mathcal{H}^n$-predictable process that, by weighting differences between $\hat{B}^n$ and $B^n(\alpha^*)$ at various times, provides great flexibility. Of course, the Y^n must be observable. For each n and k, $Z^n(k)$ is a square integrable martingale, since Y^n is bounded, with predictable variation

$$\langle Z^n(k)\rangle_t = \int_0^t Y_s^n(k)^2 \lambda_s^n(k)^{-1}\alpha_s^*(k)ds,$$

and in addition these martingales are orthogonal.

Theorem 5.13 implies the following result.

Theorem 5.23. Suppose that b_n are constants increasing to ∞ and that under P_{α^*},

$$b_n\int_0^t Y_s^n(k)^2 \lambda_s^n(k)^{-1}\alpha_s^*(k)ds \to v_t(k)$$

for each t and k, where v satisfies the properties in Theorem 5.13, and assume also that the martingales $\sqrt{b_n}Z^n(k)$ fulfill (5.20). Then under P_{α^*}, $\sqrt{b_n}Z^n \xrightarrow{d} M$, where the components of M are independent continuous Gaussian martingales with variance functions $v_t(k)$. $\square$

Given that the processes are observed over $[0,1]$, one should certainly use all the data and base the test on the values $\sqrt{b_n}Z_1^n(k)$, which are asymptotically independent and normally distributed with variances

$$v_1(k) = \lim_{n\to\infty} b_n \int_0^1 Y_s^n(k)^2 \lambda_s^n(k)^{-1} \alpha_s^*(k) ds,$$

which can be estimated consistently (see the discussion following Theorem 5.16). Therefore, one arrives at test statistics whose asymptotic distribution does not depend on α^*

Corollary 5.24. Under H_0, for each k

$$U^n(k) = Z_1^n(k)/\sqrt{\langle Z_n(k)\rangle_1} \xrightarrow{d} N(0,1).\square \qquad (5.55)$$

The test statistics $U^n(k)$ are the basic martingale tools for testing the hypothesis $H_0 : \alpha = \alpha^*$. An advantage of Theorem 5.23 is generality in terms of structure of the sequence (N^n, λ^n). One instance in which its hypotheses are satisfied is superposition of i.i.d. processes: if under each P_α, $(N^{(i)}, \lambda^{(i)})$ are i.i.d. copies of (N, λ), where N has P_α-stochastic intensity $\alpha\lambda$, then Theorem 5.23 holds for $N^n = \sum_{i=1}^n N^{(i)}$, $\lambda^n = \sum_{i=1}^n \lambda^{(i)}$ and $b_n = n$

We turn now to two-sample problems; the results apply as well to k-sample problems. For each n let $N^n = (N^n(1), N^n(2))$ be a marked point process with mark space $\{1,2\}$ and suppose that with respect to a history $\mathcal{H}^n$, $N^n(i)$ has stochastic intensity $\alpha_t(i)\lambda_t^n(i)$. The components $N^n(1)$ and $N^n(2)$ may but need not be independent; recall that both $N^n(i)$ and $\lambda^n(i)$ are observable. Our concern is the composite null hypothesis $H_0 : \alpha(1) \equiv \alpha(2)$. Although likelihood-based methods are desirable, no satisfactory results have been developed because under H_0, the common multiplier α is an infinite-dimensional nuisance parameter. By contrast, the martingale method yields asymptotically normal test statistics analogous to those for the one-sample problem, constructed by integration of predictable processes with respect to certain fundamental martingales that we now identify.

Proposition 5.25. Under H_0, for each n

a) The superposition $\bar{N}^n = N^n(1) + N^n(2)$ has $(P_\alpha, \mathcal{H}^n)$-stochastic intensity $\lambda_t^n(\alpha) = \alpha_t \bar{\lambda}_t^n$, where $\bar{\lambda}^n = \lambda^n(1) + \lambda^n(2)$.

b) The processes

$$M_t^n(i) = \int_0^t \mathbf{1}(\lambda_s^n(i) > 0)[dN_s^n(i)/\lambda_s^n(i) - d\bar{N}_s^n/\bar{\lambda}_s^n] \qquad (5.56)$$

are square integrable martingales with

$$\langle M^n(i), M^n(j)\rangle_t = \int_0^t 1(\lambda_s^n(i)\lambda_s^n(j) > 0)\left[\frac{1(i = j)}{\lambda_s^n(i)} - \frac{1}{\bar{\lambda}_s^n}\right]a_s ds. \quad (5.57)$$

Proof: Part a) is a routine calculation. For b), the first and second equalities in

$$\begin{aligned}
\int_0^t a_s 1(\lambda_s^n(i) > 0)[\bar{\lambda}_s^n/\bar{\lambda}_s^n]ds &= \int_0^t a_s 1(\lambda_s^n(i) > 0)ds \\
&= \int_0^t a_s 1(\lambda_s^n(i) > 0)[\lambda_s^n(i)/\lambda_s^n(i)]ds
\end{aligned}$$

yield $\int_0^t (\bar{\lambda}_s^n)^{-1}d\bar{N}_s^n$ and $\int_0^t (\lambda_s^n(i))^{-1}dN_s^n(i)$ as martingale estimators of the middle term. Consequently, $M_n(i)$, the difference of square integrable martingales, is itself a square integrable martingale. Derivation of (5.57) is straightforward using Theorem B.13. ∎

As in the one-sample problem there results a plethora of test statistics: given bounded predictable processes Y_n, one can define (square integrable) martingale test statistics

$$Z_t^n(i) = \int_0^t Y_s^n(i)dM_s^n(i), \quad (5.58)$$

where the $M_n(i)$ are given by (5.56). The predictable covariations are

$$\langle Z^n(i), Z^n(j)\rangle_t = \int_0^t Y_s^n(i)Y_s^n(j)d\langle M^n(i), M^n(j)\rangle_s,$$

so there is one new wrinkle: the $Z^n(i)$ are not orthogonal.

Nevertheless, a variant of Theorem 5.13 leads to a central limit theorem for the Z^n, but rather than give the most general result we present instead a specific version that seems useful for applications. Suppose that

$$Y_t^n(i) = \lambda_t^n(i)L_t^n, \quad (5.59)$$

where L^n is a predictable process depending only on $(\bar{N}^n, \bar{\lambda}^n)$ and is zero whenever $\bar{\lambda}^n$ is zero. Combining (5.56), (5.58) and (5.59) gives as test statistics the processes

$$Z_t^n(i) = \int_0^t L_s^n(i)dN_s^n(i) - \int_0^t [\lambda_s^n(i)/\bar{\lambda}_s^n]L_s^n d\bar{N}_s^n.$$

Theorem 5.26. Assume that (5.59) holds for each n and let b_n be constants increasing to infinity. Suppose that under H_0,

i) $\sqrt{b_n}L_t^n\lambda_t^n(i)/\bar{\lambda}_t^n \to 0$ in probability for each t and i;

ii) There exist functions $g(1)$ and $g(2) \in L^1[0, 1]$ such that for each t, i and j, $b_n(L_t^n)^2\lambda_t^n(i)\lambda_t^n(j) \to g_t(i)g_t(j)$ in probability;

iii) Indexed by n, t, i and j, the random variables in ii) form a uniformly integrable family.

Then $\sqrt{b_n}\, Z^n \xrightarrow{d} M$, where

$$M_t(i) = \int_0^t \sqrt{g_s(i)\bar{g}_s}\, dW_s(i) - \sum_{\ell=1}^2 \int_0^t g_s(i)\sqrt{g_s(\ell)/\bar{g}_s}\, dW_s(\ell),$$

$\bar{g} = g(1) + g(2)$, and $W(1)$ and $W(2)$ are independent Wiener processes. $\square$

We omit the proof, which is given in Andersen *et al.* (1982). Conditions ii) and iii) are jointly stronger and easier to verify than those of Theorem 5.10. In particular, $\sqrt{b_n}(Z_1^n(1), Z_1^n(2)) \xrightarrow{d} N(0, R)$ under H_0, where $R(i,j) = \int_0^1 g_s(i)[1(i = j)\bar{g}_s - g_s(j)]ds$, for which the martingale method (one of its beauties!) provides estimators

$$b_n \int_0^1 (L_s^n)^2 [\lambda_s^n(i)/\bar{\lambda}_s^n][1(i = j) - \lambda_s^n(j)/\bar{\lambda}_s^n]d\bar{N}_s^n.$$

Given the assumptions of Theorem 5.26 these estimators are weakly consistent under H_0.

We illustrate along the lines of Examples 5.7 and 5.17.

Example 5.27. (i.i.d. random variables). For $i = 1, 2$, let X_{ij} be a of i.i.d. random variables with values in $[0, 1]$ and hazard function $h(i)$, and assume that the two sequences are independent. For each n and i let $N_t^n(i) = \sum_{j=1}^n 1(X_{ij} \leq t)$; then with respect to the history $\mathcal{H}_t = \sigma(X_{ij}1(X_{ij} \leq t) : i = 1, 2; 1 \leq j \leq n)$ and with $\lambda_t^n(i) = n - N_{t-}^n(i)$, $N^n(i)$ has stochastic intensity $h_t(i)\lambda_t^n(i)$. The null hypothesis $H_0 : h_t(1) \equiv h_t(2)$ is that both sequences have the same distribution. Suppose that (5.59) holds with $L^n = \ell(\bar{\lambda}^n)$ for a fixed, deterministic function ℓ. Then with the notation $\tilde{\ell}(p) = \ell(p) + \sum_{q=p}^{2n} \ell(q)/q$ and since $\lambda_t^n(i) = \int_t^1 dN_t^n(i)$ and $\bar{\lambda}_{X_{ij}} = 2n + 1 - R_{ij}$, where R_{ij} is the rank of X_{ij} among $X_{11}, \ldots, X_{1n}, X_{21}, \ldots, X_{2n}$, we have

$$Z_1^n(i) = \sum_{j=1}^n \tilde{\ell}(2n + 1 - R_{ij}). \tag{5.60}$$

The choices $\ell(p) = p$ and $\ell(p) \equiv 1$ lead to the rank statistics of Wilcoxon and Savage, respectively. $\square$

5.4   Filtering with Point Process Observations

Stochastic intensities and martingales play important roles in state estimation as well as statistical inference, which is explored in this section. Let N be a point process on $\mathbf{R}_+$ and let $\mathcal{H}$ be a filtration satisfying Assumptions 2.10. The observations are the internal history $\mathcal{F}^N$. There is an $\mathcal{H}$-adapted *state process* Z that is of interest but not directly observable. However, Z

is partially observable to the extent that $\mathcal{F}^N$ yields information about it, which we seek to utilize in optimal fashion. The underlying probability is completely known, so the problem reduces (see Section 3.3) to computation of conditional expectations $E[J(Z)|\mathcal{F}_t^N]$, where $J(\cdot)$ is a functional of the path $s \to Z_s$.

Rather than work at this level of generality, however, we consider only the "real-time" *filtering* problem of recursive calculation of MMSE state estimators $\hat{Z}_t = E[Z_t|\mathcal{F}_t^N]$. Originally, "filtering" pertained to extraction of a signal from additive noise but now connotes, more generally, state estimation of the current state of a process, rather than a previous state (interpolation; smoothing) or a future state (prediction; extrapolation). Filtering is the most fundamental problem of the three because often a solution to it yields rather easily solutions to interpolation and prediction problems.

In what follows both N and Z are assumed to possess particular structure. Concerning the former, we suppose that there exists an $\mathcal{H}$-predictable stochastic intensity (λ_t), which (Exercise 2.16) entails existence of an $\mathcal{F}^N$-stochastic intensity $(\hat{\lambda}_t)$ satisfying $\hat{\lambda}_t = E[\lambda_t|\mathcal{F}_t^N]$ for each t. The latter is stipulated to admit a *semimartingale representation*

$$Z_t = Z_0 + \int_0^t X_s ds + M_t, \tag{5.61}$$

where

- X is an $\mathcal{H}$-progressive process

- M is a mean zero $\mathcal{H}$-martingale with $E[\int_0^t |dM_s|] < \infty$ for each t.

We further assume that Z is *bounded*. Assumptions in this paragraph are in force for the entire general discussion. Let $\hat{M}$ denote the $\mathcal{F}^N$-innovation martingale: $\hat{M}_t = N_t - \int_0^t \hat{\lambda} ds$.

The key implication of (5.61) is that one can in effect condition throughout to obtain a semimartingale representation for the estimator process $\hat{Z}$:

$$\hat{Z}_t = E[Z_0] + \int_0^t \hat{X}_s ds + \tilde{M}_t, \tag{5.62}$$

where $\hat{X}$ is $\mathcal{F}^N$-progressive (and hence computable — in principle — from the observations) and $\tilde{M}$ is a mean zero $\mathcal{F}^N$-martingale. In differential form (5.62) becomes the stochastic differential equation $d\hat{Z}_t = \hat{X}_t dt + d\tilde{M}_t$, the beginning of a method of recursive computation, but one must do more. By Theorem 2.34, the representation theorem for point process martingales, there is an $\mathcal{F}^N$-predictable process H, the *filter gain*, such that $\tilde{M}_t = \int_0^t H_s d\hat{M}_s$, leading to the stochastic differential equation

$$d\hat{Z}_t = \hat{X}_t dt + H_t(dN_t - \hat{\lambda}_t dt). \tag{5.63}$$

But while (5.63) is a principal step, two problems remain:

1. The filter gain H must be calculated, which is done in Theorem 5.30, the main result of the section.

2. One must compute $\hat{X}$ and the $\mathcal{F}^N$-stochastic intensity $\hat{\lambda}$; the former constitutes one part of Proposition 5.28, in which the semimartingale representation for Z is established, while the latter is considered following Theorem 5.30.

Now for the results, after which examples will be discussed.

Proposition 5.28. Let Z be an $\mathcal{H}$-adapted process with semimartingale representation (5.61). Then the state estimator process $\hat{Z}_t = E[Z_t|\mathcal{F}_t^N]$ has semimartingale representation (5.62), where $\hat{X}$ is the unique $\mathcal{F}^N$-progressive process such that

$$E[\int_0^\infty C_s X_s ds] = E[\int_0^\infty C_s \hat{X}_s ds] \tag{5.64}$$

for every bounded $\mathcal{F}^N$-progressive process C, and where $\tilde{M}$ is a mean zero $\mathcal{F}^N$-martingale.

Proof: In the same way that compensators are (dual) predictable projections of point processes, the process $\hat{X}$ fulfilling (5.64) is the *progressive projection* of X, realized as the Radon-Nikodym derivative, on the σ-algebra on $\mathbf{R}_+ \times \Omega$ generated by the $\mathcal{F}^N$-progressive processes, of the signed measure $X_t(\omega) dt\, P(d\omega)$ with respect to the σ-finite measure $dt\, P(d\omega)$. Now let

$$\tilde{M}_t = E[Z_0|\mathcal{F}_t^N] - E[Z_0] + E[\int_0^t X_s ds|\mathcal{F}_t^N] - \int_0^t \hat{X}_s ds + E[M_t|\mathcal{F}_t^N]; \tag{5.65}$$

then provided that $\tilde{M}$ is a martingale with respect to $\mathcal{F}^N$, (5.62) holds by algebraic manipulation. To show that $\tilde{M}$ is a martingale it suffices to show that $E[\int_0^t X_s ds|\mathcal{F}_t^N] - \int_0^t \hat{X}_s ds$ and $E[M_t|\mathcal{F}_t^N]$ are martingales. The latter is nearly immediate (Exercise 5.3). For the former, if $s < t$ and $A \in \mathcal{F}_s^N$, then the process $C_u(\omega) = \mathbf{1}(s < u \le t)\mathbf{1}(\omega \in A)$ is $\mathcal{F}^N$-progressive and hence by (5.64),

$$E\left[\left(E[\int_0^t X_u du|\mathcal{F}_t^N] - E[\int_0^s X_u du|\mathcal{F}_s^N]\right); A\right]$$
$$= E[\int_s^t X_u du; A] = E[\int C_u X_u du] = E[\int C_u \hat{X}_u du] = E[\int_s^t \hat{X}_u du; A]. \blacksquare$$

Corollary 5.29. Under the hypotheses of the proposition there is a bounded, $\mathcal{F}^N$-predictable process H satisfying (5.63). $\square$

By itself (5.64) is not easy to use for computing X; however, it does imply that for each t, $\hat{X}_t = E[X_t|\mathcal{F}_t^N]$, and if one can choose versions of these conditional expectations such that a progressive process results (as typically can be done in practice), then this is good enough.

We now calculate the filter gain more explicitly.

Theorem 5.30. Let $\tilde{M}$ be the $\mathcal{F}^N$-martingale defined by (5.65). Let J and K be the unique $\mathcal{F}^N$-predictable processes such that for every bounded, $\mathcal{F}^N$-predictable process C and each $t > 0$,

$$E\left[\int_0^t C_s Z_s \lambda_s ds\right] = E\left[\int_0^t C_s J_s \hat{\lambda}_s ds\right] \tag{5.66}$$

$$E\left[\int_0^t C_s \Delta M_s dN_s\right] = E\left[\int_0^t C_s K_s \hat{\lambda}_s ds\right], \tag{5.67}$$

where M is the martingale part of the semimartingale representation of Z. Then almost surely

$$H_t = J_t + K_t - \hat{Z}_{t-} \tag{5.68}$$

for all t.

Proof: The processes J and K are projections onto the $\mathcal{F}^N$-predictable σ-algebra $\mathcal{P}$ on $\mathbf{R}_+ \times \Omega$ and can be realized as Radon-Nikodym derivatives:

1. J is the derivative on $\mathcal{P}$ of the signed measure $Z_t(\omega)\lambda_t(\omega)dtP(d\omega)$ with respect to the σ-finite measure $\hat{\lambda}_t(\omega)dtP(d\omega)$, which agrees with the measure $\lambda_t(\omega)dtP(d\omega)$ on $\mathcal{P}$, so that absolute continuity holds;

2. K is the derivative of $\Delta M_t(\omega)dN_t(\omega)p(d\omega)$ with respect to the σ-finite measure $\hat{\lambda}_t(\omega)dtP(d\omega)$, which equals $dN_t(\omega)P(d\omega)$ on $\mathcal{P}$.

Note that the predictable process $\hat{Z}_- = (\hat{Z}_{t-})$ satisfies

$$E\left[\int_0^t C_s Z_s \hat{\lambda}_s ds\right] = E\left[\int_0^t C_s \hat{Z}_{s-} \hat{\lambda}_s ds\right] \tag{5.69}$$

for each C and t; indeed,

$$\begin{aligned} E\left[\int_0^t C_s Z_s \hat{\lambda}_s ds\right] &= \int_0^t E[C_s \hat{\lambda}_s E[Z_s | \mathcal{F}_s^N]]ds \\ &= \int_0^t E[C_s \hat{\lambda}_s \hat{Z}_s]ds = E\left[\int_0^t C_s \hat{Z}_{s-} \hat{\lambda}_s ds\right]. \end{aligned}$$

Suppose now that C is an $\mathcal{F}^N$-predictable process such that

1. The martingale $V_t = \int_0^t C_s d\hat{M}_s$ is *bounded*;

2. With H the filter gain of (5.63), $E[\int_0^t |C_s H_s| \hat{\lambda}_s ds] < \infty$ for all t.

Then from the definition of $\hat{Z}$, $E[Z_t V_t] = E[\hat{Z}_t V_t]$ for all t. We now compute both sides of this equality.

By (2.11),

$$
\begin{aligned}
Z_t V_t &= \int_0^t Z_{s-} dV_s + \int_0^t V_s dZ_s \\
&= \int_0^t Z_{s-} C_s (dN_s - \lambda_s ds) + \int_0^t Z_{s-} C_s (\lambda_s - \hat{\lambda}_s) ds \\
&\quad + \int_0^t V_s X_s ds + \int_0^t V_{s-} DM_s + \int_0^t C_s \Delta M_s dN_s.
\end{aligned}
$$

The first and fourth terms are mean zero $\mathcal{H}$-martingales (the processes Z_- and V_- are predictable) and consequently,

$$
\begin{aligned}
E[Z_t V_t] &= E\left[\int_0^t Z_{s-} C_s (\lambda_s - \hat{\lambda}_s) ds\right] + E\left[\int_0^t V_s X_s ds\right] \\
&\quad + E\left[\int_0^t C_s \Delta M_s dN_s\right] \\
&= E\left[\int_0^t V_s X_s ds\right] + E\left[\int_0^t C_s (J_s + k_s - \hat{Z}_{s-}) \hat{\lambda}_s ds\right],
\end{aligned}
$$

where we have used (5.66)–(5.69). In the same way,

$$
\hat{Z}_t V_t = \int_0^t \hat{Z}_{s-} C_s d\hat{M}_s + \int_0^t V_s X_s ds + \int_0^t V_{s-} H_s d\hat{M}_s + \int_0^t C_s H_s dN_s,
$$

with the first and third terms mean zero $\mathcal{F}^N$-martingales, so that

$$
E[\hat{Z}_t V_t] = E\left[\int_0^t V_s X_s ds\right] + E\left[\int_0^t C_s H_s \hat{\lambda}_s ds\right],
$$

since $\hat{\lambda}$ is the $\mathcal{F}^N$-stochastic intensity of N.

We conclude that

$$
E\left[\int_0^t C_s H_s \hat{\lambda}_s ds\right] = E\left[\int_0^t C_s (J_s + K_s - \hat{Z}_{s-}) \hat{\lambda}_s ds\right] \tag{5.70}
$$

for every t and each $\mathcal{F}^N$-predictable process C satisfying 1) and 2) above. The remainder of the proof is a monotone class argument to deduce from (5.70) that $H = J + K - Z_-$ on a large enough set that (5.68) holds. In particular, the conditions are satisfied for $C_s = \tilde{C}_s \mathbf{1}(s \leq S_n)$, where $\tilde{C}$ is bounded, nonnegative and $\mathcal{F}^N$-predictable and where $S_n = \inf\{t : N_t \geq n$ or $\int_0^t (1 + |H_s|) \hat{\lambda}_s ds \geq n\}$, in which case (5.70) yields

$$
E\left[\int_0^{S_n} \tilde{C}_s H_s \hat{\lambda}_s ds\right] = E\left[\int_0^{S_n} \tilde{C}_s (J_s + K_s - \hat{Z}_{s-}) \hat{\lambda}_s ds\right].
$$

From this we infer that $H_t \mathbf{1}(t \leq S_n) = (J_t + K_t - \hat{Z}_{t-}) \mathbf{1}(t \leq S_n)$ almost everywhere in (t, ω) with respect to the measure $\hat{\lambda}_t(\omega) dt P(d\omega)$ and hence also (because they agree on $\mathcal{P}$) with respect to $dN_t(\omega) P(d\omega)$. Letting $n \to \infty$ gives (5.68) since $S_n \to \infty$. ∎

The representation (5.68) simplifies if the state process Z and point process N have no jumps in common, for then $\Delta M_t dN_t \equiv 0$ (jumps in Z arise only from M) and therefore $K \equiv 0$, so that $d\tilde{M}_t = (J_t - \hat{Z}_{t-})[dN_t - \hat{\lambda}_t dt]$. But also, given integrability assumptions, by an argument similar to that leading to (5.69),

$$J_t = \widehat{Z\lambda_{t-}}/\hat{\lambda}_t. \tag{5.71}$$

Therefore, if $Z\lambda$ can be estimated simultaneously with Z, one obtains the ultimate representation

$$d\tilde{M}_t = [\widehat{Z\lambda_{t-}}/\hat{\lambda}_t - \hat{Z}_{t-}][dN_t - \hat{\lambda}_t dt], \tag{5.72}$$

but even now we are not finished.

Computation of the $\mathcal{F}^N$ stochastic intensity $\hat{\lambda}$ has been left unresolved; in general, the formula $\hat{\lambda}_t = E[\lambda_t | \mathcal{F}_t^N]$ is not useful directly. Indeed, this calculation is another filtering problem, whose solution may engender still further filtering problems.

We conclude with two examples. The first, the *disruption problem*, builds upon Exercises 2.20, 3.9, and 3.10 (see also Exercises 7.13, 7.16 and 7.23).

Example 5.31. (Disruption problem). At an unobservable random time T the intensity of N shifts from one known constant to another. The distribution of T is known and the objective is to estimate T itself, given observations of N.

Suppose that T has hazard function h and let $Z_t = 1(T \leq t)$ be the state process. Assume that there are known, distinct constants $a, b > 0$ such that with respect to $\mathcal{H}_t = \mathcal{F}_t^Z \vee \mathcal{F}_t^N$, N has stochastic intensity $\lambda_t = a + (b - a)Z_{t-}$, i.e., $\lambda = a$ on $[0, T]$ and $\lambda = b$ on (T, ∞). We seek to calculate $\hat{Z}_t = P\{T \leq t | \mathcal{F}_t^N\}$.

To start we need a semimartingale representation for Z. The process

$$M_t = Z_t - \int_0^{T \wedge t} h(s)ds = Z_t - \int_0^t (1 - Z_s)h(s)ds \tag{5.73}$$

is the $\mathcal{H}$-innovation for the point process ε_T so that $Z_t = \int_0^t h(s)(1 - Z_s)ds + M_t$ is the requisite representation. By Proposition 5.28,

$$\hat{Z}_t = \int_0^t h(s)(1 - \hat{Z}_s)ds + \tilde{M}_t;$$

it then remains to calculate (5.68) for $\tilde{M}$. Apparently, $\Delta M dN \equiv 0$, so K vanishes, while to calculate J we use (5.71) and the property that $Z^2 = Z$, which implies that $[a + (b - a)Z]Z = bZ$. Consequently,

$$J_t = b\hat{Z}_{t-}/[a + (b - a)\hat{Z}_{t-}].$$

Thus

$$
\begin{aligned}
d\hat{Z}_t \ = \ & h(t)(1 - \hat{Z}_{t-})dt \\
& + \frac{(b-a)\hat{Z}_{t-}(1 - \hat{Z}_{t-})}{a + (b-a)\hat{Z}_{t-}}[dN_t - (a + (b-a)\hat{Z}_{t-})dt],
\end{aligned}
$$

a completely self-contained representation. Note the variance factor $\hat{Z}(1 - \hat{Z})$ in the second term; analogues appear in Theorems 6.22 and 7.30.

In Section 7.2 we solve the second example using methods applicable on general spaces.

Example 5.32. (Mixed Poisson process). Let N be a mixed Poisson process on $\mathbf{R}_+$ with unobservable directing measure $M(dt) = Y\,dt$, where Y is a positive random variable with known distribution. With respect to $\mathcal{H}_t = \sigma(Y) \vee \mathcal{F}_t^N$, N has time-invariant but random stochastic intensity $\lambda_t \equiv Y$. The state process $Z_t \equiv Y$ is a martingale, so computation of MMSE state estimators $\hat{Z}_t = E[Y|\mathcal{F}_t^N]$ is merely a problem of representation of the $\mathcal{F}^N$-martingale $\hat{Z}$. Since the $\mathcal{H}$-stochastic intensity of N is the state process Z the procedure of this section yields

$$
dE[Y|\mathcal{F}_t^N] = \frac{E[Y^2|\mathcal{F}_{t-}^N] - E[Y|\mathcal{F}_{t-}^N]^2}{E[Y|\mathcal{F}_{t-}^N]}[dN_t - E[Y|\mathcal{F}_{t-}^N]dt].
$$

Unfortunately, this equation manifests a common flaw of recursive methods applied to computation of moments: the expression for one moment involves that of the next higher order, leading in general to an infinite, or open, system of simultaneous stochastic differential equations that cannot be solved meaningfully. In the case at hand it is easy to show using (5.71) that the equation for $dE[Y^2|\mathcal{F}_t^N]$ involves $E[Y^3|\mathcal{F}_t^N]$ and so on. Only in exceptional circumstances (e.g., when conditional distributions are Gaussian, as in the famous Kalman filter, or when the state process satisfies $Z^2 = Z$, as in Example 5.31 and Theorem 6.22) can the system be closed because a conditional moment of some order is a deterministic function of lower-order moments. $\square$

5.5   The Cox Regression Model

In 1972 Cox proposed a model for survival times incorporating *covariates*, that is, additional observable factors influencing the distribution of an individual's lifetime, in such a way that hazard functions for different individuals are mutually proportional. Specifically, a baseline hazard function is modified via an exponential factor containing the inner product of the covariate vector and a vector of unknown regression parameters. Known as both the *Cox regression model* and the *proportional hazards model*, it can accommodate censored survival times as well. More recently, a point process generalization, one principal topic of the section, has been devised. As will emerge, the point process model is a special case of Model 5.3; however, it is analyzed using a blend of martingale and maximum likelihood techniques, although one maximizes not the full likelihood function but only the "partial likelihood."

5.5.1 The Survival Time Model

To begin, we describe briefly the formulation, interpretation, and analysis of the Cox model for ordinary survival data. Let $T_1, \ldots, T_K$ be independent, positive random variables such that T_k, the survival time of individual k, has hazard function

$$h_t(k) = h(t)e^{\langle \beta, z_t(k) \rangle}, \qquad (5.74)$$

where h is an unknown hazard function common to all individuals, the $z(k)$ are deterministic covariate functions from $\mathbf{R}_+$ to $\mathbf{R}^p$, $\beta \in \mathbf{R}^p$ is an unknown regression parameter, and $\langle \cdot, \cdot \rangle$ denotes scalar product. When covariates do not depend on time, (5.74) becomes $h_t(k) = h(t)e^{\langle \beta, z(k) \rangle}$, so that different individuals' hazard functions are indeed mutually proportional. Returning to the general case, the observations are the survival times T_k, but keep in mind that the covariates are known.

The (full) likelihood function for the survival time model is

$$L(\beta, h) = \prod_{k=1}^{K} \left[h(T_k)e^{\langle \beta, z_{T_k}(k) \rangle} \exp\{ -\int_0^{T_k} h(t)e^{\langle \beta, z_t(k) \rangle} dt \} \right].$$

When h is unknown this functions is not only too cumbersome but also unbounded above. Moreover, as mentioned, interest often centers on β to the extent that one would like simply to treat h as an infinite-dimensional nuisance parameter. The approach proposed in Cox (1972b, 1975) is to factor the likelihood function into two terms, the first involving only β and hoped to capture most of the dependence on β, the second dependent on h and β, but not heavily on β, and then to estimate β by maximizing the first term alone. Thereafter, h is estimated from the second term, but with β there replaced by the previously derived estimator $\hat{\beta}$.

To carry out this procedure let $\lambda_t(\beta, k) = e^{\langle \beta, z_t(k) \rangle} 1(T_k \geq t)$ and define $\bar{\lambda}(\beta) = \sum_{k=1}^{K} \lambda_t(\beta, k)$, so that the likelihood function becomes

$$L(\beta, h) = [\prod_{k=1}^{K} \lambda_{T_k}(\beta, k) / \bar{\lambda}_{T_k}(\beta)] \left[\prod_{k=1}^{K} \bar{\lambda}_{T_k}(\beta) h(T_k) \right] e^{-\int_0^\infty \bar{\lambda}_s h(s) ds}. \quad (5.75)$$

Let $R_k = \{j : T_j \geq T_k\}$ be the set of individuals still at risk just before time T_k. Cox proposed that the first term,

$$C(\beta) = \prod_{k=1}^{K} \frac{\lambda_{T_k}(\beta, k)}{\bar{\lambda}_{T_k}(\beta)} = \prod_{k=1}^{K} \frac{e^{\langle \beta, zT_k(k) \rangle}}{\sum_{j \in R_k} e^{\langle \beta, zT_k(j) \rangle}}, \qquad (5.76)$$

now termed the *Cox partial likelihood,* be maximized in order to estimate β. It is not a conditional likelihood, nor does it capture all dependence of the likelihood function on β. The *Cox estimator* $\hat{\beta}$ is any value maximizing the partial likelihood.

One heuristic basis is that at time T_k the overall hazard rate from currently surviving individuals is (by independence) $h(T_k) \sum_{j \in R_k} e^{\langle \beta, zT_k(j) \rangle}$, of which $h(T_k)e^{\langle \beta, zT_k(k) \rangle}$ is contributed by individual k; hence if one thinks of T_k as known only to be the survival time of some unspecified individual still alive at T_k-, then the individual who dies at that time is k with probability $e^{\langle \beta, zT_k(k) \rangle} / \sum_{j \in R_k} e^{\langle \beta, zT_k(j) \rangle}$. Of course, the interpretation is inexact because T_k is already known to arise from individual k; moreover, there is no rigorous justification for the multiplication in (5.76).

Then, the integrated hazard function $H_t = \int_0^t h(s)ds$ is estimated by

$$\hat{H}_t = \sum_{T_i \leq t} \left(\sum_{j \in R_i} e^{\langle \hat{\beta}, zT_i(j) \rangle} \right)^{-1}. \qquad (5.77)$$

Various interpretations have been advanced, including that of a sieve-type maximum likelihood estimator relative to the likelihood function $L(\hat{\beta}, \cdot)$ (Johansen, 1983), but it seems best to think of $\hat{H}$ as a martingale estimator, especially since this strengthens the analogy to the point process model. Relative to the filtration $\mathcal{H}_t = \bigvee_{k=1}^{K} \mathcal{F}_t^{N(k)}$, the point processes $N(k) = \varepsilon_{T_k}$ have stochastic intensities $h(t)\lambda_t(\beta, k)$; consequently, $\bar{N} = \sum_{k=1}^{K} N(k)$ has stochastic intensity $h(t)\bar{\lambda}_t(\beta)$. If β were known, the martingale estimator of H would be $\hat{H}_t^* = \int_0^t \bar{\lambda}_s^{-1} d\bar{N}_s$, which with β replaced by $\hat{\beta}$ is precisely (5.77).

With the second and third terms from (5.75) written as

$$\exp\left[\sum_k \log(\bar{\lambda}_{T_k})h(T_k) - \int_0^\infty \bar{\lambda}(\beta)hds\right]$$
$$= \exp\left[\int_0^\infty \log(\bar{\lambda}h)d\bar{N}_s - \int_0^\infty \bar{\lambda}(\beta)dH_s\right],$$

upon substitution of $\hat{H}$ there ensues a quantity whose value does not depend on β, further argument why one might maximize the partial likelihood in order to estimate β. Note also that for each t, up to a constant e^t the term $\exp[-\int_0^t \bar{\lambda}hds + \int_0^t \log(\bar{\lambda}h)d\bar{N}]$ is the likelihood function for $\bar{N}$. Thus the partial likelihood is the ratio of the likelihood for $(N(1),\ldots,N(K))$ to that for the superposition $\bar{N}$, a slightly different interpretation of its allocating unmarked survival times among individuals; this interpretation carries over to the point process model.

5.5.2 The Point Process Model

We now introduce the point process model.

Model 5.33. (Intensity model with covariates). The index set I consists of all pairs (α, β), where $\alpha : [0, 1] \to \mathbf{R}_+$ is left-continuous with right-hand limits and where $\beta \in \mathbf{R}^p$ for some fixed $p \geq 1$. Let N be a marked point process with mark space $E = \{1, \ldots, k\}$. The *intensity model with covariates* has as filtration $\mathcal{H}_t = \bigvee_{k=1}^{K} \mathcal{F}_t^{N(k)}$ and $N(k)$ has $(P_{\alpha,\beta}, \mathcal{H})$-stochastic intensity

$$\lambda_t(\alpha, \beta, k) = \alpha_t e^{\langle \beta, Z_t(k) \rangle} Y_t(k), \tag{5.78}$$

where $Z(k)$ is $\mathcal{H}$-predictable with values in $\mathbf{R}^p$ and $Y(k)$ is $\mathcal{H}$-predictable and assumes only the values 0 and 1. The observations are $(N_t(k), Z_t(k), Y_t(k))$, $t \in [0, 1]$, for each k. □

One should interpret k as an index representing individuals and $Y_t(k)$ as a censoring process specifying whether individual k is under observation at time t. The function α is the *baseline intensity*. For individual k, $Z(k)$ is the *covariate process* depicting time-varying (numerical) factors influencing the event process $N(k)$. The *Cox regression parameter* β determines how covariates alter the baseline intensity; in particular, if some component of β is zero, then the corresponding component of Z has no influence. The functional form $e^{\langle \beta, Z_t(k) \rangle}$ matches that of Cox (1972b); it can be generalized (Jacobsen, 1982; Prentice and Self, 1983).

The aim is to estimate α and β from the observations $(N(k), Z(k), Y(k))$, $k = 1, \ldots, K$, and to describe asymptotics as $K \to \infty$. Especially in applications, interest focuses on the Cox regression parameter because only through it are individual-to-individual covariate differences manifested in the stochastic intensities.

Were β known, (5.78) would be a multiplicative intensity model with $\lambda_t(k) = e^{\langle \beta, Z_t \rangle} Y_t(k)$, to which techniques and results from Section 2 would apply directly. As we shall see, for the point process model as well as the Cox survival time model, one estimates β by maximizing a partial likelihood function not depending on α, then estimates α by martingale methods.

To accord with previous usage we replace K by n (note that the cardinality of the mark space now changes with n). Suppose that there are given processes $(N(k), Y(k), Z(k))$ on $[0, 1]$ such that for each n, if $\mathcal{H}_t^n = \bigvee_{k=1}^{n} \mathcal{F}_t^{N(k)}$, then for $k \leq n$, $N(k)$ has $(P_{\alpha,\beta}, \mathcal{H}^n)$-stochastic intensity (5.78). For each n let $\bar{N}(n) = \sum_{k=1}^{n} N(k)$.

Analysis commences with the partial log-likelihood function

$$C(\beta) = \sum_{k=1}^{n} \int_0^1 \langle \beta, Z_s(k) \rangle dN_s(k) - \int_0^1 \log(\sum_{j=1}^{n} Y_s(j) e^{\langle \beta, Z_s(j) \rangle}) d\bar{N}_s. \quad (5.79)$$

With identification of T_k in the Cox model with a point process $N(k)$ having stochastic intensity $h(t) e^{\langle \beta, z_t(k) \rangle} 1(T_k \geq t)$, the Cox partial likelihood (5.76) satisfies

$$
\begin{aligned}
\log C(\beta) &= \sum_{k=1}^{K} [\log \lambda_{T_k}(\beta, k) - \log \bar{\lambda}_{T_k}(\beta)] \\
&= \sum_{k=1}^{K} \int \langle \beta, z(k) \rangle dN(k) - \int \log(\sum_j e^{\langle \beta, z(j) \rangle}) d\bar{N},
\end{aligned}
$$

which is formally identical to (5.79).

The partial likelihood function has gradient (a random p-vector)

$$\nabla C(\beta) = \sum_{k=1}^{n} \int Z(k) dN(k) - \int \frac{\sum_{j=1}^{n} Y(j) e^{\langle \beta, Z(j) \rangle} Z(j)}{\sum_{j=1}^{n} Y(j) e^{\langle \beta, Z(j) \rangle}} d\bar{N}(n),$$

and the *Cox estimator* of β is any solution $\hat{\beta}$ to the likelihood equation $\nabla C(\beta) = 0$. For estimation of integrals $B_t(\alpha) = \int_0^t \alpha_s 1(\sum Y_s(j) e^{\langle \hat{\beta}, Z_s(j) \rangle} > 0) ds$ one uses what would be martingale estimators if β were known, but with β replaced by $\hat{\beta}$:

$$\hat{B}_t = \int_0^t (\sum_{j=1}^{n} Y_s(j) e^{\langle \hat{\beta}, Z_s(j) \rangle})^{-1} d\bar{N}_s(n).$$

Note the exact analogy to the estimators $\hat{H}$ of (5.77).

The central issue is asymptotic properties of $\hat{\beta}$ and $\hat{B}$ as $n \to \infty$. For the sake of economy, results are presented below for the case of i.i.d. processes, and only conditions not immediately implied by the i.i.d. hypothesis are stated as assumptions. A more general set of assumptions is formulated by Andersen and Gill (1982).

Suppose, therefore, that $(N(k), Z(k), Y(k))$ are i.i.d. copies of a process (N, Z, Y). With the true parameter values denoted by α_0 and β_0, let

$$s(\beta, t) = E_{\alpha_0, \beta_0}[Y_t e^{\langle \beta, Z_t \rangle}].$$

The conditions needed for consistency and asymptotic normality are as follows. Here and below, α_0 and β_0 are suppressed from the notation when possible.

Assumptions 5.34. a) $\int_0^1 \alpha_0(s) ds < \infty$;

b) There is a neighborhood U of β_0 such that

$$E\left[\sup_{t \in [0,1], \beta \in U} Y_t |Z_t|^2 e^{\langle \beta, Z_t \rangle}\right] < \infty;$$

c) $P\{Y_t = 1 \text{ for all } t\} > 0$;

d) The gradient ∇s and Hessian $\nabla^2 s$ of s satisfy

$$\nabla s(\beta, t) = E[Y_t e^{\langle \beta, Z_t \rangle} Z_t] \qquad (5.80)$$
$$\nabla^2 s(\beta, t) = E[Y_t e^{\langle \beta, Z_t \rangle} Z_t Z_t^T]; \qquad (5.81)$$

[We regard Z as a column vector; the "T" in (5.81) denotes transpose. Thus ∇s is a p-vector and $\nabla^2 s$ is a $p \times p$ matrix.]

e) With $u(\beta, t) = \nabla s(\beta, t)/s(\beta, t)$, the matrix

$$R = \int_0^1 [\nabla^2 s(\beta_0, t)/s(\beta_0, t) - u(\beta_0, t)u(\beta_0, t)^T]\alpha_0(t)s(\beta_0, t)dt \qquad (5.82)$$

is positive definite.

These are quite standard. The first has been assumed throughout the chapter, while the third bounds s away from zero, permitting division by it and by estimators as well. Differentiation inside the expectation to obtain (5.80) and (5.81) is justified by b), which is also used to verify a Lindeberg condition similar to (5.22). Finally, the matrix R is the inverse of the limit covariance matrix of $\sqrt{n}[\hat{\beta} - \beta_0]$ and so must be positive definite; the term "[·]" in (5.82) is a logarithmic second derivative, which shows that R plays the role of the Fisher information matrix.

Before treating asymptotic normality, we first consider weak consistency of the $\hat{\beta}$.

Proposition 5.35. If α_0 and β_0 satisfy Assumptions 5.34, then $\hat{\beta} \to \beta_0$ in probability.

Proof: For sufficiently large n the function

$$\begin{aligned} X_n(\beta) &= (1/n)[C(\beta) - C(\beta_0)] \\ &= \frac{1}{n}\left[\sum_{k=1}^n \int_0^1 \langle \beta - \beta_0, Z(k) \rangle dN(k) \right. \\ &\quad \left. - \int_0^1 \log\left(\frac{\sum_{j=1}^n Y(j)e^{\langle \beta, Z(j) \rangle}}{\sum_{j=1}^n Y(j)e^{\langle \beta_0, Z(j) \rangle}}\right) d\bar{N}(n)\right] \end{aligned}$$

is concave and uniquely maximized at $\hat{\beta}$. By the law of large numbers,

$$X_n(\beta) \to \int_0^1 [\langle \beta - \beta_0, \nabla s(\beta_0, t) \rangle - \log\{s(\beta, t)/s(\beta_0, t)\}]\alpha(t)dt$$

in probability, uniformly in a neighborhood of β_0, and evidently this function is concave and uniquely maximized at β_0. That $\hat{\beta} \to \beta_0$ follows by the analytical property that maximizers of convergent concave functions on $\mathbf{R}^p$ converge to the maximizer of the limit. ∎

Given consistency of the $\hat{\beta}$, asymptotic normality of $\sqrt{n}[\hat{B} - B]$ is a consequence of that of $\sqrt{n}[\hat{\beta} - \beta_0]$ and results for the multiplicative intensity model. Therefore, the crucial central limit theorem is the next one.

Theorem 5.36. If α_0 and β_0 fulfill Assumptions 5.34, then

$$\sqrt{n}[\hat{\beta} - \beta_0] \overset{d}{\to} N(0, R^{-1}), \tag{5.83}$$

where R is given by (5.82).

Proof: In form the argument is classical. Under Assumptions 5.34 there exists a Taylor expansion of the gradient $\nabla C(\cdot)$: for β near the true value β_0,

$$\nabla C(\beta) - \nabla C(\beta_0) = -J(\beta^*)(\beta - \beta_0), \tag{5.84}$$

where β^* is an intermediate value and

$$J(\beta) = \int_0^1 U(\beta, t) d\bar{N}_t(n), \tag{5.85}$$

with $U(\beta, t)$ the obvious empirical estimator of the term "$[\cdot]$" in (5.82): $s(\beta, t)$ is replaced by $\frac{1}{n}\sum_{k=1}^n Y_t(k)e^{\langle \beta, Z_t(k)\rangle}$ and so on. Since $\nabla C(\hat{\beta}) = 0$ by definition, (5.84) implies that

$$\frac{1}{\sqrt{n}}\nabla C(\beta_0) = \frac{J(\beta^*)}{n}\sqrt{n}[\hat{\beta} - \beta_0],$$

so that (5.83) holds provided that

$$(1/\sqrt{n})\nabla C(\beta_0) \overset{d}{\to} N(0, R) \tag{5.86}$$

and $J(\beta^*)/n \to R$ in probability.

The latter is direct: since $\hat{\beta} \to \beta_0$, we also have $\beta^* \to \beta_0$, so it is enough to show that $J(\beta_0)/n \to R$ in probability, which holds by the weak law of large numbers and the very definition of $J(\beta_0)$. The s-ratios in (5.82) are limits of the empirical estimators replacing them in (5.85), while $d\bar{N}_t(n) \to E[dN_t] = \alpha_0(s)s(\beta_0, t)dt$.

For each k, let $M_t(k) = N_t(k) - \int_0^t \alpha_0 e^{\langle \beta_0, Z(k)\rangle} ds$ be the innovation martingale associated with $N(k)$, and put $\bar{M}(n) = \sum_{k=1}^n M(k)$. Then

$$\frac{\nabla C(\beta_0)}{\sqrt{n}}$$

$$= \frac{1}{\sqrt{n}} \left[\sum_{k=1}^{n} \int Z(k) dN(k) - \int \frac{\sum_j Y(j) e^{\langle \beta_0, Z(j) \rangle} Z(j)}{\sum_j Y(j) e^{\langle \beta_0, Z(j) \rangle}} d\bar{N}(n) \right]$$

$$= \frac{1}{\sqrt{n}} \left[\sum_{k=1}^{n} \int Z(k) dM(k) - \int \frac{\sum_j Y(j) e^{\langle \beta_0, Z(j) \rangle} Z(j)}{\sum_j Y(j) e^{\langle \beta_0, Z(j) \rangle}} d\bar{M}(n) \right]$$

$$\cong \frac{1}{\sqrt{n}} \left[\sum_{k=1}^{n} \int Z(k) dN(k) - \int \frac{\nabla s(\beta_0, \cdot)}{s(\beta_0, \cdot)} d\bar{M}(n) \right]$$

$$= \frac{1}{\sqrt{n}} \sum_{k=1}^{n} \int \left[Z(k) - \frac{\nabla s(\beta_0, \cdot)}{s(\beta_0, \cdot)} \right] dM(k),$$

to which we can apply Theorem 5.11 and associated calculations to derive (5.86). ■

Asymptotic behavior of $B_t^* = \int_0^t (\sum_{k=1}^n Y(k) e^{\langle \beta_0, Z(k) \rangle})^{-1} d\bar{N}(n)$ (note that here β_0 is fixed) is an immediate consequence of Theorem 5.16.

Proposition 5.37. *If Assumptions 5.34 hold, then* $\sqrt{n}[B^* - B] \xrightarrow{d} M$, *where* M *is a continuous Gaussian martingale with variance function* $v_t(\alpha_0, \beta_0) = \int_0^t [\alpha_0(u)/s(\beta_0, u)] du$. $\square$

Taking account of the difference

$$\sqrt{n}[\hat{B}_t - B_t^*]$$

$$= \sqrt{n} \int_0^t \left[(\sum_j Y(j) e^{\langle \hat{\beta}, Z(j) \rangle})^{-1} - (\sum_j Y(j) e^{\langle \beta_0, Z(j) \rangle})^{-1} \right] d\bar{N}(n)$$

$$\cong \left\langle \hat{\beta} - \beta_0, \int_0^t \frac{\sum_j Y(j) e^{\langle \beta_0, Z(j) \rangle} Z(j)}{\sum_j Y(j) e^{\langle \beta_0, Z(j) \rangle}} d\bar{N}(n) \right\rangle$$

$$\cong -\sqrt{n} \left\langle \hat{\beta} - \beta_0, \int_0^t [\nabla s(\beta_0, u)/s(\beta_0, u)] \alpha_0(u) du \right\rangle$$

leads to the final result regarding this model.

Theorem 5.38. *If Assumptions 5.34 are fulfilled, then the vectors* $\sqrt{n}[\hat{\beta} - \beta_0]$ *and the processes*

$$\sqrt{n}[\hat{B}_t - B_t(\alpha_0)] + \sqrt{n} \left\langle \hat{\beta} - \beta_0, \int_0^t [\nabla s(\beta_0, u)/s(\beta_0, u)] \alpha_0(u) du \right\rangle$$

are asymptotically independent and the latter converges to a Gaussian martingale with variance function given in Proposition 5.37. $\square$

We omit the proof of asymptotic independence; the crux of the argument is that the prelimit martingales are orthogonal (Andersen and Gill, 1982), which becomes independence for the Gaussian limits.

5.5.3   Time-Dependent Covariate Effects

We conclude the section with a model for censored survival data allowing time-varying *covariate effects*. Conditional on the p-covariate value $\mathbf{z}$, the covariate-specific hazard function $\lambda(t|\mathbf{z})$ is given $\lambda(t|\mathbf{z}) = \lambda_0(t)e^{\langle\beta_0(t),\mathbf{z}\rangle}$, where $\beta_0(t)$ is an unknown function taking values in $\mathbf{R}^p$ and λ_0 is an unknown, nonnegative function. Note that the covariates, albeit random, are independent of time, while their effects on the hazard rate do depend on time.

Associated with each individual i is a triplet $(T_i^0, V_i, \mathbf{Z}_i)$, in which T_i^0 is a nonnegative random variable representing that individual's (potential) failure time, V_i is the (potential) censoring time, and $\mathbf{Z}_i$ is a random p-vector of covariates. The observable data for individual i are $T_i = \min\{T_i^0, V_i\}$, $D_i = 1(T_i = T_i^0)$, and $\mathbf{Z}_i$. We assume throughout that

1. The $(T_i^0, V_i, \mathbf{Z}_i)$ are i.i.d. copies of a triplet $(T^0, V, \mathbf{Z})$, in which T^0 and V are conditionally independent given the covariate $\mathbf{Z}$.

2. The distributions of T^0 and V are absolutely continuous.

3. The each component of the covariate $\mathbf{Z}$ lies in the interval $[0,1]$.

We further define $N_i(t) = 1(T_i \leq t, D_i = 1)$, the counting process of observed failures — there is at most one — for individual i; $Y_i(t) = 1(T_i \geq t)$, the "at risk" indicator process; $\lambda_i(t) = \lambda(t|\mathbf{Z}_i)Y_i(t)$, the stochastic intensity for the counting process N_i; and $M_i(t) = N_i(t) - \int_0^t \lambda_i(u)du$, the innovation martingale. Also, let $\bar{N}(t) = (1/n)\sum_1^n N_i(t)$, with $\bar{M}(t)$ defined analogously.

The main problem is to estimate the regression parameter function β_0, which is done by maximizing a "penalized" version of the partial likelihood used in the ordinary Cox regression model. More precisely, the maximum penalized partial likelihood estimator, or MPPLE, is the maximizer of

$$L(\beta) = (1/n)\sum_{i=1}^n D_i[\beta(T_i)Z_i - \log(\sum_{j=1}^n Y_j(T_i)e^{\beta(T_i)Z_j})] - (1/2)\alpha_n[\beta,\beta].$$

Here, α_n are positive numbers chosen by the statistician and, with $m \geq 3$ an integer also chosen by the statistician, $[f,g] = \int_0^1 f^{(m)}(t)g^{(m)}(t)dt$ for f and g belonging to the Sobolev space H^m of piecewise m-times differentiable functions f with $[f,f] < \infty$. The first term $L(\beta)$ is (except for the factor $1/n$) exactly the logarithm of the Cox partial likelihood. The second term is a penalty functional designed to make the estimator smooth, and thereby reduce variance. Also, without the penalty the partial likelihood is unbounded above.

We use the following notation, where $x \in \mathbf{R}$:

$$
\begin{aligned}
S_p(x; s) &= (1/n)\sum_{i=1}^{n} Y_i(s) Z_i^p e^{xZ_i}, \quad p = 0,\ldots,3 \\
A(x; s) &= S_1(x; s)/S_0(x; s) \\
V(x; s) &= (1/n)\sum_{i=1}^{n} Y_i(s)[Z_i - A(x; s)]^2 e^{xZ_i}/S_0(x; s) \\
&= S_2(x; s)/S_0(x; s) - A(x; s)^2 \\
C(x; s) &= (1/n)\sum_{i=1}^{n} Y_i(s)[Z_i - A(x; s)]^3 e^{xZ_i}/S_0(x; s) \\
s_p(x; s) &= E[Y(s) Z^p e^{xZ}], \quad p = 0,\ldots,3 \\
a(x; s) &= s_1(x; s)/s_0(x; s) \\
v(x; s) &= \frac{E[Y(s)(Z - a(x; s))^2 e^{xZ}]}{s_0(x; s)} = \frac{s_2(x; s)}{s_0(x; s)} - a(x; s)^2 \\
c(x; s) &= E[Y(s)(Z - a(x; s))^3 e^{xZ}]/s_0(x; s)
\end{aligned}
$$

Also, let $v_0 = (1/2)\inf_s v(\beta_0(s); s)$, $V_1(x; s) = \max\{V(x; s), v_0\}$, ($v_1(x; s)$ is defined analogously), and $w(s) = \lambda_0(s)s_0(\beta_0(s); s)v(\beta_0(s); s)$. Here and throughout β_0 denotes the "true" value of the unknown parameter; expectations above are taken under this value. We write $S_p(\beta, s)$ for $S_p(\beta(s); s)$, with similar abbreviations of the other quantities.

The following are basic assumptions.

Assumptions 5.39. a) The parameters α_n are deterministic;
b) $w(s) \geq w_0$ for some positive constant w_0;
c) $\beta_0 \in H^m$;
d) w is $(2m - 1)$ times continuously differentiable on $[0, 1]$.

Note that Assumption 5.39a) prevents data-dependent choice of the α_n. Assumption 5.39b) is mild, and ensures that there is adequate "action" on the entire interval $[0, 1]$, in terms of failures and covariate variability, for estimation of β_0 to be meaningful. Assumption 5.39d) is required only to prove asymptotic normality of the estimators.

Up to quantities not depending on β, the negative of the log penalized likelihood is given by

$$
H(\beta) = (1/2)\alpha[\beta, \beta] + \sum_{i=1}^{n} D_i(\log[S_0(\beta, T_i)/S_0(\beta_0, T_i)] - Z_i[\beta(T_i) - \beta_0(T_i)]),
$$

whose first- and second-order Gâteaux differentials are

$$
\begin{aligned}
\delta H(\beta; f) &= \alpha[\beta, f] + (1/n)\sum_{i=1}^{n} D_i[A(\beta, T_i) - Z_i]f(T_i) \\
\delta^2 H(\beta; f, g) &= \alpha[f, g] + (1/n)\sum_{i=1}^{n} D_i V(\beta, T_i)f(T_i)g(T_i).
\end{aligned}
$$

Existence and computation of the MPPLE are treated in Zucker and Karr (1990). The main points are that with probability one, existence conditions are fulfilled for all sufficiently large n, and that computing the MPPLE reduces to a finite-dimensional maximization.

Proceeding to consistency, we note that as $n \to \infty$,

$$
\begin{aligned}
\sup_s |S_p(\beta_0, s) - s_p(\beta_0, s)| &= O_P(n^{-1/2}) \\
\sup_s |A(\beta_0, s) - a(\beta_0, s)| &= O_P(n^{-1/2}) \\
\sup_s |V(\beta_0, s) - v(\beta_0, s)| &= O_P(n^{-1/2}) \qquad (5.87) \\
\sup_t |\bar{N}(t) - \int_0^t \lambda_0(s) s_0(\beta_0, s) ds| &= O_P(n^{-1/2}). \qquad (5.88)
\end{aligned}
$$

We now show that the MPPLE $\hat{\beta}$ converges to β_0 in probability with respect to the uniform norm on $[0, 1]$; the argument is patterned after Silverman (1982). Define

$$
\begin{aligned}
H_1(\beta) &= (1/2)\alpha[\beta, \beta] + (1/2)\int_0^1 w(s)[\beta(s) - \beta_0(s)]^2 ds \\
&\quad - (1/n)\sum_{i=1}^n \int_0^1 [Z_i - A(\beta_0, s)][\beta(s) - \beta_0(s)] dN_i(s).
\end{aligned}
$$

Then the idea of the proof is this:

- Show that the minimizer β_1 of $H_1(\cdot)$ converges to β_0 as $n \to \infty$;

- Show that for n sufficiently large, β_1 is close to $\hat{\beta}$.

The motivation for H_1 is as follows. A two-term Taylor series expansion suggests the following "approximation" to $H(\beta)$:

$$
\begin{aligned}
H(\beta) &\cong (1/2)\alpha[\beta, \beta] + (1/2)\int_0^1 V(\beta_0, s)[\beta(s) - \beta_0(s)]^2 d\bar{N}(s) \\
&\quad - (1/n)\sum_{i=1}^n \int_0^1 [Z_i - A(\beta_0, s)][\beta(s) - \beta_0(s)] dN_i(s).
\end{aligned}
$$

The results above further suggest replacing $V(\beta_0, s)$ by $v(\beta_0, s)$ and $d\bar{N}(s)$ by $\lambda_0(s) s_0(\beta_0, s) ds$. This gives $H(\beta) \doteq H_1(\beta)$.

To begin the consistency argument, define on H^m the inner products $\langle f, g \rangle_w = \int_0^1 f(s) g(s) w(s) ds$ and $\langle f, g \rangle_{H^m} = \langle f, g \rangle_w + [f, g]$. Then, by virtue of the assumption that $0 < w_0 \le w(s) \le \|w\|_\infty$ and Sobolev space theory (as in Silverman, 1982), there exist functions ϕ_ν in H^m and numbers $1 = \mu_0 \ge \mu_1 \ge \mu_2 \ge \cdots \ge 0$ such that (ϕ_ν) is an orthonormal basis for $L^2[0, 1]$ under the inner product $\langle f, g \rangle_w$ and $(\mu_\nu^{1/2} \phi_\nu)$ is an orthonormal basis for H^m under the inner product $\langle f, g \rangle_{H^m}$.

Let (b_ν) and $(b_{0\nu})$ be the coefficients in the expansions of β and β_0, respectively, in terms of (ϕ_ν). Then

$$
H_1(\beta) = (\alpha/2)\sum_0^\infty \rho_\nu b_\nu^2 + (1/2)\sum_0^\infty (b_\nu - b_{0\nu})^2 - \sum_0^\infty X_\nu(b_\nu - b_{0\nu}),
$$

where

$$X_\nu = (1/n)\sum_{i=1}^n \int_0^1 [Z_i - A(\beta_0, s)]\phi_\nu(s)dN_i(s)$$
$$= (1/n)\sum_{i=1}^n \int_0^1 [Z_i - A(\beta_0, s)]\phi_\nu(s)dM_i(s)$$

consequently, $E[X_\nu] = 0$ and $\text{Var}(X_\nu) \le 1/n$.

Evidently H_1 may be minimized by minimizing the terms in the ν-summation individually; the coefficients of β_1 are then given by $b_{1\nu} = (X_\nu + b_{0\nu})/(1 + \alpha\rho_\nu)$.

The proposition below gives bounds for the distance between β_1 and β_0.

Proposition 5.40. There exist constants $C_\epsilon^{(1)}$, $\epsilon > 0$, and $C^{(2)}$, not depending on n, such that

$$E[\|\beta_1 - \beta_0\|_\infty^2] \le C_\epsilon^{(1)}[\alpha^{-\epsilon/2m}(n^{-1}\alpha^{-1/m} + \alpha^{1-1/m})];$$
$$E[\|\beta_1 - \beta_0\|_{H^1}^2] \le C^{(2)}[\alpha^{-1/2m}(n^{-1}\alpha^{-1/m} + \alpha^{1-1/m})].\ \square$$

Before turning to the difference between β_1 and $\hat{\beta}$, further preliminaries are needed. Define

$$H_M(\beta) = (1/2)\alpha[\beta, \beta] + \int_0^1 \int_{\beta_0(s)}^{\beta(s)} \int_{\beta_0(s)}^u V_1(x; s)dx\, du\, d\bar{N}(s)$$
$$- (1/n)\sum_{i=1}^n \int_0^1 [Z_i - A(\beta_0, s)][\beta(s) - \beta_0(s)]dN_i(s).$$

The first and second-order Gâteaux differentials of H_M satisfy

$$\begin{aligned}
\delta H_M(\beta; f) &= \alpha[\beta, f] + \int_0^1 f(s)\int_{\beta_0(s)}^{\beta_1(s)} V_1(x, s)dx\, d\bar{N}(s) \\
&\quad - (1/n)\sum_{i=1}^n \int_0^1 f(s)[Z_i - A(\beta_0, s)] \\
&\quad \times [\beta(s) - \beta_0(s)]dN_i(s) \qquad (5.89) \\
\delta^2 H_M(\beta; f, g) &= \alpha[f, g] + \int_0^1 f(s)g(s)V_1(\beta, s)d\bar{N}(s) \\
&\ge \alpha[f, g] + v_0\int_0^1 f(s)g(s)d\bar{N}(s).
\end{aligned}$$

Because β_1 minimizes H_1, for each $f \in H^m$,

$$\begin{aligned}
0 = \delta H_1(\beta_1; f) &= \alpha[\beta_1, f] + \int_0^1 w(s)[\beta_1(s) - \beta_0(s)]f(s)ds \\
&\quad - (1/n)\sum_{i=1}^n \int_0^1 [Z_i - A(\beta_0, s)]dN_i(s).
\end{aligned}$$

In conjunction with (5.89), this implies that

$$\begin{aligned}
\delta H_M(\beta_1; f) &= \int_0^1 f(s)\int_{\beta_0(s)}^{\beta_1(s)} V_1(x; s)dx\, d\bar{N}(s) \\
&\quad - \int_0^1 w(s)[\beta_1(s) - \beta_0(s)]f(s)ds. \qquad (5.90)
\end{aligned}$$

The next proposition gives a bound for $\delta H_M(\beta_1; f)$.

Proposition 5.41. For $f \in H^m$ and possibly random,

$$|\delta H_M(\beta_1; f)| = O_P\left(\|\beta_1 - \beta_0\|_\infty^2 + n^{-1/2}\|\beta_1 - \beta_0\|_{H^1}\|f\|_{H^1}\right).$$

Proof: By (5.90),

$$
\begin{aligned}
\delta H_M(\beta_1; f) &= \int_0^1 f(s) \int_{\beta_0(s)}^{\beta_1(s)} [V_1(x; s) - v_1(x; s)] dx \, d\bar{N}(s) \\
&\quad + \int_0^1 \int_{\beta_1(s)}^{\beta_0(s)} [v_1(x; s) - v(\beta_0, s)] dx \, d\bar{N}(s) \\
&\quad + \int_0^1 f[\beta_1 - \beta_0] v(\beta_0, s) \\
&\quad \times [d\bar{N}(s) - s_0(\beta_0, s)\lambda_0(s)ds].
\end{aligned}
\tag{5.91}
$$

These three terms will be treated in turn.

i) For fixed s and for x between $\beta_0(s)$ and $\beta_1(s)$, the mean value theorem implies that

$$V(x; s) - v(x; s) = [V(\beta_0, s) - v(\beta_0, s)] + [C(x^*; s) - c(x^*; s)][x - \beta_0(s)]$$

for some x^* between $\beta_0(s)$ and $\beta_1(s)$. Now $|C(x^*; s) - c(x^*; s)| \le 2$. Also, by (5.87), $\sup_s |V(\beta_0, s) - v(\beta_0, s)| = O_P(n^{-1/2})$, and therefore

$$|V_1(x; s) - v_1(x; s)| \le |V(x; s) - v(x; s)| \le O_P(n^{-1/2}) + 2\|\beta_1 - \beta_0\|_\infty.$$

Consequently,

$$
\begin{aligned}
&\left|\int_0^1 f(s) \int_{\beta_0(s)}^{\beta_1(s)} [V_1(x; s) - v_1(x; s)] dx \, d\bar{N}(s)\right| \\
&= O_P\left(\left[\|\beta_1 - \beta_0\|_\infty^2 + n^{-1/2}\|\beta_1 - \beta_0\|_\infty\right]\|f\|_{H^1}\right).
\end{aligned}
$$

ii) Fix $s \in [0, 1]$ and x between $\beta_0(s)$ and $\beta_1(s)$. Then, by a mean value argument similar to that in i), $|v_1(x; s) - v(x; s)| \le 2v_0^{-1}\|\beta_1 - \beta_0\|_\infty$. Accordingly,

$$\left|\int_0^1 \int_{\beta_0(s)}^{\beta_1(s)} [v_1(x; s) - v(\beta_0, s)] dx \, d\bar{N}(s)\right| \le (2v_0^{-1} + 1)\|\beta_1 - \beta_0\|_\infty^2 \|f\|_{H^1}.$$

iii) The integrand in the third term in (5.91) is not predictable, so martingale theory cannot be applied; an alternative argument is needed. By integration by parts,

$$
\begin{aligned}
&\left|\int_0^1 v(\beta_0, s)[\beta_1(s) - \beta_0(s)][d\bar{N}(s) - s_0(\beta_0, s)\lambda_0(s)ds]\right| \\
&\le v(\beta_0, 1)|\beta_1(1) - \beta_0(1)| \, |f(1)| \left|\bar{N}(1) - \int_0^1 s_0(\beta_0, s)\lambda_0(s)ds\right| \\
&\quad + \sup_t \left|\bar{N}(t) - \int_0^t s_0(\beta_0, s)\lambda_0(s)ds\right| \\
&\quad \times \int_0^1 \left|(d/ds)[v(\beta_0, s)\lambda_0(s)[\beta_1(s) - \beta_0(s)]f(s)]\right| ds.
\end{aligned}
\tag{5.92}
$$

We have observed that $\sup_t |\bar{N}(t) - \int_0^t s_0(\beta_0, s)\lambda_0(s)ds| = O_P(n^{-1/2})$. Thus the first summand in (5.92) is $O_P(n^{-1/2}\|\beta_1 - \beta_0\|_{H^1}\|f\|_{H^1})$. The second summand is bounded by the product of a quantity that is $O_P(n^{-1/2})$ and

$$\int_0^1 \left|(d/ds)[v(\beta_0, s)[\beta_1(s) - \beta_0(s)]f(s)]\right| ds$$
$$\leq \max\{\|v'\|_\infty, 1\}\|\beta_1 - \beta_0\|_{H^1}\|f\|_{H^1}.$$

Combining i), ii) and iii) completes the proof. ∎

It follows that $\delta^2 H_M(\beta; g, g) \geq C^0\alpha^{1/m}\|g\|_{H^1}^2 - O_P(n^{-1/2}\|g\|_{H^1}^2)$ uniformly in g. Thus, if $\alpha = O(n^{-\theta})$ with $\theta < m/2$, then the probability that H_M is uniformly convex converges to 1 as $n \to \infty$. When H_M is uniformly convex, it has a unique minimizer, which we denote by β_M.

By construction, $H_M(\beta) = H(\beta)$ if, for each s, $V(x; s) \geq v_0$ for every x between $\beta_0(s)$ and $\beta(s)$. Define the (compact) interval $I = [-1 - \inf_s \beta_0(s), 1 + \sup_s \beta_0(s)]$. Then by Andersen and Gill (1982, Theorem III.1) applied with $Y_i(s)e^{xZ_i}$ viewed as a random element of the space of left-continuous, right-limited functions from $[0, 1]$ to the set of continuous functions on I, $\sup_{s\in[0,1], x\in I}|S_p(x; s) - s_p(x; s)| \stackrel{\text{a.s.}}{\to} 0$ for $p = 0, 1, 2$. Consequently, for all n sufficiently large (depending on the realization ω), $V(x; s) \geq v_0$ for all $x \in I$ and $s \in [0, 1]$.

Thus, provided $\|\beta - \beta_0\|_\infty \leq 1$, $H_M(\beta) = H(\beta)$, and similar equalities obtain for the first- and second-order Gâteaux differentials, for all n sufficiently large. Hence $\hat{\beta}$ will be equal to β_M if $\|\beta_M - \beta_0\|_\infty \leq 1$ and n is sufficiently large. Hence an analysis of the difference between β_M and β_1 provides information regarding that between $\hat{\beta}$ and β_1. This idea is implemented in the next proposition.

Proposition 5.42. Let $\epsilon > 0$ be given and suppose that $\alpha_n = O(n^{-\theta})$ with $0 < \theta < 2m/(4 + \epsilon)$. Then provided $m \geq 3$,

$$\|\hat{\beta} - \beta_1\|_\infty = O_P\left(n^{-1+[(4+\epsilon)/2m]\theta} + n^{-[1-(4+\epsilon)/2m]\theta}\right.$$
$$\left. + n^{-[1+(1-7/2m)]\theta/2}\right).$$

Proof: Put $g = \beta_1 - \beta_M$. Then, recalling that $\delta H_M(\beta_M; f) = 0$ for all $f \in H^m$, by Taylor's theorem for functionals,

$$\delta H_M(\beta_1; g) = \delta^2 H_M(\beta_M + \xi g; g, g) \tag{5.93}$$

for some $\xi \in [0, 1]$. From (5.90),

$$\delta^2 H_M(\beta_M + \xi g; g, g) \geq C^0\alpha^{1/m}\|g\|_{H^1}^2 - O_P\left(n^{-1/2}\|g\|_{H^1}^2\right). \tag{5.94}$$

On the other hand, Proposition 5.41 gives

$$|\delta H_M(\beta_1; g)| \le O_P\left(\left[\|\beta_1 - \beta_0\|_\infty^2 + n^{-1/2}\|\beta_1 - \beta_0\|_{H^1}\right]\|g\|_{H^1}\right). \quad (5.95)$$

Putting (5.94) and (5.95) into (5.93) and canceling a factor of $\|g\|_{H^1}$ yields, under the hypothesis regarding α_n,

$$\|\beta_M - \beta_1\|_{H^1} \le \alpha^{-1/m}O_P\left(\|\beta_1 - \beta_0\|_\infty^2 + n^{-1/2}\|\beta_1 - \beta_0\|_{H^1}\right). \quad (5.96)$$

Therefore, by Proposition 5.41 and the hypothesis on α_n, and the Sobolev embedding theorem, $\|\beta_1 - \beta_0\|_\infty \to 0$, (this and convergences below are in probability) and for an appropriate constant c, $\|\beta_1 - \beta_M\|_\infty \le c\|\beta_1 - \beta_M\|_{H^1} \to 0$, and so $\|\beta_M - \beta_0\|_\infty \to 0$. Accordingly, $P\{\beta_M = \hat{\beta}\} \to 1$.

Substituting the rates in Proposition 5.41 into (5.96), which now holds with β_M replaced by $\hat{\beta}$, establishes the bound for $\|\hat{\beta} - \beta_1\|_{H^1}$

By the Sobolev embedding theorem, the same holds for $\|\hat{\beta} - \beta_1\|_\infty$. ∎

Finally, the propositions combine to yield consistency.

Theorem 5.43. Let $\varepsilon > 0$ be given and suppose that $\alpha_n = O(n^{-\theta})$ with $0 < \theta < 2m/(4 + \varepsilon)$. Then

$$\begin{aligned}
\|\hat{\beta} - \beta_0\|_\infty &= O_P\left(n^{-1+[(4+\varepsilon)/2m]\theta} + n^{-1/2+[(2+\varepsilon)/4m]\theta}\right.\\
&\quad + n^{-[1+(1-7/2m)\theta]/2} + n^{-[1-(4+\varepsilon)/2m]\theta}\\
&\quad \left. + n^{-[1-(2+\varepsilon)/2m]\theta/2}\right).\ \square
\end{aligned}$$

In this expression, the first two terms represent variability and the latter three represent bias.

Finally, we consider pointwise asymptotic normality of the estimators $\hat{\beta}(t)$. Here is the necessary notation:

$$\begin{aligned}
X_\nu^* &= (1/n)\sum_{i=1}^n \int_0^1 [Z_i - a(\beta_0, s)]\phi_\nu(s)dM_i(s)\\
\beta^*(t) &= \sum_{\nu=0}^\infty [(X_\nu^* + b_{0\nu})/(1 + \alpha\rho_\nu)]\phi_\nu(t)\\
\beta_\alpha(t) &= \sum_{\nu=0}^\infty [b_{0\nu}/(1 + \alpha\rho_\nu)]\phi_\nu(t)\\
U(t) &= \sum_{\nu=0}^\infty [X_\nu^*/(1 + \alpha\rho_\nu)]\phi_\nu(t)\\
R_\alpha(s, t) &= \sum_{\nu=0}^\infty [1/(1 + \alpha\rho_\nu)]\phi_\nu(s)\phi_\nu(t)\\
r_\alpha(s, t) &= \sum_{\nu=0}^\infty [1/(1 + \alpha\rho_\nu)^2]\phi_\nu(s)\phi_\nu(t).
\end{aligned}$$

With this notation,

$$\hat{\beta}(t) - \beta_0(t) = [\hat{\beta}(t) - \beta_1(t)] + [\beta_1(t) - \beta^*(t)] + [\beta_\alpha(t) - \beta_0(t)] + U(t).$$

The idea of the proof is to show that $U(t)$, suitably standardized, is asymptotically normal, and to establish appropriate bounds on the remaining terms. A bound on $\|\hat{\beta} - \beta_1\|_\infty$ was given above. We also have

$$E\left[\|\beta_1 - \beta^*\|_\infty^2\right] \leq C_\epsilon^* n^{-2} \alpha^{-(2+\epsilon)/2m}$$
$$\|\beta_\alpha - \beta_0\|_\infty^2 \leq C_\epsilon^* \alpha^{1-1/m-\epsilon/2m}.$$

Here is the key step.

Proposition 5.44. Provided that $n^{1/2}\alpha^{1/4m} \to \infty$, as $n \to \infty$, we have $U(t)/\sqrt{\text{Var}(U(t))} \xrightarrow{d} N(0,1)$.

Proof: Put $W_{ni} = \int_0^1 R_\alpha(s,t)[Z_i - a(\beta_0, s)]dM_i(s)$. Then evidently $U(t) = (1/n)\sum_{i=1}^n W_{ni}$ and the W_{ni} are i.i.d. as i varies with n fixed, with mean 0 and (making use of orthonormality of the ϕ_ν) variance

$$\sigma_n^2(t) = E[W_{ni}^2] = r_\alpha(t,t). \tag{5.97}$$

With t fixed, because the ϕ_ν are eigenfunctions of a differential operator, there exist positive constants r_0 and r_1 (depending on t) such that

$$r_0 \alpha^{-1/2m} \leq \sigma_n^2 \leq r_1 \alpha^{-1/2m}. \tag{5.98}$$

By Chow and Teicher (1978, Corollary 12.2.2), it suffices to show that for $\xi > 0$, $nP\{|W_{ni}|/\sigma_n > \xi n^{1/2}\} \to 0$ as $n \to \infty$. By Markov's inequality, for $\Delta > 0$,

$$nP\{|W_{ni}|/\sigma_n > \xi n^{1/2}\} \leq \frac{E[|W_{ni}|^{2+\Delta}]}{\xi^{2+\Delta} n^{\Delta/2} \sigma_n^{2+\Delta}}. \tag{5.99}$$

The goal now is to show that the right-hand side of (5.99) converges to zero as $n \to \infty$.

As a preliminary, differential operator theory implies that $|\phi_\nu(t)|$ is bounded uniformly in ν and t, so that for some constant C_ϵ^*, $\sup_{s,t} |R_\alpha(s,t)| \leq C_\epsilon^* \alpha^{-1/2m}$.

Now recall that $dM_i(s) = dN_i(s) - \lambda_0(s)Y_i(s)e^{\beta_0(s)Z_i}ds$. Hence, with $d|M_i(s)|$ denoting the total variation of the signed measure $dM_i(s)$, it follows that $d|M_i(s)| \leq dN_i(s) + \|\lambda_0\|_\infty e^{\|\beta_0\|_\infty} ds$, and therefore

$$E[|W_{ni}|^{2+\Delta}] \leq 2^{2+\Delta} E\left[(\int_0^1 |R_\alpha(s,t)|dN_i(s))^{2+\Delta}\right]$$
$$+ \left(2\|\lambda_0\|_\infty e^{\|\beta_0\|_\infty}\right)^{2+\Delta} \left(\int_0^1 |R_\alpha(s,t)|ds\right)^{2+\Delta}.$$

Because the counting process N_i has at most one point,

$$E\left[(\int_0^1 |R_\alpha(s,t)|dN_i(s))^{2+\Delta}\right] = E\left[\int_0^1 |R_\alpha(s,t)|^{2+\Delta}dN_i(s)\right]$$
$$\leq K_1\sigma_n^2\alpha^{-\Delta/2m},$$

where K_1 is a constant. By similar analysis, for a constant K_2,

$$\left(\int_0^1 |R_\alpha(s,t)|ds\right)^{2+\Delta} \leq K_2\sigma_n^2\alpha^{-\Delta/2m}.$$

Thus for some κ, $E[|W_{ni}|^{2+\Delta}] \leq \kappa\sigma_n^2\alpha^{-\Delta/2m}$. By this last inequality, the right-hand side of (5.99) is bounded by $\xi^{-(2+\Delta)}\kappa(n^{1/2}\sigma_n\alpha^{1/2m})^{-\Delta}$, and this converges to zero as $n \to \infty$ by (5.98) and the hypothesis on α_n. ∎

At this point, the groundwork is laid for the final result of the chapter.

Theorem 5.45. Let $\varepsilon > 0$ be given and suppose that $\alpha_n = O(n^{-\theta})$ with $1/[1-(1+\varepsilon)/2m] < \theta < 2m/(7+\varepsilon)$. Also, suppose that $m \geq 4$. Then, as $n \to \infty$, for each fixed $t \in [0,1]$, with σ_n^2 defined as in (5.97),

$$\sqrt{n/\sigma_n^2}[\hat{\beta}(t) - \beta_0(t)] \xrightarrow{d} N(0,1).$$

Proof: It now suffices to show that $\sqrt{n/\sigma_n^2}[\hat{\beta}(t) - \beta_1(t)]$, $\sqrt{n/\sigma_n^2}[\beta_1(t) - \beta^*(t)]$ and $\sqrt{n/\sigma_n^2}[\beta_\alpha(t) - \beta_0(t)]$ converge in probability to zero as $n \to \infty$, but these have all been established. ∎

EXERCISES

5.1 Let h be the hazard function associated with a distribution function F on $\mathbf{R}_+$. Prove that for each t, $\int_0^t h(u)/[1-F(u)]du = F(t)/[1-F(t)]$.

5.2 Let $T_1, \ldots, T_k$ be independent random variables with respective hazard functions $h_1, \ldots, h_k$. Calculate the hazard function of $T = \min\{T_1, \ldots, T_k\}$.

5.3 Suppose that $\mathcal{G}$ and $\mathcal{H}$ are filtrations with $\mathcal{G}_t \subseteq \mathcal{H}_t$ for each t, and that M is an $\mathcal{H}$-martingale. Prove that the process $(E[M_t|\mathcal{G}_t])$ is a $\mathcal{G}$-martingale.

5.4 Let M be a mean zero Gaussian martingale with $E[M_t^2] = v_t$.
a) Prove that M has independent increments.
b) Calculate the covariance function $E[M_s M_t]$.
c) Suppose that $v_t = \int_0^t w_s ds$. Show that M has the same distribution as the process $\int_0^t w_s dW_s$, where W is a Wiener process.

5.5 a) Let X be a Markov process with finite state space S and generator A. Show that for each function f on S the process $M_t = f(X_t) - \int_0^t Af(X_s)ds$ is a martingale, where $Af(i) = \sum_{j \in S} A(i,j)f(j)$.
b) Let W be a Wiener process on $\mathbf{R}_+$. Prove that for each $f \in C^2(\mathbf{R})$ the process $M_t = f(W_t) - (1/2)\int_0^t f''(W_s)ds$ is a martingale.

5.6 Prove that the product limit estimator of (5.15) satisfies the differential equation (5.14).

5.7 Devise an analogue of (5.82) for the censored data model of Example 5.7 and from it derive the product limit estimator in (5.16).

5.8 a) Let N be a Poisson process on $\mathbf{R}_+$ with rate λ and let $M_t = N_t - \lambda t$ be the innovation martingale. Prove directly that $(M_t^2 - \lambda t)$ is a martingale.
b) Let W be a Wiener process on $\mathbf{R}_+$. Prove that the processes (W_t) and $(W_t^2 - t)$ are $\mathcal{F}_t^W$-martingales.
c) Relate the properties in a) and b). [*Hint:* Poisson and Wiener processes share the properties of independent and stationary increments.]

5.9 Let N be a Poisson process on $\mathbf{R}_+$ with rate 1 and for each n let $M_t(n) = (1/\sqrt{n})[N_{nt} - nt]$, $t \in [0, 1]$.
a) Prove that for each n, $M_t(n)$ is a martingale in t. [Be careful! The filtration must be specified for each n.]
b) Use Theorem 5.10 to prove that $M(n) \xrightarrow{d} W$ in $D[0, 1]$, where W is a Wiener process.

5.10 Let N be a Poisson process on $[0, 1]$ with unknown intensity function $\alpha \in L_+^1[0, 1]$.
a) Formulate this model as a multiplicative intensity model.
b) Derive the martingale estimator of $B_t(\alpha) = \int_0^t \alpha_s ds$.
c) Given i.i.d. copies of N, describe asymptotic behavior of the resultant sequence of martingale estimators.

5.11 Use Proposition 5.9 to show that (5.26) is sufficient for weak uniform consistency in the setting of Theorem 5.12.

5.12 Formulate and prove analogues of Theorems 5.12 and 5.13 for the case that the baseline stochastic intensities λ^n are *not* observable.

5.13 Let N be a point process on $[0, 1]$ that under P_θ, $\theta > 0$, has stochastic intensity $\lambda_t(\theta) = \theta \lambda_t$, where λ is an observable, predictable process.
a) Calculate the martingale estimator of the process $B_t(\theta) = \int_0^t \lambda_s(\theta) ds$.
b) For each t compute the $\mathcal{F}_t^N$-maximum likelihood estimator of θ.

5.14 a) Prove that the estimators $\hat{A}$ of (5.17) are asymptotically normal.
b) Derive asymptotic properties, as $n \to \infty$, for the estimators $\hat{A}^*$ of (5.18).
c) Calculate the asymptotic efficiency of $\hat{A}^*$ relative to $\hat{A}$.

5.15 Verify the computations in (5.49)–(5.51).

5.16 a) Prove Proposition 5.25a).
b) Verify (5.57).

5.17 Consider the two-sample multiplicative intensity model $\mathcal{P} = \{P_\theta : \theta \in \mathbf{R}_+^2\}$, where under P_θ, each point process $N^n(i)$, $n \geq 1$, $i = 1, 2$, has stochastic intensity $\theta(i)\lambda_t^n$. Formulate a test of the null hypothesis $H_0 : \theta(1) = \theta(2)$, and utilize Theorem 5.26 to derive its large-sample properties. Impose assumptions that are reasonable and defensible.

5.18 Consider the statistical model $\mathcal{P} = \{P_\lambda : \lambda > 0\}$, where under P_λ, for each n, N^n is a Poisson process on $[0, 1]$ with rate $n\lambda$. To test the null hypothesis $H_0 : \lambda = \lambda_0$ Section 5.3 yields martingales $Z_t^n = \int_0^t Y_s^n[dN_s^n/n - \lambda_0 ds]$, where the Y^n are bounded, predictable processes, and hence the test statistics U^n in (5.55).
a) Describe explicitly the statistics U^n and their asymptotic behavior when $Y^n \equiv 1$ for each n.
b) Propose and defend another choice of processes Y^n, or else argue that within the given model no other choice is more sensible.

5.19 Generalize Exercise 5.18 to the case that for $\alpha \in L_+^1[0, 1]$, under the probability P_α, the N^n are Poisson processes with intensity functions $n\alpha$.

5.20 Within the context of Example 5.27,
a) Derive (5.60).
b) Show that the Wilcoxon rank statistic $Z_1^n(i) = n(2n + 1) - \sum_{j=1}^n R_{ij}$ corresponds to the choice $\ell(p) = p$.
c) Show that the choice $\ell(p) = 1$ leads to the Savage rank statistic

$$Z_1^n(i) = n - \sum_{j=1}^n \sum_{v=2n+1-R_{1j}}^n \frac{1}{v}.$$

5.21 Construct explicitly a point process satisfying the description in Example 5.31. [*Hint:* Use a Cox process.]

5.22 In Example 5.31, give a direct proof that the process (M_t) in (5.73) is a martingale.

5.23 In the context of Example 5.31, suppose that the disruption time T has an exponential distribution with parameter λ. Show that the stochastic differential equation for $\hat{Z}_t = P\{T \le t | \mathcal{F}_t^N\}$ can be solved in closed form on each interval (T_j, T_{j+1}), where the T_j are the points of N, and calculate the solution.

5.24 Let N be a Cox process on $\mathbf{R}_+$ with directing intensity W_t^2, where W is a Wiener process. Use techniques in Section 4 to develop a stochastic differential equation fulfilled by the MMSE state estimators $E[W_t^2 | \mathcal{F}_t^N]$.

5.25 For the Cox regression model, show that when the covariate vector β is known to be zero the estimator $\hat{H}$ of (5.77) reduces to the estimator derived in Example 5.7.

5.26 Consider the Cox regression model for independent survival times $T_1, \ldots, T_K$, assuming that the common component h of the hazard functions is known.
a) Show that the likelihood function is

$$L(\beta) = \prod_{k=1}^K \left[e^{\langle \beta, z T_k \rangle} \exp\{-\int_0^{T_k} h(t) e^{\langle \beta, z_t(k) \rangle} ds\} \right].$$

b) Prove that there exists a maximum likelihood estimator of β.

5.27 Suppose that $\beta \in \mathbf{R}^p$ and that the functions $z_t(1), \ldots, z_t(K)$ from $[0, 1]$ into $\mathbf{R}^p$ satisfy $\sum_{k=1}^{K} e^{\langle \beta, z_t(k) \rangle} = 1$ for all t. Let λ be a positive, integrable function on $[0, 1]$ and let N be a Poisson process with intensity function λ. Construct independent Poisson processes $N(1), \ldots, N(K)$ by assigning a point of N at t to $N(k)$ with probability $e^{\langle \beta, z_t(k) \rangle}$, with different points allocated independently (see Definition 1.38). Calculate the conditional likelihood function of $(N(1), \ldots, N(K))$ given N and discuss its relationship to the Cox partial likelihood of (5.79).

NOTES

Broad yet detailed presentations are Gill (1980b), Jacobsen (1982), and Rebolledo (1978). The first emphasizes applications to censored data but has a good introduction to the martingale theory of inference for point processes; the second minimizes measure-theoretic technicalities by canonical constructions and algebraic arguments, and does much to reveal the true underlying structure; the third, in the French style, has influenced significantly our formulation here. Aalen (1978) (based on Aalen, 1975) is also a basic reference, while Andersen and Borgan (1985), Grigelionis (1980) and Karr (1984c) are concise surveys.

Section 5.1

The setting and stochastic intensity model (Model 5.1) are fashioned after Rebolledo (1978), while the multiplicative intensity model (Model 5.6), due to Aalen (1978), seems to be the best compromise between generality and usefulness. Aalen (1978), presaged in a sense by Nelson (1970), also introduced (in a special case) the martingale estimators of Definition 5.3 and established properties of the error processes that are described in Proposition 5.4. One should not underestimate Proposition 5.5; a sometimes decisive advantage of martingale estimators is that they are unbiased and that variances can be computed and estimated. Another, of course, is the ease with which the estimators themselves are calculated. Perhaps the most apparent disadvantage is that, in general, they estimate other random processes rather than deterministic parameters, but sometimes (see Theorems 5.15 and 5.16) this can be overcome. The likelihood representation in Theorem 5.2 (compare Theorem 2.31), due to Jacod (1975b), is analogous to the famous theorem of Girsanov (1960) for diffusion processes. Concerning the product limit estimators of Example 5.7, see Gill (1980b), Jacobsen (1982), Johansen (1978), and Nelson (1982); the key early paper on nonparametric estimation of distribution functions from censored data is Kaplan and Meier (1958). Additional discussion of the Markov process model in Example 5.8 can be found in Aalen (1978), Aalen and Johansen (1978), and Jacobsen (1982); although derived here by a martingale method the estimators $\hat{A}^*$ of (5.18) are classical.

Section 5.2

The elegant Proposition 5.9 is due to Lenglart (1977). Theorem 5.10 is not the most general central limit theorem extant for sequences of martingales, but suffices for our development; it is based on Rebolledo (1977a, 1977b). Rebolledo (1980) proves the theorem under the following "asymptotic rarefaction of jumps" (ARJ) condition, which is weaker even than the Lindeberg condition (5.22). Given a martingale M, for each $\varepsilon > 0$, let $\overline{M}(\varepsilon)$ be the martingale summing "compensated jumps" of M that exceed ε in absolute value: the quadratic variation $[\overline{M}(\varepsilon), \overline{M}(\varepsilon)]$ satisfies $[\overline{M}(\varepsilon), \overline{M}(\varepsilon)]_t = \sum_{s < t} \Delta M_s^2 1(|\Delta M_s| > \varepsilon)$. Then $M = \overline{M}(\varepsilon) + \underline{M}(\varepsilon)$, where $\underline{M}(\varepsilon)$ is a martingale with $|\Delta \underline{M}(\varepsilon)| \leq \varepsilon$. Then a sequence (M^n) of martingales satisfies ARJ if $\langle \overline{M}^n(\varepsilon), \overline{M}^n(\varepsilon) \rangle \to 0$ for each ε.

An analogous functional central limit theorem, but for semimartingales, is given in Liptser and Shiryaev (1980); see also Jacod and Shiryaev (1986). Helland (1982) contains perhaps the most accessible proof. Theorems 5.12 and 5.13 are related most directly to Rebolledo (1978), where one also finds L^p-consistency results for arbitrary values of $p > 1$. The choices $p = 2$ and $p = 4$ (the latter is used in the proof of Theorem 5.18) give particularly simple error bounds.

Section 5.3

The sufficiency assertion of Proposition 5.14 is due to Aalen (1978), where there is also a proof of completeness. In Theorems 5.15 and 5.16, new here, one can estimate deterministic integrals rather than random processes.

Theorem 5.18, due to Karr (1987a), is modeled after Grenander (1981), which along with German and Hwang (1982) can be consulted concerning related applications to nonparametric density estimation. Ramlau-Hansen (1983) studies kernel estimators of α itself:

$$\hat{\alpha}_t^* = \frac{1}{n\gamma_n} \int_0^1 Q\left(\frac{t-s}{\gamma_n}\right) \frac{J(s)}{\bar{\lambda}_s^n} dN_s^n,$$

where $J(s) = 1(\lambda_s^{(i)} = 1$ for some $i \leq n)$, $\bar{\lambda}_s^n = (1/n)\sum_{i=1}^n \lambda_s^{(i)}$, and Q is a suitable kernel.

Theorems 5.19–5.22 are also taken from Karr (1987a).

Concerning finite-dimensional versions of our results for the multiplicative intensity model, see Borgan (1984). Jacobsen (1984) discusses an approach to maximum likelihood estimation in the multiplicative intensity model that expands the model to allow processes with multiple jumps. Asymptotics for Example 5.17 are treated in Breslow and Crowley (1974), Gill (1980b), and Jacobsen (1982); the heroic covariance calculations in the first of these have been reduced by the martingale method to one line.

The material on hypothesis testing comes from Andersen *et al.* (1982), although see also Aalen (1975).

Section 5.4

The formulation and main result, Theorem 5.30, are patterned after Brémaud (1981); the theorem is due to Grigelionis (1973). Another expository treatment is given in Liptser and Shiryaev (1978). See also references cited in connection with the martingale representation of Theorem 2.34, the crucial result underlying

Theorem 5.30. Applications in later chapters include Theorems 6.22 and 7.30. Further analysis of the disruption problem of Example 5.31 appears in Chapter 7 and in Brémaud (1981), Galtchouk and Rozovskii (1971), Liptser and Shiryaev (1978) and Telksyns (1986). Example 5.32 is taken up again in Section 7.2. Extension to marked point processes is considered in Brémaud (1975b, 1981) and Hadjiev (1978).

Section 5.5

Cox (1972b) formulates the original model of survival times with covariates and shared hazard function in the functional form (5.74), and proposes the partial likelihood (5.76) for estimation of β. Cox (1975) contains further analysis and justification, while Oakes (1981) treats additional aspects. Other properties are treated in Breslow (1975), Naes (1982), Tsiatis (1981) (asymptotics), Andersen (1982) (tests for model fit), and Johansen (1983) (an extended model in which the estimators $\hat{\beta}$, $\hat{H}$ become maximum likelihood estimators).

The point process version in Model 5.33 is due to Andersen and Gill (1982), with more general risk functions considered in Jacobsen (1982) and Prentice and Self (1983); however, the fundamental results remain those of Andersen and Gill (1982). The regularity conditions in Assumptions 5.34 and their consequences — Propositions 5.35 and 5.37 and Theorems 5.36 and 5.38 — are investigated by Andersen and Gill (1982) in a setting more general than the i.i.d. case presented here.

The final model of the section, which permits time-dependent covariate *effects*, is due to Zucker and Karr (1990). The approach there is to employ penalized maximum likelihood estimation, a technique with a long history of successful application in estimation of probability density functions. Silverman (1982) is the direct antecedent of the results given for this model. Murphy and Sen (1990) applies the method of sieves in the same context.

Exercises

5.5 The principle underlying both parts is Dynkin's formula: see Çinlar (1975a) or Karlin and Taylor (1975) for elementary descriptions and Blumenthal and Getoor (1968) for a more advanced treatment. Additional applications are in Theorem 6.22 and Section 7.5.

5.6 See Gill (1980a) and Jacobsen (1982).

5.7. Comparison with Exercise 4.22 emphasizes the power of martingale techniques.

5.10 See also Section 4.6, especially Theorem 4.26.

5.12 Even when $\mathcal{H} = \mathcal{F}^N$ the processes λ^n can be unobservable. Although λ_t^n is *some* function of $\{N_u^n : u \leq t\}$, without knowledge of the law of N^n it may be impossible to determine *which* function.

5.13 See Rebolledo (1978); different approaches often coincide in special cases.

5.14 Details are given in Jacobsen (1982) (see also Aalen and Johansen, 1978). In the martingale framework temporal homogeneity is unnecessary.

5.16 See Andersen *et al.* (1982).

5.20 See Andersen *et al.* (1982); Hajek and Sidak (1967) is a general reference on rank tests.

5.24 See Section 7.5, especially Theorems 7.29 and 7.30.

5.27 This is meant to illustrate the narrowness (but nonemptiness) of the conditional likelihood interpretation of the Andersen-Gill partial likelihood (5.79).

6
Inference for Poisson
Processes on General Spaces

Beginning with this chapter we examine specific classes of point processes. The material is a mixture of general results and interesting special cases. Techniques range from martingale methods to strong approximation to sieves. The unifying thread running through Chapters 6–9 is *exploitation of stringent structural assumptions*, utilizing concepts and methods developed in Chapters 4 and 5. Justification for the restrictive assumptions is the concomitant increase in precision, specificity, and usefulness of the results.

For Poisson processes we rely principally on the properties of independent increments and conditional uniformity; the former renders many single-realization inference problems identical mathematically to multiple realization problems, while the latter allows one to make use of empirical process ideas and results. Throughout the chapter our point of view is nonparametric, with the underlying mean measure completely unknown. Parametric problems can be addressed using a combination of techniques described in Section 3.5 and *ad hoc* tools depending on the parametric model at hand, or by "general" parametric methods, as in Kutoyants (1979, 1982). Although we will not pursue such cases in detail, we note, by way of illustration, that material in Section 3.5 applies almost verbatim when the mean measure is known up to a scalar multiple (see Exercise 6.3 and Chapter 9).

Thus we have a dual viewpoint: a sequence of Poisson processes N^n with measures $n\mu$ can be interpreted either as partial superpositions of i.i.d. Poisson processes with mean measure μ or processes whose mean measures converge to infinity in a certain way. In most instances our rendition is the former, but occasionally it is useful to invoke the latter.

Contents of the sections are as follows. Section 1 is devoted to estimation of the mean measure and substitution of estimators of the mean measure to form empirical Laplace functionals and empirical zero-probability functionals specific to Poisson processes and different from those in Sections 4.2 and 4.3 (see Example 4.12 for a particular case). We also discuss application of strong approximation theorems for Poisson processes (cf. Theorems 4.25–4.26) to empirical zero-probability functionals. In Section 2 we treat nonparametric likelihood ratio tests of simple hypotheses $H_0 : \mu = \mu_0$, $H_1 : \mu = \mu_1$, but emphasize structure of likelihood ratios, absolute continuity and singularity of the probability laws of Poisson processes and local asymptotic normality of likelihood ratios. Qualitative questions of testing (e.g., for the structure of the mean measure and for equality or ordering of mean measures in the two-sample case) are mentioned briefly.

State estimation given observation of a Poisson process restricted to a subset, because of independent increments, is equally uninteresting on general spaces as on $\mathbf{R}_+$. However, for other kinds of partial observation the state estimation problem does have substance, and three such models are presented in Section 3. One is p-thinned Poisson processes as in Section 4.5, another is the stochastic integral process of Examples 3.5 and 3.10 (see also Chapter 7), and the third is a Poisson process on a product space with only one component observable, which has applications in positron emission tomography. We treat statistical estimation for these models as well. Section 4 concerns statistical inference and state estimation for Poisson processes given observation of integral data. Finally, in Section 5 we apply results in other sections to inference for random measures with independent increments; we also treat Poisson cluster processes. That ideas and techniques from earlier sections apply at all to more general classes of point processes and random measures is an important consequence of our working on general spaces, for then these kinds of extensions can be handled with little further difficulty.

6.1 Estimation

Let E be a compact set and let $N_1, N_2, \ldots$ be point processes on E. Consider the statistical model $\mathcal{P} = \{P_\mu : \mu \in \mathbf{M}_d\}$, where with respect to P_μ the N_i are i.i.d. copies of a Poisson process N with diffuse mean measure μ. Our initial concern is nonparametric estimation of μ; thereafter we examine estimators of the Laplace functional, the zero-probability functional and (only in exercises, although see also Chapter 9) the covariance measure. In each case, estimators are formed by substitution. We also discuss how the strong approximation of Poisson processes by Kiefer processes given in Theorem 4.26 produces better results for empirical zero-probability functionals and additional results in general. Discussion of estimation in the i.i.d. case concludes with application of sieve methods (see Section 5.3) to maximum likelihood estimation of an unknown intensity function. Finally, we mention some single realization results.

As estimators of μ we use the empirical averages

$$\hat{\mu}(\cdot) = (1/n)\sum_{i=1}^{n} N_i(\cdot). \tag{6.1}$$

Before proceeding, we verify that, given observation of $N_1, \ldots, N_n$, inference can be based solely on $N^n = \sum_{i=1}^{n} N_i$.

Lemma 6.1. The statistic N^n is sufficient for μ given $N_1, \ldots, N_n$.

Proof: We may suppose that $N_1, \ldots, N_n$ have been constructed as follows. Let $N^n = \sum \epsilon_{Y_k}$ be Poisson with mean $n\mu$ and let U_k be i.i.d. random variables, independent of N^n, with $P\{U_k = i\} = 1/n$, $i = 1, \ldots, n$, and put $N_i = \sum_k 1(U_k = i)\epsilon_{Y_k}$. Then $N_1, \ldots, N_n$ are independent Poisson processes (Exercise 1.16), each with mean measure μ, we have $N^n = \sum_{i=1}^{n} N_i$ by construction, and by the very nature of that construction, the conditional distribution of $N_1, \ldots, N_n$ given N^n is independent of μ. ∎

The proof highlights the dual role of N^n, which — at least for results concerning convergence in distribution — can be thought of as a superposition or simply as a single realization of a Poisson process with mean measure $n\mu$, in the same way that processes in Theorems 5.12 and 5.13 may but need not arise as superpositions of i.i.d. components. The crucial asymptotic property in both settings is that "intensities" converge to infinity.

Limit behavior of the $\hat{\mu}$ has been established already in Section 4.1; for completeness we recapitulate it.

Proposition 6.2. For each $\mu \in \mathbf{M}_d$, under P_μ
a) For each compact subset $\mathcal{K}$ of $C^+(E)$, almost surely

$$\sup_{f \in \mathcal{K}} |\hat{\mu}(f) - \mu(f)| \to 0; \tag{6.2}$$

b) The error processes $\sqrt{n}[\hat{\mu} - \mu]$, as signed random measures on E, converge in distribution to a Gaussian random measure with covariance function

$$R(f, g) = \mu(fg); \tag{6.3}$$

c) Let $\mathcal{K}$ be a compact subset of $C^+(E)$ having finite metric entropy with respect to the uniform metric. Then the error processes $\{\sqrt{n}[\hat{\mu}(f) - \mu(f)] : f \in \mathcal{K}\}$, as random elements of $C(\mathcal{K})$, converge in distribution to a Gaussian process with covariance function R given in b);

d) Almost surely the set $\{\sqrt{n}[\hat{\mu} - \mu]/\sqrt{2 \log \log n} : n \geq 3\}$ is relatively compact in the function space $C(E)^*$ with limit set $\{\eta : |\eta(f)| \leq \sqrt{\mu(f^2)}$ for all $f\}$. $\square$

For Poisson processes we define by substitution the *empirical Laplace functionals* $\hat{L}(f) = e^{-\int(1-e^{-f})d\hat{\mu}}$, with $\hat{\mu}$ given by (6.1). These have the advantage of being tailored to the Poisson model, but lack robustness. If the Poisson assumption were suspect, the general empirical Laplace functionals of Section 4.2 would be preferred.

We now summarize properties of the empirical Laplace functionals. Let $L(f) = e^{-\int(1-e^{-f})d\mu}$ be the Laplace functional of the N_i under P_μ.

Proposition 6.3. a) For each equicontinuous subset $\mathcal{K}$ of $C^+(E)$, $\sup_{f \in \mathcal{K}} |\hat{L}(f) - L(f)| \to 0$ almost surely;

b) If $\mathcal{K}$ is a compact subset of $C^+(E)$ with finite metric entropy, then the error processes $\{\sqrt{n}[\hat{L}(f) - L(f)] : f \in \mathcal{K}\}$ converge in distribution, as random elements of $C(\mathcal{K})$, to a Gaussian process with covariance function

$$R(f, g) = \mu([1 - e^{-f}][1 - e^{-g}])e^{-\mu(1-e^{-f})-\mu(1-e^{-g})}. \tag{6.4}$$

Proof: a) Equicontinuity of $\mathcal{K}$ implies that as well of the set $\mathcal{K}' = \{1 - e^{-f} : f \in \mathcal{K}\}$, which is uniformly bounded. Since $\sup_{\mathcal{K}} |\hat{L}(f) - L(f)| \leq \sup_{\mathcal{K}'} |\hat{\mu}(g) - \mu(g)|$, a) follows from (6.2).

b) For fixed $f \in \mathcal{K}$,

$$\begin{aligned}
\sqrt{n}[\hat{L}(f) - L(f)] &= \sqrt{n}[e^{-\hat{\mu}(1-e^{-f})} - e^{-\mu(1-e^{-f})}] \\
&= \sqrt{n}[\hat{\mu}(1 - e^{-f}) - \mu(1 - e^{-f})]e^{-x^*},
\end{aligned}$$

where x^* lies between $\hat{\mu}(1 - e^{-f})$ and $\mu(1 - e^{-f})$. In view of a), therefore,

$$\sup_{\mathcal{K}} \left| \sqrt{n}[\hat{L}(f) - L(f)] - \sqrt{n}[\hat{\mu}(1 - e^{-f}) - \mu(1 - e^{-f})]e^{-\mu(1-e^{-f})} \right|$$

converges to zero in P_μ-probability, and the same reasoning used to show that a continuous image of a compact set is compact confirms that $\mathcal{K}'$ has finite metric entropy. Consequently, b) ensues from Proposition 6.2c), with (6.4) obtained from (6.3) by straightforward computation. ∎

The general empirical Laplace functionals of Section 4.2, given by $\hat{L}^*(f) = (1/n)\sum_{i=1}^n e^{-N_i(f)}$, have limit covariance function

$$R^*(f, g) = e^{-\mu(1-e^{-(f+g)})} - e^{-\mu(1-e^{-f})}e^{-\mu(1-e^{-g})};$$

in particular, $\sqrt{n}[\hat{L}^*(f) - L(f)]$ has asymptotic variance

$$R^*(f, f) = e^{-\mu(1-e^{-2f})} - e^{-2\mu(1-e^{-f})}.$$

By (6.4), the asymptotic variance of $\sqrt{n}[\hat{L}(f) - L(f)]$ is

$$R(f, f) = \mu([1 - e^{-f}]^2)e^{-2\mu(1-e^{-f})},$$

so that the asymptotic efficiency of $\hat{L}^*(f)$ relative to $\hat{L}(f)$ is

$$e(\mu) = R^*(f, f)/R(f, f) = [e^{\mu(1-e^{-f})^2} - 1]/\mu([1 - e^{-f}]^2),$$

which always exceeds 1. Therefore, $\hat{L}$ is preferred to $\hat{L}^*$. (No contradiction with Example 4.12 exists because the specialized Laplace functional there is not $\hat{L}$.)

Based on the relationship $z(A) = e^{-\mu(A)}$ we define *empirical zero-probability functionals* $\hat{z}(A) = e^{-\hat{\mu}(A)}$. While nonparametric within the Poisson model, these differ nevertheless from the general empirical zero-probability functionals of Section 4.3. So far we have viewed the estimators $\hat{\mu}$ as indexed by functions rather than sets, but obviously this will not do for analysis of zero-probability functionals. Results from Chapter 4 concerning asymptotics of set-indexed empirical processes do not apply directly, although we shall discuss presently how they can be made to apply. Before doing so, however, we record elementary results.

Proposition 6.4. For each μ let $z(A) = e^{-\mu(A)} = P_\mu\{N(A) = 0\}$. Then
a) For every A, $\hat{z}(A) \to z(A)$ almost surely;
b) For $A_1, \ldots, A_k \in \mathcal{E}$,

$$\sqrt{n}\left[(\hat{z}(A_1), \ldots, \hat{z}(A_k)) - (z(A_1), \ldots, z(A_k))\right] \xrightarrow{d} N(0, R),$$

on $\mathbf{R}^k$, where $R(i, j) = \mu(A_i \cap A_j)e^{-[\mu(A_i)+\mu(A_j)]}$. $\square$

Independence of $\sqrt{n}[\hat{z}(A) - z(A)]$ and $\sqrt{n}[\hat{z}(B) - z(B)]$ for disjoint A and B carries over in the form of R. Moreover, Proposition 6.4 implies that for any sequence of error processes $\{\sqrt{n}[\hat{z}(A) - z(A)] : A \in \mathcal{C}\}$ converging in distribution, the limit covariance function must be a restriction of the function $R(A, B) = \mu(A \cap B)e^{-[\mu(A)+\mu(B)]}$.

We next explore more general version of Proposition 6.4b) based on strong approximation (see Section 4.6); the starting point is the strong approximation for homogeneous, unit rate Poisson processes on $[0,1]$ given in Theorem 4.26: asymptotic behavior of the processes $(1/\sqrt{n})\{[N^n(x) - nx] : 0 \le x \le 1\}$ is that of the Gaussian process

$$G(x) = B(x) + Zx, \tag{6.5}$$

where B is a Brownian bridge, Z has distribution $N(0,1)$, and B and Z are independent (see Theorem 4.14).

We now discuss applications to empirical zero-probability functionals, for which Theorem 4.26 must be extended to more general Poisson processes. First, if N^n has rate λn rather than n, then Theorem 4.26 applies verbatim to yield $\{(1/\sqrt{n})[N^n(x) - \lambda nx] : 0 \le x \le 1\} \overset{d}{\to} \{\lambda G(x) : 0 \le x \le 1\}$, where G is as in (6.5).

To obtain a result for zero-probability functionals we put $z(x) = e^{-\lambda x} = P_\lambda\{N(x) = 0\}$ and $z(x) = e^{-\bar{\mu}(x)}$.

Proposition 6.5. Under P_λ,

$$\sqrt{n}[\hat{z}(x) - z(x)] : 0 \le x \le 1\} \overset{d}{\to} \{\lambda G(x) : 0 \le x \le 1\}.\ \square$$

To generalize further we employ a device appearing in Kallenberg (1983) that can be viewed as a generalized quantile transformation. Suppose that μ is a diffuse measure on E, assumed without loss of generality to satisfy $\mu(E) = 1$. Then there exists an invertible mapping f_μ (not unique) of (a measure 1 subset of) $[0,1]$ into E such that $\mu = \lambda f_\mu^{-1}$, where now λ denotes Lebesgue measure. Hence a Poisson process N^n on E with mean measure $n\mu$ admits the representation

$$N^n = \sum_{i=1}^{S_n} \varepsilon_{f_\mu(X_i)}, \tag{6.6}$$

where the X_i are independent random variables uniformly distributed on $[0,1]$, and $S_n = \sum_{i=1}^{n} Y_i$, with the Y_i i.i.d. random variables independent of the X_i, each having a Poisson distribution with mean 1.

We then obtain the following result.

Proposition 6.6. Let N_n be given by (6.6), where f_μ satisfies $\mu = \lambda f_\mu^{-1}$, and let $(A_x)_{0 \le x \le 1}$ be an increasing family of subsets of E. Then with $a(x) = \mu(A_x)$, $\{\sqrt{n}[\hat{z}(A_x) - z(A_x) : 0 \le x \le 1\} \overset{d}{\to} \{e^{-a(x)}G(a(x)) : 0 \le x \le 1\}$, under P_μ, where G is the Gaussian process given by (6.5).

Proof: We may assume that $f_\mu^{-1}(A_x) = [0, a(x)]$. Let $\tilde{N}^n = \sum_{i=1}^{S_n} \varepsilon_{X_i}$, so that $N^n = \tilde{N}^n f_\mu^{-1}$ (see Section 1.5). In particular $\hat{\mu}(A_x) = (1/n)\tilde{N}^n(a(x))$ and therefore by Theorem 4.26,

$$
\begin{aligned}
\sqrt{n}[\hat{z}(A_x) - z(A_x)] \quad &= \quad \sqrt{n}[\exp\{-\tilde{N}^n(a(x))/n\} - e^{-a(x)}] \\
&\cong \quad e^{-a(x)}[\tilde{N}^n(a(x)) - na(x)]/\sqrt{n} \\
&\overset{d}{\to} \quad e^{-a(x)}G(a(x)). \quad \blacksquare
\end{aligned}
$$

In some instances an unknown mean measure is stipulated to be dominated by a known measure μ_0, but not further. For example, a Poisson process N on $\mathbf{R}_+$ may be known on physical grounds to admit an intensity function $\alpha_t = dE[N_t]/dt$, but α itself may be unknown. By sieve methods analogous to those in Theorem 5.18 — but with histogram sieves — we can estimate by maximum likelihood the unknown derivative $\alpha = d\mu/d\mu_0$ for the model $\mathcal{P} = \{P_\alpha : \alpha \in L^1_+(\mu_0)\}$, even on a general space E. Suppose that under each P_α, $N_1, N_2, \ldots$ are i.i.d. copies of a Poisson process N on E with mean measure $\mu(dx) = \alpha(x)\mu_0(dx)$, where $\mu_0 \in \mathbf{M}_d$ is *known*. With $N^n = \sum_{i=1}^n N_i$ which is remains a sufficient statistic for α given the data $N_1, \ldots, N_n$, and with P the probability under which the N_i are i.i.d. Poisson processes with mean measure μ_0, Proposition 6.10 (in Section 2) yields the log-likelihood functions

$$L_n(\alpha) = n\int_E (1 - \alpha)d\mu_0 + \int_E (\log \alpha)dN^n. \tag{6.7}$$

As is true for the multiplicative intensity model of Section 5.3, of which this model is a special case when E is a compact interval in $\mathbf{R}_+$, L_n is unbounded above as a function of α, necessitating an indirect approach to maximum likelihood estimation.

Our approach utilizes histogram sieves (see Grenander, 1981). Let $\{A_{mj} : m \geq 1, 1 \leq j \leq \ell_m\}$ be a null array of partitions of E with $\mu_0(A_{mj}) > 0$ for each m and j. For each m let I_m be the *histogram sieve* of functions $\alpha \in L^1_+(\mu_0)$ constant over each set A_{mj}. Then by straightforward computations the maximum likelihood estimator $\hat\alpha(n, m)$ relative to sample size n and I_m satisfies

$$\hat\alpha(n, m)(x) = \hat\mu(A_{mj})/\mu_0(A_{mj}), \quad x \in A_{mj}, \tag{6.8}$$

where $\hat\mu$ is given by (6.1).

The appropriate sense for estimators $\hat\alpha = \hat\alpha(n, m_n)$ defined momentarily to converge to α is with respect to the norm on $L^1_+(\mu_0)$, which we denote by $\|\cdot\|_1$. For suitable choice of m_n convergence takes place almost surely.

Theorem 6.7. If $m = m_n$ is chosen in such that, with $\tilde\ell_n = \ell_{m_n}$,

$$\sum_{n=1}^\infty \tilde\ell_n^4/n^2 < \infty, \tag{6.9}$$

then for $\hat\alpha = \hat\alpha(n, m_n)$, $\|\hat\alpha - \alpha\|_1 \to 0$ almost surely with respect to P_α.

Proof: Let n and m both be variable; then with $\mu_0(\alpha; A_{mj}) = \int_{A_{mj}} \alpha \, d\mu_0$,

$$\|\hat\alpha(n, m) - \alpha\|_1$$
$$= \sum_{j=1}^{\ell_m} \int_{A_{mj}} \left|\hat\mu(A_{mj})/\mu_0(A_{mj}) - \alpha(x)\right| \mu_0(dx)$$

$$\leq \; \sum_{j=1}^{\ell_m} \int_{A_{mj}} \left| \hat{\mu}(A_{mj})/\mu_0(A_{mj}) - \mu_0(\alpha; A_{mj})/\mu_0(A_{mj}) \right| \mu_0(dx)$$
$$+ \sum_{j=1}^{\ell_m} \int_{A_{mj}} \left| \mu_0(\alpha; A_{mj})/\mu_0(A_{mj}) - \alpha(x) \right| \mu_0(dx)$$
$$\leq \; \sum_{j=1}^{\ell_m} \left| \hat{\mu}(A_{nj}) - E_\alpha[N(A_{mj})] \right|$$
$$+ \sum_{j=1}^{\ell_m} \int_{A_{mj}} \left| \mu_0(\alpha; A_{mj})/\mu_0(A_{mj}) - \alpha(x) \right| \mu_0(dx).$$

That the second, nonrandom term converges to zero provided only that $m \to \infty$ is analytical and can be shown by a variety of techniques (see, e.g., Grenander, 1981, pp. 419–420).

We deal with the first term by methods analogous to those in Theorems 4.15 and 4.18. For $\varepsilon > 0$,

$$P_\alpha \left\{ \sum_{j=1}^{\ell_m} |\hat{\mu}(A_{mj}) - E_\alpha[N(A_{mj})]| > \varepsilon \right\}$$
$$\leq \; (1/\varepsilon^4) E_\alpha \left[\left\{ \sum_{j=1}^{\ell_m} |\hat{\mu}(A_{mj}) - E_\alpha[N(A_{mj})]| \right\}^4 \right].$$

With n and m fixed the random variables $\hat{\mu}(A_{mj}) - E_\alpha[N(A_{mj})]$ are independent in j and by brute-force expansion of the fourth power of the summation followed by repeated application of the Cauchy-Schwarz inequality, we obtain

$$P_\alpha \left\{ \sum_{j=1}^{\ell_m} |\hat{\mu}(A_{mj}) - E_\alpha[N(A_{mj})]| > \varepsilon \right\} = O(\ell_m^4/(\varepsilon^4 n^2));$$

the dominant term comes from the $O(\ell_m^4)$ summands with all indices distinct. Thus (6.9) suffices to give $P_\alpha\{\sum_{j=1}^{\ell_m} |\hat{\mu}(A_{mj}) - E_\alpha[N(A_{mj})]| > \varepsilon\} < \infty$ for every $\varepsilon > 0$, which completes the proof. ∎

In general, little can be said about estimation of the mean measure of a Poisson process in the Model 3.2 context of observation of a single realization over a noncompact set E; one can do arbitrarily poorly if the mean measure is particularly ill-chosen. However, good results obtain for stationary Poisson processes. Suppose that $E = \mathbf{R}^d$, let μ_0 denote Lebesgue measure and for each $\nu > 0$ let N be a stationary Poisson process with intensity ν under the probability P_ν. It is natural to estimate ν with estimators $\hat{\nu} = N(B)/\mu_0(B)$, where $B \uparrow E$, but if $\mu_0(B) \to \infty$ in a completely arbitrary manner, consistency and asymptotic normality can fail. For $B_n = \sum_{i=1}^{n} A_i$, where the A_i are disjoint sets with $\mu_0(A_i)$ the same for all i, the independent and stationary increments properties of N yield the following result.

Proposition 6.8. Let $A_1, A_2, \ldots$ be disjoint subsets of E with $0 < a = \mu_0(A_i) < \infty$ independently of i and for each n let $\hat{\nu} = (1/na)N(\sum_{i=1}^{n} A_i)$. Then

a) $\hat{\nu} \to \nu$ almost surely;

b) $\sqrt{n}[\hat{\nu} - \nu] \xrightarrow{d} N(0, \nu a)$. $\square$

The simplest choice of the A_i is as translates of a fixed set A_0. More generally, Theorem 1.71 and the final result of the section enable one to construct numerous consistent estimators of ν (see Chapter 9 for specific choices).

Proposition 6.9. A stationary Poisson process N on $\mathbf{R}^d$ is ergodic.

Proof: The Palm distribution $\tilde{P}$ of N, by Example 1.57, is just the P-distribution of $N + \varepsilon_0$; here we have taken $\nu = 1$ for simplicity. Therefore, $\tilde{P}$ and P agree on the invariant σ-algebra $\mathcal{I}$ (Definition 1.70) and hence by Theorem 1.71a), $E[N(f)|\mathcal{I}] = \mu_0(f)$ for all f, which proves that $\mathcal{I}$ is P-degenerate. $\blacksquare$

This concludes our general treatment of estimation, but not our consideration of the topic. Additional aspects appear in Sections 3–5.

6.2 Equivalence and Singularity; Hypothesis Testing

This section is more about properties of likelihood ratios and probability distributions of Poisson processes than about hypothesis testing *per se*, but even the less overtly statistical portions can be interpreted as addressing the simple-vs.-simple hypothesis test

$$\begin{aligned} H_0 &: \mu = \mu_0 \\ H_1 &: \mu = \mu_1 \end{aligned} \tag{6.10}$$

for a Poisson process N with unknown mean measure μ.

As in Section 1 the setting is nonparametric, but here we accentuate the single-realization case, with asymptotics pertaining to behavior of Poisson process likelihood ratios and probability laws associated with bounded sets increasing to a noncompact space E. In particular we examine equivalence and singularity of probability laws P_{μ_0} and P_{μ_1} engendering Poisson processes with mean measures μ_0 and μ_1, especially in the case that μ_0 and μ_1 are equivalent. In this case P_{μ_0} and P_{μ_1} must be either equivalent or singular; intermediate cases cannot arise. This dichotomy theorem (Theorem 6.11; see also Proposition 3.24) has analogues for Gaussian processes and diffusions (Liptser and Shiryaev, 1978), and is the main result of the section, although Proposition 6.10, which derives explicitly the form of likelihood ratios, is probably the most useful. General discussion concludes with a local asymptotic normality theorem for likelihood ratios over bounded sets.

Let E be noncompact but locally compact and let $(\Omega, \mathcal{F})$ be a measurable space supporting a (simple) point process N on E; we consider the statistical model $\mathcal{P} = \{P_\mu : \mu \in \mathbf{M}_d\}$, where under P_μ, N is a Poisson process with (diffuse) mean measure μ. For construction of likelihood ratio tests a principal interest is equivalence or singularity of probability measures P_{μ_0} and P_{μ_1} corresponding to the hypotheses (6.10) on σ-algebras $\mathcal{F}^N(B)$, where B is bounded, as well as on $\mathcal{F}^N = \mathcal{F}^N(E)$, as a function of equivalence or singularity of the mean measures μ_0 and μ_1. When $P_{\mu_0} \sim P_{\mu_1}$, there exists a positive, finite likelihood ratio, and hence a log- likelihood ratio usable for testing the hypotheses (6.10). At the other extreme, if $P_{\mu_0} \perp P_{\mu_1}$, then the test is rendered trivial: every realization of N leads without error to μ_0 or μ_1 as the "true" mean measure. Singularity of P_{μ_0} and P_{μ_1} cannot occur unless at least one of $\mu_0(E)$ and $\mu_1(E)$ is infinite; see Exercise 6.9.

Within this setting the most interesting case, mathematically and physically, is that $\mu_0 \sim \mu_1$. We commence with a criterion for equivalence of P_{μ_0} and P_{μ_1} on $\mathcal{F}^N(B)$ when B is bounded and $\mu_0 \sim \mu_1$ on B; thereafter we develop equivalence criteria for the unbounded case, including the dichotomy theorem.

Proposition 6.10. Suppose that B is bounded and that $\mu_1 \ll \mu_0$ on the σ-algebra $\mathcal{E} \cap B$. Then $P_{\mu_1} \ll P_{\mu_0}$ on $\mathcal{F}^N(B)$, with

$$\frac{dP_{\mu_1}}{dP_{\mu_0}}\bigg|_{\mathcal{F}^N(B)} = \exp\left\{\int_B \left(1 - \frac{d\mu_1}{d\mu_0}\right) d\mu_0 + \int_B \log\frac{d\mu_1}{d\mu_0} dN\right\}. \tag{6.11}$$

Proof: For $f \geq 0$ vanishing outside B,

$$E_{\mu_0}\left[e^{-N(f)}\exp\{\textstyle\int_B(1 - d\mu_1/d\mu_0)d\mu_0 + \int_B \log(d\mu_1/d\mu_0)dN\}\right]$$
$$= \exp[\textstyle\int_B(1 - d\mu_1/d\mu_0)d\mu_0]E_{\mu_0}\left[\exp\{-\int_B(f - \log[d\mu_1/d\mu_0])dN\}\right]$$
$$= \exp[\textstyle\int_B(1 - d\mu_1/d\mu_0)d\mu_0]\exp[-\int_B(1 - e^{-f}d\mu_1/d\mu_0)d\mu_0]$$
$$= \exp[-\textstyle\int_B(1 - e^{-f})d\mu_1] = E_{\mu_1}[e^{-N(f)}];$$

hence (6.11) holds by Theorem 1.12. ∎

Note the resemblance of (6.11) to the likelihood function of Section 5.3 for multiplicative intensity processes.

By interchanging μ_0 and μ_1 we establish a criterion for equivalence.

Corollary 6.11. If $\mu_0 \sim \mu_1$ on B, $P_{\mu_0} \sim P_{\mu_1}$ on $\mathcal{F}^N(B)$ and (6.11) holds. ☐

In the setting of Proposition 6.10, Neyman-Pearson critical regions for the hypotheses (6.10) have the form $\Omega_c = \{\int_B \log(d\mu_1/d\mu_0)dN \geq \delta\}$, with δ a constant depending on B, μ_0, μ_1 and the power bound under H_0 (the probability of a type I error). The null distribution of the test statistic is difficult to express in closed form; however, its Laplace transform is computable and can be used, together for example with Chebyshev's inequality, to construct approximate critical regions. Alternatively, if $\mu_0(B)$ is large then the test statistic — by independent increments — is approximately normally distributed, yielding another way to derive approximate critical regions (see M. Brown, 1972, for details).

Remaining a moment longer with the hypotheses (6.10) and observations $\mathcal{F}^N(B)$, even when μ_0 and μ_1 are not equivalent, one can formulate a likelihood ratio test by the following device. Both μ_0 and μ_1 are absolutely continuous with respect to $\mu^* = \mu_0 + \mu_1$ and Proposition 6.10 implies that on $\mathcal{F}^N(B)$

$$dP_{\mu_i}/dP_{\mu^*} = \exp\left\{\int_B(1 - d\mu_i/d\mu^*)d\mu^* + \int_B \log(d\mu_i/d\mu^*)dN\right\}$$

for $i = 0, 1$, so we may form the likelihood ratio (not a Radon-Nikodym derivative but a ratio of likelihoods notwithstanding)

$$\frac{dP_{\mu_1}/dP_{\mu^*}}{dP_{\mu_0}/dP_{\mu^*}} = \exp\left[\mu_0(B) - \mu_1(B) + \int_B \frac{d\mu_1/d\mu^*}{d\mu_0/d\mu^*}dN\right]$$

and use it to test the hypotheses (6.10).

For observation of N over the noncompact set E it is possible that $\mu_0 \sim \mu_1$ on E, and hence that $P_{\mu_0} \sim P_{\mu_1}$ on $\mathcal{F}^N(B)$ for every bounded set B, but that nevertheless $P_{\mu_0} \perp P_{\mu_1}$ on $\mathcal{F}^N$ (Exercise 6.10). Next we give a necessary and sufficient condition for $P_{\mu_0} \sim P_{\mu_1}$ in the unbounded case when $\mu_0 \sim \mu_1$ but this *dichotomy theorem* yields more: if P_{μ_0} and P_{μ_1} are not equivalent, then they must be singular.

Theorem 6.12. Suppose that $\mu_0 \sim \mu_1$ on E and that $\mu_0(E) = \mu_1(E) = \infty$. Then on $\mathcal{F}^N$ either $P_{\mu_0} \sim P_{\mu_1}$ or $P_{\mu_0} \perp P_{\mu_1}$ according as the integral

$$\int_E [1 - \sqrt{d\mu_1/d\mu_0}]^2 d\mu_0 \tag{6.12}$$

converges or diverges.

Proof: We show first that convergence in (6.12) implies that $P_{\mu_1} \ll P_{\mu_0}$; since convergence of (6.12) implies that of the corresponding integral with μ_0 and μ_1 interchanged, this proves the "convergence implies equivalence" part of the theorem. By Liptser and Shiryaev (1978, Lemma 19.13) it is enough to show that for bounded sets $B_n \uparrow E$, $\lim_n (dP_{\mu_1}/dP_{\mu_0})|_{\mathcal{F}^N(B_n)}$ exists and is finite almost surely with respect to P_{μ_1}. From (6.11),

$$dP_{\mu_1}/dP_{\mu_0}\big|_{\mathcal{F}^N(B)} = \exp\left\{\int_{B_n}(1 - d\mu_1/\mu_0)d\mu_0 + \int_{B_n}\log(d\mu_1/d\mu_0)dN\right\},$$

and by the P_{μ_1}-strong law of large numbers for N,

$$\lim_{n\to\infty} \frac{\int_{B_n}\log(d\mu_1/d\mu_0)dN}{\int_{B_n}\log(d\mu_1/d\mu_0)d\mu_1} = 1$$

almost surely with respect to P_{μ_1}, (one can choose the B_n in a manner ensuring that this holds); consequently,

$$\frac{dP_{\mu_1}}{dP_{\mu_0}} \cong \exp\left\{\int_{B_n}\left(1 - \frac{d\mu_1}{d\mu_0} + \frac{d\mu_1}{d\mu_0}\log\frac{d\mu_1}{d\mu_0}\right)d\mu_0\right\}.$$

The proof that $P_{\mu_1} \ll P_{\mu_0}$ is completed by showing that

$$\int_E \left[1 - d\mu_1/d\mu_0 + (d\mu_1/d\mu_0)\log(d\mu_1/d\mu_0)\right]d\mu_0 < \infty. \tag{6.13}$$

That (6.13) and convergence in (6.12) are equivalent is seen by first noting that the function $h(y) = 1 - y + y(\log y)$ in (6.13) can be replaced without affecting convergence of the integral by $k(y) = 1 - y + y\varphi(\log y)$, where $\varphi(x) = x$ for $|x| \leq 1$ and $x/|x|$ otherwise, and then using the property that there are constants c and c' such that $c(1 - \sqrt{y})^2 \leq k(y) \leq c'(1 - \sqrt{y})^2$ (see Grenander, 1981, Chap. 8, and Liptser and Shiryaev, 1978, Chap. 19, as well as the proof of Theorem 5.18, for details).

Conversely, given divergence in (6.12), almost surely with respect to P_{μ_1}

$$\lim_{n\to\infty} dP_{\mu_0}/dP_{\mu_1}\big|_{\mathcal{F}^N(B_n)} = \left(\lim_{n\to\infty} dP_{\mu_1}/dP_{\mu_0}\big|_{\mathcal{F}^N(B_n)}\right)^{-1} = 1/\infty = 0,$$

but by Neveu (1975, Proposition III-1-5) the limit is the Radon-Nikodym derivative of the absolutely continuous component in the Lebesgue decomposition of P_{μ_0} with respect to P_{μ_1}, and therefore $P_{\mu_0} \perp P_{\mu_1}$. ∎

Equivalent means of stating convergence in (6.12) exist. For example, M. Brown (1971, 1972) demonstrates that $P_{\mu_0} \sim P_{\mu_1}$ if and only if for some $c > 0$ (and with $g = 1 - d\mu_1/d\mu_0$)

$$\int_{\{|g|>c\}}|g|d\mu_0 + \int_{\{|g|\leq c\}}g^2 d\mu_0 < \infty.$$

An advantage of (6.12) as opposed to this criterion is that the former generalizes to point processes with nondeterministic compensators (see Liptser and Shiryaev, 1978).

Here is an extended criterion for singularity, in which it is not assumed that $\mu_0 \sim \mu_1$.

Proposition 6.13. Suppose that $\mu_1(dx) = f(x)\mu_0(dx) + \nu(dx)$ is the Lebesgue decomposition of μ_1 with respect to μ_0. Then the following are equivalent:

a) $P_{\mu_0} \perp P_{\mu_1}$ on $\mathcal{F}^N$;

b) Either $\nu(E) = \infty$ or

$$\int_E (1 - \sqrt{f})^2 d\mu_0 = \infty. \tag{6.14}$$

Proof: a) $\Rightarrow$ b). If both parts of b) fail, then there exists $A \in \mathcal{E}$ with $\mu_0(A) = \nu(A^c) = 0$ and $P_{\mu_1}\{N(A) = 0\} = e^{-\nu(A)} > 0$. Conditional on the event $\{N(A) = 0\}$, which has P_{μ_0}-probability 1, N is Poisson with mean μ_0 under P_{μ_0} and Poisson with mean $f(x)\mu_0(dx)$ under P_{μ_1}. Failure of (6.14) implies convergence in (6.12) for these conditional processes; hence by the "convergence implies equivalence" part of Theorem 6.12, $P_{\mu_1} \ll P_{\mu_0}$ on an event having positive probability for both, which makes $P_{\mu_0} \perp P_{\mu_1}$ impossible.

b) $\Rightarrow$ a). If $\nu(E) = \infty$, then there is a set A with $\mu_0(A) = 0$ (hence $P_{\mu_0}\{N(A) = 0\} = 1$) but $\nu(A) = \infty$, which implies that $P_{\mu_1}\{N(A) = 0\} = 0$; therefore, a) holds. If (6.14) is fulfilled and P_{μ_2} denotes the probability law of the Poisson process on E with mean measure $\mu_2(dx) = f(x)\mu_0(dx)$, then $P_{\mu_0} \perp P_{\mu_2}$, by Theorem 6.12. By construction, P_{μ_1} is the convolution of P_{μ_2} and P_ν, the first of which is singular with respect to P_{μ_0}, as is the second, except possibly on $\{N(E) = 0\}$; therefore $P_{\mu_0} \perp P_{\mu_1}$. ∎

From the perspective of hypothesis testing, Theorem 6.12 describes the pleasant situation that from a single realization of N there is either error-free discrimination between P_{μ_0} and P_{μ_1} (the singular case) or a positive, finite log-likelihood ratio $\int_E (1 - d\mu_1/d\mu_0)d\mu_0 + \int_E \log(d\mu_1/d\mu_0)dN$, which can be used to perform likelihood ratio tests.

Proposition 6.13 provides additional conditions for perfect discrimination between μ_0 and μ_1. If either measure is infinite, then there is no error in testing the hypotheses (6.10), while if both are finite, then every nonzero realization of N yields unerring determination of μ_0 or μ_1 but on $\{N(E) = 0\}$ flawless discrimination is not possible. However, on this event, an atom of $\mathcal{F}^N$, one can employ the log-likelihood ratio in the usual fashion.

We now examine log-likelihood function asymptotics for a sequence of processes on a bounded set with mean measures converging to infinity. As Proposition 3.25 and Theorem 5.21 suggest, the appropriate milieu for such study is local asymptotic normality under contiguous alternatives, formulated in this case as follows. Let $N_1, N_2, \ldots$ be point processes on a compact set E and let μ_0 and μ^* be elements of M_d with $\mu^* \ll \mu_0$. Suppose that under the probability measure P_0 the N_i are i.i.d. Poisson processes with mean measure μ_0 and that there are probabilities P^n under which $N_1, \ldots, N_n$ are i.i.d. Poisson processes with mean measure $\mu_0 + \mu^*/\sqrt{n}$. The contiguity theorem describes asymptotic behavior of log-likelihood ratios associated with the superpositions $N^n = \sum_{i=1}^n N_i$. Since under P_0, N^n is Poisson with mean $n\mu_0$, while under P^n its mean measure is $n\mu_0 + \sqrt{n}\mu^*$, by (6.11) the log-likelihood ratio is

$$
\begin{aligned}
L_n &= \log(dP^n/dP_0) \\
&= n\mu_0(E) - [n\mu_0(E) + \sqrt{n}\mu^*(E)] + \int_E \log \frac{d(n\mu_0 + \sqrt{n}\mu^*)}{d(n\mu_0)} dN^n \\
&= -\sqrt{n}\mu^*(E) + \int_E \log\left[1 + (1/\sqrt{n})(d\mu^*/d\mu_0)\right] dN^n.
\end{aligned}
$$

More generally, allowing for observation over a variable subset A of E we have the log-likelihood function

$$
L_n(A) = \sqrt{n}\mu^*(A) + \int_A \log\left[1 + (1/\sqrt{n})(d\mu^*/d\mu_0)\right] dN^n, \qquad (6.15)
$$

a signed random measure on E. In Theorem 5.21 we used martingale methods to analyze such processes; here we employ instead the independent increments property of Poisson processes.

Theorem 6.14. Suppose that $\int_E (d\mu^*/d\mu_0)d\mu^* < \infty$. Then with notation and hypotheses as in the preceding paragraph, under P_0, $L_n \xrightarrow{d} G$ as random measures on E, where G is a Gaussian random measure with independent increments $E[G(A)] = -\mu^*(A)/2$ and $\mathrm{Var}(G(A)) = \int_A (d\mu^*/d\mu_0)d\mu^*$.

Proof: By Theorem 1.21 and the fact that the L_n and G have independent increments (for details of the construction of G, see Neveu, 1965, pp. 84ff.), it suffices to show that $L_n(A) \xrightarrow{d} G(A)$ for each set A. Continuing from (6.15), we have

$$
\begin{aligned}
L_n(A) &\cong -\sqrt{n}\mu^*(A) + \frac{1}{\sqrt{n}} \int_A \frac{d\mu^*}{d\mu_0} dN^n - \frac{1}{2n} \int_A \frac{d\mu^*}{d\mu_0} dN^n \\
&= \frac{1}{\sqrt{n}} \sum_{i=1}^n \left(\int_A \frac{d\mu^*}{d\mu_0} dN_i - \int_A \frac{d\mu^*}{d\mu_0} d\mu_0 \right) - \frac{1}{2n} \sum_{i=1}^n \int_A \frac{d\mu^*}{d\mu_0} dN_i.
\end{aligned}
$$

The summands in the first term are i.i.d. under P_0 with mean zero and finite variance $\int_A (d\mu^*/d\mu_0)^2 d\mu_0 = \int_A (d\mu^*/d\mu_0)d\mu^*$, while by the strong law of large numbers, the second term converges to $-(1/2)\mu^*(A)$; hence $L_n(A) \xrightarrow{d} G(A)$ by the central limit theorem and Slutsky's theorem. ∎

The ease with which further testing problems can be posed for Poisson processes on general spaces is matched by the paucity of rigorous procedures available for addressing them in nonparametric settings. For nonhomogeneous Poisson processes on $\mathbf{R}_+$ there do exist techniques (some to be mentioned presently) for testing structure of rate functions or comparing rate functions. For planar Poisson processes there are "distance methods," developed mainly by British statisticians, based on distances to nearest neighbors in the point process and distances from points in $\mathbf{R}^2$ to points of N that have been applied extensively to real problems of data analysis. The reader interested in applications is urged to pursue the latter; see the chapter notes for some references.

Returning to general spaces, among testing problems of both mathematical and practical interest, but concerning which little is known, are the following.

1. *Structure of the mean measure.* The prototypical problem is to test whether μ is absolutely continuous with respect to a prescribed measure μ_0. Consistent estimators of $d\mu/d\mu_0$ under H_0 can be constructed using Theorem 6.7, but it is unclear what properties of them (intuitively, "largeness," indicative of a singular component) should lead to rejection of H_0.

2. *Equality and ordering of mean measures.* Given independent Poisson processes N_1 and N_2 with mean measures μ_1 and μ_2, except on $\mathbf{R}_+$ and with further assumptions, few generally applicable procedures are available for testing even the hypothesis that $\mu_1 = \mu_2$, let alone the hypothesis that $\mu_1 \geq \mu_2$.

3. *Whether a point process is Poisson.* The main difficulty, beyond those already present for ordinary Poisson processes on $\mathbf{R}_+$, is lack of credible, tractable alternative hypotheses.

For Poisson processes on $\mathbf{R}_+$ martingale methods are available. One could use techniques described in Section 5.3 to test, for example, whether a Poisson process N on $[0, 1]$ known to have an intensity function, has a prescribed intensity function. Two-sample tests can be effected with martingale test statistics from Section 5.3, which under $H_0 : \alpha(1) \equiv \alpha(2)$ [here $N(1)$ and $N(2)$ are independent Poisson processes with intensity functions $\alpha(1)$, $\alpha(2)$] have the advantage of removing the nuisance parameter $\alpha(1) = \alpha(2)$ from the problem. Under H_0 the process $M_t = N_t(1) - N_t(2)$ is a mean zero martingale; procedures whose asymptotics are described by Theorem 5.26 test for this property.

6.3   Partially Observed Poisson Processes

In this section we consider three models of Poisson processes that are only partially observable, and for each present results on statistical inference and state estimation. We begin with p-thinned Poisson processes, refining and specializing results appearing in Sections 4.5 and 1; in addition, we develop a maximum likelihood interpretation of the estimators $\hat{p}$ analyzed in Theorem 4.18 and establish consistency under weaker conditions. Next we examine the stochastic integral process studied in Examples 3.5 and 3.10 but in greater generality: for some results the integrand is allowed to be a semi-Markov process rather than a Markov process. For the Markov case we give an explicit application of the state estimation procedure culminating in Theorem 5.30. Finally, we treat inference for Poisson processes on product spaces such that only one component of the process is observable. The three models are illustrative rather than exhaustive; in some ways the variety of models and the techniques with which they are analyzed is the real content of the section.

6.3.1 p-Thinned Poisson Processes

We recall the setting established in Section 4.5. Let E be a compact space and let $N_1', N_2', \ldots$ be observable point processes. They are thinnings of unobservable Poisson processes N_i, so we retain the "prime" notation from Section 4.5. The statistical model is $\mathcal{P} = \{P_p\}$, where p ranges over the family of measurable functions from E to $[0, 1]$; under P_p the N_i' are i.i.d. copies of the p-thinning of the Poisson process with *known* mean measure $\mu_0 \in \mathbf{M}_d$. The model is hence a submodel of the model treated in Theorem 6.7, as well as a specialized version of Model 3.3. Our goal is to estimate the thinning function p.

Denote by P the probability measure corresponding to $p \equiv 1$ (i.e., no thinning), under which the N_i' have mean measure μ_0, and for each n let $\bar{N}_n' = \sum_{i=1}^n N_i'$. Then from (6.11) we obtain log-likelihood functions

$$L_n(p) = n \int_E (1 - p) d\mu_0 + \int_E (\log p) d\bar{N}_n'. \tag{6.16}$$

By contrast with the situation of Theorem 6.7, there do exist maximum likelihood estimators $\hat{p}$, but they are *not consistent*. Since $\log p \le 0$, the second term in the log-likelihood function is maximized by taking p to be one at each atom of $\bar{N}'_n$, and then the first term is maximized by letting p be zero everywhere else. Because $P\{\bar{N}'_n(E) < \infty\} = 1$ and μ_0 is diffuse, $\|\hat{p} - p\|_1 = \|p\|_1 = 1$, where $\|\cdot\|_1$ is the norm on $L^1(\mu_0)$; consequently, the $\hat{p}$ are not consistent. However, the histogram sieve estimators of (6.8), which coincide with the estimators of Theorem 4.18 are strongly consistent under conditions weaker in some ways than those there.

Let $\{A_{mj} : m \geq 1, 1 \leq j \leq \ell_m\}$ be a null array of partitions of E with $\mu_0(A_{mj}) > 0$ for each m and j. For each m, let I_m be the sieve of functions $p : E \to [0,1]$ constant on each A_{mj}. Then we have the following counterpart of Theorems 4.18 and 6.7.

Theorem 6.15. a) For each n and m, the n-sample maximum likelihood estimator of p relative to I_m satisfies

$$\hat{p}(n, m)(x) = \min\{\bar{N}'_n(A_{mj})/n\mu_0(A_{mj}), 1\}, \quad x \in A_{mj};$$

b) Suppose that $0 \leq \delta < 1$ and that $m = m_n$ is chosen so that with $\tilde{\ell}_m = \ell_{m_n}$,

$$\sum_{n=1}^{\infty} \tilde{\ell}_n^4/n^{2-\delta} < \infty. \tag{6.17}$$

Then with $\hat{p} = \hat{p}(n, m_n)$, for each p such that

$$\lim_{n\to\infty} n^{\delta/4} \sum_{j=1}^{\tilde{\ell}_n} \int_{A_{mj}} \left| \mu_0(p; A_{mj})/\mu_0(A_{mj}) - p(x) \right| \mu_0(dx) = 0, \tag{6.18}$$

we have $n^{\delta/4}\|\hat{p} - p\|_1 \to 0$ almost surely.

Proof: a) For $p = \sum_{j=1}^{\ell_m} a_j 1_{A_{mj}}$ belonging to I_m, (6.16) gives

$$L_n(p) = \sum_{j=1}^{\ell_m} [(1 - a_j)\mu_0(A_{nj}) + \bar{N}'_n(A_{mj})(\log a_j)].$$

Since

$$\partial L_n(p)/\partial a_j = -\mu_0(A_{mj}) + \bar{N}'_n(A_{mj})/a_j,$$

maximization with respect to a_j occurs at $\hat{a}_j = \bar{N}'_n(A_{mj})/\mu_0(A_{mj})$ provided that this quantity does not exceed 1. If it does, then the derivative is positive everywhere on $[0,1]$, so the maximizing value in $[0,1]$ is $\hat{a}_j = 1$.

b) The argument follows the proof of Theorem 6.7. One splits $n^{\delta/4}\|\hat{p}-p\|_1$ into random and nonrandom terms; for the former

$$P_p\left\{n^{\delta/4}\sum_{j=1}^{\ell_m} |\bar{N}'_n(A_{mj})/n - E_p[N'(A_{mj})]| > \varepsilon\right\}$$

$$\leq \frac{1}{\varepsilon^4 n^{\delta}} E_p\left[\left(\sum_{j=1}^{\ell_m} |\bar{N}'_n(A_{mj})/n - E_p[N'(A_{mj})]|\right)^4\right] = O\left(\frac{\ell_m^4}{\varepsilon^4 n^{2-\delta}}\right)$$

for each $\varepsilon > 0$, so that (6.17) implies that the random term converges to zero almost surely. That (6.18) suffices to dispose of the nonrandom term follows from the proof of Theorem 6.7. ∎

Unlike Theorem 4.18, Theorem 6.15 with $\delta = 0$ imposes no continuity requirement on p. A Poisson process N with mean measure μ_0 satisfies

$$E[N(A_{mj})^3] = \mu_0(A_{mj}) + 3\mu_0(A_{mj})^2 + \mu_0(A_{mj})^3,$$

so that (4.20) in this case becomes (essentially) $\tilde{l}_n^3 = O(n^\delta)$ for some $\delta < 1$, i.e., $\tilde{l}_n \cong n^{1/3-\varepsilon}$ for some $\varepsilon > 0$, while (6.17) is more restrictive: it requires that $\tilde{l}_n \cong n^{1/4-\varepsilon}$.

If each N_i', under P_p, is the p-thinning of an unobservable Poisson process N_i with mean measure μ_0, with the pairs (N_i, N_i') i.i.d., then state estimation for the underlying processes N_i is nearly trivial. By Exercise 1.16, N_i' and $N_i'' = N_i - N_i'$ are independent Poisson processes with mean measures $p(x)\mu_0(dx)$ and $[1 - p(x)]\mu_0(dx)$; consequently,

$$E[N_i|\mathcal{F}^{N_i'}] = E[N_i' + N_i''|\mathcal{F}^{N_i'}] = N_i' + (1 - p)\mu_0, \qquad (6.19)$$

where $[(1 - p)\mu_0](dx) = (1 - p(x))\mu_0(dx)$.

Of course, however, (6.19) can be implemented only if p is known. Otherwise we are in the combined statistical inference and state estimation setting of Section 3.4. With $\hat{p}$ the estimators in Theorem 6.15, we can form, guided by principles enunciated in Section 3.4, *pseudo-state estimators*

$$\hat{E}[N_{n+1}|\mathcal{F}^{N_{n+1}'}] = N_{n+1}' + (1 - \hat{p})\mu_0, \qquad (6.20)$$

which have the following consistency property.

Proposition 6.16. Under the conditions of Theorem 6.15b), for each δ and p satisfying (6.17)–(6.18),

$$n^{\delta/4} \left\| \hat{E}[N_{n+1}|\mathcal{F}^{N_{n+1}'}] - E_p[N_{n+1}|\mathcal{F}^{N_{n+1}'}] \right\| \to 0$$

almost surely, where $\|\cdot\|$ denotes the total variation norm on the set of finite signed measures on E.

Proof: From (6.19)–(6.20),

$$\left(\hat{E}[N_{n+1}|\mathcal{F}^{N_{n+1}'}] - E_p[N_{n+1}|\mathcal{F}^{N_{n+1}'}] \right)(dx) = [\hat{p}(x) - p(x)]\mu_0(dx),$$

which implies that

$$n^{\delta/4} \left\| \hat{E}[N_{n+1}|\mathcal{F}^{N_{n+1}'}] - E_p[N_{n+1}|\mathcal{F}^{N_{n+1}'}] \right\| = n^{\delta/4}\|\hat{p} - p\|_1;$$

therefore, the result follows from Theorem 6.15b). ∎

6.3.2 Stochastic Integral Process

Let N be a Poisson process on $\mathbf{R}_+$ having unknown rate λ under the probability measure P_λ, and let (X_t) be a semi-Markov process with state space $\{0,1\}$ that under each P_λ is independent of N with sojourn distributions G and F: sojourns of X in 0 having distribution G alternate with sojourns in 1 having distribution F, and all sojourns are mutually independent (see also Section 8.4). We assume that F and G are known, although if, as in Examples 3.5 and 3.10, X is a Markov process with generator

$$A = \begin{bmatrix} -a & a \\ b & -b \end{bmatrix}, \tag{6.21}$$

then the transition rates a and b can be unknown as well (see Karr, 1984a).

The observations are the stochastic integral process $(X*N)_t = \int_0^t X_s dN_s$ which is a point process, together perhaps with X itself. The interpretation is that an atom of N at u is observable in $X*N$ if and only if $X_u = 1$. When X is observable, that is, when the observed history is

$$\mathcal{H}_t = \mathcal{F}_t^{X*N} \vee \mathcal{F}_t^X, \tag{6.22}$$

inference concerning λ and state estimation for unobserved portions of N are both straightforward. Some statistical results are given in Exercise 6.16; Exercises 3.15–3.17 treat the Markov case.

After noting elementary properties of $X*N$ we examine state estimation for unobserved portions of N given the observations (6.22); thereafter we move on to the more difficult and interesting case that only $X*N$ is observable.

Proposition 6.17. Under each probability P_λ,

a) $X*N$ is a Cox process directed by the random measure $M(A) = \lambda \int_A X_u du$;

b) If F and G are nonarithmetic with finite, positive means $\bar{F}$ and $\bar{G}$, then with $q = \bar{F}/(\bar{F} + \bar{G})$, $(X*N)_t/qt \to \lambda$ almost surely;

c) If F and G are nonarithmetic with finite variances $\sigma^2(F)$ and $\sigma^2(G)$, then with $\sigma^2 = [\bar{F}^2\sigma^2(G) + \bar{G}^2\sigma^2(F)]/(\bar{F} + \bar{G})^3$, as $t \to \infty$

$$\frac{(X*N)_t - \lambda qt}{\sqrt{(1 + q\sigma)\lambda qt}} \xrightarrow{d} N(0,1). \square$$

The quantity q is the asymptotic fraction of time X spends in state 1 (i.e., the fraction of time during which N is observable). For proofs, see Karr (1982) and Grandell (1976); independence of N and X is crucial. In particular, Proposition 6.17a) implies that the $(P_\lambda, \mathcal{H})$-stochastic intensity of $X * N$ is (λX_{t-}), so that by Exercise 2.16 the $(P_\lambda, \mathcal{F}^{X*N})$-stochastic intensity is $(\lambda E_\lambda[X_t|\mathcal{F}_{t-}^{X*N}])$, whose computation is addressed in Theorem 6.22. With X observable we have a multiplicative intensity model, but asymptotics here, as $t \to \infty$, differ from those treated in Chapter 5.

Suppose now that λ is known; then state estimation for N given observations $\mathcal{H}_t$ can be performed easily.

Proposition 6.18. For each t, with $A_t = A \cap [0, t]$,

$$E_\lambda[N(A)|\mathcal{H}_t] = X * N(A_t) + \lambda \int_{A_t} (1 - X_u)du + \lambda |\mathbf{R}_+ \setminus A_t|. \qquad (6.23)$$

Proof: For fixed t and A,

$$E_\lambda[N(A)|\mathcal{H}_t]$$
$$= E_\lambda\left[X * N(A_t) + \int_{A_t}(1 - X_u)du + N(\mathbf{R}_+ \setminus A_t)\Big|\mathcal{H}_t\right]$$
$$= X * N(A_t) + \int_{A_t}(1 - X_u)E_\lambda[dN_u|\mathcal{H}_t] + E_\lambda[N(\mathbf{R}_+ \setminus A_t)|\mathcal{H}_t].$$

By independence of N and X and the independent increments property of N, $(1 - X_u)E_\lambda[dN_u|\mathcal{H}_t] = (1 - X_u)\lambda du$; similarly, $N(\mathbf{R}_+ \setminus A_t)$ is independent of $\mathcal{F}_t^N \bigvee \mathcal{F}_t^X$ and hence also of $\mathcal{H}_t$, so that

$$E_\lambda[N(\mathbf{R}_+ \setminus A_t)|\mathcal{H}_t] = E_\lambda[N(\mathbf{R}_+ \setminus A_t)] = \lambda|\mathbf{R}_+ \setminus A_t|. \quad \blacksquare$$

We now take up inference and state estimation given observation only of $X * N$; for notational simplicity let $\mathcal{F}_t = \mathcal{F}_t^{X*N}$.

Proposition 6.19. Let q and σ^2 be as in Proposition 6.17 and assume that conditions b) and c) there hold. Define estimators $\hat{\lambda} = (X * N)_t/qt$. Then under each probability P_λ,

a) $\hat{\lambda} \to \lambda$ almost surely;

b) $\sqrt{qt}[\hat{\lambda} - \lambda]/\sqrt{\lambda(1 + \lambda\sigma^2/q)} \overset{d}{\to} N(0, 1)$. $\square$

The next result not only can be used to construct likelihood ratio tests, but also serves to emphasize importance of the conditional expectations $E_\lambda[X_t|\mathcal{F}_t]$.

Proposition 6.20. For positive λ_0 and λ_1, and $t < \infty$, $P_{\lambda_0} \sim P_{\lambda_1}$ on $\mathcal{F}_t$, with

$$\log[dP_{\lambda_1}/dP_{\lambda_0}] = \int_0^t[\lambda_1\hat{X}_u(\lambda_1) - \lambda_0\hat{X}_u(\lambda_0)]du + (X * N)_t \log(\lambda_1/\lambda_0)$$
$$+ \int_0^t \log(\hat{X}_u(\lambda_1)/\hat{X}_u(\lambda_0))d(X * N)_u,$$

where $\hat{X}_t = E_\lambda[X_u|\mathcal{F}_u]\big|_{u=t-}$.

Proof: According to discussion following Proposition 6.17, the $(P_\lambda, \mathcal{F})$-stochastic intensity of $X * N$ is the process $\hat{X}_u(\lambda)$, and if P denotes a probability measure with respect to which $X * N$ is Poisson with rate 1, then by Theorem 2.31,

$$\log[dP_{\lambda_1}/dP_{\lambda_0}] = \int_0^t(1 - \hat{X}_u(\lambda))du + (X * N)_t(\log \lambda)$$
$$+ \int_0^t \log \hat{X}_u(\lambda)d(X * N)_u,$$

and from this the proposition holds by easy computations. ∎

State estimation for unobserved portions of N requires the MMSE state estimators $E_\lambda[X_t|\mathcal{F}_t]$, whose computation we consider next.

Proposition 6.21. For each t and A,

$$E_\lambda[N(A)|\mathcal{F}_t] = (X * N)_t(A) + \lambda\int_{A_t}(1 - E_\lambda[X_u|\mathcal{F}_u])du + \lambda|\mathbf{R}_+ \setminus A_t|. \qquad \square$$

Thus the crucial state estimation issue is reconstruction of the process X, to which we apply methodology developed in Section 5.4. It seems possible to obtain explicit results only when X is a Markov process, with generator (6.21), and we so assume for the rest of the discussion.

Theorem 6.22. For each λ the MMSE state estimators $\hat{X}_t = E_\lambda[X_t|\mathcal{F}_t]$ satisfy the stochastic differential equation

$$d\hat{X}_t = [-b\hat{X}_t + a(1 - \hat{X}_t)]dt + (1 - \hat{X}_{t-})[d(X * N)_t - \lambda\hat{X}_{t-}dt]. \qquad (6.24)$$

Proof: We begin with a semimartingale representation for X relative to the filtration $\mathcal{H}$ of (6.22), furnished in this case by Dynkin's formula (see also Theorem 7.30). The process $M_t = X_t - \int_0^t[a\mathbf{1}(X_u = 0) - b\mathbf{1}(X_u = 1)]du$ is a martingale; hence

$$X_t = \int_0^t[a\mathbf{1}(X_u = 0) - b\mathbf{1}(X_u = 1)]du + M_t$$

is the requisite representation. With λ fixed it follows from Proposition 5.28, (5.63) and the property $\hat{X}_t = E[X_t|\mathcal{F}_t^N]$ that there is an $\mathcal{F}$-predictable process H with

$$\hat{X}_t = \int_0^t[a(1 - \hat{X}_u) - b\hat{X}_u]du + \int_0^t H_u[d(X * N)_u - \lambda\hat{X}_{u-}du], \qquad (6.25)$$

since the binary nature of X implies that

$$P_\lambda\{X_u = 0|\mathcal{F}_u\} = 1 - P_\lambda\{X_u = 1|\mathcal{F}_u\} = 1 - E_\lambda[X_u|\mathcal{F}_u],$$

and where we have also used the property that $(\lambda \hat{X}_{t-})$ is the $(P_\lambda, \mathcal{H})$-stochastic intensity of $X * N$. By Theorem 5.30, $H_u = J_u + K_u - \hat{X}_{u-}$, where J and K are as in that Theorem, relative to the state process X. Because X is Markov and independent of N, the martingale M has no jumps in common with N; hence $K \equiv 0$ (see discussion subsequent to Theorem 5.30). By (5.71) and the crucial relationship $X^2 = X$,

$$J_u = (\lambda X^2)\hat{\,}_{u-}/\lambda \hat{X}_{u-} = (\lambda X)\hat{\,}_{u-}/\lambda \hat{X}_{u-} = 1.$$

Therefore, $H_u = 1 - \hat{X}_{u-}$; together with (6.25) this completes the proof. ∎

Using Proposition 6.21 and Theorem 6.22 it is straightforward to effect MMSE state estimation for unobserved portions of N. Note also that (6.24) provides a recursive method for computation of the state estimators $E_\lambda[X_t|\mathcal{F}_t]$: in order to calculate the increment $dE_\lambda[X_t|\mathcal{F}_t]$ one needs only the current value (so that the amount of data that must be retained in order to calculate the state estimator does not increase with time) and the innovation $dM_t = d(X * N)_t - \lambda \, E_\lambda[X_u|\mathcal{F}_u]|_{u=t-} \, dt$. The property that $X^2 = X$ is central, as will be seen more graphically in Section 7.5: it avoids introducing a state estimator of the second moment, which would have to be calculated with another stochastic differential equation.

6.3.3   Poisson Processes with One Observable Component

Let E_0 and E_1 be compact spaces (compactness of E_1 is not essential), let $E = E_0 \times E_1$, and suppose that K is a *known* transition kernel from E_0 into E_1. We consider a special case of the following situation: there is an *unobservable* point process $N^0 = \sum \epsilon_{Y_i}$ on E_0 such that each point Y_i has associated to it an observable point $X_i \in E_1$ with distribution $K(Y_i, \cdot)$ and the X_i are conditionally independent given N^0. One observes $N^1 = \sum \epsilon_{X_i}$, the second component of the two-component point process $N = \sum \epsilon_{(Y_i, X_i)}$ on E. We consider the statistical model $\mathcal{P} = \{P_\mu : \mu \in \mathbf{M}_d(E_0)\}$, where under P_μ, N^0 is Poisson with mean measure μ.

Here are elementary properties.

Lemma 6.23. Suppose that for each μ the X_i are P_μ-conditionally independent given N^0 with $P_\mu\{X_i \in B|N^0\} = K(Y_i, B)$ independently of μ. Then

a) $N = \sum \epsilon_{(Y_i, X_i)}$ is a Poisson process on E with mean measure given by $\mu(dy)K(y, dx)$;

b) $N^1 = \sum \epsilon_{X_i}$ is Poisson on E_1 with mean measure

$$\nu_\mu(dx) = \int_{E_0} \mu(dy)K(y, dx). \; \square$$

Within this setting, we consider the usual problems: estimation of μ given data $N_1^1, N_2^1, \ldots$ comprising i.i.d. copies of the second component N^1 and state estimation for N^0 given observation of N^1. As will be seen, structure of the kernel K is crucial.

We begin with estimation of μ. The extent to which observation of i.i.d. processes N_i^1 with mean measure ν_μ is informative concerning μ is determined by the kernel K. For example, if $K(y, dx) = \eta(dx)$ independently of y, where η is a probability measure on E_1, then $\nu_\mu = \mu(E_0)\eta$ and one can learn only $\mu(E_0)$. At the opposite extreme, for $K(y, dx) = \varepsilon_{f(y)}(dx)$ with f a bijection of E_0 onto a subset of E_1, then $\nu_\mu = \mu f^{-1}$, by which μ is uniquely determined.

Given the N_i^1 as data, one can do nothing more than estimate their mean measure ν_μ, using, on the basis of Proposition 6.2, the estimators $\hat{\nu} = (1/n)\sum_{i=1}^n N_i^1$. The following result summarizes what is learned thereby concerning μ itself.

Proposition 6.24. For $g \in C^+(E_1)$, let $Kg(y) = \int K(y, dx)g(x)$, and suppose that K fulfills the Feller property: $Kg \in C^+(E_0)$ for each $g \in C^+(E_1)$. Let $V = \{Kg : g \in C^+(E_1)\}$ and for $Kg \in V$ put $\hat{\mu}(Kg) = \hat{\nu}(g)$. Then Proposition 6.2 holds for compact subsets $\mathcal{K}$ of V. ▯

Of course, Proposition 6.24 fails to confront the crucial issue of how large the set V is, but especially at our level of generality this issue is difficult. Mathematically, the question is identifiability of mixtures: the mean ν_μ is the μ-mixture of the measures $K(y, \cdot)$. A mixture model $\{K(y, \cdot) : y \in E_0\}$ is *identifiable* if the mapping $\mu \to \nu_\mu$ is one-to-one. Some results are available, but we will not pursue them here; see additional comments and references in the chapter notes.

For the special case that $\mu(dy) = \alpha(y)\mu_0(dy)$ with μ_0 a known element of $\mathbf{M}_d(E_0)$ and $\alpha \in L^1(\mu_0)$, and that $K(y, dx) = k(y, x)\rho(dx)$ with ρ a probability on E_1 and $k \geq 0$ a jointly measurable function, one can show existence of maximum likelihood estimators by sieve methods. However, the estimators are only defined implicitly, although they can be calculated — as discussed momentarily — with the "EM algorithm." Consistency properties are difficult and we do not elaborate on them here.

Suppose that for each $\alpha \in L_+^1(\mu_0)$ there is a probability measure P_α under which the N_i^1 are i.i.d. copies of a Poisson process on E_1 having intensity function $\int \mu_0(dy)\alpha(y)k(y, x)$, and for each n let $\bar{N}_n^1 = \sum_{i=1}^n N_i^1$. Suppose that $E_0 = \sum_{j=1}^\ell A_j$ is a partition, which is fixed and dropped from notation, and let I be the histogram sieve of functions constant over each A_j.

Proposition 6.25. For each n there exists a maximum likelihood estimator $\hat{\alpha}_n = \sum_{j=1}^{\ell} a_j^* 1_{A_j}$, where the a_j^* satisfy the equations

$$\int_{A_i} \mu_0(dy) \left[-1 + \int_{E_1} \frac{k(y,x)}{\sum_j a_j^* \int_{A_j} \mu_0(dz) k(z,x)} \bar{N}_n^1(dx) \right] = 0 \qquad (6.26)$$

for $i = 1, \ldots, \ell$.

Proof: For general $\alpha \in L_+^1(\mu_0)$ the log-likelihood function relative to $\alpha \equiv 1$, by Proposition 6.10, is

$$\begin{aligned}
L_n(\alpha) \;=\;& n\int_{E_0}\mu_0(dy)\left[1 - \alpha(y)\int_{E_1} k(y,x)\rho(dx)\right] \\
& + \int_{E_1} \log\left[\int_{E_0}\mu_0(dx)\alpha(y)k(y,x)\right]\bar{N}_n^1(dx) \\
& - \int_{E_1} \log\left[\int_{E_0}\mu_0(dy)k(y,x)\right]\bar{N}_n^1(dx).
\end{aligned}$$

For the special case at hand, with terms not depending on α deleted, this becomes

$$\begin{aligned}
L_n(\alpha) \;=\;& -n\sum_{j=1}^{\ell}a_j\mu_0(A_j) \\
& + \int_{E_1} \log\left[\sum_{j=1}^{\ell}a_j\int_{A_j}\mu_0(dy)k(y,x)\right]\bar{N}_n^1(dx)
\end{aligned}$$

Division by n and differentiation with respect to a_i give (6.26). That a solution exists and maximizes the likelihood function follows by arguments in Vardi *et al.* (1985). ∎

The equations (6.26) are troublesome computationally: they are nonlinear, and in applications such as the positron emission tomography model of Vardi *et al.* (1985), ℓ may be on the order of 10^4. An iterative method of solution is the EM algorithm of Dempster *et al.* (1977). The algorithm leads to recursive computation of revised values a_i' from current values by means of the expression

$$a_i'\mu_0(A_i) = a_i\int_{E_1}\frac{\int_{A_i}\mu_0(dy)k(y,x)}{\sum_j a_j\int_{A_j}\mu_0(dz)k(z,x)}\frac{\bar{N}_n^1(dx)}{n}.$$

Given that the recursion converges, it is evident that the limits a_j^* satisfy (6.26).

The final result of the section examines state estimation for N^0 given observation of N^1 under the assumptions that $K(y,dx) = k(y,x)\rho(dx)$ and that μ is known. An explicit solution can be obtained for the conditional Laplace functional of N^0.

Theorem 6.26. For each μ and f

$$E_\mu[e^{-N^0(f)}|N^1] = \exp\left[-\int_{E_1} -\log\frac{\int\mu(dy)e^{-f(y)}k(y,x)}{\int\mu(dz)k(z,x)}N^1(dx)\right]. \qquad (6.27)$$

Proof: Let $\mu k(x) = \int\mu(dy)k(y,x)$. Under P_μ, N^1 is Poisson with mean measure $\mu k(x)\rho(dx)$; therefore, for $g \in C^+(E_1)$,

$$E_\mu\left[\exp\left\{-\int -\log\frac{\int\mu(dy)e^{-f(y)}k(y,x)}{\mu k(x)}N^1(dx)\right\}e^{-N_1(g)}\right]$$

$$= E_\mu\left[\exp\left\{-\left(\int -\log\frac{\int\mu(dy)e^{-f(y)}k(y,x)}{\mu k(x)} + g(x)\right)N^1(dx)\right\}\right]$$

$$= \exp\left\{-\int_{E_1}\left(1 - \frac{\int\mu(dy)e^{-f(y)}k(y,x)}{\mu k(x)}e^{-g(x)}\right)\mu k(x)\rho(dx)\right\}$$

$$= \exp\left\{-\int_{E_1}[\mu k(x) - \int_{E_0}\mu(dy)e^{-f(y)}k(y,x)e^{-g(x)}]\rho(dx)\right\}$$

$$= \exp\left\{-\int_{E_0}\int_{E_1}(1 - e^{-f(y)-g(x)})\mu(dy)k(y,x)\rho(dx)\right\}$$

$$= E_\mu[e^{-N^0(f)-N^1(g)}],$$

and an appeal to Theorem 1.12 serves to complete the proof. ∎

From (6.27) it follows that

$$E_\mu[N^0(f)|N^1] = \int_{E_1}\frac{\int\mu(dy)f(y)k(y,x)}{\int\mu(dy)k(y,x)}N^1(dx).$$

Comparison of (6.26) and (6.27) reveals the essence of the EM algorithm: it is a procedure for combined inference and state estimation, but with emphasis on inference, whereas in our other results emphasis is on state estimation. The algorithm performs an expectation step, using current parameter estimates to effect a provisional reconstruction of the unobservable Poisson process N^0, then uses this reconstruction in a maximum likelihood step to revise parameter estimates, and proceeds until convergence occurs.

6.4 Inference Given Integral Data

In this section we discuss inference for Poisson processes when observations are *integral data* $\mathcal{F}^N(g_0,\ldots,g_m) = \sigma(N(g_0),\ldots,N(g_m))$, where $g_0 \equiv 1$, $g_1,\ldots,g_m$ are continuous functions on E (assumed compact). For introductory discussion, see Section 3.1. The problems are estimation of the mean measure given observation of i.i.d. copies N_i, and state estimation, now non-trivial, for N. Moreover, we add a design element absent from other analyses: the functions g_j can be interpreted as modes of measurement that are under control of the statistician or experimenter; the design problem is to choose which measurements to perform. Although the more challenging question of choosing the g_j adaptively is not treated, we do examine criteria that may be taken into account when these functions are chosen.

A fundamental idea for either inference or state estimation is to choose $g_1, \ldots, g_m$ in such a way that every point measure $\nu = \sum_{i=1}^{\ell} \epsilon_{x_i}$ on E with $\ell \le m$ is uniquely determined by the integrals $\nu(g_0) = \nu(E)$, $\nu(g_1), \ldots, \nu(g_m)$. Then on the event $\{N(E) \le m\}$ the point process N is completely specified by $N(g_0), \ldots, N(g_m)$ and one can reconstruct the value of $N(f)$ for every $f \in C^+(E)$. Additional discussion of such sets is presented following Proposition 6.28.

In particular, for $m = 1$, g_1 must be an invertible mapping from E onto a subset of $\mathbf{R}$. For the Poisson case, by virtue of conditional uniformity, such a function and the function $g_0 \equiv 1$ are all one requires for consistent estimation of the mean measure.

Theorem 6.27. Let $N_1, N_2, \ldots$ be i.i.d. Poisson processes with mean measure μ and assume that g is a bounded, invertible mapping of E onto a subset of $\mathbf{R}$. On $\{N_i(E) = 1\}$, let $X_i = g^{-1}(N_i(g))$, and for $f \in C^+(E)$ define

$$\hat{\mu}(f) = \frac{\sum_{i=1}^{n} f(X_i)\mathbf{1}(N_i(E) = 1)}{\sum_{i=1}^{n} \mathbf{1}(N_i(E) = 0)}. \tag{6.28}$$

Then for each compact subset $\mathcal{K}$ of $C^+(E)$, almost surely

$$\sup_{f \in \mathcal{K}} |\hat{\mu}(f) - \mu(f)| \to 0. \tag{6.29}$$

Note that $\hat{\mu}$ is computable from $(N_1(E), N_1(g)), \ldots, (N_n(E), N_n(g))$. Motivation for these estimators will emerge during the proof. The key idea is that by invertibility of g, on $\{N_i(E) = 1\}$, X_i is the sole point of N_i and by conditional uniformity has distribution $\mu(\cdot)/\mu(E)$.

Proof of Theorem 6.27: With N a generic copy,

$$e^{-\mu(E)} = P\{N(E) = 0\} = \lim_{n \to \infty} (1/n)\sum_{i=1}^{n} \mathbf{1}(N_i(E) = 0), \tag{6.30}$$

while by conditional uniformity, for each f,

$$E[N(f)\mathbf{1}(N(E) = 1)] = E[N(f)|N(E) = 1]P\{N(E) = 1\} = e^{-\mu(E)}\mu(f).$$

With the X_i defined as above, invertibility of g implies that

$$N_i\mathbf{1}(N_i(E) = 1) = \epsilon_{X_i}\mathbf{1}(N_i(E) = 1),$$

and in particular $N_i(f)\mathbf{1}(N_i(E) = 1) = f(X_i)\mathbf{1}(N_i(E) = 1)$. Since $\mu(E) < \infty$, this expression and Proposition 4.8, applied to the observable point processes $\tilde{N}_i = N_i\mathbf{1}(N_i(E) = 1)$, give

$$\sup_{\mathcal{K}}|(1/n)\textstyle\sum_{i=1}^{n}f(X_i)\mathbf{1}(N_i(E) = 1) - e^{-\mu(E)}\mu(f)| \to 0 \tag{6.31}$$

almost surely, and (6.29) follows from (6.30) and (6.31). ∎

Other limit results exist as well; see Exercise 6.21 for one. When $m > 1$, more efficient methods of estimation — Theorem 6.27 uses only realizations for which $N_i(E) \le 1$ — can be devised.

Now for state estimation; we assume that the mean measure μ is known. Exact calculation of conditional expectations $E[N(f)|\mathcal{F}^N(g_0, \ldots, g_m)]$ seems difficult in generality, even for Poisson processes, but other state estimators can be computed rather easily. To illustrate, we consider optimal state estimators that are *linear* as functions of the observed data $N(g_0), \ldots, N(g_m)$. Although it is not essential, we continue to assume that $g_0 \equiv 1$.

The pleasant solution to this problem is that the optimal linear estimator is given by $\hat{N}(f) = a(f) + N(\hat{f})$ where $a(f)$ is a constant and $\hat{f}$ is the projection of f onto the span of $\{g_0, \ldots, g_m\}$ in the Hilbert space $L^2(\mu)$.

Proposition 6.28. Let $V = \text{sp}(g_0, \ldots, g_m)$. Then for $f \in C^+(E)$ the solution to the problem

$$\begin{array}{cc} \text{minimize} & E\left[(N(f) - a - N(g))^2\right] \\ \text{s.t. } a \in \mathbf{R}, g \in V \end{array}$$

is $g^* = \hat{f}$, the orthogonal projection of f onto V, and $a^* = \mu(f) - \mu(\hat{f})$.

Proof: We minimize $h(a, a_0, \ldots, a_m) = E[(N(f) - a - \sum_j a_j N(g_j))^2]$ by brute force. Setting the derivative $\partial h/\partial a$ equal to zero implies that

$$a^* = \mu(f) - \textstyle\sum_{j=0}^{m}a_j\mu(g_j). \tag{6.32}$$

Similarly, setting $\partial h/\partial a_i$ equal to zero gives

$$E[N(f)N(g_i)] = a\mu(g_i) + \textstyle\sum_{j=0}^{m}a_jE[N(g_j)N(g_i)],$$

into which we substitute $E[N(f)N(g)] = \mu(fg) + \mu(f)\mu(g)$ [see Example 1.15 and (6.32)] to obtain

$$\begin{aligned} \mu(fg_i) + \mu(f)\mu(g_i) &= \mu(f)\mu(g_i) - \mu(g_i) \\ &\quad + \textstyle\sum_{j=0}^{m}a_j\mu(g_ig_j) + \sum_{j=0}^{m}a_j\mu(g_j)\mu(g_i), \quad (6.33) \end{aligned}$$

which simplifies to

$$\mu(fg_i) = \sum_{j=0}^{m} a_j \mu(g_j g_i), \quad i = 0, \ldots, m, \tag{6.34}$$

precisely the "normal equations" characterizing the projection $\hat{f}$. That $\hat{f}$ minimizes $\mu((f-g)^2)$ over $g \in V$ implies that the solution to (6.32) and (6.33) minimizes $E\left[(N(f) - a - N(g))^2\right]$. ∎

Computation of the state estimator $\hat{N}(f) = \mu(f) - \mu(\hat{f}) - N(\hat{f})$ arising from Proposition 6.28 is no more difficult than calculation of the projection $\hat{f}$, which is ordinarily effected by solving the normal equations (6.34). A design aspect is that if $g_0, \ldots, g_m$ are chosen to be an orthonormal subset of $L^2(\mu)$ [i.e., $\mu(g_j g_k) = 1(k = j)$], then the normal equations are solvable in closed form: $a_i^* = \mu(fg_i)$ for each i.

An alternative approach to state estimation disregards the Poisson structure of N and chooses $g_0, \ldots, g_m$ so that each point measure $\nu = \sum_{i=1}^{\ell} \epsilon_{x_i}$ on E with $\ell \leq m$ is fully determined by $\nu(g_0) \ldots, \nu(g_m)$. Such sets of functions are known as *Tchebycheff systems*; the simplest example is the functions $g_j(x) = x^j$ on $[0,1]$. In this case, whatever the law of N, the point process $\tilde{N} = N 1(N(E) \leq m)$ is reconstructible without error from $N(g_0), \ldots, N(g_m)$, and the only problem remaining is how to estimate N on $\{N(E) > m\}$. One approach is to suppose that the g_j comprise a *complete* Tchebycheff system, that is, $\{g_0, \ldots, g_m\}$ is a Tchebycheff system for each m. For data $\mathcal{F}^N(g_0, \ldots, g_m)$, exact reconstruction of N is then possible on $\{N(E) \leq m\}$ while on $\{N(E) = k\}$, $k > m$, one can reconstruct N under the approximation that $N(g_i) = E[N(g_i)]$ for $i = m + 1, \ldots, k$.

6.5 Random Measures with Independent Increments; Poisson Cluster Processes

This section is a combination of results presented previously; its purpose is to emphasize that inference procedures for Poisson processes on general spaces — this is an important reason for our working in a general context — apply also to random measures and point processes that admit Poisson cluster representations. In particular we consider random measures with independent increments, whose Poisson cluster representation is given by Theorem 1.34 and for which reasonably specific results can be developed. For infinitely divisible point processes, with Poisson cluster representation derived in Theorem 1.32, the situation is less tractable, although the third model of Section 3, of a Poisson process with an unobservable component, is germane for some problems. Our development is sketchy; as stated before, one intention is to stress breadth of applicability of methods developed earlier in the chapter.

We begin with (purely atomic) random measures with independent increments but no fixed atoms. According to Theorem 1.34, such a random measure M on a (compact) space E admits the Poisson cluster representation

$$M = \sum_j U_j \epsilon_{X_j}, \tag{6.35}$$

where $N = \sum \epsilon_{(X_j, U_j)}$, the "measure of jumps," is a Poisson process on $E \times (0, \infty)$ with mean measure μ satisfying

$$\int_{(0,\infty)} (1 - e^{-u}) \mu(E \times du) < \infty. \tag{6.36}$$

Since observation of M over $A \subseteq E$ is equivalent to observation of N over $A \times (0, \infty)$, provided the mean measure of N is locally finite, material in Sections 1–4 applies. The key difference from infinitely divisible point processes is that here the Poisson cluster representation is compatible with natural forms of observation. It is important to remember that local finiteness of μ is defined relative to $E \times (0, \infty)$ rather than $E \times [0, \infty)$ because in many interesting cases, for example most Levy processes on $\mathbf{R}_+$, (increasing processes with independent and stationary increments), $\mu(E \times (0, \varepsilon)) = \infty$ for every $\varepsilon > 0$ (there are many small jumps) even though $\mu(E \times [\varepsilon, \infty)) < \infty$ (there cannot be too many large jumps).

For P_μ the probability under which N is Poisson with mean μ and with M given by (6.35),

$$
\begin{aligned}
E_\mu[e^{-M(f)}] &= E_\mu \left[\exp(-\sum_j U_j f(X_j)) \right] \\
&= E_\mu \left[\exp \left\{ -\int_{E \times (0,\infty)} u f(x) N(dx, du) \right\} \right] \\
&= \exp \left\{ -\int_{E \times (0,\infty)} [1 - e^{-u f(x)}] \mu(dx, du) \right\}; \tag{6.37}
\end{aligned}
$$

therefore, the law of M is determined by μ.

Estimation of Laplace functionals of truncations of M, given i.i.d. copies, can be performed in the following manner.

Proposition 6.29. Suppose that $\mu \in \mathbf{M}_d(E \times (0, \infty))$ satisfies (6.36) and that P_μ is a probability with respect to which the random measures $M_1, M_2, \ldots$ are i.i.d. with Laplace functional (6.37), and let $N_1, N_2, \ldots$ be the associated Poisson processes on $E \times (0, \infty)$. Given the representation $M_i = \sum_j U_{ij} \epsilon_{X_{ij}}$, for each $\delta > 0$, define the thinned random measure

$$M_i^\delta = \sum_j U_{ij} \mathbf{1}(U_{ij} > \delta) \epsilon_{X_{ij}},$$

and let $\hat{\mu} = (1/n) \sum_{i=1}^{n} N_i$ be the usual estimators of the mean measure of the N_i. Then for the empirical Laplace functionals

$$\hat{L}^{\delta}(f) = \exp\left\{-\int_{E\times(\delta,\infty)}[1 - e^{-uf(x)}]\hat{\mu}(dx, du)\right\}$$

and each compact subset $\mathcal{K}$ of $C^+(E)$, almost surely

$$\sup_{f\in\mathcal{K}} \left|\hat{L}^{\delta}(f) - E_{\mu}[e^{-M^{\delta}(f)}]\right| \to 0. \tag{6.38}$$

Proof: Observe first that on $E \times (\delta, \infty)$ one can apply Proposition 6.3a) to the Poisson processes N_i, to deduce that for $\mathcal{K}'$ a compact subset of $C^+(E \times (\delta, \infty))$,

$$\sup_{g\in\mathcal{K}'} \left|e^{-\hat{\mu}(g)} - \exp\left\{-\int_{E\times(\delta,\infty)}[1 - e^{-g(x,u)}]\mu(dx, du)\right\}\right| \to 0 \tag{6.39}$$

almost surely. For $f \in C^+(E)$ define $\tilde{f} \in C^+(E \times (\delta, \infty))$ by $\tilde{f}(x, u) = uf(x)$; then for each $a > 0$, $\{\tilde{f} : f \in \mathcal{K}\}$ is a compact subset of $C^+(E \times (\delta, a))$ and (6.39) yields

$$\sup_{f\in\mathcal{K}} \left|e^{-\hat{\mu}(\tilde{f}\mathbf{1}_{E\times(\delta,a)})} - \exp\left\{-\int_{E\times(\delta,a)}[1 - e^{-uf(x)}]\mu(dx, du)\right\}\right| \to 0.$$

For large a, (6.36) implies that the difference between $\exp[-\hat{\mu}(\tilde{f}\mathbf{1}_{E\times(\delta,a)})]$ and $\exp[-\hat{\mu}(\tilde{f}\mathbf{1}_{E\times(\delta,\infty)})]$ is small uniformly in $f \in \mathcal{K}$, as is that between $\exp\{-\int_{E\times(\delta,a)}[1 - e^{-uf(x)}]d\mu\}$ and $\exp\{-\int_{E\times(\delta,\infty)}[1 - e^{-uf(x)}]d\mu\}$, so that (6.38) holds. ∎

For likelihood ratios we obtain a rather precise result by invoking the independent increments property of the Poisson process N that represents M: for δ_n decreasing to zero the point processes $N_n(\cdot) = N(\cdot \cap (E \times (\delta_n, \delta_{n-1})))$ are independent Poisson processes under P_μ with *finite* mean measures $\mu(\cdot \cap (E \times (\delta_n, \delta_{n-1})))$. Using Theorem 6.12 we derive a sufficient condition for equivalence, together with an expression for the likelihood ratio. The sample space is $\Omega = \mathbf{M}_a(E)$, the set of purely atomic measures on E, with coordinate mapping $M(\omega) = \omega$ and σ-algebra $\mathcal{F}^M = \sigma(M)$.

Theorem 6.30. Suppose that μ_0 and μ_1 are Radon measures on $E \times (0, \infty)$, each fulfilling (6.36), and that in addition $\mu_0 \sim \mu_1$ on $E \times (0, \infty)$ and

$$\int_{E\times(0,\infty)}[1 - \sqrt{d\mu_1/d\mu_0}]^2 d\mu_0 < \infty.$$

Then $P_{\mu_0} \sim P_{\mu_1}$ on $\mathcal{F}^M$ and with $\delta_n \to 0$, $\delta_0 = \infty$ and $B_n = (\delta_n, \delta_{n-1}]$,

$$\frac{dP_{\mu_1}}{dP_{\mu_0}} = \prod_{n=1}^{\infty} \exp\left\{(\mu_1 - \mu_0)(E \times B_n) + \sum_{j:U_j\in B_n} \log \frac{d\mu_1}{d\mu_0}(X_j, U_j)\right\}. \tag{6.40}$$

Proof: For the Poisson process N, Theorem 6.12, which applies by virtue of the assumptions here, gives, with $C_n = (\delta_n, \infty)$,

$$dP_{\mu_1}/dP_{\mu_0}\big|_{\mathcal{F}^N(E \times C_n)} = \exp\left\{(\mu_1 - \mu_0)(E \times C_n) + \int_{E \times C_n} \log(d\mu_1/d\mu_0)dN\right\}.$$

Utilizing the martingale convergence theorem (as in Liptser and Shiryaev, 1978, Theorem 19.6) to take limits, along with invertibility of the mapping $\sum_j u_j \varepsilon_{x_j} \to \sum_j \varepsilon_{(x_j, u_j)}$, we arrive at (6.40). ∎

Note that unlike the analogue (6.11) for Poisson processes, the log-likelihood ratio (6.40) is not a linear function of M, but it is, of course, linear in N. Theorem 6.30 has application to problems ranging from L^1-maximum likelihood estimation via sieves (as in Theorem 6.7) to likelihood ratio tests to state estimation for random measures with conditionally independent increments; concerning the latter, see Karr (1983).

Moving on, we recall from Theorem 1.32 that a point process N on E is infinitely divisible if and only if N is a Poisson cluster process:

$$N \overset{d}{=} \sum_j N_j, \tag{6.41}$$

where $\tilde{N} = \sum_j \varepsilon_{N_j}$ is a Poisson process on $\mathbf{M}_p(E)$ with mean measure μ satisfying $\int_{\mathbf{M}_p}[1 - e^{-\nu(E)}]\mu(d\nu) < \infty$.

Potentially, methods introduced in the chapter are applicable, but there is a major difficulty: (6.41) gives only equality in distribution, as contrasted with the pointwise representation (6.35). There is no way in general to reconstruct the components N_j from their superposition N, which prevents one's exploiting the Poisson cluster representation (6.41). Empirical methods from Section 4.2 can be applied to estimate from i.i.d. copies of N the Laplace functional

$$E_\lambda[e^{-N(f)}] = \exp\{-\int[1 - e^{-\nu(f)}]\lambda(d\nu)\},$$

but usually it is impossible to estimate λ directly as the mean of the processes $\tilde{N}_i = \sum_j \varepsilon_{N_{ij}}$ because the N_{ij} cannot be calculated from N_i.

Of course, for processes constructed explicitly as Poisson cluster processes in the sense of Definition 1.30, things are more hopeful. A Poisson cluster process N on E has the form (6.41), where there is an underlying Poisson process $\tilde{N} = \sum_j \varepsilon_{(X_j, N_j)}$ on a space $E_0 \times E$, where $N^0 = \sum_j \varepsilon_{X_j}$, the cluster center process, is Poisson on E_0 with mean measure μ_0, and where $\tilde{N}$ is obtained from N^0 by position-dependent marking (Example 1.28) using a transition kernel $K : E_0 \to \mathbf{M}_p(E)$. Hence $\tilde{N}$ is Poisson with mean measure $\mu_0(dx)K(x, d\nu)$, and while $\mathbf{M}_p(E)$ is not compact, the general theory applies as long as E_0 is compact. If the process $N_1 = \sum_j \varepsilon_{N_j}$ is observable, as in some applications it is even though none of the X_i is observable, then methods from Section 3 apply provided that the kernel K is known. In particular, Theorem 6.26 allows MMSE reconstruction of the cluster center process N^0 from observation of N^0. The *really* interesting state estimation problem of estimating the cluster center process N^0 from the Poisson cluster process alone has yet to be solved, however.

EXERCISES

6.1 Let N be a Poisson process on a general space E with mean measure μ. Show that for each k the cumulant measure γ^k (Definition 1.62) is given by
$$\gamma^k(f_1 \odot \cdots \odot f_k) = \mu(\prod_{i=1}^{k} f_i)$$

6.2 Let N be a Poisson process on $[1, \infty)$ with mean measure $\mu(ds) = \lambda s^a ds$, where $\lambda > 0$ and $a \in \mathbf{R}$ are unknown.
a) Show that if $a < -1$, then consistent estimation of λ and a given one realization of N is impossible.
b) Given that $a > 0$, show that the estimators $\hat{a} = \log_2(N_t/N_{t/2}) - 1$ and $\hat{\lambda} = (\hat{a} + 1)N_t/t^{\hat{a}+1}$ are strongly consistent as $t \to \infty$.
c) Still with $a > 0$, discuss asymptotic normality of the estimators $\hat{a}$ and $\hat{\lambda}$.

6.3 Suppose that under P_λ, $N_1, N_2, \ldots$ are i.i.d. Poisson processes on a compact set E with mean measure $\mu = \lambda \mu^*$ where λ^* is a *known* element of $\mathbf{M}_d$.
a) Construct estimators $\hat{\lambda}$ given observations $N_1, \ldots, N_n$, that are strongly consistent and asymptotically normal and verify these properties.
b) Examine asymptotic behavior of the empirical Laplace functionals $\hat{L}(f) = \exp[-\hat{\lambda}\mu^*(1 - e^{-f})]$ and calculate their asymptotic efficiency relative to the general empirical Laplace functionals of (4.13).

6.4 (Continuation of Exercise 6.3) Assume that E is noncompact and that $\mu^*(E)$ is infinite. Suppose that under P_λ, $\lambda > 0$, N is a Poisson process with mean measure $\lambda\mu^*$. Repeat Exercise 6.3 relative to observation of a single realization of N over bounded sets $B_n \uparrow E$. If necessary, specify the B_n to have additional properties, but impose as few restrictions as possible.

6.5 Analyze consistency and asymptotic normality of the estimators $\hat{\rho}(A \times B) = \hat{\mu}(A \cap B)$ of the covariance measure ρ of a Poisson process, given as data i.i.d. copies of the process.

6.6 Verify that $R(A, B) = \mu(A \cap B)e^{-[\mu(A)+\mu(B)]}$ is the correct covariance function in the context of Proposition 6.4.

6.7 Let N be a Poisson process on $[0, 1]$ with unknown intensity function α.
a) Derive the martingale estimator of the process $B_t(\alpha) = \int_0^t \alpha_s ds$.
b) Discuss asymptotic behavior of estimators $\hat{B}^n$ associated with Poisson N^n having intensity functions $\alpha^n = n\alpha$.

6.8 Let N be a Poisson process on $[0, 1]$ with unknown *decreasing* intensity function α, which is further assumed to be continuous and strictly positive. Show that within the statistical model specified by these restrictions the maximum likelihood estimator of α is the derivative of the least concave majorant of the function $t \to N_t$.

6.9 Suppose that μ_0 and μ_1 are *finite* measures on E and let N be a Poisson process that under P_i has mean μ_i.
a) Prove that P_1 and P_2 cannot be singular on $\mathcal{F}^N(E)$.
b) Given the Lebesgue decomposition $dP_1 = Z dP_0 + dQ$, where $Q \perp P_0$, provide an upper bound on the mass of Q.
c) Discuss implications for hypothesis testing.

6.10 Give an explicit example of a space E (necessarily not compact — why?) and measures μ_0 and μ_1 on E such that (for P_μ the law of the Poisson process with mean measure μ), $P_{\mu_0} \sim P_{\mu_1}$ on $\mathcal{F}^N(B)$ for every bounded set B but $P_{\mu_0} \perp P_{\mu_1}$ on $\mathcal{F}^N(E)$.

6.11 Let P_0 and P_1 be probabilities with respect to which the Poisson process N on $\mathbf{R}_+$ has equivalent mean measures μ_0, μ_1.
a) Prove directly that the likelihood ratio process

$$Z_t = \exp\left\{ \int_0^t (1 - d\mu_1/d\mu_0)ds + \int_0^t \log(d\mu_1/d\mu_0)dN \right\}$$

is a $(P_0, \mathcal{F}^N)$-martingale.
b) Derive a stochastic differential equation satisfied by Z and discuss its application to recursive computation of likelihood ratios.

6.12 a) Let N be a Poisson process on $\mathbf{R}_+$ with unknown rate λ, let $L_t(\lambda) = -\lambda t + N_t(\log \lambda)$ be the log-likelihood process (with a constant term deleted), and let $\hat{\lambda} = N_t/t$ be the $\mathcal{F}_t^N$-maximum likelihood estimator. Prove that under P_λ, $2[L_t(\hat{\lambda}) - L_t(\lambda)] \overset{d}{\to} \chi_1^2$, a χ^2-distribution with 1 degree of freedom, as $t \to \infty$.
b) Generalize to the case of Exercise 6.3.

6.13 Let $N_1, N_2, \ldots$ be i.i.d. copies of a Poisson process on a compact space E with unknown mean measure μ, let μ^* be a fixed element of M_d, and consider testing the hypothesis $H_0 : \mu \ll \mu^*$. Given observations $N_1, \ldots, N_n$ and a finite partition $E = \sum_{j=1}^J A_j$, under H_0 (6.8) provides a restricted maximum likelihood estimator of $d\mu/d\mu^*$. A plausible criterion for testing H_0 is to reject if the test statistic $T = \max_j\{N^n(A_j)/\mu^*(A_j)\}$ is too large, where $N^n = \sum_{i=1}^n N_i$, the motivation being that if H_0 fails, then there exists a component of μ singular with respect to μ^* and if successively finer partitions were employed, then the test statistics would diverge to ∞.
a) Given fixed $\mu \ll \mu^*$, sample size n, and partition (A_j), calculate the distribution of T.
b) Analyze large-sample behavior under the assumptions of Theorem 6.7.

6.14 Let N_1 and N_2 be independent Poisson processes on $[0,1]$ with mean measures μ_1 and μ_2 satisfying (in distribution function notation) $\mu_1(1) = \mu_2(1) = 1$. Propose, motivate, and analyze a test of the hypothesis $H_0 : \mu_1(t) \geq \mu_2(t)$ for all t. [*Hint:* Use conditional uniformity; see the discussion of stochastic orderings in Section 8.2.]

6.15 Prove Proposition 6.20.

6.16 Formulate and prove an analogue of Proposition 6.20 for the observations $\mathcal{H}_t$ of (6.22).

6.17 Prove Lemma 6.23.

6.18 Let $N_1 = \sum \varepsilon_{T_i}$ and $N_2 = \sum \varepsilon_{S_i}$ be independent Poisson processes on $\mathbf{R}_+$ with rates λ and μ and suppose that one observes only the Poisson samples $N_1(S_i)$, $i = 1, 2, \ldots$. That is, observations are the cumulative numbers of arrivals in N_1 at each arrival time of N_2. Suppose that μ is known but that λ is not.
a) Show that the maximum likelihood estimator of λ given the observations $N_1(S_1), \ldots, N_1(S_n)$, is $\hat{\lambda} = \mu N_1(S_n)/n$.
b) Prove that the estimators are strongly consistent and asymptotically normal.
c) Construct a likelihood ratio test of the hypotheses $H_0 : \lambda = \lambda_0$, $H_1 : \lambda = \lambda_1 > \lambda_0$.

6.19 (Continuation of Exercise 6.18) Suppose that the sampling times S_n are also observable. Develop asymptotic properties of the estimators $\hat{\lambda}^* = N_1(S_n)/S_n$ corresponding to the data $(N_1(S_1), S_1), \ldots, (N_1(S_n), S_n)$ and calculate their asymptotic efficiency relative to the estimators $\hat{\lambda}$ of Exercise 6.18.

6.20 (Continuation of Exercise 6.18) Suppose now that λ and μ are both known. Given as data the pairs $(N_1(S_n), S_n)$, let $\mathcal{H}_t = \sigma(\mathbf{1}(S_n \leq t)(N_1(S_n), S_n) : n \geq 1)$ be the σ-algebra describing observations over $[0, t]$.
a) Calculate explicitly the MMSE state estimators $E[N_1(t)|\mathcal{H}_t]$.
b) Derive a stochastic differential equation satisfied by these estimators.

6.21 Formulate and prove a central limit theorem for the estimators $\hat{\mu}$ given by (6.28).

6.22 Let N be a Poisson process on a compact space E with known mean measure μ and let the observations be integral data $N(g_0) = N(E)$, $N(g_1), \ldots, N(g_m)$, where $g_0 \equiv 1$, $g_1, \ldots, g_m$ are continuous functions on E. Let V be the linear span of $\{g_0, \ldots, g_m\}$. Show that for fixed $f \in C^+(E)$, the MMSE linear predictor of $e^{-N(f)}$ given these observations, which corresponds to the values of $g \in V$ and $a \in \mathbf{R}$ minimizing $E\left[(e^{-N(f)} - a - N(g))^2\right]$, is given by g^*, the orthogonal projection of $(e^{-f} - 1)\exp[-\mu(1 - e^{-f})]$ onto V, and $a^* = E[e^{-N(f)}] - \mu(g^*)$.

6.23 Let $N_1, N_2, \ldots$ be i.i.d. Poisson processes on $[0,1]$ with unknown rate $\lambda > 0$ and let g be a positive function. Suppose that one can observe only the integral data $N_1(g), N_2(g), \ldots$.

a) Develop asymptotic properties of the estimators $\hat{\lambda} = [1/(n \int_0^1 g)] \sum_{i=1}^n N_i(g)$.
b) Suppose that g can be chosen (but only *a priori*) by the experimenter. Propose and defend an optimality criterion and calculate the optimal choice of g relative to it.

6.24 Let $N = \sum \varepsilon_{T_i}$ be a Poisson process on $\mathbf{R}$ with known rate λ, let $(S_i)_{i \in \mathbf{Z}}$ be i.i.d. random variables independent of N with distribution function G, and let $\tilde{N} = \sum \varepsilon_{T_i + S_i}$ be the point process formed by displacing each point T_i of N by the distance S_i. Exercise 1.20 shows that estimation of G from observation of N alone is impossible; here we examine estimation of G given observation of N and $\tilde{N}$ separately. (One does not know the links between the two processes.)
a) Prove that $N^* = \sum \varepsilon_{(T_i, T_i + S_i)}$ is a Poisson process with

$$E[e^{-N^*(h)}] = \exp\{-\lambda \int \int [1 - e^{-h(x, x+y)}] dx\, G(dy)\}.$$

b) Show that for $z \in \mathbf{R}$ and $t, \tau > 0$,

$$P\{N((z, t+z]) = \tilde{N}((z, \tau + z]) = 0\} = \exp\{-\lambda \int_0^t [G(\tau - x) - G(x)] dx\},$$

and prove that G is uniquely determined by these integrals.
c) For fixed t and τ devise strongly consistent estimators of the probability $P\{N((0, t]) = 0, \tilde{N}((0, \tau]) = 0\}$ given observation of N and $\tilde{N}$ over increasingly large sets.

6.25 Suppose that there is an unknown number m of sources transmitting signals according to mutually independent Poisson processes $N_1, \ldots, N_m$ with known rate λ. When a signal is received one can discern from which source it emanated. Let $K(t) = \sum_{j=1}^m \mathbf{1}(N_j(t) \geq 1)$ be the number of sources that have transmitted in $[0, t]$ and let $N(t) = \sum_{j=1}^m N_j(t)$ be the total number of signals received during the same interval.
a) Show that $K(t)$ is a sufficient statistic for m.
b) Show that the estimators $\hat{m} = K(t)/(1 - e^{-\lambda t})$ are unbiased.
c) Suppose that there is a observation cost $c > 0$ per unit of time and that the loss function is $\ell(t) = E[(\hat{m} - m)^2] + ct$. Show that as $m \to \infty$ the optimal observation time t^* minimizing ℓ satisfies $t^* = \log(m\lambda/c)/\lambda + O(1/m)$.

6.26 (Continuation of Exercise 6.25) Suppose that the individual transmission rate λ is also unknown.
a) Prove that given observation over $[0, t]$, the statistic $(K(t), N(t))$ is sufficient for (m, λ).
b) Let a be a function on $[1, \infty)$ that is bounded away from zero, has four continuous derivatives on $[1, 3]$ and satisfies $a(x)/(1 - e^{-a(x)}) = x$ for $x \geq 2$. Show that as $m \to \infty$ and $\lambda t \to \infty$ the estimators $\hat{m} = K(t)/(1 - e^{-a[N(t)/K(t)]})$ satisfy

$$
\begin{aligned}
E[\hat{m} - m] &= -(1/2)\lambda t e^{-\lambda t}[1 + o(1)] + O(m^{-k}) \\
\text{Var}(\hat{m}) &= m e^{-\lambda t}[1 + o(1)] + O(m^{-k})
\end{aligned}
$$

for every $k > 0$. [The intuitive basis for this is that $a(N(t)/K(t))$ is an estimator of λt, which one then substitutes into the estimators of Exercise 6.25b).]

NOTES

References for theory of nonhomogeneous Poisson processes on $\mathbf{R}_+$ are Khinchin (1956b, 1960) and Snyder (1975), while for Poisson processes on general spaces, in addition to the still readable early works of Dobrushin (1956) and Doob (1953) there are more modern expositions in Kallenberg (1983) and Matthes et al. (1978). In this book we have omitted the recently emerged central role of Poisson processes as building blocks for construction of Markov (and other) processes; works treating this role include Çinlar and Jacod (1981), El-Karoui and Lepeltier (1977), Watanabe (1977) and Yor (1976). To our knowledge there exists no treatment other than this of inference for Poisson processes on general spaces.

Section 6.1

Without restriction on the form of the mean measure, single-realization inference for Poisson processes can be arbitrarily uninformative even on $\mathbf{R}_+$; thus one is led in the nonparametric case to the multiple-realization formulation of this section. The simplest parametric case, homogeneous Poisson processes on $\mathbf{R}_+$, is treated in Section 3.5; see the associated notes for additional references. In Chapter 9 homogeneous Poisson process on $\mathbf{R}^d$, $d \geq 1$, are analyzed as stationary point processes. In Exercise 6.3 the mean measure is presumed known within a scalar multiplier. Many other sources address special estimation problems or applicability of particular methods. Prominent among them are Kutoyants (1979, 1982), by far the most comprehensive examination of parametric maximum likelihood estimation, Bartoszynski et al. (1981) on penalized maximum likelihood estimation of intensity functions, and Krickeberg (1974, 1982) on estimation of moment measures. Nonparametric Bayesian estimation is treated by Lo (1982); sequential methods are described by Dvoretsky et al. (1953b) and in a more modern setting permitting censored observations by Vardi (1979). "Linear" inference is discussed by Clevenson and Zidek (1977).

Ripley (1981, 1988) and Diggle (1983) contain expository development of "distance methods" for planar Poisson processes. These techniques, which focus on distances from points in $\mathbf{R}^2$ to the nearest point of the process or from a point of the process to its nearest neighbor, have received widespread and productive application despite incomplete understanding of their theoretical properties.

The main theorem of the section, Theorem 6.7, appears here for the first time. It employs ideas based on Grenander (1981, Chapter 8) and Karr (1987a).

Section 6.2

Of results here, Theorem 6.12, taken from Karr (1983), is an extension to general spaces of Liptser and Shiryaev (1978, Theorem 19.7). Related versions are given by M. Brown (1971, 1972)— the latter is perhaps the most detailed treatment available of likelihood ratios and likelihood ratio tests for Poisson processes on $\mathbf{R}_+$, as well as the source of Proposition 6.13. Weiss (1979) shows that in the noncompact case singularity of Poisson process laws on $\mathcal{F}^N$ is equivalent to singularity on an appropriately defined tail σ-algebra. Albeit new, the contiguity property in Theorem 6.14 is not especially deep. Variations and special cases can be found in Boswell (1966), Cox and Lewis (1966), and Saw (1975), all concerning tests of hypotheses of "trend"(see also Section 3.5); in Dvoretsky *et al.* (1953a), Epstein (1960), and Kiefer and Wolfowitz (1957), which examine sequential testing of hypotheses for Poisson processes on $\mathbf{R}_+$; and in Lewis (1965), which deals with the problem of testing whether a renewal process is Poisson, given asynchronous observations (see also Section 8.2).

Boswell (1966), M. Brown (1972), and Saw (1975) examine other aspects of testing for nonhomogeneous Poisson processes on $\mathbf{R}_+$. Brown treats mainly Neyman-Pearson tests of simple hypotheses for models $\mathcal{P} = \{P_\alpha\}$, where under P_α the Poisson process has mean measure $\alpha_t \mu_0(dt)$ with μ_0 known but not necessarily Lebesgue measure; normal approximations for test statistics $\int_0^t \log[\alpha_s(1)/\alpha_s(0)]dN_s$ as $t \to \infty$ are among the results. Boswell and Saw consider trend and monotonicity properties of several kinds.

Section 6.3

As mentioned in the text, even though each model treated here is of particular interest for applications, an important message of the section is the variety of problems that can be formulated and of techniques available for their solution. Theorem 6.15 has not appeared previously; it is a closer relative of Theorem 6.7 in terms of hypotheses and method of proof than of Theorem 4.18 even though the latter also addresses, but in a more general context, estimation of thinning functions.

With the exception of Propositions 6.20 and 6.21, which are elementary, results on the stochastic integral process $X * N$ come from Karr (1982), but proofs here are simplified by use of martingale techniques. Theorem 6.22 is a very special case of Theorem 7.30; the property that $X^2 = X$ not only makes $X * N$ a Cox process (see Karr, 1982) but also yields the single recursive equation (6.24) rather than the infinite system of equations in Theorem 7.31. Konecny (1984) applies filtering techniques to calculate likelihood functions for use in maximum likelihood estimation; see Smith and Karr (1985) and also Section 7.5 for application to models of precipitation. Further analysis of the process as a renewal process (Exercise 3.15 and Theorem 8.24) can be found in Chapter 8. Freed and Shepp (1982) examine calculation of state estimators $P\{X_0 = 1|\mathcal{F}_t^{X*N}\}$; even this seemingly innocuous problem admits no truly "simple" solution.

Results concerning Poisson processes with one observable component stem from Karr *et al.* (1990), motivated by Vardi *et al.* (1985). In the latter the EM algorithm is utilized to calculate the maximum likelihood estimators in the context of Proposition 6.25; good descriptions of the algorithm itself and of applications to inference when there is incomplete data are the original paper of Dempster *et al.* (1977), the survey paper of Redner and Walker (1984) and Wu (1983). The mixture identifiability problem arising in Proposition 6.24 is discussed by Barndorff-Nielsen (1965) and Teicher (1961) (see also Lindsay, 1983a, 1983b). In view of Theorem 1.32, some problems of inference and state estimation for Poisson cluster processes fall within this setting (see also Section 5).

Section 6.4

See Section 3.1 for introductory discussion of inference given integral data. Karlin and Studden (1966) treats complete Tchebycheff systems, which are important in approximation theory and description of compact, convex sets of measures.

Section 6.5

Brief as it is, this section exhibits rather clearly two kinds of problems: easy problems, for example inference for random measures with independent increments, for which the Poisson cluster representation is compatible with natural forms of observation and to which previously derived results apply in a straightforward manner, and difficult problems, whose solution is not facilitated by the point process's admitting a Poisson cluster representation, because the representation cannot be reconstructed from observation of the process. For such point processes other methods remain to be developed. Of results here, Theorem 6.30 comes from Karr (1983) (see also Akritas and Johnson, 1981 and Kailath and Segall, 1975), while Proposition 6.29 is new. There exist several papers treating special topics in inference for Poisson cluster processes; among them are Baudin (1981) on nearest-neighbor methods in $\mathbf{R}^2$; Brown and Silverman (1978) on tests based on distance methods; Kryscio and Saunders (1983) on nearest neighbors, interpoint distances, and discrimination between Poisson and Poisson cluster processes; and Strauss (1975) on tests for clustering.

Exercises

6.5 See also Section 9.1.

6.8 Grenander (1956) derives analogues in the context of estimation of density functions.

6.12 This property corresponds to a result for the classical case of i.i.d. real data (see Bickel and Doksum, 1977, or Wilks, 1962). Fundamentally a consequence of asymptotic normality of derivatives of log-likelihood functions ("score" functions), it also appears, for example, in Theorem 9.23.

6.13 The suggested technique is related to methods proposed by Saw (1975).

6.17 See the discussion of position-dependent marking in Section 1.5 and also Exercise 1.11.

6.18 See Kingman (1963) and Section 8.1 concerning Poisson sampling of renewal processes. Additional Poisson sampling problems are examined in Chapter 10.

6.24 Random translations of Poisson processes are discussed in M. Brown (1969), Milne (1970), and Thédeen (1964, 1967a); the second of the four is most oriented to statistical inference.

6.25 See Starr (1974), to whom the model seems due, and Vardi (1980).

6.26 See Hall (1982).

7
Inference for Cox
Processes on General Spaces

With their conditionally Poisson structure, Cox processes are uniquely suited to application in a variety of situations, many of which entail state estimation. Lundberg (1940) used Pólya processes, a class of mixed Poisson processes, to model accident statistics. Cox (1955), a seminal work, presented a model of breakages of thread as it is fed into looms. The thread is composed of long, internally homogeneous segments tied together; within each, weak points causing breakages can be represented as an ordinary Poisson process, but the rate varies randomly from segment to segment. If the rate is modeled as a pure jump random process (X_t), then the point process of breakages is a Cox process directed by the random measure $M(B) = \int_B X_t dt$. Snyder (1975) treats communications applications in which the directing intensity is an unobservable information process modulating the observable Cox process. A Cox process model of precipitation occurrences is formulated in Smith and Karr (1983) (see also Rohde and Grandell, 1981) using the stochastic integral $X * N$ of Examples 3.5 and 3.10 and Section 6.3; the 0-1 Markov intensity process represents alternating climatological states, a "dry" state during which, for meteorological reasons, precipitation cannot arise, and a "wet" state during which precipitation events constitute a Poisson process.

The unifying element in these applications is a physical interpretation for the directing random measure, that is, the structure of a Cox process mirrors the physical nature of the system being modeled. Thus a binary Markov process may represent an on/off signal transmitted optically through space or through a fiber, which is not observed directly. Instead (see Snyder, 1975), observations are photon counts, some of them noise, for which there is significant theoretical and empirical evidence substantiating the assumption that they are conditionally Poisson.

Another shared trait is that typically only the Cox process is observable: the directing measure cannot be observed directly even though it may be of greater physical importance. Thus arises the state estimation problem for Cox processes: optimal reconstruction, realization by realization, of the directing measure from Cox process observations. It will occupy much of the chapter. The key property appears in Theorem 7.6, and expresses MMSE state estimators in terms of Palm distributions of the directing measure, which are expressible in turn using those of the Cox process by means of Proposition 1.55.

Even for statistical inference one ordinarily wishes to deal with probability laws of directing measures because it is they that are dictated by physical considerations, but must do so with observations solely of the Cox processes. Special structure of Cox processes is exploited for statistical inference through distributional relationships that make possible estimation and hypothesis testing for unobservable directing measures.

We pause to recollect definitions and results from Chapter 1. Given a point process N and random measure M on the space E and defined over the same probability space, N is a *Cox process directed by M* if conditional on M, N is a Poisson process with mean measure M. We say also that (N, M) is a *Cox pair*. When M has the form $M(B) = \int_B X(y)\nu(dy)$ with X a positive, measurable process and ν a diffuse measure on E, we call X the *ν-directing intensity* of N. The following result collects computational relationships derived in Chapter 1.

Lemma 7.1. Let (N, M) be a Cox pair. Then
a) The probability laws $\mathcal{L}_N$ and $\mathcal{L}_M$ determine each other uniquely;
b) N is simple if and only if M is diffuse;
c) $L_N(f) = L_M(1 - e^{-f})$;
d) $\mu_N = \mu_M$. $\square$

Throughout the chapter we assume that *all directing measures are diffuse*, so that the Cox processes are simple.

It follows that the family of Cox processes is as large as that of diffuse random measures; not surprisingly, for statistical estimation, the topic of Section 1, one must rely primarily on the general methods developed in Chapter 4, as opposed to highly specialized techniques such as those in Chapter 6. In some cases martingale methods from Chapter 5 apply, but it seems difficult to exploit Cox and martingale structures simultaneously.

One class of Cox processes for which specialized estimation techniques *have* been developed is the mixed Poisson processes, treated earlier in Examples 4.22 and 5.31. A *mixed Poisson process* has directing measure $M = Y\nu$, where $\nu \in \mathbf{M}_d$ and Y is a positive random variable. There are two unknown, infinite-dimensional parameters, ν and the distribution F of Y, and special structure can be utilized effectively not only for statistical estimation (Section 1) and state estimation (Section 2) but also for combined statistical inference and state estimation (Section 4).

Figuratively and literally, the heart of the chapter is Section 2, on state estimation. There we develop a variety of techniques for MMSE reconstruction of unobservable directing measures from Cox process observations, in settings ranging from completely general to mixed Poisson processes. Explicit and recursive computation of MMSE state estimators $E[e^{-M(f)}|\mathcal{F}^N(A)]$ are addressed. Of course, only on $\mathbf{R}_+$ is recursive calculation of MMSE state estimators treated meaningfully. Optimal state estimators are nonlinear functions of the observations $N_A(\cdot) = N(\cdot \cap A)$, which is a principal reason for their being very difficult to calculate; consequently, we also consider minimum error linear state estimators (Theorems 7.16 and 7.17). Applications of state estimation results appear in Sections 2 and 5.

Properties of likelihood ratios and associated issues of hypothesis testing are examined in Section 3, where state estimation techniques are again central, this time to representation and calculation of likelihood ratios. We also derive a separation theorem for Cox processes on $\mathbf{R}_+$, which allows recursive computation of likelihood ratios by substitution of state estimators of the directing intensity into Poisson process likelihood ratios.

Section 4 treats in detail the problem of combined statistical inference and state estimation for mixed Poisson processes, with briefer analysis of more difficult techniques available for general Cox processes.

A Cox process N on $\mathbf{R}_+$ with directing intensity X has $\mathcal{F}^N$-stochastic intensity $\lambda_t = E[X_t|\mathcal{F}_{t-}^N]$, and martingale tools, from Theorem 2.31 for calculation of likelihood ratios to the state estimation procedure of Section 5.4, apply, but only (it seems) at the expense of one's more or less ignoring the Cox process structure. Such matters are treated in Section 5, which also contains a state estimation theorem for Cox processes on $\mathbf{R}_+$ whose directing intensity is a function of a Markov process.

7.1   Estimation

Let $(N_1, M_1), (N_2, M_2), \ldots$ be i.i.d. copies of a Cox pair (N, M) on a compact space E and assume that only the Cox processes N_i are observable. To estimate the Laplace functional $L_N(f) = E[e^{-N(f)}]$ in the absence of additional assumptions on the law $\mathcal{L}_N$, one would employ empirical Laplace functionals introduced in Section 4.2

$$\hat{L}_N(f) = (1/n)\sum_{i=1}^{n} e^{-N_i(f)}, \tag{7.1}$$

to which Theorems 4.9 and 4.10 apply directly, with the limiting covariance function given by the latter.

In many cases, however, as examples and applications throughout the chapter confirm, principal interest focuses on the law of the directing measures since typically it is specified initially. Consequently, it is more germane to estimate the Laplace functional $L_M(f) = E[e^{-M(f)}]$. By Theorem 1.12, L_N and L_M determine each other via Lemma 7.1c, which in inverted form becomes

$$L_M(g) = L_N(-\log(1 - g)), \tag{7.2}$$

provided that $0 \leq g < 1$. For statistical estimation this restriction is harmless, but it does cause difficulties for state estimation (see Section 4).

Amalgamation of (7.1) and (7.2) yields the empirical Laplace functionals

$$\hat{L}_M(g) = \hat{L}_N(-\log(1 - g)) = (1/n)\sum_{i=1}^{n} e^{-N_i(-\log(1-g))}, \tag{7.3}$$

defined for functions $g \in C^+(E)$ satisfying $0 \leq g < 1$. Without additional restrictions, (7.3) is the extent to which special structure of Cox processes is exploited.

Properties of $\hat{L}_M$, all straightforward consequences of Theorems 4.9 and 4.10, are now summarized.

Proposition 7.2. a) For each equicontinuous subset $\mathcal{K}$ of $C^+(E)$ such that

$$\sup_{g \in \mathcal{K}} \|g\|_\infty < 1, \tag{7.4}$$

$\sup_{g \in \mathcal{K}} |\hat{L}_M(g) - L_M(g)| \to 0$ almost surely;

b) If $\mathcal{K}$ is a compact subset of $C^+(E)$ fulfilling (7.4) and having finite metric entropy, then the error processes $\{\sqrt{n}[\hat{L}_M(g) - L_M(g)] : g \in \mathcal{K}\}$ converge in distribution, as random elements of $C(\mathcal{K})$, to a continuous Gaussian process G with covariance function

$$R(g_1, g_2) = L_M(g_1 + g_2 - g_1 g_2) - L_M(g_1)L_M(g_2). \; \square \tag{7.5}$$

The covariance function R differs from that of the empirical Laplace functionals $\hat{L}_M^*(g) = (1/n)\sum_{i=1}^{n} e^{-M_i(g)}$ that would be used if the M_i were observable. The latter, by Theorem 4.10, is

$$R^*(g_1, g_2) = L_M(g_1 + g_2) - L_M(g_1)L_M(g_2),$$

and in particular, for $g \in C^+(E)$ with $g < 1$, the asymptotic efficiency of $\hat{L}_M(g)$ relative to $\hat{L}_M^*(g)$ is

$$e(\hat{L}_M(g), \hat{L}_M^*(g)) = \frac{R^*(g, g)}{R(g, g)} = \frac{L_M(2g) - L_M(g)^2}{L_M(2g - g^2) - L_M(g)^2} < 1.$$

Hence there is, not unexpectedly, loss of efficiency in consequence of one's inability to observe the M_i directly.

By contrast with the Poisson case, superpositions $N^n = \sum_{i=1}^n N_i$ are *not* sufficient statistics for the unknown law of the M_i, even for mixed Poisson processes, and statistical inference cannot be based on N^n alone. In Section 3 we confirm this assertion by calculation of likelihood functions, but another justification can be deduced from the following result.

Lemma 7.3. For each n the superposition N^n is a Cox process with directing measure $M^n = \sum_{i=1}^n M_i$. □

That Lemma 7.3 implies that N^n is not sufficient can be argued heuristically as follows: if it were, then $L_M(f) = \lim L_{M^n}(f)^{1/n}$ could be estimated from the N^n alone, but by applying (7.3) to the Cox pair (N^n, M^n), we obtain the estimators $\hat{L}_{M^n}(g) = \exp\{-N^n(-\log(1 - g))\}$, whose nth roots converge not to L_M but only to $\exp\{-\mu_N(-\log(1 - g))\}$.

In view of Lemma 7.1d, estimation of the mean measure $\mu_M = \mu_N$ is routine: Lemma 7.3 shows that N^n *is* a sufficient statistic for μ_M, and the estimators $\hat{\mu}_M = (1/n)N^n = (1/n)\sum_{i=1}^n N_i$ have the limiting behavior described in Proposition 4.8, provided only that moment hypotheses be fulfilled. Variance properties, expectedly, differ from those of the estimators $\hat{\mu}_M^* = \frac{1}{n}\sum_{i=1}^n M_i$ that would be utilized if the M_i were observable, and there is concomitant loss of efficiency (Exercise 7.7).

Depending on one's point of view, for estimation in the context of general Cox processes, there is either little need or little hope to go beyond what we have presented already, but given appropriate structural assumptions one can develop effective specialized techniques. The one class of Cox processes (other than Poisson processes, obviously) for which this has been done not only in reasonable detail but also in (relative) generality is the mixed Poisson processes.

Given a diffuse measure ν on E and a probability distribution F on $\mathbf{R}_+$, let $P_{\nu, F}$ be a probability measure under which $(N_1, Y_1), (N_2, Y_2), \ldots$ are i.i.d. copies of a pair (N, Y) such that Y is a random variable with distribution F and N is a Cox process with directing measure $M = Y\nu$. While the methods above apply to estimation of the Laplace functional $L_N(g) = \exp\{-\int F(du)e^{-u\nu(1-e^{-g})}\}$, the goal now is to estimate ν and F separately, assuming that only the N_i are observable.

We assume that

$$F(0) = 0 \tag{7.6}$$

$$\nu(E) = 1 \tag{7.7}$$

$$m_F = \int F(du)u < \infty. \tag{7.8}$$

The assumption (7.8) implies that the mean measure $\mu_N = \mu_M = m_F\nu$ is finite. The condition (7.7) or an alternative is needed in order to ensure that the full nonparametric model $\mathcal{P} = \{P_{\nu,F}\}$ be identifiable: otherwise, a multiplicative factor can be shifted between ν and the Y_i while affecting neither the distribution of the M_i nor that of the N_i. Alternatives to (7.7) have been employed (for estimation of ν alone it is convenient to suppose instead that $m_F = 1$, because then ν is the mean of the N_i), but the form we have chosen permits use of conditional uniformity of mixed Poisson processes (Exercise 7.3), allows straightforward estimation of ν and F, and is suited to estimation of more complicated objects required in the combined statistical inference and state estimation setting of Section 4.

We work below only with the full nonparametric model, with the N_i as the sole observations. If ν is known or the Y_i are observable, further specialization is possible. The latter problem can be approached directly (Exercise 7.8). The former, which ceases to be a point process problem, has been treated in the literature; some references are given in the chapter notes. See also the discussion relative to (7.10).

Estimation of ν and F is direct. In view of Exercise 7.2, N_i has the structure $\sum_{j=1}^{N_i(E)} \varepsilon_{X_{ij}}$, where $N_i(E)$ has the mixed Poisson distribution

$$Q_F(k) = P_{\nu,F}\{N(E) = k\} = \int F(du)e^{-u}u^k/k!, \tag{7.9}$$

which by (7.7) is not dependent on ν, and where the X_{ij} are i.i.d. random elements of E, independent of $N_i(E)$ and with distribution ν. The latter independence manifests itself as asymptotic independence of the estimators we are about to define.

The estimators

$$\hat{\nu}(\cdot) = \sum_{i=1}^n N_i(\cdot)/\sum_{i=1}^n N_i(E)$$

are simply normalizations of empirical processes with random sample sizes. (For similar analysis of Poisson processes see Theorem 4.26 and related discussion.) The distribution F is determined uniquely by the integrals $Q_F(k)$, which we estimate with empirical averages:

$$\hat{Q}(k) = (1/n)\sum_{i=1}^n \mathbf{1}(N_i(E) = k). \tag{7.10}$$

There are disadvantages to this choice, for characteristics of F such as moments are not easily calculated from Q_F, and must instead be estimated independently rather than by substitution as functionals of Q_F. Concerning alternatives, Karr (1984b) presents techniques for estimation of ν and the Laplace transform of F, and Simar (1976) addresses maximum likelihood estimation of F itself, given i.i.d. mixed Poisson random variables as data. In the latter context estimation of moments as functionals is direct.

Properties of $\hat{\nu}$ and $\hat{Q}$ can be developed without difficulty.

Theorem 7.4. Assume that (7.6)–(7.8) hold. Then almost surely with respect to $P_{\nu,F}$,

a) For each compact subset $\mathcal{K}$ of $C^+(E)$, $\sup_{f \in \mathcal{K}} |\hat{\nu}(f) - \nu(f)| \to 0$;

b) $\sup_{k \in \mathbf{N}} |\hat{Q}(k) - Q(k)| \to 0$.

Proof: Of course, b) is simply the Glivenko-Cantelli theorem (see Gaenssler, 1984) applied to the empirical distributions $\hat{Q}$ of the random variables $N_i(E)$, and requires no further comment.

As for a),

$$\sup_{f \in \mathcal{K}} |\hat{\nu}(f) - \nu(f)|$$

$$\leq \frac{1}{(1/n)\sum_{i=1}^n N_i(E)} \sup_{f \in \mathcal{K}} |(1/n)\sum_{i=1}^n N_i(f) - m_F\nu(f)|$$
$$+ \sup_{f \in \mathcal{K}} \nu(f) \, |1/[(1/n)\sum_{i=1}^n N_i(E)] - 1/m_F|$$

By the strong law of large numbers, almost surely $\frac{1}{n}\sum_{i=1}^n N_i(E) \to m_F$, and hence the second term converges to zero. By Proposition 4.8, since under $P_{\nu,F}$ the N_i have finite mean measure $m_F\nu$,

$$\sup_{f \in \mathcal{K}} |(1/n)\sum_{i=1}^n N_i(f) - m_F\nu(f)| \to 0. \blacksquare$$

Asymptotic normality is hardly more difficult; however, we present only a sketch of the proof, because more complicated results, in particular Theorem 7.25, are proved using arguments easily adapted to this case.

Theorem 7.5. Assume that (7.6)–(7.8) hold and that $E_{\nu,F}[Y^2] < \infty$. Then

$$\begin{bmatrix} \{\sqrt{n}[\hat{\nu}(f) - \nu(f)] : f \in C^+(E)\} \\ \{\sqrt{n}[\hat{Q}(k) - Q_F(k)] : k \in \mathbf{N}\} \end{bmatrix} \xrightarrow{d} \begin{bmatrix} \{G(f) : f \in C^+(E)\} \\ \{Z(k) : k \in \mathbf{N}\} \end{bmatrix}, \quad (7.11)$$

with the first components viewed as random measures on E and the second as ordinary sequences of random variables, where G is a mean zero Gaussian random measure with covariance function

$$R_G(f, g) = [\nu(fg) - \nu(f)\nu(g)]/m_F, \quad (7.12)$$

Z is a mean zero Gaussian sequence with covariance function

$$R_Z(k,j) = 1(k=j)Q_F(k) - Q_F(k)Q_F(j), \qquad (7.13)$$

and G and Z are independent.

Proof: (Sketch). For each f

$$\sqrt{n}[\hat{\nu}(f) - \nu(f)] = \frac{\sqrt{n}}{(1/n)\sum_{i=1}^n N_i(E)}(1/n)\sum_{i=1}^n [N_i(f) - \nu(f)N_i(E)]$$
$$\cong (1/m_F\sqrt{n})\sum_{i=1}^n \int[f - \nu(f)]dN_i$$

by Slutsky's theorem. Asymptotic normality is, ultimately, a consequence of the central limit theorem (we view the first components as random measures). Covariance computations are routine: (7.13) follows from the role of the $\hat{Q}$ as empirical processes on $\mathbf{N}$, (7.12) comes from

$$\text{Cov}_{\nu,F}(N(g), N(h)) = m_F\nu(gh) + \text{Var}_{\nu,F}(Y)\nu(g)\nu(h)$$

(Exercise 4.16), and finally independence of G and Z (i.e., the covariance is zero) is a consequence of conditional uniformity: for each k

$$E_{\nu,F}\left[\int\{f - \nu(f)\}dN \cdot 1(N(E) = k)\right] = 0. \blacksquare$$

As promised, more complicated analogues are proved more carefully in Section 4, where in order to perform approximate state estimation for mixed Poisson processes whose probability law is unknown, we estimate integrals $\int F(du)e^{-u\nu(A)}u^k = k!P_{\nu,F}\{N(A) = k\}/\nu(A)^k$ for arbitrary subsets A of E. For now, though, Theorem 7.5 concludes our discussion of estimation for mixed Poisson processes.

7.2 State Estimation

Let (N, M) be a Cox pair on a locally compact space E. This section is devoted to the state estimation problem of minimum mean squared error (MMSE) reconstruction of the directing measure M from observation of the Cox process N over a subset A of E, under the assumption that the probability law of N is known. We begin with a very general result, which is then specialized to produce MMSE state estimators for specific classes of Cox processes.

The main result is based on Palm distributions, concerning which we recall basic properties (see Section 1.7 for details). Denote by $Q_N(\mu, d\nu)$, $\mu \in \mathbf{M}_p$, the *reduced* Palm distributions of N, with heuristic interpretation

$$Q_N(\mu, d\nu) = P\{N - \mu \in d\nu | \mu \subseteq \operatorname{supp} N\},$$

and let $Q_M(\mu, d\nu)$ be the *unreduced* Palm distributions associated with M. For $\mu \in \mathbf{M}_p$ we recall the *reduced* Palm process N_μ with probability law $Q_N(\mu, \cdot)$, and similarly, let M_μ be the *unreduced* Palm process of M, with law $Q_M(\mu, \cdot)$. Proposition 1.55 relates these processes: for each μ, N_μ has the distribution of a Cox process directed by M_μ.

With this preparation we can formulate the main theorem on state estimation, in which we calculate explicitly the MMSE state estimators $E[e^{-M(f)} | \mathcal{F}^N(A)]$.

Theorem 7.6. Let (N, M) be a Cox pair. Then for each set A with $E[M(A)] < \infty$ and each $f \in C^+(E)$,

$$E[e^{-M(f)} | \mathcal{F}^N(A)] = \left. \frac{E[e^{-M_\mu(A)} e^{-M_\mu(f)}]}{E[e^{-M_\mu(A)}]} \right|_{\mu = N_A}, \qquad (7.14)$$

where $N_A(\cdot) = N(\cdot \cap A)$ is the restriction of N to A (i.e., the observations).

Proof: Let $\mathcal{L}$ denote the law of M and let $L(\mu, f) = E[e^{-M_\mu(f)}] = \int Q_M(\mu, d\nu) e^{-\nu(f)}$ denote the Laplace functional of the Palm process M_μ. To prove (7.14) it suffices by Theorem 1.12 to show that for g a positive function vanishing on A^c and each k,

$$E\left[e^{-N(g)} \mathbf{1}(N(A) = k) \frac{L(N, f + \mathbf{1}_A)}{L_N(\mathbf{1}_A)} \right] = E[e^{-N(g)} \mathbf{1}(N(A) = k) e^{-M(f)}],$$

which we do by calculating the two sides separately. Since N is conditionally Poisson given M and by conditional uniformity of Poisson processes,

$$E\left[e^{-N(g)} \mathbf{1}(N(A) = k) L(N, f + \mathbf{1}_A) / L_N(\mathbf{1}_A) \right]$$
$$= (1/k!) \int \mathcal{L}(d\nu) \int_A \nu(dx_1) \cdots \int_A \nu(dx_k) e^{-\sum_{i=1}^k g(x_i)}$$
$$\times L(\textstyle\sum_{i=1}^k \varepsilon_{x_i}, f + \mathbf{1}_A) / L(\textstyle\sum_{i=1}^k \varepsilon_{x_i}, \mathbf{1}_A). \qquad (7.15)$$

Consider now a random measure $\tilde{M}$ with law $\tilde{\mathcal{L}}(d\nu) = (1/C) e^{-\nu(A)} \mathcal{L}(d\nu)$, where $C = \int \mathcal{L}(d\nu) e^{-\nu(A)}$, which is positive and finite. Then, it is straightforward (Exercise 7.10) to verify that the Palm Laplace functionals $\tilde{L}(\mu, f) = E[e^{-\tilde{M}(f)}]$ are related to those for M by

$$\tilde{L}(\mu, f) = L(\mu, f + \mathbf{1}_A) / L(\mu, \mathbf{1}_A).$$

Substitution into (7.15) yields

$$E\left[e^{-N(g)}\mathbf{1}(N(A)=k)L(N,f+\mathbf{1}_A)/L_N(\mathbf{1}_A)\right]$$

$$= (C/k!)\int L(d\nu)\int_A \nu(dx_1)\cdots\int_A\nu(dx_k)e^{-\sum_{i=1}^k g(x_i)}\tilde{L}(\sum_{i=1}^k \varepsilon_{x_i},f)$$

$$= (C/k!)E\left[\int_A\tilde{M}(dx_1)\cdots\int_A\tilde{M}(dx_k)e^{-\sum_{i=1}^k g(x_i)}e^{-\tilde{M}(f)}\right]$$

$$= (1/k!)E\left[\int_A M(dx_1)\cdots\int_A M(dx_k)e^{-\sum_{i=1}^k g(x_i)}e^{-M(A)-M(f)}\right].(7.16)$$

On the other hand, suppose that for each ν, $\tilde{N}^\nu$ is Poisson with mean measure ν; then once more by conditional uniformity

$$E[e^{-\tilde{N}^\nu(g)}\mathbf{1}(\tilde{N}^\nu(A)=k)]$$

$$= P\{\tilde{N}^\nu(A)=k\}E[e^{-\tilde{N}^\nu(g)}|\tilde{N}^\nu(A)=k]$$

$$= (e^{-\nu(A)}/k!)\int_A\nu(dx_1)\cdots\int_A\nu(dx_k)e^{-\sum_{i=1}^k g(x_i)};$$

consequently,

$$E[e^{-N(g)}\mathbf{1}(N(A)=k)]$$

$$= \int L(d\nu)e^{-\nu(f)}E[e^{-\tilde{N}^\nu(g)}\mathbf{1}(\tilde{N}^\nu(A)=k)]$$

$$= (1/k!)E\left[\int_A M(dx_1)\cdots\int_A M(dx_k)e^{-\sum_{i=1}^k g(x_i)}e^{-M(A)-M(f)}\right].(7.17)$$

Comparison of (7.16) and (7.17) indicates that the proof is complete. ∎

Alternative forms of (7.14) will be used elsewhere in this section and in other sections; we give them as corollaries.

Corollary 7.7. With the notation and hypotheses of Theorem 7.6, for each set Γ,

$$P\{M\in\Gamma|\mathcal{F}^N(A)\} = \left.\frac{E[e^{-M_\mu(A)}\mathbf{1}(M_\mu\in\Gamma)]}{E\left[e^{-M_\mu(A)}\right]}\right|_{\mu=N_A}. \qquad (7.18)$$

Proof: By (7.14) the two sides of (7.18), viewed with N_A fixed as probability measures on $\mathbf{M}_d$, have the same Laplace functional; hence they are equal by Theorem 1.12. ∎

Corollary 7.8. Under the hypotheses of Theorem 7.6, for each function $f\in C^+(E)$,

$$E[M(f)|\mathcal{F}^N(A)] = \left.\frac{E[e^{-M_\mu(A)}M_\mu(f)]}{E\left[e^{-M_\mu(A)}\right]}\right|_{\mu=N_A}. \qquad (7.19)$$

Proof: For f such that $E[M(f)] < \infty$, (7.19) is a direct consequence of (7.18); validity for general $f \geq 0$ holds by approximation arguments. ∎

Note that (7.18) provides explicitly a regular version of the conditional distribution $P\{M \in (\cdot)|\mathcal{F}^N(A)\}$. The Cox relationship between M_μ and the reduced Palm process N_μ allows one to convert (7.18) to the very intuitive expression

$$P\{M \in \Gamma|\mathcal{F}^N(A)\} = P\{M_\mu \in \Gamma|N_\mu(A) = 0\}\Big|_{\mu=N_A}.$$

That is, conditioning on $\{N_\mu(A) = 0\}$ simply assures, in effect, that N has no points in A other than at the atoms of μ.

One can hardly call Theorem 7.6 truly "useful" for calculation of MMSE state estimators; it is too general. However, specialized versions such as Theorem 7.9 for Cox processes admitting directing intensities and Proposition 7.10 for mixed Poisson processes certainly *are* useful, among other things for derivation of recursive methods for computation of state estimators and likelihood ratios. As discussed in Chapter 1 and also in Section 3.3 (see in particular Example 3.17) the *raison d'être* for recursive methods is to avoid recalculating state estimators from scratch as additional observations are obtained, by instead computing only incremental changes in the estimators.

Only for point processes on $\mathbf{R}_+$ can one unambiguously make sense of observations increasing infinitesimally; in this context Theorem 6.22 and the more general Theorem 7.30 are prime examples. The former gives a recursive method for calculation of state estimators $E_\alpha[X_t|\mathcal{F}_t^{X*N}]$: the change at time t in the value of the estimator is determined by previously calculated values and the innovation $d(X*N)_t - \alpha E_\alpha[X_t|\mathcal{F}_{t-}^{X*N}]dt$, and consists of two terms. The first, a predictable change, involves only a dt-differential, results solely from one's having observed the process over a larger set, and is functionally independent of the innovation at time t. The second, expressed in Theorem 6.22 in the canonical form "filter gain times innovation," vanishes if there is no innovation, that is, if one's knowledge does not increase beyond what can be predicted from past observations. It is quite feasible to utilize Theorem 6.22 for real-time calculations. Related computational methods are developed in Theorems 7.15 and 7.30.

We now specialize, here by restricting the structure of the directing measure and in Section 5 by assuming the underlying space to be $\mathbf{R}_+$. The next result concerns Cox processes with directing intensities; however, rather than introduce explicitly the law of the intensity, we continue to work with that of the directing measure M.

Theorem 7.9. Let (N, M) be a Cox pair and suppose that there is a measure $\nu \in \mathbf{M}_d$ such that $P\{M \ll \nu\} = 1$. Let $\mathcal{L}$ denote the probability law of M. Then for each set A such that $E[M(A)] < \infty$ and each $f \in C^+(E)$,

$$E[e^{-M(f)}|\mathcal{F}^N(A)]$$
$$= \frac{\int \mathcal{L}(d\eta)\exp\{-\eta(A) + \int_A \log(d\eta/d\nu)dN\}e^{-\eta(f)}}{\int \mathcal{L}(d\eta)\exp\{-\eta(A) + \int_A \log(d\eta/d\nu)dN\}} \qquad (7.20)$$

Proof: In view of (7.14) it is enough to show that for $\mu = \sum_{i=1}^k \varepsilon_{x_i}$,

$$E[e^{-M_\mu(g)}] = \frac{E[e^{-M(g)}\prod_{i=1}^k (dM/d\nu)(x_i)]}{E[\prod_{i=1}^k (dM/d\nu)(x_i)]}, \qquad (7.21)$$

where the M_μ are the unreduced Palm processes of M. Heuristically, this is obvious from the interpretation

$$E[e^{-M_\mu(g)}] = \frac{E[e^{-M(g)}M(dx_1)\cdots M(dx_k)]}{E[M(dx_1)\cdots M(dx_k)]}$$

and the assumed form of M. To make the reasoning rigorous we use results from Section 1.7: with μ_M^k the k-variate mean measure of M,

$$\int_{A_1}\nu(dx_1)\cdots\int_{A_k}\nu(dx_k)E\left[e^{-M(g)}\prod_{i=1}^k (dM/d\nu)(x_i)\right]$$
$$= E[e^{-M(g)}M(A_1)\cdots M(A_k)]$$
$$= \int_{A_1\times\cdots\times A_k}\mu_M^k(dx_1,\ldots,dx_k)L_M(\sum_{i=1}^k \varepsilon_{x_i},g)$$
$$= \int_{A_1}\nu(dx_1)\cdots\int_{A_k}\nu(dx_k)E\left[\prod_{i=1}^k (dM/d\nu)(x_i)\right]L_M(\sum_{i=1}^k \varepsilon_{x_i},g)$$

for all choices of $A_1,\ldots,A_k$, which proves (7.21). ∎

We next examine the most important special case of Theorem 7.9, namely mixed Poisson processes.

Proposition 7.10. Let N be a mixed Poisson process on E with directing measure $M = Y\nu$, where $\nu \in \mathbf{M}_d$ and Y is a positive random variable with distribution F satisfying (7.8). Then for each set A with $\nu(A) < \infty$ and each f

$$E[e^{-M(f)}|\mathcal{F}^N(A)] = \frac{\int F(du)e^{-u\nu(A)}u^{N(A)}e^{-u\nu(f)}}{\int F(du)e^{-u\nu(A)}u^{N(A)}}. \qquad (7.22)$$

Proof: Define $h : [0,\infty) \to \mathbf{M}_d$ by $h(u) = u\nu$, so that the law $\mathcal{L}$ of M satisfies $\mathcal{L} = Fh^{-1}$. By (7.20) and change of variables

$$E[e^{-M(f)}|\mathcal{F}^N(A)]$$

$$= \frac{\int F(du)\exp\{-u\nu(A) + \int_A \log(d[u\nu]/d\nu)dN\}e^{-u\nu(f)}}{\int F(du)\exp\{-u\nu(A) + \int_A \log(d[u\nu]/d\nu)dN\}}$$

$$= \frac{\int F(du)\exp\{-u\nu(A) + N(A)(\log u)\}e^{-u\nu(f)}}{\int F(du)\exp\{-u\nu(A) + N(A)(\log u)\}}. \blacksquare$$

In particular, we have the following consequence, which is central in Section 4 for analysis of combined inference and state estimation of mixed Poisson processes.

Corollary 7.11. Suppose that (N, M), ν and F satisfy the hypotheses of Proposition 7.10. Then for each set A with $\nu(A) < \infty$,

$$E[M(\cdot)|\mathcal{F}^N(A)] = \frac{\int F(du)e^{-u\nu(A)}u^{N(A)+1}}{\int F(du)e^{-u\nu(A)}u^{N(A)}}\nu(\cdot).\ \square \qquad (7.23)$$

Obviously, since $M = Y\nu$, (7.23) implies that

$$E[Y|\mathcal{F}^N(A)] = \frac{\int F(du)e^{-u\nu(A)}u^{N(A)+1}}{\int F(du)e^{-u\nu(A)}u^{N(A)}}. \qquad (7.24)$$

Note also in (7.22)–(7.24) that $N(A)$ is a "sufficient statistic" for state estimation given the observations $\mathcal{F}^N(A)$; this is another manifestation of conditional uniformity and, indeed, characterizes mixed Poisson processes.

When $\nu(E) = \infty$ it is possible, given observations over bounded sets B_n increasing to E, to recover the multiplier Y exactly. In some ways, though, this is a negative result; it demonstrates that mixed Poisson processes are not ergodic, a sometimes serious limitation in inference and modeling applications.

Proposition 7.12. Let (N, M), ν and F satisfy the hypotheses of Proposition 7.10, suppose that $\nu(E) = \infty$, and let $B_1 \subseteq B_2 \subseteq \cdots$ be bounded sets increasing to E such that for $\tilde{N}$ a Poisson process with mean measure ν, $\lim \tilde{N}(B_n)/\nu(B_n) = 1$ almost surely. Then $\lim E[Y|\mathcal{F}^N(B_n)] = Y$ almost surely.

Proof: By the martingale convergence theorem, $E[Y|\mathcal{F}^N(B_n)]$ converges to $E[Y|\mathcal{F}^N]$ a.s., so it is enough to show that Y is $\mathcal{F}^N$-measurable, but by the strong law of large numbers and the conditionally Poisson structure of N, we have $\lim N(B_n)/\nu(B_n) = Y$ almost surely. $\blacksquare$

We next solve the state estimation problem for p-thinned Cox processes by reducing it to the principal problem of the section, so that methods already described apply. Comparable results for Poisson processes appear in Section 6.3.

Suppose that (N, M) is a Cox pair with known law and that $p : E \to (0, 1]$ is a known function. Neither the directing measure nor the Cox process N is observable; instead, one observes only the p-thinning N' of N, possibly only over a subset A of E (see Definition 1.38 and Section 4.5). By Proposition 1.39, the observed process N' and the random measure $M'(dx) = p(x)M(dx)$ form a Cox pair; this relationship together with the following result enables one to reconstruct the underlying process N from the thinned process N'. (See also Theorem 4.23, a general result formulated using Palm distributions.)

Theorem 7.13. For each A with $E[M(A)] < \infty$ and each $f \in C^+(E)$,

$$E[N(f)|\mathcal{F}^{N'}(A)] \;=\; \int_A f \, dN' + E\left[\int_A f(1 - p)dM \,\middle|\, \mathcal{F}^{N'}(A)\right]$$
$$+ E\left[\int_{A^c} f \, dM \,\middle|\, \mathcal{F}^{N'}(A)\right]. \tag{7.25}$$

We prove the theorem after interpreting (7.25) and indicating how, in conjunction, for example, with (7.19) or (7.22), it would be applied. First the interpretation: write N as $N = N'_A + N''_A + N_{A^c}$, where $N'' = N - N'$. Then the three terms on the right-hand side of (7.25) are MMSE estimators of $N'_A(f)$, $N''_A(f)$, and $N_{A^c}(f)$, given the observations $\mathcal{F}^{N'}(A)$. The first is clear: N'_A is observable, while the second results from the property that N'_A and N''_A are conditionally independent given M, with mean measures $1_A(x)p(x)M(dx)$ and $1_A(x)(1 - p(x))M(dx)$, so that the best estimator of $N''_A(f)$ is that of its conditional mean $\int_A f(1 - p)dM$. Similarly, N_{A^c} is conditionally independent of N'_A given M, so $N_{A^c}(f)$ can be estimated only to the extent that one can estimate $\int_{A^c} f \, dM$.

To utilize (7.25) we write it as

$$E[N(f)|\mathcal{F}^{N'}(A)] \;=\; \int_A f \, dN' + E\left[\int_A \{f(1 - p)/p\}dM' \,\middle|\, \mathcal{F}^{N'}(A)\right]$$
$$+ E\left[\int_{A^c} (f/p)dM' \,\middle|\, \mathcal{F}^{N'}(A)\right],$$

where M' is the directing measure of the observable Cox process N'. Results elsewhere in the section can then be applied to calculate the second and third terms.

Proof of Theorem 7.13: We may and do prove (7.25) separately for functions f vanishing on A^c, then for functions f vanishing on A.

Suppose that $f, g \geq 0$ vanish on A^c. Then on the one hand, conditional independence of N' and N'' given M implies that

$$E[N(f)e^{-N'(g)}|M = \nu]$$

$$
\begin{aligned}
=\ & E[N'(f)e^{-N'(g)}|M=\nu] + E[N''(f)|M=\nu]E[e^{-N'(g)}|M=\nu]\\
=\ & \nu(pfe^{-g})e^{-\nu(p(1-e^{-g}))} + \nu(f(1-p))e^{-\nu(p(1-e^{-g}))}
\end{aligned}
$$

(cf. Exercise 7.9); consequently,

$$
\begin{aligned}
E[N(f)e^{-N'(g)}] =\ & E[M(pfe^{-g})e^{-M(p(1-e^{-g}))}]\\
& + E[M(f(1-p))e^{-M(p(1-e^{-g}))}]. \qquad (7.26)
\end{aligned}
$$

On the other hand,

$$
\begin{aligned}
E&\left[\left\{N'(f) + E[\textstyle\int_A f(1-p)dM|\mathcal{F}^{N'}(A)]\right\}e^{-N'(g)}\right]\\
=\ & E[N'(f)e^{-N'(g)}] + E\left[\{\textstyle\int_A f(1-p)dM\}e^{-N'(g)}\right]\\
=\ & E[N'(f)e^{-N'(g)}] + E\left[\{\textstyle\int_A f(1-p)dM\}E[e^{-N'(g)}|M]\right]\\
=\ & E[N'(f)e^{-N'(g)}] + E[M(f(1-p))e^{-M(p(1-e^{-g}))}]
\end{aligned}
$$

which is identical to (7.26) and establishes the first part of (7.25).

For f vanishing on A and g on A^c the argument is easier:

$$
\begin{aligned}
E[N(f)e^{-N'(g)}] =\ & E\left[E[N(f)e^{-N'(g)}|\mathcal{F}^M \vee \mathcal{F}^{N'}(A)]\right]\\
=\ & E[M(f)e^{-N'(g)}]\\
=\ & E[E[M(f)|\mathcal{F}^{N'}(A)]e^{-N'(g)}],
\end{aligned}
$$

and thus the proof is complete. ∎

This ends for the moment our study of state estimation on general spaces; we return to the topic at the end of the section, in Section 3 for computation of likelihood ratios and in Section 4 in the context of combined statistical inference and state estimation.

Next we explore consequences of the assumption that the underlying space is $\mathbf{R}_+$, but in a context different from that of Section 5, where we assess applicability of martingale methods of inference and state estimation for Cox processes on $\mathbf{R}_+$ with directing measures $M(A) = \int_A X_u du$. Here, more generally, directing measures have the form $M(A) = \int_A X_u d\nu(u)$, where ν is a diffuse measure on $\mathbf{R}_+$ and X is a positive, measurable stochastic process. It follows that Theorem 7.9, and for mixed Poisson processes, Proposition 7.10 apply. The latter, especially, yields useful methods for recursive calculation of state estimators. Nonetheless, explicit techniques of computation are also important, one of which we illustrate for a specific case.

Example 7.14. Let N be a Cox process on $\mathbf{R}_+$ with directing intensity a Markov process X with state space $[0, \infty)$. Let (P_t) be the transition function of X and let (Q_t) be the subordinate transition function generated by the multiplicative functional $Z_t = e^{-\int_0^t X_u du}$; that is, $Q_t g(x) = E_x[g(X_t)Z_t]$, where P_x is the probability law of X corresponding to the condition $X_0 = x$ (see Blumenthal and Getoor, 1968, for details). For $r \in \mathbf{N}$ and $0 \leq t_1 < \cdots < t_r \leq t$, let

$$g_t(t_1, \ldots, t_r) = \int \pi(dx_0) \left[\prod_{i=1}^r \int Q_{t_i - t_{i-1}}(x_{i-1}, dx_i)x_i \right] \int Q_{t-t_r}(x_r, dy),$$

where π is the distribution of X_0, and further define

$$
\begin{aligned}
h_t(t_1, \ldots, t_r; v) &= g_t(t_1, \ldots, t_j, v, t_{j+1}, \ldots, t_r), & v &\in (t_j, t_{j+1}) \\
&= g_t(t_1, \ldots, t_r, v), & v &\in (t+r, t) \\
&= \int \pi(dx_0)\left[\prod_{i=1}^r \int Q_{t_i - t_{i-1}}(x_{i-1}, dx_i)x_i\right] \\
&\quad \times \int P_{v-t_r}(x_r, dx_{r+1})x_{r+1}, & v &> t.
\end{aligned}
$$

Theorem 7.9 applies and shows that if on $[0, t]$, $N = \sum_{i=1}^k \varepsilon_{t_i}$, then for $f \in C^+(\mathbf{R}_+)$,

$$E[M(f)|\mathcal{F}_t^N] = \int_0^t f(v) h_t(t_1, \ldots, t_k; v) dv / g_t(t_1, \ldots, t_k).$$

Verifications and computational details are given in Karr (1983). For the special case that X has state space $\{i, j\}$ and positive, finite transition rates q_i $(i \to j)$ and q_j $(j \to i)$, the transition function can be calculated in closed form:

$$Q_t = \begin{bmatrix} p(t) + (j + q_j)r(t) & q_i r(t) \\ q_j r(t) & p(t) + (i + q_i)r(t) \end{bmatrix}, \tag{7.27}$$

where $p(t) = (r_1 e^{r_1 t} - r_2 e^{r_2 t})/(r_1 - r_2)$ and $r(t) = (e^{r_1 t} - e^{r_2 t})/(r_1 - r_2)$, with r_1 and r_2 the (real and negative) roots of the equation

$$x^2 + (i + j + q_i + q_j)x + (iq_i + jq_j + ij) = 0. \tag{7.28}$$

When $i = 0$, $j = \alpha$, $q_0 = a$ and $q_1 = b$, we have the process of Theorem 6.22; in this case (7.28) can be solved in closed form, there is a simplified form of (7.27), and one obtains an explicit calculation of state estimators as a complement to the recursive method of Theorem 6.22. $\square$

We next analyze recursive versions of (7.24) for estimation of the directing multiplier of a mixed Poisson process on $\mathbf{R}_+$. We assume that the measure ν is diffuse, although not necessarily Lebesgue measure, and use distribution function notation: $\nu(t) = \nu([0, t])$, $N(t) = N([0, t])$, $N(t-) = N([0, t))$. According to (7.24),

$$E[Y|\mathcal{F}_t^N] = \int F(du)e^{-u\nu(t)} u^{N(t)+1} / \int F(du)e^{-u\nu(t)} u^{N(t)}.$$

This expression is satisfactory for many purposes, but the state estimators can also be calculated recursively.

Theorem 7.15. Define the $\mathcal{F}^N$-predictable processes

$$
\begin{aligned}
H(t) &= \left[\int F(du)e^{-u\nu(t)}u^{1+N(t-)}\right]^2 \\
&\quad - \left[\int F(du)e^{-u\nu(t)}u^{2+N(t-)}\right]\left[\int F(du)e^{-u\nu(t)}u^{N(t-)}\right] \\
J(t) &= \left[\int F(du)e^{-u\nu(t)}u^{1+N(t-)}\right]\left[\int F(du)e^{-u\nu(t)}u^{N(t-)}\right].
\end{aligned}
$$

Then the MMSE state estimators $E[Y|\mathcal{F}_t^N]$ satisfy the stochastic differential equation

$$
dE[Y|\mathcal{F}_t^N] = -[H(t)/J(t)][dN_t - E[Y|\mathcal{F}_{t-}^N]d\nu(t)]. \quad\square
$$

For the proof we refer to Karr (1983).

By arguments given in Chapter 2, the $\mathcal{F}^N$-stochastic intensity of N is $E[Y|\mathcal{F}_{t-}^N]$, and therefore this equation is in the canonical form "filter gain times innovation," with no predictable term.

For some applications even recursive methods for computation of state estimators are too cumbersome, in part because, as made manifest in Section 5, there often results an infinite, or "open" system of simultaneous stochastic differential equations, each introducing successively higher-order moments, to which no solution can sensibly be said to exist. (In many ways, Theorem 6.22, with its single equation, is the exceptional case.) Therefore, it is of interest to develop simplified techniques for state estimation.

One approach is to impose *a priori* restrictions on the form of state estimators as functions of the observations, the most obvious of which, of course, is linearity. We end the section with two results: one for Cox processes on $\mathbf{R}_+$, the other for general spaces. Proposition 6.28 is analogous in motivation (both underlying state estimation problems are difficult to solve in full generality) and in the mathematical form of the results.

Let N be a Cox process on $\mathbf{R}_+$ with directing measure $\int_A X_u d\nu(u)$, where ν is a diffuse measure on $\mathbf{R}_+$ and X is a positive process. For each t it is desired to estimate X_t by a *linear* state estimator

$$
\hat{X}_t = a + \int_0^t h(u)dN_u, \tag{7.29}
$$

where a is a constant and h a deterministic function, so that the estimator $\hat{X}_t$ is a *nonrandom* linear functional of the observations. We wish to calculate the MMSE state estimator of the form (7.29). Even though in some applications the time t may be fixed, it is easier to think of calculating $\hat{X}_t$

for all values of t, in the same way that a recursive computational technique such as that of Theorem 7.15 does not calculate a state estimator for only a single t-value but provides instead a recipe valid for every value. Thus in (7.29) the constant a becomes a function $a(t)$ and h becomes a function of two variables.

As might be suspected, the process $dN_t - E[X_t]d\nu(t)$ plays a role analogous to that of the innovation martingale in, for example, Theorems 5.30 and 6.22. Although deterministic, equation (7.30) is analogous to stochastic differential equations derived previously.

Theorem 7.16. Let N be a Cox process on $\mathbf{R}_+$ admitting directing intensity X with respect to the diffuse measure ν. Assume that $E[X_t^2] < \infty$ for each t, and let $R(t, u) = \mathrm{Cov}(X_t, X_u)$ be the covariance function of X. Then for each t the MMSE linear state estimator of X_t, given observations $\mathcal{F}_t^N$, is

$$\hat{X}_t = E[X_t] + \int_0^t h^*(t, u)\{dN_u - E[X_u]d\nu(u)\}, \qquad (7.30)$$

where h^* satisfies the integral equation

$$h^*(t, u)E[X_u] + \int_0^t h^*(t, s)R(s, u)d\nu(s) = R(t, u) \qquad (7.31)$$

for $0 \le u \le t$. The optimal estimation error is $E[(\hat{X}_t - X_t)^2] = h^*(t, t)E[X_t]$.

Proof: For optimality it is enough to show that with

$$\tilde{X}_t = a(t) + \int_0^t [h^*(t, u) + g(t, u)]\{dN_u - E[X_u]d\nu(u)\},$$

the mean squared error $E[(\tilde{X}_t - X_t)^2]$ is minimized by $a(t) = E[X_t]$ and $g \equiv 0$. By straightforward computations making use of (7.31) (and relegated to Exercise 7.15)

$$
\begin{aligned}
E[(\tilde{X}_t - X_t)^2] = {} & R(t, t) + (a(t) - E[X_t])^2 - \int_0^t h^*(t, u)R(u, t)d\nu(u) \\
& + \int_0^t \int_0^t g(t, u)R(u, s)g(t, s)d\nu(u)d\nu(s) \\
& + \int_0^t E[X_u]g(t, u)^2 d\nu(u). \qquad (7.32)
\end{aligned}
$$

In this expression, only the second term, which is evidently minimized by choosing $a(t) = E[X_t]$, and the fourth and fifth terms depend on a or g. The latter two are nonnegative, the first because as a covariance function R is positive definite, and each is minimized by taking $g \equiv 0$. Consequently, the minimal error is

$$E[(\tilde{X}_t - X_t)^2] = R(t, t) - \int_0^t h^*(t, u)R(u, t)d\nu(u) = h^*(t, t)E[X_t]$$

by another application of (7.31). ∎

Solving (7.31) can be easy or difficult depending on the form of the functions $t \to E[X_t]$ and R; for an illustration of the former, see Exercise 7.24. In less fortuitous instances, however, the same escalation as in Theorem 7.29 can arise: one equation may only engender others. However, note one additional advantage of linear state estimation: only first and second moments of the directing intensity are required.

We revert now to Cox processes on general spaces and derive an analogue of Theorem 7.16. Let (N, M) be a Cox pair on a locally compact space E, let μ be the mean measure of N and M, and let $f \in C^+(E)$ satisfy $E[M(f)^2] < \infty$. Suppose that N has been observed over the bounded set A. By analogy with (7.30), we seek that element of $C^+(E)$ minimizing the mean squared error $E\left[\{M(f) - \int_A g(dN - d\mu)\}^2\right]$; the minimizer $\hat{f}$ leads to the MMSE linear state estimator of $M(f)$, given the observations $\mathcal{F}^N(A)$, of Theorem 7.17. Not only is the problem analogous to that treated in Theorem 7.16, but so is the solution.

Theorem 7.17. The MMSE linear state estimator of $M(f)$, given the observations $\mathcal{F}^N(A)$, is $\hat{M}(f) = \mu(f) + \int_A \hat{f}[dN - d\mu]$, where $\hat{f}$ satisfies the equation

$$E[M(f)M(h)] + \mu(f)\mu(h) = \mu(\hat{f}h) + E[M(\hat{f})M(h)]\mu(\hat{f})\mu(h) \qquad (7.33)$$

for every nonnegative function h on A such that $E[M(h)^2] < \infty$. Also,

$$E[(\hat{M}(f) - M(f))^2] = \text{Var}(M(f)) - E[M(f)M(\hat{f})] + \mu(\hat{f})\mu(f). \qquad (7.34)$$

Proof: The optimal linear estimator $\hat{M}(f)$ is characterized by the property that, for each function h on A with $E[N(h)^2] < \infty$, $\hat{M}(f) - M(f)$ is orthogonal to $N(h)$ in the Hilbert space $L^2(P)$; that is, $\hat{M}(f)$ is the unique solution of the "normal equations"

$$
\begin{aligned}
0 &= E[\{M(f) - (\mu(f) + N(\hat{f})\mu(\hat{f}))\}N(h)] \\
&= E[M(f)N(h)] - \mu(f)\mu(h) - E[N(\hat{f})N(h)] + \mu(\hat{f})\mu(h) \\
&= E[M(f)N(h)] - \mu(f)\mu(h) - E[M(\hat{f}h) + M(\hat{f})M(h)] + \mu(\hat{f})\mu(h),
\end{aligned}
$$

with $\hat{f}$ assumed to vanish on A^c, and where we have used the property that $E[N(\hat{f})N(h)|M] = M(\hat{f}h) + M(\hat{f})M(h)$. The last expression above gives (7.33) as characterization of the optimal integrand $\hat{f}$. Then, (7.34) follows by routine Hilbert space reasoning (see also Grandell, 1976). ∎

The mapping $f \to \hat{f}$ defined by (7.33) is linear, so for estimation of $M(f)$ for functions f belonging to a given class V, taken without loss of generality to be a vector space, it is necessary to estimate only for functions comprising a basis of V. One should regard (7.33) as an equation for measures on A, as the following example illustrates.

Example 7.18. Let N be a mixed Poisson process on E with directing measure $M = Y\nu$, where $\nu \in \mathbf{M}_d$ and $E[Y^2] < \infty$. We assume for simplicity that $E[Y] = 1$, so that $\mu_N = \mu_M = \nu$. Then (7.33) becomes

$$E[Y^2]\nu(f)\nu(h) - \nu(f)\nu(h) = \nu(\hat{f}h) + E[Y^2]\nu(\hat{f})\nu(h) - \nu(\hat{f})\nu(h),$$

which simplifies to $\text{Var}(Y)[\nu(f) - \nu(\hat{f})]\nu(\cdot) = \nu(\hat{f} \times \cdot)$, whose solution is $\hat{f} = \text{Var}(Y)\nu(f)/[1 + \text{Var}(Y)\nu(A)]$, a constant function. □

While challenging and intriguing in its own right, state estimation is also important to calculation of likelihood ratios, the topic of next section.

7.3 Likelihood Ratios and Hypothesis Tests

In form, this section parallels Section 2 and, although less clearly, Section 6.2. It is in the spirit of the latter in emphasizing structure of likelihood ratios rather than details of hypothesis testing. Nevertheless, at some level much of it can be interpreted as addressing construction of likelihood ratio tests for simple hypotheses when the data are observations of a Cox process.

Let (N, M) be a Cox pair on a locally compact space E with only the Cox process N observable, and consider the test

$$\begin{aligned} H_0 &: \mathcal{L}_M = \mathcal{L}_0 \\ H_1 &: \mathcal{L}_M = \mathcal{L}_1. \end{aligned} \tag{7.35}$$

Although one could formulate an equivalent problem in terms of the law of N, not only the mathematics of Cox processes but also the physics of systems they are used to model dictate that the law of M be regarded as the fundamental object defining the distribution of the pair (N, M). But since the observations are the Cox process rather than the directing measure, a likelihood ratio test for the hypotheses (7.35) must not be based on the likelihood ratio $(d\mathcal{L}_1/d\mathcal{L}_0)(M)$, which cannot be calculated from the observations, but instead on the N-likelihood ratio $(dP_1/dP_0)|_{\mathcal{F}^N}$, where under P_i the directing measure has law $\mathcal{L}_i$. Moreover, the likelihood ratio must be calculated either explicitly or recursively; we consider both, beginning with the former.

Concerning the N-likelihood ratio, its key feature is that it is the conditional expectation of the M-likelihood ratio given the observations of N, and from this, together with Theorems 7.6 and 7.9 and Proposition 7.10, we derive a number of expressions and computational techniques. The main result establishes this property along with a sufficient condition for equivalence of P_0 and P_1 on $\mathcal{F}^N$, namely that of $\mathcal{L}_0$ and $\mathcal{L}_1$.

Proposition 7.19. Let (N, M) be a Cox pair on E such that under the probability P_i, the directing measure has law $\mathcal{L}_i$, $i = 0, 1$. If $\mathcal{L}_1 \ll \mathcal{L}_0$, then $P_1 \ll P_0$ on $\mathcal{F}^N$ and for each set A

$$dP_1/dP_0\big|_{\mathcal{F}^N(A)} = E_0[(d\mathcal{L}_1/d\mathcal{L}_0)(M)|\mathcal{F}^N(A)]. \tag{7.36}$$

Proof: For each function f vanishing outside A,

$$
\begin{aligned}
E_1[e^{-N(f)}] &= \int \mathcal{L}_1(d\nu)e^{-\nu(1-e^{-f})} \\
&= \int \mathcal{L}_0(d\nu)(d\mathcal{L}_1/d\mathcal{L}_0)(\nu)e^{-\nu(1-e^{-f})} \\
&= E_0\left[(d\mathcal{L}_1/\mathcal{L}_0)(M)E_0[e^{-N(f)}|M]\right] \\
&= E_0\left[E_0[(d\mathcal{L}_1/\mathcal{L}_0)(M)|\mathcal{F}^N(A)]e^{-N(f)}\right],
\end{aligned}
$$

which together with Theorem 1.12 verifies (7.36). ∎

Thus we have several explicit expressions for likelihood ratios.

Corollary 7.20. Suppose that under P_i, (N, M) is a Cox pair with the directing measure having law $\mathcal{L}_i$. If $\mathcal{L}_1 \ll \mathcal{L}_0$, then $P_1 \ll P_0$ on $\mathcal{F}^N$ and for each set A with $E_0[M(A)] < \infty$,

$$\frac{dP_1}{dP_0}\bigg|_{\mathcal{F}^N(A)} = \frac{E[e^{-M_\mu^0(A)}(d\mathcal{L}_1/d\mathcal{L}_0)(M_\mu^0)]}{E[e^{-M_\mu^0(A)}]}\Bigg|_{\mu=N_A},$$

where the M_μ^0 are Palm processes of M relative to $\mathcal{L}_0$. □

Corollary 7.21. Let (N, M) be a Cox pair such that under the probability P_i, N has law $\mathcal{L}_i$. Suppose that $\mathcal{L}_1 \ll \mathcal{L}_0$ and that there is $\nu \in \mathbf{M}_d$ such that $P_0\{M \ll \nu\} = 1$. Then $P_1 \ll P_0$ on $\mathcal{F}^N$ and for each set A with $E_0[M(A)] < \infty$,

$$\frac{dP_1}{dP_0}\bigg|_{\mathcal{F}^N(A)} = \frac{\int \mathcal{L}_1(d\eta)\exp\{-\eta(A) + \int_A \log(d\eta/d\nu)dN\}}{\int \mathcal{L}_0(d\eta)\exp\{-\eta(A) + \int_A \log(d\eta/d\nu)dN\}}. \ □ \tag{7.37}$$

Note the continuing primacy of state estimation: to calculate likelihood ratios one utilizes techniques for state estimation of an unobservable directing measure.

If in the setting of Corollary 7.21, $\mathcal{L}_i = \varepsilon_{\nu_i}$, so that under P_i, N is Poisson with mean measure ν_i, then with $\nu = \nu_0 + \nu_1$ one obtains by formal substitution into (7.37) the expression

$$\left. \frac{dP_1}{dP_0} \right|_{\mathcal{F}^N(A)} = \exp\left[\nu_0(A) - \nu_1(A) + \int_A \log \frac{d\nu_1/d\nu}{d\nu_0/d\nu} dN \right],$$

which was proposed in Section 6.2 for likelihood ratio tests for Poisson processes with not necessarily equivalent mean measures. In particular, when $\nu_1 \ll \nu_0$, it assumes the form

$$dP_1/dP_0 \big|_{\mathcal{F}^N(A)} = \exp\left\{ \int_A (1 - d\nu_1/d\nu_0) d\nu_0 + \int_A \log(d\nu_1/d\nu_0) dN \right\}, \quad (7.38)$$

the principal Poisson likelihood ratio of Proposition 6.10. However, none of this is justified in terms of Corollary 7.21 because for $\mathcal{L}_i = \varepsilon_{\nu_i}$; we have $\mathcal{L}_0 \perp \mathcal{L}_1$, the extreme opposite of the absolute continuity assumed in the theorem. Nonetheless, (7.38) is a valid Poisson likelihood ratio and thus for Cox processes it is possible that $P_0 \sim P_1$ on $\mathcal{F}^N$ but $P_0 \perp P_1$ on $\mathcal{F}^M$. Further comments about singularity will be given momentarily, but first we conclude the discussion of absolute continuity by considering mixed Poisson processes.

Given probability distributions F_0 and F_1 on $\mathbf{R}_+$ and diffuse measures ν_0 and ν_1 on E [for identifiability, assume that $\int F_0(du)u = \int F_1(du)u = 1$] we need a criterion for absolute continuity of the distribution $\mathcal{L}_1$ of $M = Y\nu_1$ under a probability P_1 such that $P_1\{Y \in (\cdot)\} = F_1(\cdot)$ with respect to the corresponding law $\mathcal{L}_0$ constructed from v, ν_0 and F_0. This is provided by the following result.

Lemma 7.22. With the notation and hypotheses of the preceding paragraph, the following assertions are equivalent:

a) $\mathcal{L}_1 \ll \mathcal{L}_0$;

b) There is $a \in (0, \infty)$ such that $\nu_1 = a\nu_0$ and such that the measure $B \to F_1(a^{-1}B)$ is absolutely continuous with respect to F_0. $\square$

Thus there is no loss of generality in supposing that $\nu_0 = \nu_1$, whose common value we denote by ν, and dropping the restriction that $\int F_i(du)u = 1$. From Propositions 7.10 and 7.19 we then derive a representation of likelihood ratios for mixed Poisson processes.

Proposition 7.23. Suppose that under P_i, N is a mixed Poisson process on E with directing measure $M = Y\nu$, where $\nu \in \mathbf{M}_d$ does not depend on i, while Y has P_i-distribution F_i. Assume that $E_0[Y] < \infty$. If $F_1 \ll F_0$ then

$P_1 \ll P_0$ on $\mathcal{F}^N$ and for each bounded set A,

$$\left. \frac{dP_1}{dP_0} \right|_{\mathcal{F}^N(A)} = \frac{\int F_1(du)e^{-u\nu(A)}u^{N(A)}}{\int F_0(du)e^{-u\nu(A)}u^{N(A)}} \cdot \square$$

Proposition 7.23 confirms even in the mixed Poisson case that the superposition of i.i.d. Cox processes is not a sufficient statistic for an unknown probability law: the n-sample likelihood function obtained by multiplying terms above cannot be expressed as a function of only the superposition $N^n = \sum_{i=1}^n N_i$.

Attention so far has focused mainly on absolute continuity relationships between probability laws, and for a reason: singularity of laws of directing measures, as already noted, does not imply that of the associated Cox processes. Thus there is no analogue for Cox processes of Theorem 6.12 or Proposition 6.13, nor is there a dichotomy theorem. If under P_0, N is Poisson with mean measure μ_0, and under P_1, N is a Cox process whose directing measure M has distribution $P\{M = \mu_0\} = P\{M = \mu_1\} = 1/2$, where $\mu_0 \perp \mu_1$ then the law of N under P_1 is neither equivalent to nor singular with respect to the law under P_0.

It is, however, rather easy (Exercise 7.18) to show that if $P_0 \perp P_1$ on $\mathcal{F}^N$, then $P_0 \perp P_1$ on $\mathcal{F}^M$; except for this and the absolute continuity result given in general form in Corollary 7.20, one can say little more.

Recursive computation is as important for likelihood ratios as for any state estimation problem. In Theorem 7.24 we provide, for a large class of Cox processes on $\mathbf{R}_+$, not only a recursive method for calculation but also a *separation theorem*, whose content is that because a Cox process is conditionally Poisson, one can calculate a likelihood ratio for Cox processes by taking a Poisson likelihood ratio

$$\frac{dP_{\nu_1}}{dP_{\nu_0}} = \exp\left[\int_A \left(\frac{d\nu_0}{d\nu} - \frac{d\nu_1}{d\nu} \right) d\nu + \int_A \log \frac{d\nu_1/d\nu}{d\nu_0/d\nu} dN \right]$$

and simply replacing $d\nu_0/d\nu$ and $d\nu_1/d\nu$ by conditional expectations of the directing intensity calculated under the candidate probability laws for M. Thus hypothesis testing (in the engineering literature, "detection") is separated from (state) estimation.

We assume a Cox pair (N, M) on $\mathbf{R}_+$ such that $M(A) = \int_A X_t d\nu(t)$, with ν stipulated only to be diffuse. The case that ν is Lebesgue measure is pursued further in Section 5 using stochastic intensity methods; there, a separation theorem is proved directly from Theorem 2.31.

Theorem 7.24. Let (N, M) be a Cox pair on $\mathbf{R}_+$ with the structure stipulated above, and assume that $\mathcal{L}_1 \ll \mathcal{L}_0$ where $\mathcal{L}_i$ is the P_i-law of M.

Then $P_1 \ll P_0$ on $\mathcal{F}^N$ and for each t, with $\hat{X}_t^i = E_i[X_t|\mathcal{F}_t^N]$,

$$\log (dP_1/dP_0)\big|_{\mathcal{F}_t^N} = \int_0^t(\hat{X}_u^0 - \hat{X}_u^1)d\nu(u) + \int_0^t \log(\hat{X}_{u-}^0/\hat{X}_{u-}^1)dN_u. \quad (7.39)$$

Proof: A rigorous proof proceeds from the property that the $(P_i, \mathcal{F}^N)$-compensator of N is $A_t^i = \int_0^t \hat{X}_{u-}^i \, d\nu(u)$, then appeals to representation theorems for likelihood ratios that generalize Theorem 2.31 (see, e.g., Jacod, 1975b, Theorem 4.5; Jacod and Shiryaev, 1986; and Liptser and Shiryaev, 1978).

Instead, we give an heuristic argument based on calculation of a stochastic differential. Specifically, with distribution function notation used as convenient,

$$d\left(\log \int \mathcal{L}_0(d\eta) \exp\{-\eta(t) + \int_0^t \log(d\eta/d\nu)dN\}\right)$$

$$= -\frac{\int \mathcal{L}_0(d\eta) \exp[-\eta(t) + \int_0^t \log(d\eta/d\nu)dN](d\eta/d\nu)(t)}{\int \mathcal{L}_0(d\eta) \exp[-\eta(t) + \int_0^t \log(d\eta/d\nu)dN]}d\nu(t)$$

$$+ \left\{\log \int \mathcal{L}_0(d\eta) \exp[-\eta(t) + \int_0^t \log(d\eta/d\nu)dN](d\eta/d\nu)(t)\right.$$

$$\left. - \log \int \mathcal{L}_0(d\eta) \exp[-\eta(t) + \int_0^t \log(d\eta/d\nu)dN]dN_t\right\}$$

$$= -\hat{X}_t^0 d\nu(t) + \log \hat{X}_{t-}^0 dN_t,$$

since (7.20) implies that

$$\hat{X}_t^0 = E_0[(dM/d\nu)(M)|\mathcal{F}_t^N]$$

$$= -\frac{\int \mathcal{L}_0(d\eta) \exp[-\eta(t) + \int_0^t \log(d\eta/d\nu)dN](d\eta/d\nu)(t)}{\int \mathcal{L}_0(d\eta) \exp[-\eta(t) + \int_0^t \log(d\eta/d\nu)dN]}d\nu(t).$$

Hence, by (7.37),

$$d\left(\log[dP_1/P_0]\big|_{\mathcal{F}_t^N}\right) = (\hat{X}_t^0 - \hat{X}_t^1)d\nu(t) + \log(\hat{X}_{t-}^1/\hat{X}_{t-}^0)dN_t. \blacksquare \quad (7.40)$$

The formula (7.39), especially its differential form (7.40), is of computational as well as intuitive value: it provides a technique for recursive calculation of the likelihood ratios $(dP_1/dP_0)|_{\mathcal{F}_t^N}$ as random functions of the observation time t. One should not view Theorem 7.24 as an end in itself, but rather as an instrument for computation, which merely reduces calculation of likelihood ratios to that of state estimators for the directing intensity. Even in rigidly structured situations the latter problem is not trivial; we take it up again in Section 5.

We turn briefly to testing hypotheses other than the simple pair (7.35). There are two fundamental difficulties (at least). The first is the size of the class of Cox processes: there are as many Cox processes as diffuse random measures. It is inescapable that restriction of the class of candidate Cox processes in the statistical model must precede any serious attempt to test hypotheses, which leads to the second difficulty, of specifying subclasses of Cox processes that are sufficiently restricted to permit reasonably precise tests, yet broad enough not to render the test meaningless in the sense, for example, that physical grounds suggest that the model itself is inappropriate. Even among Cox processes on $\mathbf{R}_+$ only the following subclasses seem to admit the right balance of restrictive structure and breadth:

- Poisson processes

- Mixed Poisson processes

- Cox processes that are also renewal processes

- Cox processes that are also Poisson cluster processes.

The third class includes the stochastic integral process discussed in Sections 6.3 and 5 (see also Chapter 8). Unfortunately (in a sense), in each case the additional structure is exploited more intensively than the Cox structure, which becomes superfluous. One can rarely implement the Neyman-Pearson theory and, especially if parameters are estimated simultaneously, tests are often decided only according as a log-likelihood ratio is positive or negative.

Even for the simple-vs.-simple test (7.35) the outlook for analysis à la Neyman-Pearson can be grim. Consider the mixed Poisson case with likelihood ratio given by Proposition 7.23. Its distribution, even though dependent on the data $\mathcal{F}^N(A)$ only through $N(A)$, is intractable. Normal approximation is impossible since a mixed Poisson process, unless it is actually Poisson, is not ergodic.

Tests with qualitative flavor are even more difficult. Consider the problem of testing whether a mixed Poisson process is Poisson, where the underlying measure ν is unknown. This problem is hopeless in the single-realization case. Suppose even that $E = \mathbf{R}_+$ and that ν is proportional to Lebesgue measure. Given observation of a single realization over $[0, t]$, tests such as uniform conditional tests (Section 3.5) cannot distinguish between Poisson and mixed Poisson processes: both are conditionally uniform. A single realization of the process over $\mathbf{R}_+$ is no more useful; if one conditions on the value of $\lim_{t \to \infty} N_t/t$ (the intensity in the Poisson case and the random value of the directing multiplier in the mixed Poisson case), then again the two classes are indistinguishable. If the null hypothesis were "Poisson with pre-scribed intensity ν_0" the outlook would be less bleak, but in applications there is rarely justification for hypotheses this narrow.

For data representing i.i.d. realizations there are more possibilities but few have been worked out in detail. Indeed, so little has been developed that one cannot identify with confidence the principal difficulties that arise.

7.4 Combined Inference and State Estimation

Suppose that $(N_1, M_1), (N_2, M_2), \ldots$ are i.i.d. Cox pairs on a compact space E, with the law $\mathcal{L}$ of the M_i unknown, and that only the N_i are observable. Calculation of state estimators $E[M_{n+1}|\mathcal{F}^{N_{n+1}}(A)]$ derived in Section 2 is impossible without effectively full knowledge of the law $\mathcal{L}$. Nonetheless, there are situations in which state estimation remains the primary goal, lack of knowledge of $\mathcal{L}$ notwithstanding. In this section we solve the prob-lem of combined statistical inference and state estimation for mixed Poisson processes. We also mention, albeit briefly, extensions to more general Cox processes.

For mixed Poisson processes the true state estimators are sufficiently simple in structure, and attributes of $\mathcal{L}$ appearing in them sufficiently facile to estimate, that without further restrictions on E pseudo-state estimators $\hat{E}[M_{n+1}|\mathcal{F}^{N_{n+1}}(A)]$ defined in (7.46) approximate the true state estimators, as $n \to \infty$, at a rate that is best possible given our assumption that the parameters F and ν defining $\mathcal{L}$ are entirely unknown. Moreover, for some purposes the processes N_i may be observed over varying sets A_i.

By contrast, results available concerning combined nonparametric infer-ence and state estimation for general Cox processes are much less satisfac-tory, for reasons indicated in discussion concluding the section.

Thus we establish the following setting. Let $(N_1, Y_1), (N_2, Y_2), \ldots$ be i.i.d. copies of a pair consisting of a simple point process N on E and a positive random variable Y; for $\nu \in \mathbf{M}_d$ and F a probability distribution on $\mathbf{R}_+$, let $P_{\nu,F}$ be a probability under which Y has distribution F and N is a mixed Poisson process with directing measure $M = Y\nu$. We work with the full nonparametric model $\mathcal{P} = \{P_{\nu,F}\}$, except that throughout ν is assumed to be diffuse and the regularity conditions (7.6)–(7.8) are presumed to hold.

Suppose that $N_1, \ldots, N_n$ have been observed previously over all of E, that N_{n+1} is observed over a subset A, and that we seek to reconstruct $M_{n+1} = Y_{n+1}\nu$. To this end we devise *pseudo-state estimators* approximating the true MMSE state estimators

$$E_{\nu,F}[M_{n+1}|\mathcal{F}^{N_{n+1}}(A)] = \frac{\int F(du)e^{-u\nu(A)}u^{N_{n+1}(A)+1}}{\int F(du)e^{-u\nu(A)}u^{N_{n+1}(A)}}\nu, \qquad (7.41)$$

and examine the behavior as $n \to \infty$ of the difference between the pseudo- and the true state estimators.

We embark on this path by introducing the integrals

$$K_A(k) = \int F(du)e^{-u\nu(A)}u^k = k!P_{\nu,F}\{N(A) = k\}/\nu(A)^k \qquad (7.42)$$

and rephrasing (7.41) as

$$E_{\nu,F}[M_{n+1}|\mathcal{F}^{N_{n+1}}(A)] = \frac{K_A(N_{n+1}(A) + 1)}{K_A(N_{n+1}(A))}\nu. \qquad (7.43)$$

To create pseudo-state estimators, guided by the principle of separation discussed in Chapters 1 and 3, we utilize the previous observations $N_1, \ldots, N_n$ to construct estimators $\hat{\nu}$ and $\hat{K}_A$, and then form the pseudo-state estimator by substituting them for ν and K_A in (7.43). The estimator $\hat{\nu}$ is that used in Section 1:

$$\hat{\nu}(\cdot) = \sum_{i=1}^n N_i(\cdot)/\sum_{i=1}^n N_i(E). \qquad (7.44)$$

At the expense of increased variance, since $(1/n)\sum_{i=1}^n N_i(E) \to 1$ almost surely by (7.7), we could use instead $\hat{\nu} = (1/n)\sum_{i=1}^n N_i$, which we do below when the N_i are observed over differing sets and only the multipliers Y_i are to be estimated. On the basis of (7.42) we take

$$\hat{K}_A(k) = (k!/\hat{\nu}(A)^k)(1/n)\sum_{i=1}^n \mathbf{1}(N_i(A) = k), \qquad (7.45)$$

and finally the $(n + 1)$-sample pseudo-state estimator is

$$\hat{E}[M_{n+1}|\mathcal{F}^{N_{n+1}}(A)] = \frac{\hat{K}_A(N_{n+1}(A) + 1)}{\hat{K}_A(N_{n+1}(A))}\hat{\nu}. \qquad (7.46)$$

Note that "current" observations N_{n+1} are *not* used in forming the estimators $\hat{\nu}$ and $\hat{K}_A$. In practice they should be used, at least in (7.45), in order not to discard additional information they contain, but for proving theorems it is more convenient to exclude them, for then in (7.46) the estimators $\hat{\nu}$, $\hat{K}_A$ and the random variable $N_{n+1}(A)$ are independent. Asymptotically, it is irrelevant which option one selects.

The main result concerning asymptotic behavior of differences between pseudo- and true state estimators, viewed as signed random measures on

E, is Theorem 7.26, to the effect that $\sqrt{n}$ times the difference converges in distribution to a mixture of Gaussian random measures. In view of Theorems 7.5 and 7.25 this rate of convergence is best possible because the same rate applies to the estimation error processes $\hat{\nu} - \nu$ and $\hat{K}_A - K_A$ and there is no possibility that pseudo-state estimators can approximate true at a rate faster than the law of the directing measures can be estimated.

While Theorem 7.26 is the *main* result, the next theorem is the *key* result. Not only does it yield Theorem 7.26 by a direct, albeit laborious, transformation argument, but it also provides the details needed to flesh out the skeletal proof given for Theorem 7.5. The $Q_F(k)$ and $\hat{Q}(k)$ of (7.9) and (7.10) are intimately related to (indeed, simpler than) the $K_A(k)$ and $\hat{K}_A(k)$ of this section. We stress that for now the set A is fixed.

Theorem 7.25. In addition to the hypotheses set forth above, assume that $\nu(A) > 0$ and that $E_{\nu,F}[Y^2] < \infty$. Then

$$
\begin{bmatrix} \{\sqrt{n}[\hat{\nu}(f) - \nu(f)] : f \in C(E)\} \\ \{\sqrt{n}[\hat{K}_A(k) - K_A(k)] : k \in \mathbf{N}\} \end{bmatrix} \xrightarrow{d} \begin{bmatrix} \{G(f) : f \in C(E)\} \\ \{Z(k) : k \in \mathbf{N}\} \end{bmatrix} \quad (7.47)
$$

under $P_{\nu,F}$, the first components as random measures on E and the second as sequences of random variables, where G is a mean zero Gaussian random measure and Z is a mean zero Gaussian sequence. The process (G, Z) is Gaussian with covariance function (where $m = E_{\nu,F}[Y]$)

$$
\text{Cov}(G(f), G(g)) = [\nu(fg) - \nu(f)\nu(g)]/m \quad (7.48)
$$

$$
\text{Cov}(Z(k), Z(j)) = 1(k = j)\frac{K_A(k)}{\nu(A)^k} - K_A(k)K_A(j)
$$
$$
+ \frac{kjK_A(k)K_A(j)}{\nu(A)}(1 - \nu(A))(2 + 1/m) \quad (7.49)
$$

$$
\text{Cov}(G(f), Z(k)) = kK_A(k)\frac{\nu(f; A)}{\nu(A)} + \nu(f; A^c)K_A(k + 1)
$$
$$
- \nu(f)[kK_A(k) + \nu(A^c)K_A(k + 1)]
$$
$$
+ kK_A(k)[\nu(f; A) - \nu(f)\nu(A)]/m. \quad (7.50)
$$

Proof: Let $\hat{Q}(k) = (1/n)\sum_{i=1}^n 1(N_i(A) = k)$ and (with dependence on ν and F suppressed) let $Q(k) = P_{\nu,F}\{N(A) = k\} = (\nu(A)^k/k!)K_A(k)$. We show first the central limit theorem

$$
\begin{bmatrix} \{\sqrt{n}[\hat{\nu}(f) - \nu(f)] : f \in C(E)\} \\ \sqrt{n}[\hat{\nu}(a) - \nu(A)] \\ \{\sqrt{n}[\hat{Q}(k) - Q(k)] : k \in \mathbf{N}\} \end{bmatrix} \xrightarrow{d} \begin{bmatrix} \{G^0(f) : f \in C(E)\} \\ W^0 \\ \{Z^0(k) : k \in \mathbf{N}\} \end{bmatrix}, \quad (7.51)
$$

with G^0 a Gaussian random measure, W^0 a normally distributed random variable and Z^0 a Gaussian sequence. By Theorem 1.12 and Billingsley (1968, Theorem 2.2), it is enough to show that for fixed f and J, the random vectors

$$\sqrt{n}\left([\hat{\nu}(f) - \nu(f)], [\hat{\nu}(A) - \nu(A)], [\hat{Q}(j) - Q(j)], j = 0, \ldots, J\right) \qquad (7.52)$$

have, asymptotically, a joint normal distribution, and by the Cramér-Wold device this follows from asymptotic normality of summations

$$\sqrt{n}\left([\hat{\nu}(f) - \nu(f)] + a[\hat{\nu}(A) - \nu(A)] + \sum_{\ell=0}^{J} a_\ell[\hat{Q}(\ell) - Q(\ell)]\right), \qquad (7.53)$$

where $a, a_0, \ldots, a_J$ are constants. As in Theorem 7.5, within error converging in probability to zero,

$$\sqrt{n}[\hat{\nu}(f) - \nu(f)] = (1/m\sqrt{n})\sum_{i=1}^{n} \int[f - \nu(f)]dN_i,$$

and in the same way,

$$\sqrt{n}[\hat{\nu}(A) - \nu(A)] = (1/m\sqrt{n})\sum_{i=1}^{n}[N_i(A) - \nu(A)N_i(E)],$$

which enables us to write (7.53) as

$$(1/\sqrt{n})\sum_{i=1}^{n}\left\{[N_i(f) - \nu(f)] + a[N_i(A) - \nu(A)N_i(E)]\right.$$
$$\left. + \sum_{\ell=0}^{J} a_\ell[1(N_i(A) = \ell) - Q(\ell)]\right\}. \qquad (7.54)$$

But (7.54) is just a sum of i.i.d. random variables with mean zero and finite variance, so asymptotic normality holds by the central limit theorem. (A simplified version of the foregoing argument proves Theorem 7.5.)

It remains at the moment to calculate covariances associated with the limit process in (7.51). The computations are straightforward and more boring than interesting; hence some of them are given as exercises! Here we simply record the results:

$$\text{Cov}(G^0(f), G^0(g)) = [\nu(fg) - \nu(f)\nu(g)]/m \qquad (7.55)$$

$$\text{Var}(W^0) = \nu(a)(1 - \nu(a))/m \qquad (7.56)$$

$$\text{Cov}(Z^0(k), Z^0(j)) = 1(j = k)Q(k) - Q(k)Q(j) \qquad (7.57)$$

$$\text{Cov}(G^0(f), W^0) = [\nu(f; A) - \nu(f)\nu(A)]/m \qquad (7.58)$$

$$\text{Cov}(G^0(f), Z^0(k)) = kQ(k)\frac{\nu(f; A)}{\nu(A)} + \frac{\nu(f; A^c)\nu(A)^k K_A(k+1)}{k!}$$
$$- \frac{\nu(f)}{k!}[kQ(k) + \nu(A^c)\nu(A)^k K_A(k+1)] \qquad (7.59)$$

$$\text{Cov}(W^0, Z^0(k)) = kQ(k)Q(k)[1 - \nu(A)]. \qquad (7.60)$$

Note that (7.56) arises from taking $f = g = 1_A$ in (7.55); that (7.57) is related closely to (7.13); that (7.60) corresponds to $f = 1_A$ in (7.59); and that the covariances in (7.58) and (7.59) are zero when $A = E$.

To proceed from (7.51) to (7.47) we use transformation theory (the "delta method") and another liberal dose of covariance computations. By reasoning similar to that just gone through, (7.47) holds if for each f and J the asymptotic distribution of the random vectors

$$
\begin{bmatrix}
\sqrt{n}[\hat{\nu}(f) - \nu(f)] \\
\sqrt{n}[\hat{K}_A(0) - K_A(0)] \\
\sqrt{n}[\hat{K}_A(1) - K_A(1)] \\
\vdots \\
\sqrt{n}[\hat{K}_A(J) - K_A(J)]
\end{bmatrix}
=
\begin{bmatrix}
\sqrt{n}[\hat{\nu}(f) - \nu(f)] \\
\sqrt{n}[\hat{Q}_{(}0) - Q(0)] \\
\sqrt{n}[\hat{Q}(1)/\hat{\nu}(A) - Q(1)/\nu(A)] \\
\vdots \\
\sqrt{n}[J!\hat{Q}(J)/\hat{\nu}(A)^J - J!Q(J)/\nu(A)^J]
\end{bmatrix}
$$

is normal, but this follows from applying the delta method to (7.52) and the transformation $(a, b, c_0, \ldots, c_J) \to (a, c_0, c_1/b, \ldots, J!c_J/b^J)$. The covariance function (7.48)–(7.50) follows at once from (7.55)–(7.60). ∎

With one more transformation argument we obtain the main result.

Theorem 7.26. Under the hypotheses of Theorem 7.25,

$$
\sqrt{n}\left[\hat{E}[M_{n+1}|\mathcal{F}^{N_{n+1}}(A)] - E_{\nu,F}[M_{n+1}|\mathcal{F}^{N_{n+1}}(A)]\right] \overset{d}{\to} H \tag{7.61}
$$

as signed random measures on E, where the distribution of the random measure H is a mixture, with mixing distribution $P_{\nu,F}\{N(A) \in (\cdot)\}$, of those of the sequence (G_k) of Gaussian random measures whose covariance function is given by (7.62) below.

Proof: By Slutsky's theorem, for fixed k and f

$$
\sqrt{n}\left[\frac{\hat{K}_A(k+1)}{\hat{K}_A(k)}\hat{\nu}(f) - \frac{K_A(k+1)}{K_A(k)}\nu(f)\right]
$$

$$
\cong \; \nu(f)\sqrt{n}\left[\frac{\hat{K}_A(k+1)}{\hat{K}_A(k)} - \frac{K_A(k+1)}{K_A(k)}\right] + \frac{K_A(k+1)}{K_A(k)}\sqrt{n}[\hat{\nu}(f) - \nu(f)]
$$

$$
\cong \; (\nu(f)/K_A(k))\sqrt{n}[\hat{K}_A(k+1) - K_A(k+1)]
$$
$$
- (\nu(f)K_A(k+1)/K_A(k)^2)\sqrt{n}[\hat{K}_A(k) - K_A(k)]
$$
$$
+ (K_A(k+1)/K_A(k))\sqrt{n}[\hat{\nu}(f) - \nu(f)].
$$

Hence by reasoning similar to that in the proof of Theorem 7.25 and by the conclusion to that theorem, the sequence of signed random measures

$$
\left\{\left\{\sqrt{n}\left[\frac{\hat{K}_A(k+1)}{\hat{K}_A(k)}\hat{\nu}(f) - \frac{K_A(k+1)}{K_A(k)}\nu(f)\right] : f \in C(E)\right\} : k \in \mathbf{N}\right\}
$$

converges in distribution to a jointly Gaussian sequence (G_k) of Gaussian random measures whose covariance function, by routine calculations, is

$$
\begin{aligned}
&\text{Cov}(G_k(f), G_j(g)) \\
&= \ \nu(f)\nu(g)\left[\frac{R(k+1,j+1)}{K_A(k)K_A(j)} - \frac{K_A(k+1)R(k,j+1)}{K_A(k)^2 K_A(j)}\right. \\
&\quad\left. - \frac{K_A(k+1)R(k+1,j)}{K_A(k)K_A(j)^2} + \frac{K_A(k+1)K_A(j+1)R(k,j)}{K_A(k)^2 K_A(j)^2}\right] \\
&\quad + \frac{\nu(f)}{K_A(j)^2}\left[K_A(j+1)R(f,j+1) - K_A(j+1)^2 R(f,j)\right] \\
&\quad + \frac{\nu(g)}{K_A(k)^2}\left[K_A(k+1)R(g,k+1) - K_A(k+1)^2 R(g,k)\right] \\
&\quad + \frac{K_A(k+1)K_A(j+1)}{K_A(k)K_A(j)}R(f,g), \qquad\qquad (7.62)
\end{aligned}
$$

where R is the covariance function of (7.48)–(7.50).

For each n, $N_{n+1}(A)$ is independent of $N_1, \ldots, N_n$, and hence of the n-sample estimators $\hat{\nu}$ and $\hat{K}_A$, and from this, the continuous mapping theorem and Billingsley (1968, Theorem 3.2), (7.61) ensues. ∎

In particular, taking the suppressed function in (7.61) to be $f \equiv 1$ and using the relationship

$$
E_{\nu,F}[M_{n+1}(E)|\mathcal{F}^{N_{n+1}}(A)] = E_{\nu,F}[Y_{n+1}|\mathcal{F}^{N_{n+1}}(A)],
$$

we infer that for the pseudo-state estimators

$$
\hat{E}[Y_{n+1}|\mathcal{F}^{N_{n+1}}(A)] = \frac{\hat{K}_A(N_{n+1}(A)+1)}{\hat{K}_A(N_{n+1}(A))} \qquad\qquad (7.63)
$$

of the directing multipliers the following limit behavior obtains.

Proposition 7.27. Under the hypotheses of Theorem 7.25,

$$
\sqrt{n}\left[\hat{E}[Y_{n+1}|\mathcal{F}^{N_{n+1}}(A)] - E_{\nu,F}[Y_{n+1}|\mathcal{F}^{N_{n+1}}(A)]\right] \xrightarrow{d} Y,
$$

where Y has the mixed normal distribution of $H(1)$ in Theorem 7.26. ☐

Note that computation of the pseudo-state estimator (7.63) requires knowledge only of $N_1(A), \ldots, N_n(A)$ and $N_1(E), \ldots, N_n(E)$, and that, as mentioned previously, even the latter would be dispensable if instead of (7.44) one utilized the estimators $\hat{\nu}(A) = (1/n)\sum_{i=1}^n N_i(A)$. An analogue of Proposition 7.27 holds, with a different covariance function (see Karr,

1984b). But the distribution of the $N_i(A)$ depends on A only through $\nu(A)$ (Exercise 1.3), raising hope that a result comparable to Proposition 7.27 holds if each process N_i is observed over a set A_i depending deterministically on i. Indeed, this is so, provided that the values $\nu(A_i)$ satisfy a fairly stringent stability condition.

In this setting, we can approximate the true state estimators

$$E_{\nu,F}[Y_{n+1}|\mathcal{F}^{N_{n+1}}(A)] = \frac{K_{A_{n+1}}(N_{n+1}(A_{n+1}) + 1)}{K_{A_{n+1}}(N_{n+1}(A_{n+1}))}$$

by pseudo-state estimators

$$\hat{E}[Y_{n+1}|\mathcal{F}^{N_{n+1}}(A)] = \frac{\hat{K}(N_{n+1}(A_{n+1}) + 1)}{\hat{K}(N_{n+1}(A_{n+1}))}$$

where

$$\hat{K}(k) = k! \sum_{i=1}^n N_i(A_i) / \sum_{i=1}^n 1(N_i(A_i) = k),$$

which can be calculated from only the data $N_1(A_1), \ldots, N_n(A_n), N_{n+1}(A_{n+1})$. In Karr (1984b) the following limit theorem is proved.

Theorem 7.28. Assume that
a) There is $c \in (0,1)$ such that $(1/\sqrt{n}) \sum_{i=1}^n |\nu(A_i) - c| \to 0$;
b) $\int F(du)u^k < \infty$ for each k.
Then under $P_{\nu,F}$,

$$\sqrt{n} \left[\hat{E}[Y_{n+1}|\mathcal{F}^{N_{n+1}}(A_{n+1})] - E_{\nu,F}[Y_{n+1}|\mathcal{F}^{N_{n+1}}(A_{n+1})] \right] \xrightarrow{d} Z,$$

where Z has a mixed normal distribution. $\square$

The proof is little more difficult than those above; where before appeal was made to the central limit theorem for i.i.d. random variables, one uses instead the Lindeberg-Feller theorem, whose applicability is justified by a) and b). See Karr (1984b) for additional discussion.

An altogether different procedure applies to extremely general — but still i.i.d. — Cox pairs (N_i, M_i) with no structure other than diffuseness imposed on the directing measures. We omit a detailed presentation, but inasmuch as the methodology combines several diverse topics, an outline is merited.

1. For simplicity suppose that E is compact and that $A = E$. Assuming that $E[M(E)^k] < \infty$ for each k, the true state estimators of

$$E[e^{-M_{n+1}(f)}|N_{n+1}] = E[e^{-M_\mu(1+f)}]/E[e^{-M_\mu(1)}]\Big|_{\mu=N_{n+1}}$$

are expressible via Laplace functionals $L_M(\mu, f) = E[e^{-M_\mu(f)}]$ of the unreduced Palm processes M_μ of a generic directing measure M:

$$E[e^{-M_{n+1}(f)}|N_{n+1}] = L_M(N_{n+1}, 1 + f)/L_M(N_{n+1}, 1). \qquad (7.64)$$

In order to form pseudo-state estimators, one must estimate these Palm Laplace functionals. Were the M_i observable this could be done directly, but only the Cox processes are observable, so additional analysis is needed.

2. By Proposition 1.55 for each μ the reduced Palm process N_μ of N, with Palm Laplace functional $L_N(\mu, f)$, has the distribution of a Cox process directed by M_μ, so that for functions g satisfying $0 \le g < 1$,

$$L_M(\mu, g) = L_N(\mu, -\log(1 - g)). \qquad (7.65)$$

Consequently, one can estimate the $L_M(\mu, \cdot)$ by applying (7.65) to estimators of the $L_N(\mu, \cdot)$. Since the N_i are observable i.i.d. point processes, methods of Section 4.4 can be used to construct estimators $\hat{L}_N(\mu, g)$ that are asymptotically normal, but only in the integrated sense of Theorem 4.16.

3. There is further complication: neither of the functions $g = 1 + f$ and $g \equiv 1$ in (7.64) satisfies $0 \le g < 1$, and therefore (7.65) does not apply. One can circumvent this difficulty by passing to moment-generating functionals rather than Laplace functionals, but at a penalty: it becomes necessary to assume that $E[e^{tM(E)}] < \infty$ for some $t > 0$. Granted this assumption, for functions f satisfying $0 \le f < 1$ the true state estimators

$$E[e^{M_{n+1}(f)}|N_{n+1}] = \frac{L_M(N_{n+1}, 1 - f)}{L_M(N_{n+1}, 1)} = \frac{L_N(N_{n+1}, -\log f)}{L_N(N_{n+1}, \infty)}$$

[where "∞" denotes the function identically equal to ∞, but $L_N(\mu, \infty)$ is shorthand for $P\{N_\mu(E) = 0\} = \lim_{t\to\infty} \int Q_N(\mu, d\nu)e^{-t\nu(E)}$] can be approximated by pseudo-state estimators

$$\hat{E}[e^{M_{n+1}(f)}|N_{n+1}] = \hat{L}_N(N_{n+1}, -\log f)/\hat{L}_N(N_{n+1}, \infty).$$

However, the degree of approximation is much worse than in the mixed Poisson case, and the culprit is the integrated convergence in distribution in Theorem 4.16.

Theorem 7.29. For f a function satisfying $0 < f \leq 1$ and $\delta > 0$,

$$\lim_{n \to \infty} n^{1/2-\delta} E\left[(\hat{E}[e^{M_{n+1}(f)}|N_{n+1}] - E[e^{M_{n+1}(f)}|N_{n+1}])^2\right] = 0.\,\square$$

While very general, this theorem is disappointing in the sense that the rate of (L^2-)convergence, $n^{-1/4+\delta}$, is distinctly less than one would wish (based on central limit theory, the hoped-for rate of convergence is $n^{-1/2+\delta}$).

7.5 Martingale Inference for Cox Processes

Suppose that N is a Cox process on $\mathbf{R}_+$ with directing measure

$$M(A) = \int_A X_t dt. \tag{7.66}$$

With respect to the filtration $\mathcal{H}_t = \sigma(X) \vee \mathcal{F}_t^N$, N has stochastic intensity (X_t) (this process is $\mathcal{H}$-predictable because $X_t \in \mathcal{H}_0$ for every t) and hence N has $\mathcal{F}^N$-stochastic intensity $(\hat{X}_{t-})$, where $\hat{X}_t = E[X_t|\mathcal{F}_t^N]$. In this section we examine applicability of martingale methods from Chapters 2 and 5 to inference and state estimation for such Cox processes. That the methods do apply in principle is evident; the key issue is to make use, within the martingale context, of the Cox process structure, which is manifested only in that $X_t \in \mathcal{H}_0$ for each t. It seems difficult to do so; indeed, because X is not observable one is often better off simply using either only methods developed elsewhere in this chapter or only martingale techniques.

In connection with statistical estimation, if the directing intensity is neither restricted to have special structure nor at least partially observable, use of martingale methods seems precluded and one is left with the Laplace functional techniques described in Section 1 and Chapter 4. However, the consequences are less than ideal in terms of what can be inferred about the directing intensity viewed as a stochastic process rather than as the random measure of (7.66). For (N, M) a Cox pair with M given by (7.66), Lemma 7.1 yields $E[\exp\{-\int g(t)X_t dt\}] = L_N(-\log(1-g))$ provided that $0 \leq g < 1$, but whereas in principle from this one can recover information or estimate finite-dimensional distributions, moments, ... of X, in practice computation seems difficult. Of course, integrals $\mu_M(f) = \int f(t)E[X_t]dt$, the mean measure of M, can be estimated by empirical averages $(1/n)\sum_{i=1}^{n} N_i(f)$, but provide only "integrated" information about the function $t \to E[X_t]$. For statistical estimation given i.i.d. copies, there seems to be no way (with or without martingale methods) to exploit systematically the special structure inherent in (7.66). Of course, if N is a mixed Poisson process, then Theorems 7.5 and 7.15 apply, but then there is so much special structure that martingale techniques are unnecessary.

On the other hand, with additional specialization beyond (7.66) and modification of the observability structure, one achieves a multiplicative intensity model, to which martingale methods apply but for which it is hard to exploit the Cox process structure. Specifically, if N is a Cox process on $[0, 1]$ directed by the random measure $M(A) = \int_A \alpha_t X_t dt$, with α an unknown function, and if the process X is observable, then we have a special case of the multiplicative intensity model of Section 5.1. Methods and results from Sections 5.2 and 5.3 apply to estimation of α; in particular, given data comprising i.i.d. pairs (N_i, X_i), then with $N^n = \sum_{i=1}^n N_i$ and $X^n = \sum_{i=1}^n X_i$, Theorems 5.12 and 5.13 apply to the martingale estimators $\hat{B}_t^n = \int_0^t (X_u^n)^{-1} dN_u^n$ of the processes $B_t^n(\alpha) = \int_0^t \alpha_u \mathbf{1}(X_u^n > 0) du$. However, in this setting one seemingly cannot refine or sharpen martingale results by using Cox process structure.

Thus for statistical estimation one seemingly cannot have it both ways; it is possible to exploit Cox process structure or stochastic intensity structure, but not both simultaneously.

For state estimation the situation is similar. In the martingale context, Cox structure is usable to the extent that it implies that the $\mathcal{F}^N$-stochastic intensity is the conditional expectation of the directing intensity, but not much further. Given a Cox process N with directing measure (7.66), the central state estimation problem is to reconstruct the directing intensity; we now explore the consequences of Theorem 5.30 concerning it. The main difficulty is derivation of a semimartingale representation for the directing intensity. If anything, the Cox structure becomes a hindrance because it is more difficult to construct a useful representation for the filtration $\mathcal{H}_t = \sigma(X) \vee \mathcal{F}_t^N$ than for the smaller filtration $\mathcal{G}_t = \mathcal{F}_t^X \vee \mathcal{F}_t^N$, with respect to which N still has stochastic intensity (X_{t-}). Once one passes to $\mathcal{G}$, though, the Cox structure of N is lost.

However, if that structure is restricted sufficiently, useful conclusions can be obtained. While valid more generally than for Cox processes, the next result is given here because its principal applications have been to Cox processes and because in the Cox case its hypotheses are fulfilled. It treats the case that the stochastic intensity is a function of a Markov process; in the engineering literature such processes are termed Markov-modulated Poisson processes. Special cases are the stochastic integral of Section 6.3 and the Cox processes of Example 7.14.

Theorem 7.30. Let N be an integrable point process on $\mathbf{R}_+$ admitting $\mathcal{H}$-stochastic intensity $\lambda_t = f(Y_{t-})$, where Y is a standard Markov process (with respect to $\mathcal{H}$) with generator A, and f is a bounded, positive function belonging to the domain of A. Then the MMSE state estimators $\hat{\lambda}_t =$

$E[f(Y_t)|\mathcal{F}_t^N]$ satisfy the stochastic differential equation

$$d\hat{\lambda}_t = E[Af(Y_t)|\mathcal{F}_t^N]dt + \{[\widehat{\lambda^2}]_{t-} - (\hat{\lambda}_{t-})^2]/\hat{\lambda}_{t-}\}dM_t, \qquad (7.67)$$

where $dM_t = dN_t - \hat{\lambda}_t dt$ is the $\mathcal{F}^N$-innovation martingale and $\widehat{\lambda^2}$ is the process of MMSE state estimators of λ_t^2. The initial condition is $\hat{\lambda}_0 = E[f(Y_0)]$.

Proof: Formally, the proof is identical to that of the special case in Theorem 6.22, except that the latter contains simplifications (because there $\lambda^2 = \lambda$) not valid in general. One works through the computations associated with Theorem 5.30 and its antecedents. The semimartingale representation of λ is $\lambda_t = \lambda_0 + \int_0^t Af(Y_u)du + \tilde{M}_t$ by Dynkin's formula, where $\tilde{M}$ is a mean zero $\mathcal{H}$-martingale, and hence by Proposition 5.28 and the property that $\hat{X}_t = E[X_t|\mathcal{F}_t^N]$, $\hat{\lambda}$ satisfies the stochastic differential equation

$$d\hat{\lambda}_t = E[Af(Y_t)|\mathcal{F}_t^N]dt + H_t dM_t.$$

By Theorem 5.30 the filter gain H has the form $H_t = J_t + K_t - \hat{\lambda}_{t-}$, where J are K are predictable processes that remain to be calculated. Since the martingale $\tilde{M}$ and the point process N have no jumps in common almost surely (this uses the Markov property of Y), $K \equiv 0$ by discussion following Theorem 5.30, while $J_t = (\widehat{\lambda^2})_{t-}/\hat{\lambda}_{t-}$. Substitution now yields (7.67). ∎

Several comments are in order. First, note that the filter gain in (7.67) is the conditional variance $E[\lambda_t^2|\mathcal{F}_t^N] - E[\lambda_t|\mathcal{F}_t^N]^2$ divided by the conditional mean. Second, save in special cases such as Theorem 6.22, the equation (7.67) does not involve $(\hat{\lambda}_t)$ alone and hence is not self-sufficient for recursive computation of $\hat{\lambda}$. Indeed, (7.67) introduces two additional terms, $E[Af(Y_t)|\mathcal{F}_t^N] = \widehat{Af(Y)}_t$ and $(\widehat{\lambda^2})_t$, neither of which is expressible in general as function of $\hat{\lambda}$. Estimation of them using two equations analogous to (7.67) (see also Theorem 7.31) introduces four further equations, and so on, resulting in an infinite system of stochastic differential equations that can be "solved" only by recourse to (almost always unjustified) approximations or simplifications.

Alternatively, instead of estimating moments one can allow measure-valued state processes and estimate distributions themselves, as we describe briefly. We remain in the setting of Theorem 7.30; in this discussion important technical details are ignored. The measure-valued state process $\eta_t = \varepsilon_{Y_t}$, given appropriate assumptions, has semimartingale representation

$$\eta_t = \int_0^t A(Y_u)du + \tilde{M}_t,$$

where $\tilde{M}$ is a martingale taking values in the space of signed measures on E and $A(y)$ is the measure against which a function g belonging to the domain of A integrates to $Ag(y)$. It follows by arguments in the proof of Theorem 7.30 that the state estimators $\hat{\eta}_t = E[\eta_t|\mathcal{F}_t^N]$, where the conditional expectations are random measures on E, have semimartingale representation

$$\hat{\eta}(g)_t = \int_0^t \widehat{Ag(Y)}_u du + \int_0^t H_u(g)[dN_u - \hat{\lambda}_{u-} du],$$

with $\hat{\lambda}$ as in Theorem 7.30. But

$$\int_0^t \widehat{Ag(Y)}_u = \int_0^t \int_E P\{Y_u \in dy|\mathcal{F}_u^N\} Ag(y) du = \int_0^t \int_E \hat{\eta}_u(dy) Ag(y) du,$$

while since $\lambda_t = \int \eta_t(dy) f(y)$, $\hat{\lambda}_t = \int \hat{\eta}(dy) f(y)$. Two more computations then suffice to give the following result.

Theorem 7.31. The measure-valued process $\hat{\eta}_t = E[\eta_t|\mathcal{F}_t^N]$ satisfies the stochastic differential equation

$$d\hat{\eta}_t(\cdot) = [\int_E \hat{\eta}_t(dy) A(y)(\cdot)]dt + \{[\hat{\eta}_{t-}(f \times (\cdot)) - \hat{\lambda}_{t-} \hat{\eta}_{t-}(\cdot)]/\hat{\lambda}_{t-}\} dM_t,$$

where $dM_t = dN_t - \hat{\lambda}_{t-} dt$ as in Theorem 7.30.

Proof: By Theorem 5.30 the measure-valued filter gain is

$$H_t = [(\widehat{\lambda\eta})_{t-} - \hat{\lambda}_{t-} \hat{\eta}_{t-}]/\hat{\lambda}_{t-}$$

(the K-term again vanishes), but for g in the domain of the generator, $\lambda_t \eta_t(g) = f(Y_t) = \eta_t(fg)$, and consequently $(\widehat{\lambda\eta})_{t-}(\cdot) = \hat{\eta}_{t-}(f \times (\cdot))$. ∎

We have digressed from Cox processes on $\mathbf{R}_+$, although it should be borne in mind that a Cox process with directing intensity is one case in which the hypotheses of Theorem 7.30 are known to be verified. Concerning likelihood ratios, Theorem 2.31 yields the following separation theorem, which contains a special case of Theorem 7.24.

Proposition 7.32. Suppose that under the probability P, (N, M) is a Cox pair with M of the form (7.66), and let $\tilde{P}$ be a probability under which N is a Poisson process with rate 1. Then for each t, $P \ll \tilde{P}$ on $\mathcal{F}_t^N$ with

$$dP/d\tilde{P}\Big|_{\mathcal{F}_t^N} = \exp\left\{\int_0^t (1 - \hat{X}_u)du + \int_0^t (\log \hat{X}_{u-})dN_u\right\}, \qquad (7.68)$$

where $\hat{X}_u = E[X_u|\mathcal{F}_u^N]$. □

In general, $P \perp \tilde{P}$ on $\mathcal{F}^M$, so Proposition 7.32 furnishes yet another example in which $P \ll \tilde{P}$ on $\mathcal{F}^N$ despite their being singular on $\mathcal{F}^M$.

In particular, if P_0 and P_1 are probabilities with respect to each of which N is a Cox process with directing measure of the form (7.66), then (7.68) implies that

$$\left.\frac{dP_1/d\tilde{P}}{dP_0/d\tilde{P}}\right|_{\mathcal{F}_t^N} = \exp\left\{\int_0^t(\hat{X}_u^1 - \hat{X}_u^0)du + \int_0^t \log[\hat{X}_{u-}^1/\hat{X}_{u-}^0]dN_u\right\}, \quad (7.69)$$

where $\hat{X}_u^i = E_i[X_u|\mathcal{F}_u^N]$. Whenever $P_1 \ll P_0$ on $\mathcal{F}^N$, (7.69) gives the likelihood ratio dP_1/dP_0 and, provided that state estimators $\hat{X}_u^i$ can be calculated, is usable for likelihood ratio tests. The log-likelihood function

$$L_t(P_0, P_1) = \int_0^t(\hat{X}_u^1 - \hat{X}_u^0)du + \int_0^t \log(\hat{X}_{u-}^1/\hat{X}_{u-}^0)dN_u \quad (7.70)$$

can be used more generally, although without absolute continuity or equivalence it may assume infinite values, but still the state estimators must be computed. Theorem 7.30 presents a set of conditions under which this might be possible, as well as a computational method. We conclude the section with a specific application in which computations have actually been done.

Contained in Smith and Karr (1985) is a statistical analysis of some point process models for times and amounts of summer season precipitation events at a single site. Three Cox process models are considered:

1. Poisson processes with constant intensity

2. The stochastic integral $X * N$ of Section 6.3, with X a Markov process

3. A class of Neyman-Scott cluster processes.

The first two have already been described in detail; the third will be treated momentarily. Each entails unknown parameters that must be estimated before one can perform likelihood ratio tests, so that the procedure is not a straightforward application of (7.70). Specifics are discussed following description of the cluster processes.

The Neyman-Scott cluster processes considered as models of rainfall are Poisson cluster processes in the sense of Definition 1.30, but are more easily understood via an explicit construction. Let $\tilde{N} = \sum \varepsilon_{T_i}$, the *cluster center process*, be a Poisson process on $\mathbf{R}_+$ with rate c, and to each cluster center T_i let there be associated a random number X_i of secondary points with locations T_{ij}, $j = 1, \ldots, X_i$, in such a manner that

- X_i has a Poisson distribution with mean a

- The "offsets" $T_{ij} - T_i$ are independent and identically exponentially distributed with mean $1/b$

- X_i and the $T_{ij} - T_i$ are independent of one another, of $\tilde{N}$ and of the X_k and $T_{k\ell}$ for $k \neq i$.

The Poisson cluster process $N = \sum_i \sum_j \varepsilon_{T_{ij}}$ consisting of all the secondary points, is termed a *Neyman-Scott cluster process*, with the general interpretation that the cluster center T_i engenders succeeding events at times T_{ij}. A specific interpretation will be given momentarily.

First, here are basic properties.

Proposition 7.33. The Neyman-Scott cluster process N is a Cox process with directing intensity

$$X_t = ab \int_0^t e^{-b(t-u)} d\tilde{N}_u \qquad (7.71)$$

and a Markov process. ☐

The proof is straightforward: intuitively, the Markov property of X results from the stochastic differential equation $dX_t = -bX_t dt + ab d\tilde{N}_t$, a consequence of (7.71), and from the independent increments property of $\tilde{N}$.

As models of summer season (July–October) rainfall in the Potomac River basin in the eastern United States, each class of processes carries a distinctive physical interpretation. The Poisson model represents a constant rate of occurrence of rainfall events. For the stochastic integral, the directing intensity represents a climatological process, with $X = 0$ a "dry" state (cool, high-pressure conditions) during which there is no precipitation and $X = 1$ a "wet" state during which events occur in Poisson fashion. Not only is this interpretation consistent qualitatively with climatological data but also (Exercise 3.15), $X * N$ is a renewal process, which accords with data analysis in Smith (1981) but lacks clear physical justification. In the Neyman-Scott cluster process, cluster centers correspond to passages of cold and warm fronts, each having associated with it some (or possibly no) subsequent rainfall events (see Kavvas and Delleur, 1981, for details).

Statistical analysis was performed as follows. By Proposition 7.33, Proposition 7.32 applies to the Neyman-Scott cluster model and to the stochastic integral $X * N$ for calculation of log-likelihood functions. However, it was necessary first to estimate unknown parameters: for Poisson processes the rate c, for $X * N$ the rate c of events during periods when $X = 1$ and the transition rates a and b of X, and for the cluster process the rate c of the cluster center process, the mean number a of cluster members per cluster center, and b, the exponential parameter of the cluster center-cluster member offsets. For the Poisson model this was done using $\hat{c} = N_t/t$, while for the other two classes it was done by maximizing numerically the likelihood functions (7.68) over a grid of (a, b, c) values using a nonlinear optimization algorithm, choosing as maximum likelihood estimators $\hat{a}$, $\hat{b}$ and $\hat{c}$ (the parameters have different interpretations in the two models) the values for which the likelihood was greatest. To compute state estimators of the directing intensity, Theorem 6.22 was utilized for $X * N$ and an approximation to (7.67) — third moments were declared to be zero — for the cluster process. For computational purposes, time discretization was used to convert stochastic differential equations to difference equations.

Likelihood ratio tests were performed using the log-likelihood functions (7.70), but with the probability measures P_0 and P_1 there replaced by measures associated with previously calculated maximum likelihood estimators. Then, given two classes, one was deemed a better description of the data if the maximum value of the likelihood function was greater for it. In terms of (7.70) and likelihood ratio tests, this amounts to choosing P_1 over P_0 if $L_t(P_1, P_0) > 0$ and choosing P_0 over P_1 otherwise. The hypotheses $H_0 : P = P_0$ and $H_1 : P = P_1$ are thus treated symmetrically.

Computational difficulties precluded complete analysis of the Neyman-Scott alternative. Between the Poisson model and the stochastic integral, for small values of t the Poisson hypothesis was accepted, but for larger values the stochastic integral was the better model. For simulated realizations of $X * N$ convergence of maximum likelihood estimators to true parameter values was exceedingly slow, especially for ill-chosen but physically reasonable combinations of the parameters. See Smith and Karr (1985) for details and numerical results.

In addition to its other properties, the stochastic integral $X * N$ is a renewal process, so it forms a fitting bridge to the next chapter, where it will be analyzed yet again.

EXERCISES

7.1 Let (N, M) be a Cox pair. Prove that if N is simple, then M is diffuse almost surely.

7.2 Let N be a mixed Poisson process with parameters (ν, F) satisfying (7.6)–(7.8). Prove that N admits a representation $N = \sum_{i=1}^{N(E)} \varepsilon_{X_i}$, where $N(E)$ has the mixed Poisson distribution (7.9) and the X_i are i.i.d. with distribution ν and independent of $N(E)$.

7.3 Deduce from Exercise 7.2 a mixed Poisson process N satisfying (7.6)–(7.8) is conditionally uniform with respect to ν: given that $N(E) = k$, N has the same distribution as the empirical process $\tilde{N} = \sum_{i=1}^{k} \varepsilon_{Z_i}$, where the Z_i are i.i.d. random elements of E, each with distribution ν.

7.4 Let N be a mixed Poisson process on $\mathbf{R}_+$. Show that if N is a renewal process, then N is Poisson.

7.5 Derive the covariance function R of (7.5).

7.6 Prove Lemma 7.3.

7.7 Let (N_i, M_i) be i.i.d. copies of a Cox pair on a compact space E. Calculate the asymptotic efficiency of the estimators $\hat{\mu}_M = (1/n)\sum_{i=1}^n N_i$ of the mean measure of M, which are used if only the N_i are observable, relative to the estimators $\hat{\mu}_M^* = (1/n)\sum_{i=1}^n M_i$, which would be used if the M_i were observable.

7.8 Let (N_i, Y_i) be i.i.d. copies of the pair (N, Y) in which N is a mixed Poisson process directed by the random measure $M = Y\nu$, where the Y_i satisfy $E[Y_i] = 1$. Assume that the underlying space is compact and suppose that both the N_i and the Y_i are observable.
a) Develop asymptotic properties of the estimators $\hat{\nu}^* = \sum_{i=1}^n N_i / \sum_{i=1}^n Y_i$ of the mean measure ν of the N_i.
b) Compare with those of the estimators $\hat{\nu} = (1/n)\sum_{i=1}^n N_i$ that would be used if the Y_i were not observable. In particular, calculate the asymptotic relative efficiency.

7.9 Let N be a Poisson process with mean measure μ. Show that for $g, h \geq 0$,
$$E[e^{-N(g)}N(h)] = \exp[-\mu(1 - e^{-g})]\mu(he^{-g}).$$

7.10 Let M be a diffuse random measure with probability law $\mathcal{L}$ and let $\tilde{M}$ have law $\tilde{\mathcal{L}}(d\nu) = (1/C)e^{-\nu(A)}\mathcal{L}(d\nu)$, where A is a fixed set and $C = \int \mathcal{L}(d\nu)e^{-\nu(A)}$. Show that the Palm Laplace functionals $L(\mu, f) = E[e^{-M_\mu(f)}]$ and $\tilde{L}(\mu, f) = E[e^{-\tilde{M}_\mu(f)}]$ satisfy $\tilde{L}(\mu, f) = L(\mu, f + 1_A)/L(\mu, 1_A)$.

7.11 Let N be a mixed Poisson process on $\mathbf{R}_+$ with directing measure $M = Y\nu$.
a) Prove that for each t,
$$E[Y^2|\mathcal{F}_t^N] = \int F(du)e^{-u\nu(t)}u^{N_t+2} / \int F(du)e^{-u\nu(t)}u^{N_t}.$$
b) Derive the expression in Example 5.31 directly from this.

7.12 Consider the Markov-modulated Poisson process of Example 7.14.
a) Solve (7.28) for the case $i = 0$, $j = \alpha$, $q_0 = a$ and $q_\alpha = b$, where $a, b > 0$.
b) Obtain the corresponding explicit version of (7.27).
c) Contrast this method of calculation with that of Theorem 6.22.

7.13 Let $a, b > 0$ be known and distinct, let $T > 0$ be an unobservable random variable with known, continuous distribution G, let X be the process $X_t = a1(t < T) + b1(t \geq T)$, and suppose that the observations are a Cox process N with directing intensity X. [This is the "disruption problem;" see also Exercise 5.23.]
a) Derive a stochastic differential equation satisfied by the MMSE state estimators $P\{T \leq t|\mathcal{F}_t^N\} = P\{X_t = b|\mathcal{F}_t^N\}$.
b) Solve the equation explicitly for the case that G is an exponential distribution.

7.14 Provide details missing from the proof of Theorem 7.13.

7.15 Give a detailed derivation of (7.32).

7.16 For the disruption problem of Exercise 7.13, for each t calculate the MMSE linear estimator of $\mathbf{1}(T \le t)$ given the observations $\mathcal{F}_t^N$.

7.17 Work through the following heuristic derivation of (7.33).
a) Argue that $\hat{M}(f) - M(f)$ is characterized by its being orthogonal to $N(dx)$ for every $x \in A$.
b) Deduce from a) that for $x \in A$, $\hat{f}$ satisfies

$$E[M(f)M(dx)] + \mu(f)\mu(dx) = \hat{f}(x)\mu(dx) + E[M(\hat{f})M(dx)] - \mu(\hat{f})\mu(dx),$$

the differential version of (7.33).

7.18 Let P_0 and P_1 be probabilities under each of which (N, M) is a Cox pair on E. Prove that if $P_0 \perp P_1$ on $\mathcal{F}^N$, then $P_0 \perp P_1$ on $\mathcal{F}^M$.

7.19 Let E be a compact set, let $\mathcal{L}_0$ and $\mathcal{L}_1$ be equivalent probabilities on $\mathbf{M}_d$, let M be the coordinate random measure, and let M_μ^i be the unreduced Palm processes of M under P_i. Prove that

$$\frac{E[e^{-M_\mu^0(E)}(d\mathcal{L}_1/d\mathcal{L}_0)(M_\mu^0)]}{E[e^{-M_\mu^0(E)}]} = \frac{E[e^{-M_\mu^1(E)}]}{E[e^{-M_\mu^1(E)}(d\mathcal{L}_0/d\mathcal{L}_1)(M_\mu^1)]}.$$

7.20 Prove Lemma 7.22.

7.21 Consider once more the disruption problem of Exercise 7.13. For fixed $\alpha > 0$, develop for each t level-α likelihood ratio test of the "hypothesis" $H_0 : T \le t$, given the observations $\mathcal{F}_t^N$. That is, define a critical region $\Gamma \in \mathcal{F}_t^N$ such that on the event Γ, $P\{T > t | \mathcal{F}_t^N\} \le \alpha$ almost surely.

7.22 In the context of Theorem 7.25, calculate the covariances in (7.55)–(7.60).

7.23 (Continuation of Exercise 7.13) Consider the disruption problem of Exercise 7.13 with a and b known but G an exponential distribution with unknown parameter λ.
a) Show that consistent estimation of λ from a single realization of a generic copy N, even observed over $[0, \infty)$, is not possible.
b) Devise strongly consistent and asymptotically normal estimators $\hat{\lambda}$ (as $n \to \infty$) given as data i.i.d. copies N_i, each observed over $[0, 1]$. [*Hint:* One choice uses the identity $E[N_i(1)] = b + (b - a)(1 - e^{-\lambda})/\lambda.$]
c) Construct pseudo-state estimators $\hat{P}\{T_{n+1} \le t | \mathcal{F}_t^{N_{n+1}}\}$ by replacing λ in the stochastic differential equation derived in Exercise 7.13 by $\hat{\lambda}$ (based on $N_1, \ldots, N_n$) and examine asymptotics as $n \to \infty$.

7.24 Work explicitly through the calculations associated with Theorem 7.16 under the assumption that the directing intensity is a Markov process with finite state space and generator A.

7.25 a) Prove Proposition 7.33.

b) Calculate the Laplace functional of the Neyman-Scott cluster process N of that Proposition.

7.26 Let N be a Cox process on $\mathbf{R}_+$ with directing intensity X, a Markov process with state space $\{0, \alpha\}$, where $\alpha > 0$ is unknown, and with known, positive transition rates a $(0 \to \alpha)$ and b $(\alpha \to 0)$. Construct strongly consistent and asymptotically normal estimators of α given the observations $\mathcal{F}_t^N$, $t \geq 0$, and verify these properties.

NOTES

Expository accounts of Cox processes are given by Grandell (1976), Krickeberg (1972), Matthes *et al.* (1978), and Snyder (1975); only the latter contains applications. As described in Chapter 1, a fundamental theoretical role of Cox processes concerns thinnings (see Kallenberg, 1975a; Mecke, 1968). A plethora of modeling applications — central to each is a physical interpretation of the directing measure or intensity — includes aerosol pollutants (Rohde and Grandell, 1981), optical signal transmission (Karp and Clark, 1970; Snyder, 1975), and precipitation (Smith and Karr, 1983). Previous treatments of inference in some generality are Grandell (1976) and Krickeberg (1982); each concentrates on state estimation.

Section 7.1

As intimated above, there are no previous works on statistical estimation for Cox processes on general spaces, in part because Proposition 7.2 [which merely combines (7.2) with Theorems 4.9 and 4.10] is as far as Cox process structure can be exploited without introducing additional assumptions. Except for Poisson processes, only mixed Poisson processes have been studied in detail; much remains to be done to identify and analyze other worthwhile special cases. Theorems 7.4 and 7.5, taken from Karr (1984b), are the most general results available in the i.i.d. case; the estimators have been modified slightly to yield easier proofs and less complicated covariances. Albrecht (1982b), Simar (1976) and Tucker (1964) treat special cases of estimation for mixed Poisson processes. Lemma 7.3 highlights a key difference between Poisson and Cox processes.

Section 7.2

State estimation, motivated originally by applications, is *the* fundamental inference problem for Cox processes from a theoretical viewpoint as well. The main result, Theorem 7.6, is due to Karr (1985a); along with Theorem 4.23 it emphasizes the central role of Palm distributions for state estimation. However, despite its elegance it is in some sense too general to be really useful. The special cases of Theorems 7.9 and 7.15, Propositions 7.10 and 7.12 and Example 7.14, which are more immediately applicable, appear in Karr (1983). There, Theorem 7.9 is the most general result and is proved by appeal to the Bayes' theorem of Kallianpur and Striebel (1968) (see also Krickeberg, 1982). Special versions of Proposition 7.10 and Theorem 7.15

are in Grandell (1976), Liptser and Shiryaev (1978) and Snyder (1975). Maziotto and Szpirglas (1980) analyze recursive state estimation for planar mixed Poisson processes. Theorem 7.16 is due to Grandell (1976); the formally new Theorem 7.17 uses exactly the same proof. An infinitesimal/recursive version for general spaces is Karr (1983, Theorem 3.11). Further special cases of state estimation are considered in Boel and Beñes (1980) (diffusion process intensities); Freed and Shepp (1982) (estimation of X_0 for the process $X * N$ of Section 6.3); Grandell (1976), Karp and Clark (1970), and Macchi and Picinbono (1972) (photon counts); and Snyder (1972a, 1975) (engineering applications). Snyder (1972a) is the seminal early work.

Section 7.3

Even though none of the results here is especially deep, to our knowledge this is the first systematic study of likelihood ratios and absolute continuity/singularity for Cox processes on general spaces. Analogues of Theorem 7.24, the separation theorem, appear in Brémaud (1981), Brémaud and Jacod (1077), and Snyder (1975); in the first and third one can locate interpretations and implications of "separation" (see also Proposition 7.32). The intractable problem of discriminating between Poisson and mixed Poisson processes is discussed by Albrecht (1982a) and Sundt (1982); the root of the difficulties is nonergodicity of mixed Poisson processes (see Proposition 7.12).

Section 7.4

All of the results pertaining to mixed Poisson processes are due to Karr (1984b), but, as in Theorems 7.4 and 7.5, estimators and covariances have been simplified. In the proof of Theorem 7.25 are details needed to flesh out that of Theorem 7.5. The only extant treatment of combined inference and state estimation for general Cox processes is Karr (1985a).

Section 7.5

As observed in the text, the message of the section is negative: it seems impossible to exploit Cox structure within the context of martingale inference. The relevant characterization of Cox processes, $\mathcal{H}_0$-measurability of the stochastic intensity or compensator (Brémaud, 1975a, 1981; see also Exercise 2.10) simply does not help. Indeed, historical development of the theory confirms this. Results originally proved for Cox processes using specialized methods (e.g., Snyder, 1972a) have been extended to point processes with much more general stochastic intensities, and in some sense one cannot hope to exploit what has become an unessential hypothesis. As a consequence, Section 5.4, particularly Theorem 5.30, is the main antecedent of this section. Theorem 7.30, in some ways the culmination of the pre-martingale era, is due to for Cox processes Snyder (1972a). Dynkin's formula, which yields the semimartingale representation there, is an important result for Markov processes and is explained, for example, in Karlin and Taylor (1975). That Theorem 6.22 is a special case of Theorem 7.30 is apparent; just *how* special is initially less so. Closing an infinite system of stochastic differential equations by writing it as a single equation for a measure-valued process, as in Theorem 7.31, is elegant mathematically but of dubious practical value. Kunita (1971) and Fuisaki *et al.* (1972) are related works.

Neyman and Scott (1972) contains a discussion of the cluster process bearing their names; applications to modeling of precipitation is studied in Kavvas and Delleur (1981) and Smith and Karr (1985).

Exercises

7.2 For more on this characteristic feature of mixed Poisson processes, see Feigin (1979), Kallenberg (1975b, 1983) (where "conditionally uniform" is termed "symmetrically distributed"), Matthes *et al.* (1978), Mecke (1976), and Puri (1982).

7.4 See Theorem 8.24 and Kingman (1963), the original source.

7.7 Lost efficiency is an inevitable consequence of having "inappropriate" observations.

7.10 See Karr (1985a).

7.12 See Karr (1983, 1984a) for additional aspects.

7.13 Among references on this venerable problem are Brémaud (1980), Davis *et al.* (1975), Galtchouk and Rozovskii (1971), Liptser and Shiryaev (1978), Telksnys (1986) and Wan and Davis (1977).

7.26 See also Exercise 3.17.

8
Nonparametric Inference for Renewal Processes

Although of considerable intrinsic interest, renewal processes are important as well because of regenerative processes, that is, continuous-time stochastic processes that begin anew at embedded random times forming a renewal process. Usefulness of renewal processes stems principally from asymptotic properties, not only their own but also of solutions to renewal equations. Hence in this chapter there is some emphasis on functionals of renewal processes (e.g., recurrence time processes) and of interarrival time distributions (e.g., renewal measures).

Statistical inference for renewal processes is based on the property that interarrival times are i.i.d., and requires few tools beyond those developed in earlier chapters. For state estimation, as will emerge, central roles are played by the backward and forward recurrence time processes.

We begin with a prefatory discussion of key concepts and results, but it is in no sense complete; consult the chapter notes for references. Let $N = \sum_{i=0}^{\infty} \varepsilon_{T_i}$ be a simple point process on $\mathbf{R}_+$ with arrival times $T_1 < T_2 < \cdots$, and let $U_i = T_i - T_{i-1}$, $i \geq 1$, be the interarrival times. From Definition 1.6, N is a *renewal process* if the U_i are i.i.d. In a renewal process, at each arrival time T_i the arrival counting process $N_t = N((0,t]) = \sum_{i=1}^{\infty} \mathbf{1}(T_i \leq t)$ renews itself probabilistically: its future increments form another renewal process. (Therefore, the T_i are sometimes called renewal times.) The distribution F of the interarrival times is termed the *interarrival distribution*, and is the sole determinant the law of N. Because N is simple, $F(0) = 0$, and we assume throughout that $F(\infty) = 1$, although often neither of these assumptions is imposed when probabilistic properties of renewal processes are studied. The mean and variance of F are denoted by m and σ^2, and when it exists the hazard function $F'/(1-F)$ by h.

Given the i.i.d. structure of the interarrival times, asymptotic properties of the arrival time sequence are manifested as well in the counting process.

Proposition 8.1. Let N be a renewal process with interarrival distribution F.

a) If $m < \infty$, then $\lim_{t\to\infty} N_t/t = 1/m$ almost surely;

b) If $\sigma^2 < \infty$, then $\sqrt{t}[N_t/t - 1/m] \overset{d}{\to} N(0, \sigma^2/m^3)$. $\square$

In particular, Proposition 8.1 provides apparatus to estimate the mean interarrival time from observation of the counting process (N_t), although if it were used, for example, to construct hypothesis tests or confidence intervals, one would also have to estimate the variance (Exercise 8.9).

Associated with a renewal process are many functionals, the two most fundamental of which are the backward and forward recurrence time processes, whose values at t are the time elapsed since the last arrival before t and the time from t until the first arrival thereafter. More precisely, the *backward recurrence time* at t is

$$V_t = t - T_{N_t}, \tag{8.1}$$

which is zero if $N_t = 0$, and the *forward recurrence time* at t is

$$W_t = T_{N_t+1} - t. \tag{8.2}$$

Pictorially, as shown in Figure 8.1, $t \to V_t$ is an "ascending sawtooth" and $t \to W_t$ a "descending sawtooth." For state estimation their roles are crucial and complementary: to predict the future of a renewal process one needs concerning the past only the cumulative number of arrivals and the current value of the backward recurrence time (i.e., N_t and V_t), while once the forward recurrence time is predicted, the remainder of the future can be predicted deterministically.

The function (also viewed as a measure) $R(t) = 1 + E[N_t]$ is the *renewal function*, and is expressible in terms of the interarrival distribution F as

$$R(t) = \sum_{k=0}^{\infty} F^k(t) = \sum_{k=0}^{\infty} P\{T_k \le t\}, \tag{8.3}$$

where F^k denotes the k-fold convolution of F with itself and $F^0 \equiv 1$ is the distribution function of $T_0 \equiv 0$. The renewal function is analogous to the potential measure of a Markov process, but its most important role is in the solution of renewal equations (Theorem 8.4).

Here are elementary properties.

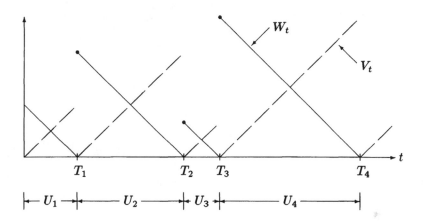

Figure 8.1: Recurrence Time Processes

Proposition 8.2. Let R be the renewal function associated with the renewal process N. Then $R(t) < \infty$ for all $t < \infty$ and $\lim_{t \to \infty} R(t)/t = 1/m$, even if $m = \infty$, in which case $1/m = 0$. $\square$

We next consider renewal equations. Denote by **B** the set of functions on $\mathbf{R}_+$ that are bounded over bounded intervals. Given a locally finite measure ν on $\mathbf{R}_+$ and $f \in \mathbf{B}$, the convolution $\nu * f(t) = \int_0^t \nu(du) f(t - u)$ also belongs to **B**. A *renewal equation* associated with F is an equation

$$f = g + F * f, \tag{8.4}$$

where $g \in \mathbf{B}$ is known and $f \in \mathbf{B}$ is unknown. The following example not only illustrates the typical genesis of a renewal equation but also points the way to the general solution.

Example 8.3. Let W be the forward recurrence time process defined by (8.2); we wish to calculate the distribution of W_t for each t. Let y be fixed and put $f(t) = P\{W_t > y\}$. Then

$$
\begin{aligned}
P\{W_t > y\} &= P\{W_t > y, T_1 > t\} + P\{W_t > y, T_1 \leq t\} \\
&= P\{T_1 > t + y\} + E\left[P\{W_t > y | T_1\} \mathbf{1}(T_1 \leq t)\right],
\end{aligned}
$$

since once $T_1 > t$, $W_t > y$ if and only if $T_1 > t + y$. Now we make the crucial *renewal argument*: $\tilde{N} = \sum_{k=2}^{\infty} \varepsilon_{T_k - T_1}$ is a renewal process independent of T_1, with the same law as N. Therefore (with $\tilde{W}$ the forward recurrence time for $\tilde{N}$), on the event $\{T_1 \leq t\}$, $P\{W_t > y|T_1\} = P\{\tilde{W}_{t-T_1} > y\} = f(t - T_1)$, and hence f satisfies the renewal equation (8.4) with $g(t) = 1 - F(t + y)$.

But we may calculate f directly: using (8.3),

$$
\begin{aligned}
P\{W_t > y\} &= P\{N_{t+y} - N_t = 0\} \\
&= \sum_{k=0}^{\infty} P\{T_k \leq t, T_{k+1} > t + y\} \\
&= \sum_{k=0}^{\infty} \int_0^t F^k(du)[1 - F(t + y - u)] \\
&= \int_0^t R(du)[1 - F(t + y - u)] = R * g(t). \qquad (8.5)
\end{aligned}
$$

Thus for this specific case the renewal equation is solved by $f = R * g$. ☐

The same is true in general.

Theorem 8.4. Given $g \in \mathbf{B}$, the unique solution in $\mathbf{B}$ to the renewal equation (8.4) is $f = R * g$. ☐

For a Poisson process with rate λ, $R(t) = 1 + \lambda t$, and substitution in (8.5) — in such computations one must not neglect the atom of the renewal measure at the origin — gives $P\{W_t > y\} = e^{-\lambda y}$ regardless of t, a property that characterizes Poisson processes among renewal processes (Exercise 8.24).

In most cases, though, the renewal function cannot be calculated explicitly, so one is led to hope, based also on "nice" asymptotic behavior of renewal processes, that solutions of renewal equations behave similarly.

The hope is not unfounded, as the *key renewal theorem* demonstrates.

Theorem 8.5. Assume that $g \in \mathbf{B}$ is directly Riemann integrable and that F is nonarithmetic. Then

$$
\lim_{t \to \infty} R * g(t) = (1/m) \int_0^\infty g(u)du; \qquad (8.6)
$$

the possibility $m = \infty$ is not excluded. ☐

Choosing $g = \mathbf{1}_{[0,h]}$ in (8.6) yields the *Blackwell renewal theorem*:

$$
\lim_{t \to \infty} [R(t + h) - R(t)] = h/m, \qquad (8.7)
$$

a strengthened form of the second statement in Proposition 8.2. The equations (8.6) and (8.7) are in fact equivalent.

The function $g(t) = 1 - F(t + y)$ in (8.5) is directly Riemann integrable provided that $m < \infty$, and if F is nonarithmetic, then

$$
\lim_{t \to \infty} P\{W_t > y\} = (1/m) \int_y^\infty [1 - F(u)]du = H_F(y). \qquad (8.8)
$$

This distribution arises (Exercise 1.31) in connection with stationary renewal processes: if N is modified so that U_1 has distribution H_F, then $N^* = \sum_{n=0}^{\infty} \varepsilon_{T_n}$ (a special case of a *delayed* renewal process) is a stationary point process on $\mathbf{R}_+$, additional aspects of which are treated in Chapter 9.

Finally, we introduce the notion of a regenerative process.

Definition 8.6. Let X be a (càdlàg) process taking values in a LCCB space E. Then X is a *regenerative process* if there exist $\mathcal{F}^X$-stopping times $0 = T_0 < T_1 < T_2 < \cdots$ such that

a) $N = \sum \varepsilon_{T_n}$ is a renewal process;

b) For each n, k, $t_1, \ldots, t_k \geq 0$ and (bounded, continuous) function f on E^k, $E[f(X_{T_n+t_1}, \ldots, X_{T_n+t_k})|\mathcal{F}^X_{T_n}] = E[f(X_{t_1}, \ldots, X_{t_k})]$.

At each *regeneration time* T_n, the future of X becomes a probabilistic replica independent of the past. Given a suitable definition of the excursions $Z_{n+1}(t) = X_{T_n+t}$, $0 \leq t \leq U_{n+1}$, the pairs (U_n, Z_n) are i.i.d., and hence we can study X using the marked point process $\bar{N} = \sum \varepsilon_{(T_n, Z_n)}$.

Other than a renewal process the simplest example of a regenerative process is an irreducible Markov process with finite state space, which is regenerative with respect to the times of return to its initial state (Exercise 8.6).

Renewal theorems apply to regenerative processes. For example, with A a fixed subset of E, the function $t \to P\{X_t \in A\}$ satisfies the renewal equation

$$P\{X_t \in A\} = P\{X_0 \in A, T_1 > t\} + \int_0^t F(du)P\{X_{t-u} \in A\},$$

where F is the distribution of the times between regenerations. Consequently, if F fulfills the hypotheses of Theorem 8.5, then with m the mean of F,

$$\lim_{t \to \infty} P\{X_t \in A\} = (1/m)\int_0^\infty P\{X_0 \in A, T_1 > u\}du, \tag{8.9}$$

a limit theorem of remarkable power and generality.

This concludes our survey of renewal and regenerative processes. The remainder of the chapter treats inference and is organized in the following manner. Section 1 is devoted to estimation of the interarrival distribution and, by substitution, functionals such as the renewal function and the distribution H_F of (8.8), given as data different kinds of observations of a single realization of a renewal process. For synchronous observations, empirical process methods from Chapter 4 apply; we do not dwell on this case. Mainly, we treat asynchronous data representing "real-time" observation; here empirical process techniques apply with minor modifications. Finally, partly leading to Chapter 10, we consider Poisson samples of a renewal process: observations are the values of the counting process (N_t) at arrival times in a Poisson process independent of N. We also examine martingale methods; when the hazard function h exists, N admits $\mathcal{F}^N$-stochastic intensity $\lambda_t = h(V_{t-})$.

Section 2 is a study of hypothesis testing, with analysis of absolute continuity, likelihood functions and local asymptotic normality. Thereafter we discuss some fairly specific tests concerning structural characteristics of the interarrival distribution. Tests whether a renewal process is a Poisson or Cox process are also treated; they lead to interesting characterizations.

State estimation and combined statistical inference and state estimation form the topic of Section 3. Finally, Section 4 describes techniques for statistical estimation for Markov renewal processes based on complete or censored observations.

8.1   Estimation of the Interarrival Distribution

Suppose that N is a renewal process under P_F with interarrival distribution F, which we take to be completely unknown, yielding the full nonparametric model $\mathcal{P} = \{P_F\}$. Here our primary focus is on nonparametric estimation of F given various kinds of observations of N. Parametric models, for example where the interarrival times have a gamma distribution (Examples 3.4 and 3.9) are ordinarily analyzed using specialized techniques; the reliability literature (see the chapter notes) is a good source for specifics. Because interarrival times are positive, classical procedures for normally distributed data are not applicable.

For synchronous data, represented by σ-algebras $\mathcal{F}^N_{T_n} = \sigma(T_1, \ldots, T_n) = \sigma(U_1, \ldots, U_n)$ and corresponding to observation of N until the nth arrival, the empirical distribution functions

$$\hat{F}(x) = (1/n)\textstyle\sum_{i=1}^n \mathbf{1}(U_i \leq x) \tag{8.10}$$

are nonparametric maximum likelihood estimators of F, and standard results apply. (See Chapter 4 or, e.g., Gaenssler, 1984, or Pollard, 1984.)

In particular, we have the following asymptotic properties.

Proposition 8.7. Let $\hat{F}$ be given by (8.10). Then under P_F,
a) $\sup_{x \in \mathbf{R}_+} |\hat{F}(x) - F(x)| \to 0$;
b) As random elements of $D[0, \infty)$,

$$\{\sqrt{n}[\hat{F}(x) - F(x)] : x \in \mathbf{R}_+\} \overset{d}{\to} \{G(x) : x \in \mathbf{R}_+\}, \tag{8.11}$$

where G is a Gaussian process with covariance function $R(x, y) = F(x)[1 - F(y)]$ for $0 \leq x \leq y$. $\square$

Rarely, however, does one observe synchronous data; rather, the arrival counting process N is observed in "real time," resulting in the asynchronous data $\mathcal{F}_t^N = \sigma(N_u : 0 \leq u \leq t)$. The obvious analogues of the estimators (8.10) are the empirical distribution functions

$$\hat{F}(x) = (1/N_t)\sum_{i=1}^{N_t} \mathbf{1}(U_i \leq x), \tag{8.12}$$

defined to be identically 1 on $\{N_t = 0\}$. We can establish an analogue of Proposition 8.7 without difficulty, but first we remark on some additional aspects of (8.12).

Not all of the information contained in $\mathcal{F}_t^N$ is used in forming the estimator $\hat{F}$: the ongoing interarrival interval is known to exceed the current backward recurrence time V_t, but this is ignored in (8.12). Therefore, in particular, the $\hat{F}$ are not maximum likelihood estimators. Partial information about U_{N_t+1} *is* incorporated in the modified estimators

$$
\begin{aligned}
\hat{F}^*(x) &= \frac{N_t}{N_t + 1}\hat{F}(x), \quad x \leq V_t \\
&= \hat{F}(x), \quad\quad\quad x > V_t,
\end{aligned}
\tag{8.13}
$$

whose asymptotic behavior is identical to that of the $\hat{F}$. (In Section 8.4 partial observations of Markov renewal processes are incorporated similarly.) A more serious — and inescapable — shortcoming is the *length bias* in the estimators $\hat{F}$: since N has been observed over an interval of length t, there are no observed U_i with values exceeding t, so that $\hat{F}(x) = 1$ for all $x \geq t$.

Concerning asymptotics, strong uniform consistency is nearly immediate given Proposition 8.7a).

Proposition 8.8. Let $\hat{F}$ be given by (8.12). Then almost surely under each P_F, $\sup_{x \in \mathbf{R}_+} |\hat{F}(x) - F(x)| \to 0$. $\square$

To prove a central limit theorem, we must reconcile the relative rates of growth of t in (8.12) and $n \ (= N_t)$ in (8.10). This is done using the strong law of large numbers, which entails an assumption that the mean interarrival time be finite.

Theorem 8.9. Assume that $m = E_F[U_i] < \infty$. Then under P_F,

$$\{\sqrt{t}[\hat{F}(x) - F(x)] : x \in \mathbf{R}_+\} \xrightarrow{d} \{G(x) : x \in \mathbf{R}_+\} \tag{8.14}$$

as random elements of $D[0, \infty)$, where G is a Gaussian process with covariance function $R(x, y) = mF(x)[1 - F(y)]$ for $0 \leq x \leq y$.

Proof: In view of finiteness of m and Proposition 8.1,

$$\sqrt{t}[\hat{F} - F](x) = \sqrt{t/N_t}(1/\sqrt{N_t})\sum_{i=1}^{N_t}[\mathbf{1}(U_i \leq x) - F(x)]$$
$$\cong \sqrt{m}(1/\sqrt{N_t})\sum_{i=1}^{N_t}[\mathbf{1}(U_i \leq x) - F(x)]. \qquad (8.15)$$

On the interval $[T_n, T_{n+1})$, the process $A_t(x) = (1/\sqrt{N_t})\sum_{i=1}^{N_t}[\mathbf{1}(U_i \leq x) - F(x)]$ satisfies $A_t(x) = (1/\sqrt{n})\sum_{i=1}^{n}[\mathbf{1}(U_i \leq x) - F(x)]$, and hence (Serfozo, 1975), utilizing also the fact that $T_n/n \to m$,

$$\lim_{t\to\infty} A_t(\cdot) \overset{d}{=} \lim_{n\to\infty}(1/\sqrt{n})\sum_{i=1}^{n}[\mathbf{1}(U_i \leq (\cdot)) - F(\cdot)].$$

Appealing to Proposition 8.7b) and remembering the factor $\sqrt{m}$ in (8.15), we arrive at (8.14). ∎

If F admits hazard function h, then martingale concepts are useful for interpretation of the estimators even though results from Chapter 5 do not apply in our single-realization setting. Let $L_t = V_{t-}$ be the *left*-continuous backward recurrence time process, which is therefore $\mathcal{F}^N$-predictable. The functions $t \to L_t$ and $t \to V_t$ differ only at the points T_n, where $L_t = U_n$ but $V_t = 0$. In some contexts the difference is immaterial but in others, for example Theorem 8.17, it is crucial.

The stochastic intensity of N is then derived easily.

Proposition 8.10. If F admits hazard function h, then the $(P_F, \mathcal{F}^N)$-stochastic intensity of N is $\lambda_t = h(L_t)$. □

Consider now the identity

$$F(x) = E\left[\int_0^{U_i \wedge x} h(y)dy\right]. \qquad (8.16)$$

With x fixed the process $(\hat{F}_t(x))_{t\geq 0}$ satisfies

$$N_t\hat{F}_t(x) = \sum_{i=1}^{N_t}\mathbf{1}(U_i \leq x) = \int_0^t \mathbf{1}(\lambda_u \leq x)dN_u, \qquad (8.17)$$

so that $N_t\hat{F}_t(x)$ is the martingale estimator of the process

$$B_t(F, x) = \int_0^t \mathbf{1}(\lambda_u \leq x)du = \int_0^{L_t \wedge x} h(u)du + \sum_{i=i}^{N_t}\int_0^{U_i \wedge x} h(u)du.$$

Therefore, $\hat{F}_t(x)$ estimates $B_t(F, x)/N_t$. By (8.16) the latter converges almost surely to $F(x)$, and hence we have a martingale interpretation of the estimators.

We next consider estimation of functionals of F by corresponding functionals of $\hat{F}$, remaining with asynchronous data. The mean of the empirical distribution function $\hat{F}$ is

$$\hat{m} = (1/N_t)\sum_{i=1}^{N_t}U_i = (t - V_t)/N_t, \qquad (8.18)$$

whose asymptotic properties are those of the estimators t/N_t, which are in turn straightforward consequences of Proposition 8.1.

For estimation of the renewal function the natural choice is $\hat{R} = \sum_{k=0}^{\infty} \hat{F}^k$, with $\hat{F}$ given by (8.12); these estimators have the following properties.

Proposition 8.11. For each $x_0 < \infty$, $\sup_{x \leq x_0} |\hat{R}(x) - R(x)| \to 0$ almost surely with respect to P_F.

Proof: Proposition 8.8 reduces this proposition to the analytical assertion that if F_n, F are distribution functions on $(0, \infty)$ with renewal functions R_n, R and if $F_n \to F$ uniformly, then $R_n \to R$ uniformly on compact sets. We outline one of several arguments. Let $\tilde{F}_n(u) = \int F_n(dx)e^{-ux}$, $\tilde{F}$, $\tilde{R}_n$ and $\tilde{R}$ be the Laplace transforms. Then by (8.3) and the convergence $F_n \to F,$, for each u, $\tilde{R}(u) = 1/[1 - \tilde{F}(u)] = \lim_{n \to \infty} 1/[1 - \tilde{F}_n(u)] = \lim_{n \to \infty} \tilde{R}_n(u)$; therefore, $R_n \to R$ vaguely. One then applies standard reasoning used to deduce locally uniform convergence of monotone functions from pointwise convergence. ∎

Unfortunately, this line of attack does not yield a useful central limit theorem. Thus while solutions to renewal equations can be estimated consistently, less can be said about the estimation error.

A compromise is to estimate limits of solutions to renewal equations. Under the assumptions of Theorem 8.5, if f is the solution of the renewal equation (8.4), then $\lim_{t \to \infty} f(t) = (1/m) \int_0^{\infty} g$, so since m can be estimated using Proposition 8.1 or (8.18), whenever g, which usually depends on f, and may depend as well on the interarrival distribution F, can be estimated, so can the functional $(1/m) \int_0^{\infty} g$. Because of the variety of functions involved, it is difficult to develop a comprehensive theory; instead, we illustrate with a specific case that arises again in Section 3.

Let H_F be the P_F-limit distribution of the forward recurrence time W_t of N. By (8.8), provided that m is finite and F nonarithmetic, and given the observations $\mathcal{F}_t^N$, H_F can be estimated by substitution:

$$\hat{H}(y) = (1/\hat{m})\int_y^{\infty}[1 - \hat{F}(u)]du, \tag{8.19}$$

with $\hat{F}$ given by (8.12) and $\hat{m}$ by (8.18). In the following result we impose rather stringent hypotheses on F in order to simplify the proof; the practical impact is minor, however.

Theorem 8.12. Assume that F is continuous and that there is $x_0 < \infty$ such that $F(x_0) = 1$. Then
a) Almost surely with respect to P_F, $\sup_{y \in \mathbf{R}_+} |\hat{H}(y) - H_F(y)| \to 0$;

b) As $t \to \infty$,

$$\{\sqrt{t}[\hat{H}(y) - H_F(y)] : 0 \le y \le x_0\} \xrightarrow{d} \{J(y) : 0 \le y \le x_0\},$$

where

$$J(y) = (1/m^2)\int_y^{x_0}[1 - F(u)]du \int_0^{x_0} G(v)dv - (1/m)\int_y^{x_0} G(u)du,$$

with G the Gaussian process of Theorem 8.9.

Proof: a) By Proposition 8.1 and (8.18), $\hat{m} \to m$ almost surely, while

$$\sup_{y \in \mathbf{R}_+} \left| \int_y^\infty [1 - \hat{F}(u)]du - \int_y^\infty [1 - F(u)]du \right|$$

$$= \sup_{y \in x_0} \left| \int_y^{x_0}[1 - \hat{F}(u)]du - \int_y^{x_0}[1 - F(u)]du \right|$$

$$\le x_0 \sup_{x \le x_0} |\hat{F}(x) - F(x)|,$$

which converges to zero by Proposition 8.8.

b) In view of a), within error converging in probability to zero uniformly in y,

$$\sqrt{t}[\hat{H}(y) - H_F(y)]$$

$$= \sqrt{t}\left\{(1/\hat{m} - 1/m)\int_y^{x_0}[1 - F] + (1/\hat{m})\int_y^{x_0}[\hat{F} - F]\right\}$$

$$\cong (1/m^2)\int_y^{x_0}[1 - F]\int_0^{x_0}\sqrt{t}[\hat{F} - F] + (1/m)\int_y^{x_0}\sqrt{t}[\hat{F} - F]$$

$$\xrightarrow{d} (1/m^2)\int_y^{x_0}[1 - F]\int_0^{x_0} G - (1/m)\int_y^{x_0} G$$

by Theorem 8.9 and the continuous mapping theorem. It follows by continuity of F that G is continuous [in fact, $G(x) = B(F(x))$, where B is a Brownian bridge] and hence the integrals involving it are well defined. ∎

Our analysis of asynchronous data concludes with application of marked point processes and empirical process methods to regenerative processes. Let X be a regenerative process with state space E and renewal process $N = \sum \epsilon_{T_n}$ of regeneration times. Let a death state Δ be adjoined to E (Blumenthal and Getoor, 1968) and for each n define a random element Z_{n+1} of the function space $D = D(E \cup \{\Delta\}, [0, \infty))$ by

$$\begin{aligned} Z_{n+1}(t) &= X_{T_n+t} & t < U_{n+1} \\ &= \Delta & t \ge U_{n+1}. \end{aligned}$$

The function Z_{n+1} is the *excursion* of X between the regeneration times T_n and T_{n+1}. The Z_n determine X; in particular, $U_{n+1} = \inf\{t : Z_{n+1}(t) = \Delta\}$.

That the Z_n are i.i.d. random elements of D, which is certainly plausible, is confirmed by the following result.

Proposition 8.13. Let X be regenerative, with excursions Z_n. Then the Z_n are i.i.d. random elements of D whose distribution determines the law of X. Conversely, given i.i.d. random elements Z_n of D such that the death times $U_n = \inf\{t : Z_n(t) = \Delta\}$ are finite almost surely, then with $T_n = \sum_{i=1}^{n} U_i$ and $X_t = Z_{n+1}(t - T_n)$, $t \in [T_n, T_{n+1})$, X is regenerative with regeneration times T_n. $\square$

We study X via the marked point process $\bar{N} = \sum_{n=0}^{\infty} \epsilon_{(T_n, Z_n)}$, in which T_n is marked by the just-completed excursion Z_n. Observation of X over $[0, t]$ is essentially equivalent to that of $\bar{N}_t = \sum_{n=0}^{\infty} \mathbf{1}(T_n \le t)\epsilon_{(T_n, Z_n)}$, and so a plausible estimator of the law $\mathcal{L}$ of the excursions is

$$\dot{\mathcal{L}} = (1/N_t)\sum_{n=1}^{N_t} \epsilon_{Z_n}.$$

As with the estimators (8.12), information about "in progress" excursions is discarded.

The space D is a complete, separable metric space and empirical process asymptotics hold for the $\hat{\mathcal{L}}$, although in the central limit theorem one must again reconcile different sampling rates.

Theorem 8.14. Let X be a regenerative process with state space E and excursion sequence (Z_n), and let $\mathcal{L}$ be the law of the Z_n. Then

a) If $\mathcal{C}$ is a Vapnik-Chervonenkis class of (Borel) subsets of D, then $\sup_{C \in \mathcal{C}} |\hat{\mathcal{L}}(C) - \mathcal{L}(C)| \to 0$ almost surely;

b) If $m = E_F[U_i]$ is finite, and if $\mathcal{C}$ is a class of subsets of D satisfying appropriate regularity conditions, then there is a continuous Gaussian process $\{G(C) : C \in \mathcal{C}\}$ such that $\sqrt{t}[\hat{\mathcal{L}} - \mathcal{L}] \xrightarrow{d} G$ as random elements of the metric space V described in Section 4.1; G has covariance function $R(C_1, C_2) = m[\mathcal{L}(C_1 \cap C_2) - \mathcal{L}(C_1)\mathcal{L}(C_2)]$. $\square$

Our discussion of estimation concludes with Poisson sampling of renewal processes, a special case of which appeared in Exercise 6.18. Suppose that under P_F, $N = \sum_{n=1}^{\infty} \epsilon_{T_n}$ is a renewal process with interarrival distribution F, that $\tilde{N} = \sum_{n=1}^{\infty} \epsilon_{S_n}$ is a Poisson process with (known) rate 1, and that N and $\tilde{N}$ are independent. The observations are the values of the counting process (N_t) at the arrival times S_n of $\tilde{N}$, that is, the *Poisson samples* $N_{S_1}, N_{S_2}, \ldots$, or equivalently the numbers $Y_k = N_{S_k} - N_{S_{k-1}}$ of arrivals in N during interarrival intervals in $\tilde{N}$. (Additional models of this ilk are analyzed in Chapter 10.)

The goal is to estimate the interarrival distribution F from the Y_k observed in real time, so that the σ-algebra corresponding to observation over

$[0, t]$ is $\mathcal{H}_t = \sigma((S_k, Y_k)\mathbf{1}(S_k \leq t) : k \geq 0)$. But since $\tilde{N}$ is independent of N and has rate 1, the real-time sampling rate is equal to 1, and properties established for estimators based on $\mathcal{G}_k = \sigma(Y_1, \ldots, Y_k)$ carry over with no change to analogous estimators based on $\mathcal{H}_t$. Consequently, we pursue only estimators based on the $\mathcal{G}_k$, beginning with a description of the probabilistic structure of Y_k.

Proposition 8.15. Let N, $\tilde{N}$ and (Y_k) be as described, and for each k let

$$Q(k) = \int F(du)e^{-u}u^k/k!. \tag{8.20}$$

Then under P_F, $(Y_k) \overset{d}{=} (\delta_k X_k)$, where

a) The X_k are i.i.d., with geometric distribution $Q(0)^{n-1}[1 - Q(0)]$;

b) For each k, $\delta_k = \mathbf{1}(R_n = k$ for some $n)$, where (R_n) is a discrete-time delayed renewal process: $R_1, R_2 - R_1, R_3 - R_2, \ldots$ are independent; $P\{R_1 = \ell\} = Q(\ell - 1)$, and for $j \geq 2$, $P\{R_j - R_{j-1} = \ell\} = Q(\ell)/[1 - Q(0)]$;

c) The sequences (X_k) and (δ_k) are independent.

One can interpret the construction via point processes: if $N^* = \sum_{n=1}^{\infty} \epsilon_{R_n}$ is viewed as a point process on $\mathbf{N}$, then $\delta_k = N^*(\{k\})$ and Y is the integer-valued random measure obtained by integrating X with respect to N^*.

Proof of Proposition 8.15: Introduce the dual process $\tilde{Y}_k = \tilde{N}_{T_k} - \tilde{N}_{T_{k-1}}$, the number of points of $\tilde{N}$ in the interarrival interval $(T_{k-1}, T_k]$ of N. Because $\tilde{N}$ is Poisson and independent of N, the $\tilde{Y}_k$ are i.i.d. with the mixed Poisson distribution Q of (8.20). The sequences (Y_k) and $(\tilde{Y}_k)$ determine each other pathwise (a picture is recommended!), and in particular, given positive integers $q, r_1, \ldots, r_q$ and $k_1, \ldots, k_q$, and with $R_\ell = \sum_{i=1}^{\ell} r_i$ and $K_\ell = \sum_{i=1}^{\ell} k_i$,

$$P\{Y(R_\ell) = k_\ell, \ell = 1, \ldots, q; Y_j = 0 \text{ for all other } j < R_q\}$$
$$= P\{\tilde{Y}_1 = r_1 - 1; \tilde{Y}(K_\ell + 1) = r_{\ell+1}, \ell = 1, \ldots, q - 1;$$
$$\tilde{Y}(K_q + 1) \geq 1; \tilde{Y}_j = 0 \text{ for all other } j \leq K_q\}$$
$$= Q(r_1 - 1)Q(r_2)\cdots Q(r_q)[1 - Q(0)]Q(0)^{K_q - q}.$$

Summed over $k_1, \ldots, k_q$, this becomes

$$P\{Y(R_\ell) \neq 0, \ell = 1, \ldots, q; Y_j = 0 \text{ for all other } j < R_q\}$$
$$= Q(r_1 - 1)[\prod_{j=1}^{q} Q(j)]/(1 - Q(0)),$$

which shows that $\delta_k = \mathbf{1}(Y_k \neq 0)$ has the structure indicated in b).

These two formulas further imply that

$$P\left\{Y(R_\ell) = k_\ell, \ell = 1, \ldots, q | \delta_j = 1 \Leftrightarrow j \in \{R_1, \ldots, R_q\}\right\}$$
$$= [1 - Q(0)]^q Q(0)^{K_q - q} = \prod_{j=1}^q Q(0)^{k_j - 1}[1 - Q(0)],$$

which proves both a) and c). ∎

To estimate F, we estimate the mixed Poisson distribution Q of (8.20), by which F is uniquely determined (see the discussion preceding Theorem 7.4). Given observations $Y_1, \ldots, Y_n$, the estimator of $Q(k)$ is

$$\hat{Q}(k) = (1/N(S_n))\sum_{\ell=1}^{N(S_n)} \mathbf{1}(\tilde{Y}_\ell = k),$$

since having observed $Y_1, \ldots, Y_n$ we can reconstruct $\tilde{Y}_\ell$ for $\ell = 1, \ldots, N(S_n)$. Relative to $(\tilde{Y}_k)$, these estimators are asynchronous samples; reasoning used already in the chapter, along with parts of Theorems 7.4 and 7.5, produces the final result of the section.

Theorem 8.16. a) Almost surely, $\sup_{k \in \mathbb{N}} |\hat{Q}(k) - Q(k)| \to 0$;
b) If $m = E_F[U_i]$ is finite, then as sequences of random variables

$$\{\sqrt{n}[\hat{Q}(k) - Q(k)] : k \in \mathbb{N}\} \overset{d}{\to} \{G(k) : k \in \mathbb{N}\},$$

where G is Gaussian with covariance function $R(k,j) = m[\mathbf{1}(k = j)Q(k) - Q(k)Q(j)]$. □

This ends our discussion of estimation; a key point is that no new tools were needed, even for Poisson samples. What is new are the renewal process context, the need to adjust for sampling rate differentials caused by asynchronous or Poisson sampling, and interest in functionals such as renewal functions and recurrence time distributions.

8.2 Likelihood Ratios and Hypothesis Tests

Like its counterparts in earlier chapters, this section does not emphasize specifics of hypothesis testing, although there are more details here than in the others. We do consider structure and asymptotic properties of likelihood ratios, structural characteristics of the interarrival distribution, and characterizations of Poisson and Cox processes among renewal processes. Most of the tests mentioned concerning the interarrival distribution address specific qualitative characteristics and stochastic orderings studied mainly in the reliability literature. These include hypotheses that the interarrival times (which in reliability are lifetimes of i.i.d. systems, each, when it fails, replaced instantaneously) are stochastically larger than those in a renewal process with interarrival distribution F_0, and hypotheses concerning structure of the hazard function of F, for example that it is increasing (the systems deteriorate with age) or constant (the process is Poisson).

We begin with analysis of likelihood functions and ratios. As in earlier chapters, the results, in practical terms, are directed at construction of likelihood ratio tests for simple hypotheses

$$
\begin{aligned}
H_0 &: F = F_0 \\
H_1 &: F = F_1,
\end{aligned}
\tag{8.21}
$$

but also have inherent interest. Our statistical model consists of those P_F for which F admits density F' and hence hazard function $h = F'/(1 - F)$. Only likelihood ratios associated with asynchronous observations are treated.

Theorem 8.17. Let P denote the probability with respect to which N is Poisson with rate 1. Then for each $t < \infty$, $P_F \ll P$ on $\mathcal{F}_t^N$ with

$$
dP_F/dP\big|_{\mathcal{F}_t^N} = e^t \left\{ \prod_{i=1}^{N_t} F'(U_i) \right\} [1 - F(L_t)],
\tag{8.22}
$$

where (L_t) is the left-continuous backward recurrence time process.

Proof: We apply Theorem 2.31 even though direct arguments are available, because the need to use predictable stochastic intensities is made strikingly clear. By Proposition 8.10, the $(P, \mathcal{F}^N)$-stochastic intensity of N is $\lambda_t = h(L_t)$, so by Theorem 2.31 $P_F \ll P$ on $\mathcal{F}_t^N$ for each t and the derivative satisfies

$$
\begin{aligned}
\log(dP_F/dP) &= \int_0^t (1 - \lambda_u)du + \int_0^t (\log \lambda_u)dN_u \\
&= t - \int_0^t h(L_u)du + \int_0^t (\log h(L_u))dN_u \\
&= t - \sum_{i=1}^{N_t} \int_0^{U_i} h(u)du + \int_0^{L_t} h(u)du + \sum_{i=1}^{N_t} \log h(U_i) \\
&= t + \sum_{i=1}^{N_t} \log F'(U_i) + \log[1 - F(L_t)]. \blacksquare
\end{aligned}
$$

Note what would have happened had we attempted to use $h(V_t)$ rather than $h(L_t)$ as stochastic intensity: since $V_{T_i} = 0$ for each t the dN-integral would have diverged to $-\infty$ and no likelihood function would have arisen; singularity of P_F and P on $\mathcal{F}_t^N$ might have been concluded erroneously.

To test the hypotheses (8.21) given observations $\mathcal{F}_t^N$ one can use the likelihood ratio

$$
L_t(F_0, F_1) = \frac{1 - F_1(L_t)}{1 - F_0(L_t)} \prod_{i=1}^{N_t} \frac{F_1'(U_i)}{F_0'(U_i)},
\tag{8.23}
$$

additional properties of which will be derived momentarily. First we observe that because there exist strongly consistent estimators of F, singularity on $\mathcal{F}_\infty^N$ always occurs for different interarrival distributions.

Proposition 8.18. Given probability distributions F_0 and F_1, either $P_{F_0} = P_{F_1}$ on $\mathcal{F}_\infty^N$ or $P_{F_0} \perp P_{F_1}$, according as $F_0 = F_1$ or $F_0 \neq F_1$. $\square$

Hence interesting behavior of likelihood ratios as $t \to \infty$ occurs only locally. Intuitively, it is clear that one should deal with a fixed null distribution F and t-dependent alternatives converging to it at rate $1/\sqrt{t}$, but it is not clear in what sense the convergence should take place. The following contiguity theorem reveals that the difference between the hazard functions should converge to zero at rate $1/\sqrt{t}$, which is not altogether surprising in view of Theorems 2.31 and 8.17. As it turns out, (8.23) is not convenient in this context and we employ a modification of (8.22) instead.

Theorem 8.19. Let F and F^* be distributions on $\mathbf{R}_+$ admitting hazard functions h and h^*, such that $h^*(x) = 0$ whenever $h(x) = 0$ (which implies that $F^* \ll F$). For each $t > 1$, let F_t have hazard function $h_t = h + (1/\sqrt{t})[h^* - h]$, and assume that $\sigma^2 = E_F[\{h^*(U_i)/h(U_i) - 1\}^2] < \infty$, and that $m = E_F[U_i]$ is finite as well. Then for each t, $P_{F_t} \ll P_F$ on $\mathcal{F}_t^N$; moreover, under P_F,

$$\log \, dP_{F_t}/dP_F \big|_{\mathcal{F}_t^N} \overset{d}{\to} N(-\sigma^2/2m, \sigma^2/m).$$

Proof: Formally, by dividing dP_{F_t}/dP as given by (8.22) by dP_F/dP and using Proposition 8.10 (more rigorously, by mimicking the proof of Theorem 8.17 with Theorem 2.31 replaced by an analogue of Theorem 5.2), we infer that absolute continuity holds, with

$$
\begin{aligned}
\log(dP_{F_t}&/dP_F)\big|_{\mathcal{F}_t^N} \\
&= \textstyle\int_0^t [h(L_u) - h_t(L_u)]du + \int_0^t \log[h_t(L_u)/h(L_u)]dN_u \\
&= \textstyle(1/\sqrt{t})\int_0^t [h(L_u) - h^*(L_u)]du \\
&\quad + \textstyle\int_0^t \log(1 + (1/\sqrt{t})[h^*(L_u)/h(L_u) - 1])dN_u \\
&\cong \textstyle(1/\sqrt{t})\left\{ \int_0^t [h(L_u) - h^*(L_u)]du + \int_0^t [h^*(L_u)/h(L_u) - 1]dN_u \right\} \\
&\quad - \textstyle(1/2t)\int_0^t [h^*(L_u)/h(L_u) - 1]^2 dN_u \\
&= \textstyle(1/\sqrt{t})\int_0^t [h^*(L_u)/h(L_u) - 1][dN_u - h(L_u)du] \\
&\quad - \textstyle(1/2t)\int_0^t [h^*(L_u)/h(L_u) - 1]^2 dN_u.
\end{aligned}
$$

By the strong law of large numbers,

$$(1/t)\int_0^t [h^*(L_u)/h(L_u) - 1]^2 dN_u = (N_t/t)(1/N_t)\sum_{i=1}^{N_t}[h^*(U_i)/h(U_i) - 1]^2$$
$$\xrightarrow{d} \sigma^2/m.$$

This expression, Proposition 8.10, Theorem B.21 and ergodicity of renewal processes imply that

$$(1/\sqrt{t})\int_0^t [h^*(L_u)/h(L_u) - 1][dN_u - h(L_u)du] \xrightarrow{d} N(0, \sigma^2/m),$$

which completes the proof. ∎

We now take up some specific questions of hypothesis testing. The methods below are based on the i.i.d. structure of the interarrival times, regardless of whether observations are synchronous or asynchronous; moreover, as we have seen, asymptotics in the latter case differ from more directly deduced properties of the former only in the ubiquitous variance adjustment factor $m = E_F[U_i]$. Hence for simplicity we shall work only with synchronous data. Given observations $U_1, \ldots, U_n$, except that the U_i are positive and hence cannot be normally distributed, the mathematics is classical. The contribution of renewal processes lies principally in the *kinds of questions* that are posed.

As intimated previously, reliability theory is the source of many such questions; here the renewal process describes failures of i.i.d. objects tested sequentially, with instantaneous replacements at times of failures. Consider, for example, the following question. Suppose that a newly developed system is compared with an extant system in order to ascertain whether the new system possesses a "longer lifetime"; the data are lifetimes of i.i.d. copies of the new system. Although this might be viewed as a two-sample problem, but often there are enough data concerning the old system that its lifetime distribution F_0 is taken as known.

Of several interpretations of "longer lifetime," one is the relatively strong condition that the F-distributed lifetimes of the new system be *stochastically larger* than F_0-lifetimes in the sense that $F(x) \leq F_0(x)$ for every $x \geq 0$. (One says also that F is stochastically larger than F_0; for elementary properties, see Exercise 8.15.) If this is so, then at least in terms of longevity the new system is superior.

The hypothesis

$$H_0 : F \leq F_0, \quad F \neq F_0 \tag{8.24}$$

can be tested nonparametrically with the one-sided Kolmogorov-Smirnov statistic $D^- = \sup_{x \in \mathbf{R}_+}[F_0(x) - \hat{F}(x)]$, where $\hat{F}$ is given by (8.10). Like the

Kolmogorov-Smirnov statistic, its two-sided analogue, D^- is distribution-free provided that F_0 is continuous, and has known and rather simple limit distribution.

Suppose now that F_0 and F_1 are distribution functions with F_1 stochastically larger than F_0. If N^0 and N^1 are associated renewal processes, then it seems plausible that for each t, N_t^0 should be stochastically larger than N_t^1, i.e., the process with shorter lifetimes should have more failures. Were it possible to construct N^0 and N^1 on the same probability space in such a manner that $N_t^0 \geq N_t^1$ for all t, this and other distributional relationships would hold simply as expectations of pointwise inequalities. We achieve this by a coupling construction.

Proposition 8.20. Let F_0 and F_1 be distribution functions on $\mathbf{R}_+$ such that F_0 is stochastically *smaller* than F_1. Then there exists a probability space on which are defined counting processes (N_t^0) and (N_t^1) such that

a) N^i is a renewal process with interarrival distribution F_i;

b) Almost surely, $N_t^0 \geq N_t^1$ for all t.

Proof: It suffices to construct random variables U^0 and U^1 such that U^i has distribution F_i and $U^0 \leq U^1$ almost surely, for then i.i.d. copies of U^i can be taken as the interarrival times for N^i, and it is evident that a) and b) hold. But the construction is easy: let Z be uniformly distributed on $[0, 1]$ and let $(U^0, U^1) = (F_0^{-1}(Z), F_1^{-1}(Z))$. ∎

Proposition 8.20 is mainly a tool for study of renewal processes and derivation of approximations and bounds for quantities that are difficult to compute explicitly but important in inference nevertheless.

To illustrate, the distribution of N_t, needed for example to calculate critical regions for hypothesis tests and to compute power, is expressible in terms of the interarrival distribution F: $P\{N_t < n\} = P\{T_n > t\} = 1 - F^n(t)$, but save in a handful of cases cannot be calculated explicitly. If F^* is a more tractable distribution than F satisfying $F^* \leq F$, then by Proposition 8.20, for each n and t, $P_F\{N_t < n\} \leq P_{F^*}\{N_t < n\}$, which can be used to derive bounds. A specific example appears in Exercise 8.19.

In reliability it is often of interest to study aging properties of systems. One would expect a typical physical system to undergo a "burn-in" period during which failure is relatively more likely, that systems surviving the burn-in would exhibit a period during which the failure rate is essentially constant, and that finally, systems would age, becoming more and more likely to fail. We introduce three formalizations of the simpler notion of systems that merely grow increasingly less reliable with age, and discuss briefly some procedures that have been developed to test for them, given i.i.d. samples $U_1, \ldots, U_n$, against the null hypothesis that the U_i are exponentially distributed. The tests, therefore, examine whether a renewal process is Poisson.

First we define the properties to be tested.

Definition 8.21. Let F be a distribution function on $\mathbf{R}_+$ and let $\bar{F}(x) = 1 - F(x)$. Then

a) F has *increasing failure rate* if the function $x \to -\log \bar{F}(x)$ is convex on the support of F;

b) F has *increasing failure rate average* if $x \to -(1/x)\log \bar{F}(x)$ is increasing on the support of F;

c) F has the property *new better than used* if $\bar{F}(x+y) \geq \bar{F}(x)\bar{F}(y)$ for all x and y.

It is conventional to write $F \in$ IFR, $F \in$ IFRA, $F \in$ NBU when F has the indicated property. Analogues such as decreasing failure rate,..., which describe systems that improve with age, can be formulated and studied using similar methods.

If F admits hazard function h (in reliability, the *failure rate function*), then $F \in$ IFR if and only if h is increasing, and $F \in$ IFRA if and only if the function $x \to (1/x)\int_0^x h$ is increasing. The less obviously motivated class of IFRA distributions, which contains the IFR distributions, is important because it arises in the reliability of multicomponent systems and systems that fail from the cumulative effect of randomly occurring shocks (Exercise 8.22). Weaker still than IFRA is NBU, whose interpretation is that a single system is less likely to remain in operation at time $x + y$ than if it had been replaced at time x with a new copy.

With respect to all three classes exponential distributions are the borderline case: if F is exponential, then $-\log \bar{F}$ is linear, $-(1/x)\log \bar{F}(x)$ is constant, and $\bar{F}(x+y) = \bar{F}(x)\bar{F}(x)$ for all x and y. In fact, each of these properties characterizes exponential distributions. The exponential case, consequently, is the natural null hypothesis to be tested against the increasingly broad alternatives

$$H_1 : F \in \text{IFR}, \ F \text{ not exponential} \qquad (8.25)$$

$$H_1 : F \in \text{IFRA}, \ F \text{ not exponential} \qquad (8.26)$$

$$H_1 : F \in \text{NBU}, \ F \text{ not exponential}, \qquad (8.27)$$

which represent particular classes of non-Poisson renewal processes.

The literature on tests and maximum likelihood estimation for distributions with monotone failure rate is extensive; here we only mention selected aspects. Tests for the IFR alternative (8.25) are based (see also Example 3.14) on normalized spacings $D_i = (n - i + 1)[U_{(i)} - U_{(i-1)}]$, where $U_{(0)} = 0 < U_{(1)} < \cdots < U_{(n)}$ are the order statistics engendered by $U_1, \ldots, U_n$. Behavior under the null hypothesis of exponentiality has been noted in Section 3.3 and Exercise 3.34.

Behavior under the IFR alternative has also been characterized.

Proposition 8.22. If $F \in$ IFR, then the spacings D_i are stochastically decreasing in i. $\square$

Test statistics based on spacings include the *time on test*, defined by $T = [\sum_{i=1}^{n} \sum_{j=1}^{i} D_j] / \sum_{i=1}^{n} D_i$, which has asymptotic minimax properties (Barlow *et al.*, 1972), and rank statistics $\sum_{i=1}^{n} i \log[1 - R_i/(n + 1)]$, where R_i is the rank of D_i among $D_1, \ldots, D_n$. The latter are locally most powerful (Bickel and Doksum, 1968). Similar methods have been applied to the IFRA alternative (8.26) (see the chapter notes).

The NBU alternative (8.27) has been examined by Hollander and Proschan (1972). Deviation of $F \in$ NBU from exponentiality is described using the functional

$$\gamma(F) = \int \int [\bar{F}(x)\bar{F}(y) - \bar{F}(x + y)] F(dx) F(dy),$$

which is zero for F an exponential distribution and positive for all other $F \in$ NBU. The test statistic is a U-statistic approximation to the corresponding functional $\gamma(\hat{F})$ of the empirical distribution function of (8.10):

$$U = [1/2n(n - 1)(n - 2)] \sum 1(U_i > U_j + U_k),$$

where the summation is over all triples of distinct integers i, j and k with $j < k$. Limit theory for U-statistics (see Serfling, 1980) yields consistency, asymptotic normality, and characteristics of the null distribution.

Narrower alternatives are usually formulated parametrically; see Exercise 8.26 for the most important case, Weibull distributions.

In the opposite direction of broader alternatives, the situation is less understood. There exist "nonparametric" characterizations of Poisson processes among renewal processes, each reducing ultimately to a characterization of exponential distributions, but none seems to have been investigated as a tool for inference. Nevertheless, they are interesting in their own right, so we present one of the simplest.

Proposition 8.23. Let N be a renewal process with interarrival distribution F and forward recurrence time process W. Then the following are equivalent for each $c > 0$:

a) N is a Poisson process with rate $1/c$;

b) $E[W_t] = c$ for each $t > 0$.

Proof: Integration with respect to y of the renewal equation

$$P\{W_t > y\} = 1 - F(t + y) + \int_0^t F(du)P\{W_{t-u} > y\},$$

satisfied by the function $t \to P\{W_t > y\}$ (see Example 8.3), and application of b) yield

$$c = E[W_t] = \int_t^\infty [1 - F(u)]du + cF(t),$$

so that F is differentiable, with hazard function $h \equiv 1/c$. ∎

Related characterizations of Poisson processes among renewal processes are mentioned in Exercises 8.24 and 8.25 and in the chapter notes.

There exists also a characterization of Cox processes that are renewal processes. The stochastic integral $X * N$ of Example 3.5 and Section 6.3 is a renewal process provided that the sojourn distribution in state 1 is exponential. The content of the following theorems is that every Cox process that is also a renewal process has this form.

The crucial properties are two. Let N be a renewal process with interarrival hazard function h and suppose that N is also a Cox process with directing intensity X. Then first of all, X can assume only one nonzero value, by the following argument. In consequence of Proposition 8.10, N has $\mathcal{F}^N$-stochastic intensity $\lambda_t = h(L_t)$, and therefore by Exercise 2.16 and Theorem 7.9, with $Z_t = \exp\{-\int_0^t X_u du\}$,

$$h(L_t) = E[X_t|\mathcal{F}_t^N] = E[Z_t\{\textstyle\prod_{i=1}^n X_{t_i}\}X_t]/E[Z_t\textstyle\prod_{i=1}^n X_{t_i}]$$

evaluated at $n = N_t, t_1 = T_1, \ldots, t_n = T_n$. That is, for each n and for $t_1 < \cdots < t_n < t$,

$$h(t - t_n) = E[Z_t\{\textstyle\prod_{i=1}^n X_{t_i}\}X_t]/E[Z_t\textstyle\prod_{i=1}^n X_{t_i}],$$

and letting all the t_i increase to t gives $h(0) = E[Z_t X_t^{n+1}]/E[Z_t h(0)^n]$, or $E[Z_t X_t^n] = E[Z_t]h(0)^n$, which implies that the only nonzero value that X can assume is $\lambda = h(0)$.

Second, sojourns of X in λ must be exponentially distributed, intuitively because unless the hazard function were constant, times between arrivals in N would not be independent: short interarrival times will cluster as X remains in λ, to be followed by a longer interarrival time during which X spends time in zero.

Here is the characterization.

Theorem 8.24. Let N be a point process on $\mathbf{R}_+$ that is simultaneously a stationary renewal process and a Cox process with stationary directing intensity. Then the Laplace transform $\ell(u) = \int F(dx)e^{-ux}$ has the form

$$\ell(u) = \lambda/[\lambda + u\int(1 - e^{-uz})K(dz)], \qquad (8.28)$$

where $\lambda > 0$ and K is a measure on $(0, \infty)$ satisfying $\int K(dz)z < \infty$.

Proof: Let $\Lambda_t = \int_0^t X_u du$ and let Λ^{-1} denote the right-continuous inverse. By Exercise 2.11 N admits the representation $N = \sum_{n=0}^{\infty} \epsilon_{T_n}$, where $T_n = \Lambda^{-1}(S_n)$ with $\tilde{N} = \sum \epsilon_{S_n}$ a Poisson process with rate 1 and independent of X. It follows that $E[e^{-u\Lambda^{-1}(u)}] = [1 - \ell(u)]/mu\ell(u)$ (here m is the mean interarrival time) and that for $s, t > 0$,

$$E[e^{-u\{\Lambda^{-1}(t\mid s)\ \Lambda^{-1}(t)\}}] = e^{\ s[1-\ell(u)]/\ell(u)}. \qquad (8.29)$$

These properties are consequences of a theorem of Kingman (1963) concerning the way in which the law of the Poisson sample process (T_n) determines that of the process Λ^{-1} being sampled (see Proposition 10.1).

Also by Kingman (1963), the independent increments property of (T_n) and stationarity of the increments $T_2 - T_1, T_3 - T_2, \ldots$ are shared by Λ^{-1}. Application of the Lévy-Khintchine representation theorem (see Blumenthal and Getoor, 1968, and Theorem 1.34) provides existence of $\beta \geq 0$ and a measure H with $\int H(dz)z < \infty$ such that

$$E[e^{-u\{\Lambda^{-1}(t+s)-\Lambda^{-1}(t)\}}] = \exp\left\{-s[\beta u + \int_0^\infty (1 - e^{-uz})H(dz)]\right\}.$$

Together with (8.29) this implies that

$$\ell(u) = 1/[1 + u\beta + \int_0^\infty (1 - e^{-uz})H(dz)], \qquad (8.30)$$

and that $m = \beta + \int H(dz)z$. But β cannot be zero, for if it were then (8.30) would imply that $P\{\Lambda^{-1}(0) = 0\} = \lim_{u\to\infty} E[e^{-u\Lambda^{-1}(0)}] = \beta/m = 0$, and hence that $\Lambda \equiv 0$, an obvious contradiction. Thus, (8.28) holds with $\lambda = 1/\beta$ and $K(dz) = \lambda H(dz)$. ∎

When the measure K in Theorem 8.24 is finite, so that Λ^{-1} is a compound Poisson process, then the directing intensity has the structure described before the theorem.

Theorem 8.25. In the context of Theorem 8.24, suppose that $k = K(\infty) - K(0)$ is finite. Then X is a semi-Markov process with state space $\{0, \lambda\}$, sojourns in λ are exponentially distributed with parameter k, and sojourns in 0 have distribution $G(z) = [K(z) - K(0)]/k$. ☐

It may also be wondered which renewal processes are infinitely divisible. Haberland (1975) shows that only Poisson processes are simultaneously renewal processes and Poisson cluster processes.

8.3 State Estimation; Combined Inference and State Estimation

Let N be a renewal process with known interarrival distribution F. We first consider the prediction problem of estimating future behavior of N given observations $\mathcal{F}_t^N$. Intuitively, the forward recurrence time W_t should be a "sufficient statistic" for state estimation of the process $N_u^t = N_{t+u} - N_t$, in the sense that once W_t is estimated the remainder of the future is straightforward, because after $T_{N_t+1} = t + W_t$ the renewal process is a probabilistic replica of itself. Confirmation is contained in the following result, which shows that the situation is symmetric: once the backward recurrence time V_t is known, additional information in $\mathcal{F}_t^N$ does not improve predictions of N^t. Let $\bar{F}(x) = 1 - F(x)$ be the survivor function of F.

Theorem 8.26. Let $t > 0$ be fixed and let (N_u^t) be as above. Then for each set Γ,

$$
\begin{aligned}
P\{N^t \in \Gamma | \mathcal{F}_t^N\} &= P\{N^t \in \Gamma | V_t\} \\
&= (1/\bar{F}(V_t)) \int F(V_t + dy) P\{N \in \Gamma | T_1 = y\}. \quad (8.31)
\end{aligned}
$$

Proof: It suffices to establish equality of the first and third members in (8.31) when $\{N^t \in \Gamma\} = \{N_{u_1}^t = j_1, \ldots, N_{u_k}^t = j_k\} = \{N_{t+u_1} - N_t = j_1, \ldots, N_{t+u_k} - N_t = j_k\}$ for $0 \le u_1 < \cdots < u_k$ and $0 \le j_1 \le \cdots \le j_k$. Given $\Lambda = \{N_{t_1} = i_1, \ldots, N_{t_\ell} = i_\ell\}$ with $0 \le t_1 < \cdots < t_\ell \le t$ and $0 \le i_1 \le \cdots \le i_\ell$,

$$
\begin{aligned}
P\{\Gamma \cap \Lambda\} &= \sum_{q=0}^{\infty} P\{\Lambda, N_t = i_\ell + q, N_{t+u_j} = i_\ell + q + i_j, j = 1, \ldots, k\} \\
&= \sum_{q=0}^{\infty} \int_0^t P\{\Lambda, T_{i_\ell+q} \in dz\} \int_{t-z}^{\infty} F(dx) P\{N \in \Gamma | T_1 = z + x - t\} \\
&= \sum_{q=0}^{\infty} \int_0^t P\{\Lambda, T_{i_\ell+q} \in dz\} \int F(dy + t - z) P\{N \in \Gamma | T_1 = y\} \\
&= \int_0^t P\{\Lambda, V_t \in t - dz\} \frac{\int F(dy + t - z) P\{N \in \Gamma | T_1 = y\}}{1 - F(t - z)} \\
&= E\left[(1/\bar{F}(V_t)) \int F(dy + V_t) P\{N \in \Gamma | T_1 = y\}; \Lambda\right],
\end{aligned}
$$

which confirms (8.31). ∎

In particular, we have the following consequence.

Proposition 8.27. For each t and z,

$$P\{W_t > z|V_t\} = \bar{F}(V_t + z)/\bar{F}(V_t).\qquad(8.32)$$

Proof: In (8.31) choose Γ so that $\{N^t \in \Gamma\} = \{N_z^t = 0\} = \{N_{t+z} - N_t = 0\} = \{W_t > z\}$. Since $P\{N \in \Gamma|T_1 > y\} = P\{T_1 > z|T_1 > y\} = \mathbf{1}(y > z)$, (8.31) then gives $P\{W_t > z|V_t\} = \int F(V_t + dy)\mathbf{1}(y > z)/\bar{F}(V_t)$. ∎

Therefore,

$$P\{N^t \in \Gamma|\mathcal{F}_t^N\} = \int P\{W_t \in dy|V_t\}P\{N \in \Gamma|T_1 = y\},\qquad(8.33)$$

with the dependence on $\mathcal{F}_t^N$ encapsulated in the conditional distribution $P\{W_t > z|V_t\}$. For many calculations the explicit form of (8.32) is adequate, but it is also possible to derive a recursive method of computation in the canonical form of predictable part plus filter gain times innovation.

Proposition 8.28. Let N be a renewal process with interarrival hazard function h. Then for each z the MMSE state estimators $\hat{W}_t = P\{W_t > z|\mathcal{F}_t^N\} = P\{W_t > z|V_t\}$ satisfy the stochastic differential equation

$$d\hat{W}_t = \{\bar{F}(z)h(V_t) - \hat{W}_t h(V_t + z)\}dt + [\bar{F}(z) - \hat{W}_{t-}][dN_t - h(L_t)dt].\qquad(8.34)$$

Proof: Recall that (L_t) is the left-continuous backward recurrence time process. We first observe that from (8.32)

$$P\{W_t > z|V_t\} = \bar{F}(V_t + z)/\bar{F}(V_t) = \exp\left\{-\int_{V_t}^{V_t+z} h(s)ds\right\}.$$

Between points of N, $t \to V_t$ is increasing linearly; thus if $N_{t+\Delta} - N_t = 0$, then

$$\begin{aligned}
d\hat{W}_t &= \exp\left\{-\int_{V_t+\Delta}^{V_t+z+\Delta} h(s)ds\right\} - \exp\left\{-\int_{V_t}^{V_t+z} h(s)ds\right\}\\
&= \exp\left\{-\int_{V_t}^{V_t+z} h(s)ds\right\}\left[\exp\left\{-\int_{V_t}^{V_t+\Delta} h(s)ds - \int_{V_t+\Delta}^{V_t+z+\Delta} h(s)ds\right\} - 1\right]\\
&\cong \hat{W}_t[h(V_t) - h(V_t + z)]\Delta
\end{aligned}$$

as $\Delta \to 0$. Hence between points of N the "ordinary" differential equation

$$d\hat{W}_t = \hat{W}_t[h(V_t) - h(V_t + z)]dt\qquad(8.35)$$

holds.

At a point t of N, there occurs the discrete correction

$$d\hat{W}_t = \bar{F}(z) - \hat{W}_{t-} = \bar{F}(z) - \bar{F}(L_t + z)/\bar{F}(L_t).\qquad(8.36)$$

To obtain (8.34), combine (8.35) and (8.36), add and subtract the term $\{\bar{F}(z)\bar{F}(L_t + z)/\bar{F}(L_t)\}h(L_t)dt$, and finally appeal to the property that almost surely $L_t = V_t$ for almost every t (with respect to Lebesgue measure) in order to replace L_t by V_t in the predictable part of (8.34). ∎

For fixed z the equation (8.34) is entirely recursive, involving no other state estimators. Combined with (8.33) it provides a recursive method for calculation of $P\{N^t \in \Gamma | \mathcal{F}_t^N\}$.

Our analysis of state estimation *per se* concludes with p-thinned renewal processes, where p is a constant. Let $N = \sum_{n=1}^{\infty} \epsilon_{T_n}$ be a renewal process with interarrival distribution F and let $Z_1, Z_2, \ldots$ be i.i.d. random variables, independent of N, with $P\{Z_n = 1\} = p \in (0,1]$, $P\{Z_n = 0\} = q = 1 - p$. The p-thinning of N, the point process $N' = \sum Z_n \epsilon_{T_n}$, can be characterized in the following manner.

Proposition 8.29. The p-thinning N' is a renewal process with interarrival distribution $\tilde{F} = p\sum_{k=1}^{\infty} q^{k-1} F_k$. □

Some aspects of inference are examined in Exercises 8.29 and 8.30.

The state estimation problem of interest is to reconstruct N from asynchronous observations of N', which entails calculating MMSE state estimators $E[N_t | \mathcal{F}_t^{N'}]$. General methods developed in Section 5.4 yield the solution.

Theorem 8.30. Suppose that F admits hazard function h. Then the state estimators $E[N_t | \mathcal{F}_t^{N'}]$ satisfy the stochastic differential equation

$$
\begin{aligned}
dE[N_t | \mathcal{F}_t^{N'}] = \ &(1/p)\tilde{h}(L_t')dt \\
&+ \left\{ 1 + (pE[N_u h(L_u) | \mathcal{F}_u^{N'}]/\tilde{h}(L_u') - E[N_u | \mathcal{F}_u^{N'}])\big|_{u=t-} \right\} \\
&\times [dN_t' - \tilde{h}(L_t')dt],
\end{aligned} \tag{8.37}
$$

where $\tilde{h}$ is the hazard function of $\tilde{F}$ and L and L' are the left-continuous backward recurrence time processes for N and N'.

Although it is easy to show that the hazard function $\tilde{h}$ exists, it does not seem to be related in a simple manner to the hazard function of F.

Proof of Theorem 8.30: We work through calculations associated with Theorem 5.30. Notation matches that of Section 5.4 only partially, however: in particular, the observed point process denoted there by N is in this case the p-thinned process N'. By Proposition 8.10 the state process $Z_t = N_t$, with respect to the filtration $\mathcal{H}_t = \mathcal{F}_t^N \vee \mathcal{F}_t^{N'}$, has semimartingale representation

$$
Z_t = \int_0^t h(L_u)du + M_t, \tag{8.38}
$$

where M is an $\mathcal{H}$-martingale. The $\mathcal{H}$-stochastic intensity of N' is

$$\lambda_t = ph(L_t), \tag{8.39}$$

so there exists an $\mathcal{F}^{N'}$-stochastic intensity $\hat{\lambda}$ satisfying $\hat{\lambda}_t = pE[h(L_t)|\mathcal{F}_t^{N'}]$ almost surely for each t. By Proposition 8.29 and another appeal to Proposition 8.10, $\hat{\lambda}_t = \tilde{h}(L_t')$, and therefore

$$E[h(L_t)|\mathcal{F}_t^{N'}] = \tilde{h}(L_t')/p \tag{8.40}$$

almost surely for each t.

From Proposition 5.28, Theorem 5.30, (8.39) and (8.40),

$$dE[N_t|\mathcal{F}_t^{N'}] = \tilde{h}(L_t')/p + H_t[dN_t' - \tilde{h}(L_t')dt], \tag{8.41}$$

where the filter gain H is given by $H_t = J_t + K_t - E[N_s|\mathcal{F}_s^{N'}]\big|_{s=t-}$, with J and K the $\mathcal{F}^{N'}$-predictable processes of Theorem 5.29. We compute the former using (8.40):

$$J_t = (1/\hat{\lambda}_t)\, E[N_u h(L_u)|\mathcal{F}_u^{N'}]\big|_{u=t-} = (1/\tilde{h}(L_t'))\, E[N_u h(L_u)|\mathcal{F}_u^{N'}]\big|_{u=t-}.$$

By contrast with previous cases, the process K is not zero. With M the martingale in (8.38), $\Delta M_t = 1$ whenever $dN_t' \neq 0$ and thus K must satisfy

$$E\left[\int_0^t CK\lambda ds\right] = E\left[\int_0^t C(\Delta M)dN'\right] = E\left[\int_0^t CdN'\right] = E\left[\int_0^t C\lambda ds\right]$$

for each bounded, $\mathcal{F}^{N'}$-predictable process C and each $t > 0$. Therefore, $K \equiv 1$ [the solution to (5.67) is unique] and now only substitution into (8.41) remains in order to complete the proof. ∎

We conclude the section with brief analysis of combined statistical inference and state estimation. In view of Proposition 8.27 the key state estimators for which approximating pseudo-state estimators must be produced are the conditional distributions $P\{W_t > z|V_t\}$. Our setting, now different from Sections 3.4 and 7.4, is asynchronous observation of a single realization of a renewal process N with unknown interarrival distribution F. Invoking once again the principle of separation, given observation $\mathcal{F}_t^N$ we construct from the estimator $\hat{F}$ of (8.12), by (8.32), the pseudo-state estimators

$$\hat{P}\{W_t > z|V_t\} = [1 - \hat{F}(V_t + z)]/[1 - \hat{F}(V_t)], \tag{8.42}$$

where $0/0 = 0$. [The denominator on the right-hand side of (8.42) cannot vanish unless the numerator does also.] These pseudo-state estimators converge to the true state estimators at an optimal rate, provided only that the mean interarrival time be finite.

Theorem 8.31. Assume that $m = E_F[U_i]$ is finite and let the pseudo-state estimators $\hat{P}\{W_t > z|V_t\}$ be given by (8.42). Then

$$\left(\sqrt{t}[\hat{P}\{W_t > z|V_t\} - P\{W_t > z|V_t\}]\right)_{z \in \mathbf{R}_+}$$

$$\overset{d}{\rightarrow} \left(G(V_\infty + z)/\bar{F}(V_\infty) - \bar{F}(V_\infty + z)G(V_\infty)/\bar{F}(V_\infty)^2\right)_{z \in \mathbf{R}_+}, (8.43)$$

where

i) G is the Gaussian process of Theorem 8.9;
ii) $P\{V_\infty \leq y\} = (1/m) \int_0^y [1 - F(x)]dx$;
iii) G and V_∞ are independent.

Proof: The error process $\{\sqrt{t}[\hat{F}(x) - F(x)] : x \in \mathbf{R}_+\}$ and the backward recurrence time V_t are asymptotically independent (Exercise 8.32); therefore, $(\sqrt{t}[\hat{F} - F], V_t) \overset{d}{\rightarrow} (G, V_\infty)$, where G and V_∞ satisfy i) – iii). From this, (8.43) follows by the "delta method" and the continuous mapping theorem. ∎

Note that Theorems 8.31 and 7.26 are similar structurally as well as in their relationships to Theorems 8.9 and 7.5, even though one deals with single-realization asymptotics and the other with asymptotics of i.i.d. copies.

To cope with more general state estimators

$$P_F\{N^t \in \Gamma|\mathcal{F}_t^N\} = (1/\bar{F}(V_t)) \int F(V_t + dy)P_F\{N \in \Gamma|T_1 = y\}$$

one must also estimate $P_F\{N \in \Gamma|T_1 = y\}$; whether this is feasible depends on Γ. Whenever it is, in the sense that one can devise asymptotically normal estimators $\hat{P}\{N \in \Gamma|T_1 = y\}$, an analogue of Theorem 8.31 holds, but with more complicated covariance relationships. In other cases there are directly defined pseudo-state estimators $\hat{P}\{N^t \in \Gamma|V_t\}$ (see Exercise 8.33 for an example).

8.4   Estimation for Markov Renewal Processes

A Markov renewal process describes by times of jumps and states visited the evolution of a finite-state, continuous-time process whose structure generalizes that of a Markov process: the sequence of states visited remains a Markov chain, but time spent in a state need not be exponentially distributed and may depend on the next state entered. A Markov renewal process is also a collection of simultaneously evolving renewal processes, representing times of entrances to various states, with a particular interdependence, and tools developed for statistical inference for renewal processes — empirical and martingale — can be applied, even in the presence of severely censored observations. In this section we treat estimation for two models of Markov renewal processes, one with complete observation of a single realization and the second with i.i.d. copies of a heavily censored process.

We begin with a short description of Markov renewal and semi-Markov processes. Let S be a finite set, let $(X, T) = ((X_n, T_n))_{n \geq 0}$ be a process with $X_n \in S$ for each n, and suppose that $0 = T_0 < T_1 < \cdots$. Then (X, T) is a *Markov renewal process* if

$$P\{X_{n+1} = j, T_{n+1} - T_n \leq x | X_0, \ldots, X_n, ; T_0, \ldots, T_n\}$$
$$= P\{X_{n+1} = j, T_{n+1} - T_n \leq x | X_n\}$$

for each n, j and x. We assume homogeneity, so that the transition probabilities

$$Q(i, j, x) = P\{X_{n+1} - j, T_{n+1} - T_n \leq x | X_n = i\}$$

known as the *semi-Markov kernel* of (X, T), do not depend on n.

The Markov renewal process is a Markov chain with state space $S \times \mathbf{R}_+$, but with additional structure.

Proposition 8.32. Let (X, T) be a Markov renewal process with semi-Markov kernel Q. Then

a) $X = (X_n)$ is a Markov chain with transition matrix

$$P(i, j) = Q(i, j, \infty) = \lim_{x \to \infty} Q(i, j, x); \tag{8.44}$$

b) $T_1, T_2 - T_1, \ldots$ are conditionally independent given X;

c) Times of entrances to a fixed state constitute a (possibly delayed or terminating) renewal process. $\square$

The law of a Markov renewal process is determined by the semi-Markov kernel Q, although typically (as for Markov processes) it is specified by giving the transition probabilities $P(i, j)$ for the embedded Markov chain (X_n) and the conditional sojourn distributions

$$G(i, j, x) = P\{T_{n+1} - T_n \leq x | X_n = i, X_{n+1} = j\},$$

with Q then *defined* by

$$Q(i, j, x) = P(i, j)G(i, j, x), \tag{8.45}$$

so that P and G are the fundamental parameters. Below we estimate Q by first estimating unconditional sojourn distributions

$$H(i, x) = P\{T_{n+1} - T_n \leq x | X_n = i\} = \sum_j Q(i, j, x). \tag{8.46}$$

The continuous time process $Y_t = X_n$, $t \in [T_n, T_{n+1})$, is the *semi-Markov process* engendered by (X, T). Its structure generalizes that of a Markov process; indeed, in the special case that $Q(i, j, x) = P(i, j)[1 - e^{-\lambda(i)x}]$, where $0 < \lambda(i) < \infty$ and $P(i, i) = 0$ for each i, Y is a Markov process with generator $A(i, j) = \lambda(i)P(i, j)$, $i \neq j$. At the opposite extreme, if $S = \{i\}$ is a singleton, then (T_n) is an ordinary renewal process with interarrival distribution $F(\cdot) = Q(i, i, \cdot)$.

Asymptotic behavior of Markov renewal and semi-Markov processes is determined by the renewal structure together with limit properties of the Markov chain (X_n). If X is irreducible, so that (since S is finite) there exists a probability distribution ν on S satisfying $\nu(j) = \sum_i \nu(i)P(i, j)$ for all $j \in S$, where P is given by (8.44), if (X, T) is aperiodic [this is unrelated to aperiodicity of X and connected instead with the distribution functions $G(i, j, \cdot)$], and if $m(i) = E[T_1 | X_0 = i]$ is finite for each i, then

$$\lim_{t \to \infty} P\{Y_t = k\} = \nu(k)m(k) / \sum_{j \in S} \nu(j)m(j);$$

see Çinlar (1975a, 1975b) for details, as well as (8.9) and associated discussion. Under the same hypotheses there hold strong laws of large numbers and central limit theorems for functionals $Z_t = \sum_{T_n \leq t} f(X_{n-1}, X_n, T_n - T_{n-1})$.

One can describe the Markov renewal process (X, T) not only by the semi-Markov process Y but also, and more completely, by the marked point process

$$N^0 = \sum_n \epsilon_{(T_n, X_{n-1}, X_n)} \tag{8.47}$$

with mark space $E = S \times S$. The time T_n is marked by the states X_{n-1} and X_n from and to which the semi-Markov process jumps. Our inference procedures are formulated in terms of N^0.

The statistical problem is: given observations of the semi-Markov process Y (equivalently, of the marked point process N^0), estimate the semi-Markov kernel Q. Below we examine two models, the former involving complete observation of one realization and the latter having i.i.d. copies of N^0, each only partially observed, with a censoring mechanism similar to that in Example 5.7.

8.4.1   Single Realization Inference

For complete observation of single realizations, although one must use limit theorems for Markov renewal processes, results parallel those in Section 1.

If for all i, j and x,

$$G(i, j, x) = H(i, x), \tag{8.48}$$

that is, although while not necessarily exponential, the distribution of a sojourn in state i does not depend on the next state to be entered [this assumption always holds for the Markov renewal process $(\tilde{X}, T)$ defined by $\tilde{X}_n = (X_n, X_{n+1})$], then $P(i, j)$ and $H(i, x)$ can be estimated as follows. Let (compare Examples 2.33 and 5.8)

$$N_t^0(i, j) = \sum_n \mathbf{1}(T_n \leq t, X_{n-1} = i, X_n = j) \qquad (8.49)$$

be the number of i-to-j transitions in $[0, t]$ and let $N_t^0(i) = \sum_{j \in S} N_t^0(i, j)$ be the number of sojourns in state i completed before time t. The estimators

$$\hat{H}(i, x) = (1/N_t^0(i))\sum_{i=1}^{N_t^0(i)} \mathbf{1}(X_{n-1} = i, T_n - T_{n-1} \leq x)$$

are analogous to the empirical distribution functions $\hat{F}$ of (8.12) and have similar asymptotic properties, while $\hat{P}(i, j) = N_t^0(i, j)/N_t^0(i)$, as estimators of the transition matrix of the embedded Markov chain, except for the random sample size, are nonparametric maximum likelihood estimators, and are strongly consistent and asymptotically normal.

Finally, by substitution into (8.45), one arrives at

$$\hat{Q}(i, j, x) = \hat{P}(i, j)\hat{H}(i, x),$$

for which the following properties hold.

Theorem 8.33. Assume that the Markov renewal process (X, T) is irreducible and aperiodic, that $E[T_1|X_0 = i] < \infty$ for each i, and that (8.48) holds. Then for each i and j

a) $\sup_{x \in \mathbf{R}_+} |\hat{Q}(i, j, x) - Q(i, j, x)| \to 0$ almost surely;

b) The error processes $\{\sqrt{t}[\hat{Q}(i, j, x) - Q(i, j, x)] : x \in \mathbf{R}_+\}$ converge in distribution to a Gaussian process.

Proof: a) That $\sup |\hat{H}(i, x) - H(i, x)| \to 0$ almost surely follows by the argument used to prove Proposition 8.8, while $|\hat{P}(i, j) - P(i, j)| \to 0$ by Basawa and Prakasa Rao (1980, Sect. 4.2).

b) Let $\tilde{m}(i)$ be the mean recurrence time for state i. Then by Basawa and Prakasa Rao (1980, Sect. 4.2) and the sampling rate reconciliation made throughout Section 1,

$$\sqrt{t}[\hat{P}(i, j) - P(i, j)] \overset{d}{\to} N(0, \tilde{m}(i)P(i, j)[1 - P(i, j)]),$$

while Theorem 8.9 establishes that

$$\{\sqrt{t}[\hat{H}(i, x) - H(i, x)] : x \in \mathbf{R}_+\} \overset{d}{\to} \{W(i, x) : x \in \mathbf{R}_+\},$$

where $W(i)$ is Gaussian with covariance $R(x,y) = \tilde{m}(i)H(i,x)[1 - H(i,y)]$, $0 \leq x \leq y$. These limit processes are independent, and hence b) holds. ∎

Moreover, the processes $W(i)$ are mutually independent and independent of the process $Z(i,j) = \lim_{t\to\infty} \sqrt{t}[\hat{P}(i,j) - P(i,j)]$, which has covariance function

$$R((i,j),(k,\ell)) = \mathbf{1}(i = k)\tilde{m}(i)[\mathbf{1}(j = \ell)P(i,j) - P(i,j)P(k,\ell)],$$

and thus further covariance relationships can be calculated.

8.4.2   Multiple Realization Inference

Here martingale methods are used to derive estimators; limit theorems are demonstrated by direct appeal to the i.i.d. structure, but with martingale ideas employed to deduce covariance structure of Gaussian limit processes.

Both the model for partially observed Markov renewal processes and results concerning it are due to Gill (1980a). Let (X,T) be a Markov renewal process, described by the marked point process N^0 of (8.47), for each t let $N_t^0 = \sum_{i\in S} N_t^0(i) = \sum_n \mathbf{1}(T_n \leq t)$ and let (L_t^0) be the left-continuous backward recurrence time process of (N_t^0).

The underlying process is not completely observable: each sojourn interval $T_{n+1} - T_n$ is subject to (perhaps total) censoring. More precisely, there are censoring variables V_n satisfying $T_n \leq V_n \leq T_{n+1}$ (and additional restrictions) such that for each n and each t in the possibly empty interval $(T_n, V_n]$, the values of Y_{t-}, L_t^0 and all $\Delta N_t^0(i,j)$ are observed. Thus about the sojourn in the state X_n one learns nothing, partial information, or everything according as V_n is equal to T_n, strictly between T_n and T_{n+1}, or equal to T_{n+1}. In detail,

- If $T_n = V_n$, the observation interval $(T_n, V_n]$ is empty and the sojourn — both its length $T_{n+1} - T_n$ and the state X_n occupied — is entirely invisible.

- If $T_n < V_n < T_{n+1}$, the sojourn is partially observable: the state X_n is known, it is known that $T_{n+1} - T_n$ exceeds the observation span $V_n - T_n$, and it is known when observability ceases at V_n that V_n differs from T_{n+1} (In Example 5.7 the indicator variables D_i convey corresponding information.)

- If $V_n = T_{n+1}$, the length $T_{n+1} - T_n$ of the sojourn, the current state X_n *and* the state X_{n+1} entered at T_{n+1} are all known.

Observation, to the extent possible, is over $[0, \infty)$. However, assumptions below imply that $V_n = T_n$ for all but finitely many values of n.

By analogy with Example 5.7 and the description of the underlying semi-Markov process using the marked point process N^0, observations are also described by marked point processes; note that in these expressions x is not a time variable, but instead the argument of distribution functions. Let

$$N(i, j, x) = \sum_n \mathbf{1}(X_n = i, X_{n+1} = j, T_{n+1} - T_n \leq x, V_n = T_{n+1}) \quad (8.50)$$

be the number of completely observed sojourns in i that last no longer than x and terminate with a jump of the semi-Markov process to j, and let

$$K(i, x) = \sum_n \mathbf{1}(X_n = i, T_{n+1} - T_n \geq x, V_n - T_n > x) \quad (8.51)$$

be the number of sojourns in i *observed* to last at least as long as x. Also, put $N(i, x) = \sum_j N(i, j, x)$ and $K(x) = \sum_i K(i, x)$.

Before constructing estimators of the semi-Markov kernel Q we give the compensators of the point processes $N^0(i, j)$. They are not in general integrated stochastic intensities because the $Q(i, j, \cdot)$ need not be absolutely continuous; nonetheless, the following result is related closely to Proposition 8.10.

Proposition 8.34. For each i and j, the $\mathcal{F}^N$-compensator of $N^0(i, j)$ is

$$A_t(i, j) = \int_0^t \frac{\mathbf{1}(Y_{u-} = i)}{1 - H(i, L_u^0-)} dQ(i, j, L_u^0),$$

where H is given by (8.46). $\square$

The observability process can depend on the Markov renewal process, but only through the initial state, with all additional randomness conditionally independent of N^0 given X_0. In order that quantities in (8.50)–(8.51) be finite, most of the semi-Markov process must be unobservable. The following assumptions are in force for the remainder of the section.

Assumptions 8.35. a) There exists a σ-algebra $\mathcal{H}$, conditionally independent of $\mathcal{F}^{N^0}(\mathbf{R}_+ \times E)$ given X_0, such that the V_n are stopping times of the filtration $\mathcal{G}_t = \mathcal{H} \vee \sigma(X_0) \vee_{i,j \in S} \mathcal{F}_t^{N(i,j)}$.
b) $E[\sum_n \mathbf{1}(V_n > T_n)] < \infty$.

Suppose now that we are given i.i.d. copies of both Markov renewal process and observability process, each defined on $\mathbf{R}_+$ and fulfilling Assumptions 8.35. For each n, let $N^n(i, j, x)$, $K^n(i, x)$, $N^n(i, x)$ and $K^n(x)$ denote the sums (unnormalized) over the first n processes of the functionals defined by (8.50)–(8.51). We finally construct estimators of the unconditional so-journ distributions H and the semi-Markov kernel Q. For the former we use product limit estimators, as in Example 5.7:

$$\hat{H}(i, x) = 1 - \prod_{u \le t} \left[1 - \frac{\Delta N^n(i, u)}{K^n(i, u)} \right] = \int_0^x \frac{1 - \hat{H}(i, u-)}{K^n(i, u)} dN^n(i, u); \quad (8.52)$$

the second expression is the recursive method of calculation. To estimate Q we unravel $\hat{H}$:

$$\hat{Q}(i, j, x) = \int_0^x \frac{1 - \hat{H}(i, u-)}{K^n(i, u)} dN^n(i, j, u). \quad (8.53)$$

A substitution justification is given in Exercise 8.36.

Note that full use is made of partially observed sojourns. If desired, one can estimate the transition probabilities $P(i, j)$ by $\hat{P}(i, j) = \hat{Q}(i, j, \infty)$, but the censoring may prevent these estimators from being consistent.

The estimators $\hat{H}$ are weakly uniformly consistent and asymptotically normal. We sketch a proof of the first assertion, but omit that of the latter. Of course, one must restrict attention to x-values for which the censoring does not preclude observed sojourns of length x or longer. For each i, let $\tau_i = \sup\{x : E[K(i, x)] > 0\}$; estimation of $H(i, x)$ for $x > \tau_i$ is impossible.

Theorem 8.36. Suppose that Assumptions 8.35 are satisfied. Then for each i and j, $\sup_{x < \tau_i} |\hat{Q}(i, j, x) - Q(i, j, x)| \to 0$ in probability; consequently, $\sup_{x < \tau_i} |\hat{H}(i, x) - H(i, x)| \to 0$ in probability.

Proof: Introduce processes

$$Z^n(i, j, x) = N^n(i, j, x) - \int_0^x \frac{K^n(i, u)}{1 - H(i, u-)} dQ(i, j, u) \quad (8.54)$$

(not calculable from the observations), put $Z^n(i, x) = \sum_j Z^n(i, j, x)$, and let $Z(i, j, x)$ and $Z(i, x)$ be associated with a generic copy of the process.

By the weak law of large numbers,

$$\sup_{x < \tau_i} |(1/n)N^n(i, j, x) - E[N(i, j, x)]| \to 0 \quad (8.55)$$

and

$$\sup_{x < \tau_i} \left| \int_0^x \frac{(1/n)K^n(i, u)}{[1 - H(i, u-)]} dQ(i, j, u) - \int_0^x \frac{E[K(i, u)]}{1 - H(i, u-)} dQ(i, j, u) \right| \to 0, \quad (8.56)$$

both in the sense of convergence in probability. Laborious but not deep calculations reveal that for each i, j and x,

$$
\begin{aligned}
\hat{Q}(i,j,x) &- Q(i,j,x) \\
&= \int_0^x \frac{1 - \hat{H}(i,u-)}{(1/n)K^n(i,u)}(1/n)dZ^n(i,j,u) \\
&\quad - Q(i,j,x)\int_0^x \frac{1 - \hat{H}(i,u-)}{[1 - H(i,u)](1/n)K^n(i,u)}(1/n)dZ^n(i,u) \\
&\quad + \int_0^x Q(i,j,u)\frac{1 - \hat{H}(i,u-)}{[1 - H(i,u)](1/n)K^n(i,u)}(1/n)dZ^n(i,u), \quad (8.57)
\end{aligned}
$$

in view of which it suffices to show that $(1/n)\sup_x |Z^n(i,j,x)| \to 0$ for each i and j. By (8.55)–(8.56),

$$
\sup \left|(1/n)Z^n(i,j,x) - \left\{ E[N(i,j,x)] - \int_0^x \frac{E[K(i,u)]}{1 - H(i,u-)}dQ(i,j,u)\right\}\right| \to 0
$$

in probability, so it remains only to establish the identity

$$
E[N(i,j,x)] = \int_0^x \frac{E[K(i,u)]}{1 - H(i,u-)}dQ(i,j,u),
$$

which is done using Proposition 8.34 and martingale techniques; see Gill (1980a) for details. ∎

We conclude with a central limit theorem.

Theorem 8.37. Suppose that Assumption 8.35a) holds and that there is $\epsilon > 0$ such that $E[\{\sum_j 1(V_j > T_j)\}^{7+\epsilon}] < \infty$, and for each i, let τ_i be such that $E[K(i,\tau_i)] > 0$. Then as random elements of $\prod_{i \in S}(D[0,\tau_i])^{|S|}$ the error processes $\{\{\sqrt{n}\hat{Q}(i,j,x) - Q(i,j,x)] : x \le \tau_i\} : i,j \in S\}$ converge in distribution to a Gaussian process $G = G(i,j,x)$ with representation

$$
\begin{aligned}
G(i,j,x) &= \int_0^x \frac{1 - H(i,u-)}{E[K(i,u)]}dM(i,j,u) \\
&\quad - Q(i,j,x)\int_0^x \frac{1 - H(i,u-)}{[1 - H(i,u)]E[K(i,u)]}dM(i,u) \\
&\quad + \int_0^x Q(i,j,u)\frac{1 - H(i,u-)}{[1 - H(i,u)]E[K(i,u)]}dM(i,u),
\end{aligned}
$$

where
i) For each i and j, $M(i,j,\cdot)$ is a mean zero Gaussian martingale;
ii) The families $\{M(i,j,\cdot) : j \in S\}$ are independent as i varies;

iii) $M(i) = \sum_j M(i,j)$;

iv) The following covariance relationships hold:

$$\langle M(i,j)\rangle_x = \int_0^x \frac{E[K(i,u)][1 - \Delta Q(i,j,u)]}{1 - H(i,u-)} dQ(i,j,u),$$

while for $j \neq j'$,

$$\langle M(i,j), M(i,j')\rangle_x = -\int_0^x \frac{E[K(i,u)]\Delta Q(i,j,u)}{1 - H(i,u-)} dQ(i,j',u). \square$$

In fact, the error processes $\sqrt{n}[\hat{Q} - Q]$ and $\sqrt{n}[\hat{H} - H]$ are jointly asymptotically Gaussian since

$$\sqrt{n}[\hat{H}(i,x) - H(i,x)] = \sqrt{n}[\sum_j \hat{Q}(i,j,x) - Q(i,j,x)].$$

The key point of the proof (Gill, 1980a) is to show asymptotic normality of the processes $\sqrt{n}Z^n$ introduced in (8.54). Except for tightness this follows from the ordinary central limit theorem, although it must be verified that the limit covariance is correct. In Gill (1980a) tightness is established by a Skorohod embedding construction and the covariance function is computed using martingale techniques. Note that Theorems 8.36 and 8.37 resemble results that would ensue from Theorems 5.12 and 5.13 if the processes Z^n were martingales, but this is not the case.

EXERCISES

8.1 Prove that a homogeneous Poisson process on $\mathbf{R}_+$ is a renewal process.

8.2 Prove Proposition 8.1.

8.3 Let W and V be the forward and backward recurrence time processes in a renewal process with interarrival distribution F.
a) For fixed y and x, exhibit $P\{W_t > y, V_t > x\}$ as the solution of a renewal equation.
b) Assuming F to be nonarithmetic, calculate the limit of this probability as $t \to \infty$.

8.4 (Continuation of Exercise 8.3) Let $I_t = V_t + W_t$ be the length of the interarrival interval containing t.
a) Derive a renewal equation for $P\{I_t > x\}$.
b) Calculate $\lim_{t\to\infty} P\{I_t > x\}$.

8.5 Let N be a renewal process with nonarithmetic interarrival distribution F and let h be a bounded, continuous function satisfying $h(x) = \int_{-\infty}^{\infty} h(x-y)dF(y)$ for all $x \in \mathbf{R}_+$.
a) Let $X_n = h(x - T_n)$, where x is fixed. Prove that $(X_n, \mathcal{F}_{T_n}^N)$ is a martingale.
b) With the aid of a) deduce that h must be constant. [*Hint:* Use the martingale convergence theorem and the Hewitt-Savage zero-one law.]

8.6 Let (X_t) be an irreducible Markov process with finite state space S and generator A satisfying $A(i, i) = -\lambda(i)$ and $A(i, j) = \lambda(i)Q(i, j)$, $i \neq j$, where $\lambda(i)$ is the exponential parameter for sojourns in state i and Q is the transition matrix of the embedded Markov chain.
a) Verify that X is regenerative with respect to the sequence of times T_n of returns to its initial state X_0.
b) Use (8.9) to deduce that for each j,

$$\lim_{t \to \infty} P_j\{X_t = j\} = \frac{\nu(j)/\lambda(j)}{\sum_{k \in S} \nu(k)/\lambda(k)},$$

where ν is the unique invariant distribution of Q and P_j is the law of X under the condition $X_0 = j$.

8.7 Prove Proposition 8.8.

8.8 Let N be a renewal process with unknown interarrival distribution F, observed asynchronously. Calculate the bias of the estimators $\hat{F}$ and $\hat{F}^*$ of (8.12) and (8.13).

8.9 Let N be a renewal process with unknown interarrival distribution F, observed asynchronously. Propose $\mathcal{F}_t^N$-estimators of the interarrival variance σ^2 and establish their asymptotic properties as $t \to \infty$

8.10 Let N be an asynchronously observed renewal process with unknown interarrival distribution F and backward recurrence time process V.
a) Prove that $V_t/N_t \to 0$ almost surely.
b) Deduce from this a strong law of large numbers for the estimators $\hat{m}$ of (8.18).
c) Repeat b) for the mean $\hat{m}^*$ of the estimators $\hat{F}^*$ of (8.13).

8.11 Verify the identity (8.16).

8.12 A *Weibull* distribution with parameters $\lambda, \alpha > 0$ has distribution function $F_{\lambda,\alpha}(x) = 1 - e^{-\lambda x^\alpha}$, $x > 0$.
a) Calculate the hazard function and describe its monotonicity properties.
b) Let N be a renewal process with Weibull interarrival distribution F. Analyze the problem of maximum likelihood estimation of λ and α from asynchronous observations $\mathcal{F}_t^N$.
c) In the context of b), derive $\mathcal{F}_t^N$-estimators that are strongly consistent and asymptotically normal; verify these properties.

8.13 Provide a direct (i.e., non-martingale) proof of Theorem 8.17.

8.14 Generalize Theorem 8.17 to probabilities P_0 and P_1 under which N is a renewal process with interarrival distributions F_0, F_1 satisfying $F_1 \ll F_0$.

8.15 a) Let T and S be random variables with distributions F and G. Prove that F is stochastically larger than G if and only if $E[h(T)] > E[h(S)]$ for every positive, increasing function h.
b) Prove that the property of "stochastically larger" is not preserved under

addition; that is, there exist random variables T, $\tilde{T}$, S and $\tilde{S}$ with T ($\tilde{T}$) stochastically larger than S ($\tilde{S}$) for which $T + \tilde{T}$ is *not* stochastically larger than $S + \tilde{S}$.

8.16 Let F be a distribution function on $\mathbf{R}_+$ with finite mean m and let $H_F(t) = (1/m)\int_0^t[1 - F(u)]du$ be the associated limiting forward recurrence time distribution [see (8.8)].
a) Show that $H_F = F$ if and only if F is an exponential distribution.
b) Explore stochastic order relationships, if any, between F and H_F.

8.17 Let F be a distribution function on $\mathbf{R}_+$ with finite mean m and let $G(dv) = (1/m)vF(dv)$, which according to Exercise 1.31 is the distribution, in a stationary renewal process with interarrival distribution F, of the interarrival interval U_1, which is constrained to contain the origin. Analyze stochastic order relationships between F and G.

8.18 Let F and G be distribution functions on $\mathbf{R}_+$ with hazard functions h_F and h_G, and let λ be a function satisfying

$$\lambda(t) \geq \max\left\{\sup_{s \leq t} h_F(s), \sup_{s \leq t} h_G(s)\right\}, \qquad t \geq 0,$$

and let $\tilde{N} = \sum \varepsilon_{T_n}$ be a Poisson process with intensity function λ. Construct indicator variables δ_k, $k \geq 1$, as follows: $P\{\delta_1 = 1\} = h_G(T_1)/\lambda(T_1)$, while for $k > 1$,

$$\begin{aligned} P\{\delta_k = 1|\delta_1,\ldots,\delta_{k-1}\} &= h_G(T_k)/\lambda(T_k) && \text{if } \delta_1 = \cdots = \delta_{k-1} = 0 \\ &= h_F(T_k - T_j)/\lambda(T_k) && \text{otherwise,} \end{aligned}$$

where $j = \max\{i < k : \delta_i = 1\}$.
a) Prove that $N = \sum_{k=1}^\infty \delta_k\varepsilon_{T_k}$ is a delayed renewal process in which the first arrival time $T_1 = U_1$, has distribution G and the remaining interarrival times have distribution F.
b) Under the further assumption that h_F is bounded and bounded away from zero, use this construction, with $G(dv) = (1/m)[1 - F(v)]dv$, to prove the Blackwell renewal theorem (8.7) for F.

8.19 Discuss how the construction in Exercise 8.18 can be used to produce computable upper bounds on probabilities $P_F\{N_t \geq k\}$, which would be used, for example, to construct hypothesis tests.

8.20 Prove the (strict) inclusions IFR $\subset$ IFRA $\subset$ NBU (see Definition 8.21).

8.21 Let $T \geq 0$ have distribution F, and for each t, let $X_t = (T - t)^+$ be the remaining life at t of an object with lifetime T. Prove that $F \in$ IFR if and only if X_t is stochastically decreasing in t.

8.22 Consider a device subject to shocks that constitute a Poisson process $N = \sum \varepsilon_{T_n}$ with rate $\lambda > 0$; the shock at T_n causes a random damage X_n. Assume that the X_i are i.i.d. with distribution F and independent of N. The device fails when the cumulative damage $Y_t = \sum_{i=1}^{N_t} X_i$ exceeds a threshold a. Prove that distribution of the failure time $T = \inf\{t : Y_t \geq a\}$ of the device belongs to the class IFRA.

8.23 (Continuation of Exercise 8.22) Suppose that F and a are known but that λ is not. Discuss estimation of λ from observation only of the failure times T_j of i.i.d. devices.

8.24 Let N be a renewal process with interarrival distribution F whose forward recurrence time process satisfies $P\{W_t > y\} = 1 - F(y)$ for all t and y. Prove that N is a Poisson process.

8.25 Let N be a renewal process with interarrival distribution F and backward recurrence time process V. For each t, let $F_t(x)$ be $F(x)$ if $x \leq t$ and 1 otherwise. Prove that if V_t has distribution F_t for all t, then N is Poisson.

8.26 Let N be a renewal process with Weibull interarrival distribution F.
a) Propose and analyze a test of the hypotheses $H_0 : \alpha = 0$; $H_1 : \alpha > 0$ given synchronous observations $\mathcal{F}^N_{T_n}$.
b) Repeat a) for asynchronous observations $\mathcal{F}^N_t$.

8.27 Derive (8.37) and (8.38).

8.28 Prove Proposition 8.29.

8.29 Let N' be the p-thinning of a renewal process N with known, absolutely continuous interarrival distribution F. Only N' is observable; only p is unknown.
a) Show that the $\mathcal{F}^{N'}_t$-estimators $\hat{p} = N'_t/R(t)$ are unbiased, where R is the renewal function of F.
b) Describe the large-sample behavior of these estimators.

8.30 (Continuation of Exercise 8.29) Suppose now that N is also observable.
a) Prove that (N_t, N'_t) is a sufficient statistic for p given observation over $[0, t]$.
b) Compute the asymptotic efficiency of the estimators $\hat{p}$ of Exercise 8.29 relative to the estimators $\hat{p}^* = N'_t/N_t$, where $0/0 = 0$.

8.31 Work through the calculations in Theorem 8.30 for the case that N is a Poisson process.

8.32 Verify carefully the asymptotic independence used in the proof of Theorem 8.31.

8.33 Let N be a renewal process with interarrival distribution F and backward recurrence time process V. For each i, let $N^t_u = N_{t+u} - N_t$.
a) Show that $P\{N^t_u = k|V_t\} = F^k(u + V_t) - F^{k+1}(u + V_t)$ for each k.
b) Suppose now that F is not known. Construct pseudo-state estimators $\hat{P}\{N^t_u = k|V_t\}$ given observations $\mathcal{F}^N_t$ and examine their properties as $t \to \infty$.

8.34 Let Y be a semi-Markov process with finite state space S and semi-Markov kernel $Q(i, j, x) = P(i, j)G(i, j, x)$ as in (8.45). Let N^0 be the marked point process given by (8.49). Assuming that $G(i, j, \cdot)$ has hazard function $h(i, j, \cdot)$, show that $N^0(i, j)$ has $\mathcal{F}^Y$-stochastic intensity

$$\lambda_t(i, j) = \mathbf{1}(Y_{t-} = i)P(i, j)h(i, j, V_{t-}),$$

where V_t is the time since the most recent transition of Y before t.

8.35 Prove Proposition 8.34.

8.36 In the notation of Section 4, prove that for each i and t,

$$H(i, t) = \int_0^t \{(1 - H(i, s-))/E[K(i, s)]\} dE[N(i, s)].$$

8.37 Let N be a renewal process with backward recurrence time process V.
a) Prove that the bivariate process (N_t, V_t) is Markov.
b) Interpret Theorem 8.26 in terms of part a).

NOTES

For a number of years, Feller (1971, Chap. XI) was *the* treatment of renewal theory, but because regenerative processes are omitted it must be considered incomplete. More modern accounts [e.g., Çinlar (1975a) and Ross (1982)] do include regenerative processes; the former also presents semiregenerative processes. Other general sources are Cox (1962), which contains distributional calculations possibly appearing nowhere else, and Smith (1958). Applications of renewal processes are too numerous to describe. Perhaps the most important are queueing and reliability (see Section 2); Daley and Milne (1973) provides access to others.

The introduction surveys the fundamental results; proofs can be found, for example, in Çinlar (1975a) and Feller (1971). Theorem 8.5, the main result, is due in the form stated to Smith (1958); the simpler but equivalent form (8.7) was proved by Blackwell (1948). Recall that the distribution F is *arithmetic* if there exists $a > 0$ such that $F(\{na : n \in N\}) = 1$ and *nonarithmetic* otherwise. An analogue of (8.6) holds in the arithmetic case. Direct Riemann integrability is also treated in Feller (1971); roughly speaking, f is directly Riemann integrable if the integral $\int_0^\infty f(s)ds$ can be defined directly rather than in the improper sense, as the limit as $t \to \infty$ of integrals over $[0, t]$.

Feller (1971) seems to be the first expository treatment of renewal equations *per se* and their systematic application to computations involving functionals of renewal processes. As (8.4) and the proof of Theorem 10.7 show, however, "sample path" arguments can render renewal equations superfluous. Concerning Proposition 8.1, see Cox (1962) or Feller (1971).

Section 8.1

Synchronous observation of renewal processes neither introduces new problems nor requires tools beyond those discussed in Chapter 4. Transition to asynchronous observation entails only having to reconcile real- and discrete-time sampling rates, as in Theorem 8.9, although it must also be assumed that the mean interarrival time is finite. Our approach utilizes results of Serfozo (1975) that deduce asymptotic properties — including functional limit theorems — of a continuous-time process Y from those of an embedded sequence Y_{T_n}, where the T_n are stopping times satisfying a weak law of large numbers. The key additional assumption, that fluctuations of Y between successive T_n not be too wild, is trivially satisfied in our case. Theorem 8.12, although possibly new, is not deep and is but one of many variants. The

Poisson sampling model and Proposition 8.15 are from Kingman (1963) (see also Exercise 6.18 and Chapter 10). Theorem 8.16 is similar to Theorems 7.4 and 7.5.

Parametric inference for renewal processes amounts in practice to inference for gamma- or Weibull-distributed i.i.d. data and is discussed, for example, in Barlow and Proschan (1975) and Cox and Lewis (1966). Sources with more emphasis on survival analysis and estimation of hazard functions are Kalbfleisch and Prentice (1980), Lawless (1982) and Nelson (1982). Vardi (1982a, 1982b) examines length bias in samples of i.i.d. renewal processes over a finite time interval, possibly with additional censoring.

Section 8.2

The "theoretical" properties of likelihoods, Theorems 8.17 and 8.19, are the main results. Evidently, the former is not new, although the martingale proof, with its striking confirmation of the need to use predictable stochastic intensities, may be; direct proofs are also possible (Exercise 8.13). Theorem 8.19 appears here for the first time.

Except for queueing theory, reliability is the oldest area of application of point processes. In its literature one can often find specifics of inference that are glossed over in statistics books. Previously mentioned sources, such as Barlow and Proschan (1975), Kalbfleisch and Prentice (1980) and Nelson (1982), along with Mann et al. (1974), provide entry to the voluminous literature. Barlow et al. (1972) is a nice treatment of inference problems, many but not all of reliability origin, under order restrictions. Qualitative concepts such as stochastic orderings and monotonicity of hazard functions arose mainly in reliability.

Ross (1982) and Stoyan (1983) are expository developments of stochastic orderings, not only the "stochastically larger than" ordering $F \leq F_0$ but also others involving hazard functions and likelihood ratios. The former includes simple coupling constructions such as that in Proposition 8.20; the latter explores extensively implications concerning comparison of stochastic processes. Lindvall (1981, 1982) uses couplings in order to prove renewal theorems.

The families IFR, IFRA and NBU are increasingly broad interpretations of the intuitive idea of "older is worse"; recently, even weaker notions have been devised. Tests for them (and "older is better" counterparts) constitute a central part of reliability theory. In addition to the books mentioned above, one can consult key papers, among them Barlow (1968), Bickel and Doksum (1968), Bickel (1969), and Proschan and Pyke (1967) on tests for IFR; Barlow and Proschan (1969) on tests for IFRA; and Hollander and Proschan (1972) and Chen et al. (1983) on tests for NBU. All treat i.i.d. lifetimes; some allow censoring. Testing for "U-shaped" hazard functions has also been considered; this problem is akin to testing for unimodality of a density function.

Proposition 8.23, possibly the simplest of several related characterizations of Poisson processes among renewal processes, is due to Çinlar and Jagers (1973) (see also Exercises 8.24 and 8.25). Kingman (1963) is the source of Theorems 8.24 and 8.25.

Testing whether a point process is a renewal process has been addressed using serial correlation coefficients and spectra of counts and interarrival times (see, e.g.,

Cox and Lewis, 1966). Difficulties arise with alternative hypotheses. For example, some procedures take as an alternative the point process of Wold (1948), whose interarrival times form a Markov chain, even though this model has never found plausible application.

Section 8.3

Nearly all of this section is new; Theorem 8.30 is the "most novel" result, whereas Theorem 8.26 should really be regarded as part of the folklore and Theorem 8.31 is an amalgamation of other results. Thinned renewal processes have also been studied by Mogyorodi (1971, 1972, 1973a, 1973b) and Räde (1972a, 1972b). Poisson limit theorems for superpositions of independent, sparse renewal processes are given in Çinlar (1972); the renewal structure of the components leads to simplified conditions for validity of general theorems.

Section 8.4

General exposition of Markov renewal and semi-Markov processes is given by Çinlar (1975a, 1975b); Pyke (1961a, 1961b), and Pyke and Schaufele (1964), the latter treating limit theorems, are key early works, although the idea stems from Lévy (1954).
   Estimation for Markov renewal processes divides into pre- and post-martingale eras, each exemplified by a single paper. Moore and Pyke (1968), the source of Theorem 8.33, is the pre-martingale paper. Under (8.48), which holds without loss of generality by expansion of the state space and redefinition of the X_n, one simply combines results for empirical distribution functions (Chapter 4) and those for Markov chains (see, e.g., Basawa and Prakasa Rao, 1980, or Billingsley, 1961a, 1961b). The post-martingale results are from Gill (1980a).

Exercises

8.5 This property, known in analysis as the Choquet-Dény theorem, is a key step in the proof of the renewal theorem given by Feller (1971).

8.6 See Çinlar (1975a).

8.12 Cox and Lewis (1966) contains this and additional aspects, but with a different parameterization.

8.15 Ross (1982) presents related properties.

8.17 Spurious waiting-time "paradoxes" have been inferred in this context.

8.19 M. Brown (1980) presents related properties.

8.22 See Barlow and Proschan (1975).

8.25 Chung (1972) establishes the result under the weaker hypothesis that V_t has distribution F_t only for $t \leq t_0$.

8.26 This is a parametric test whether F has increasing failure rate.

8.34 See also Proposition 8.34.

8.36 See Gill (1980a); this is the substitution motivation for (8.52).

9
Inference for Stationary Point Processes

In earlier chapters we emphasized "multiple realization" inference (Model 3.1), in which setting the structure of the underlying space was not restricted severely. By contrast, in this chapter we address only single-realization inference (Model 3.2) in a context entailing strong assumptions on the space and the point processes. Stationarity replaces the "identically distributed" part of "i.i.d."; to derive asymptotic properties of estimators we substitute for "independent," hypotheses of ergodicity and finiteness of reduced cumulant measures.

Of course, in order that stationarity even make sense the underlying space must admit a group structure; for concreteness and simplicity we shall suppose that $E = \mathbf{R}^d$ for some $d \geq 1$.

Perhaps the most severe restriction imposed by the setting concerns *what* can be estimated easily and effectively. Whereas given i.i.d. multiple realizations of a point process N, one can estimate in a straightforward manner almost any aspect of the probability law $\mathcal{L}_N$ (see Chapter 4), here we focus on estimation of Palm measures, which do determine $\mathcal{L}_N$, albeit indirectly, but from which estimators of $\mathcal{L}_N$ and many functionals, notably reduced moment measures, can be constructed by substitution.

We pause to establish notation, and to recall concepts and results from Section 1.8. Let $E = \mathbf{R}^d$, and for $x \in E$, let τ_x be the translation operator $y \to \tau_x y = y - x$. Lebesgue measure on E is denoted simply by dx, or if necessary by λ. A point process N on E, defined over the probability space $(\Omega, \mathcal{F}, P)$, is *stationary* with respect to shift operators $(\theta_x)_{x \in E}$ satisfying conditions set forth in Section 1.8, provided that $N\tau_x^{-1} = N \circ \theta_x$ for each x, and that (θ_x) be measure-preserving for P: $P\theta_x^{-1} = P$ for each x. In

348

consequence, N is *stationary in law*:

$$N\tau_x^{-1} \stackrel{d}{=} N \tag{9.1}$$

for every x; conversely, if (9.1) holds, then there exists a canonical realization of N that is stationary.

By (9.1), for each x the mean measure μ_N satisfies $\mu_N \tau_x^{-1} = \mu_N$, and hence has the form $\mu_N = \nu_N \lambda$; the constant ν_N is the *intensity* of N. We assume throughout the chapter, often without explicit mention, that ν_N is *finite and positive*; usually we impose stronger moment conditions.

The Palm measure P^* of N has the interpretation $P^*(\Gamma)/P^*(\Omega) = P\{N \in \Gamma | N(\{0\}) = 1\}$, and satisfies

$$E\left[\int N(\omega, dx) H(\theta_x, \omega)\right] = E^*\left[\int H(\omega, x) dx\right], \tag{9.2}$$

where $H : \Omega \times E \to \mathbf{R}$ is bounded and measurable. Palm measures play a central role in estimation procedures developed in Section 1, mainly through a special case of (9.2): for H a function from $\mathbf{M}_p$ to $\mathbf{R}$ and K a compact subset of E,

$$E^*[H(N)] = (1/\lambda(K))E\left[\int_K N(dx) H(N\tau_x^{-1})\right]. \tag{9.3}$$

Thus we use integrals $(1/\lambda(K)) \int_K N(dx) H(N\tau_x^{-1})$, to estimate $E^*[H(N)]$. Here K denotes the set over which N is observed.

We now present a "zero-infinity" law for stationary point processes: for a stationary point process, any finitely defined configuration of points (e.g., pairs of points within a prescribed distance of one another) occurs either not at all or infinitely often. Exercise 1.28 is a special case: for N stationary point process on $\mathbf{R}$, $P\{N = 0\} + P\{N(-\infty, 0) = N(0, \infty) = \infty\} = 1$. Such a result is required in order to ensure that for sets $K \uparrow E$ there is enough data (in particular, infinitely many points of N) for consistency and asymptotic normality to hold. We recall the notation

$$N^k(dx_1, \ldots, dx_k) = N(dx_1) \cdots N(dx_k),$$

so that for $A \in \mathcal{E}^k$, $N^k(A)/k!$ is the number of k-configurations of points of N falling in A.

Proposition 9.1. Let N be a stationary point process and let A be a subset of E^k invariant under the transformation $(y_1, \ldots, y_k) \to (y_1 - x, \ldots, y_k - x)$ for each $x \in E$. Then $P\{N^k(A) = 0\} + P\{N^k(A) = \infty\} = 1$.

Proof: Let $\Gamma = \{0 < N^k(A) < \infty\}$. If $P\{\Gamma\} > 0$, then the formula $Q\{\cdot\} = P\{\cdot|\Gamma\}$ defines a probability with respect to which N remains stationary, but under which the E-valued functional

$$Y = (1/N^k(A))\int_A[(1/k)\sum_{i=1}^k x_i]N^k(dx)$$

has a translation invariant distribution, which would be the nonexistent uniform distribution on E. ∎

The remainder of the chapter concentrates on statistical inference for stationary point processes, and is organized in the following manner. Section 1 concerns estimation of Palm integrals (9.3). Among these are the intensity of N as well as reduced moment and cumulant measures. We assume stationarity and ergodicity, but the setting is nonparametric.

Concepts and techniques of spectral analysis appear in Section 2; here one exploits stationarity of the increments of a stationary point process in the frequency domain rather than the time domain. A spectral representation theorem is derived, spectral measures and density functions are defined, and methods for estimation are described. Likelihood ratios and parametric maximum likelihood estimation are examined in Section 3, along with the structure of the stochastic intensity of a stationary point process on **R**. Finally, in Section 4 we discuss state estimation for stationary point processes, concentrating on linear state estimators but considering both time- and frequency-domain approaches.

9.1 Estimation of Palm and Moment Measures

We pause to recall further notation from Chapter 1.

The point process N admits a moment of order k with respect to P if the measure

$$\mu^k(dx_1,\ldots,dx_k) = E[N^k(dx_1,\ldots,dx_k)]$$

belongs to $\mathbf{M}(E^k)$, in which case it is termed the *moment measure* of order k. For $z \in E^{k-1}$ let λ_z be the image of Lebesgue measure λ under the mapping $x \to (z_1 + x,\ldots,z_{k-1}+x,x)$ of E into E^k. By Proposition 1.60, there exists a disintegration

$$\mu^k(\cdot) = \int_{E^{k-1}}\lambda_z(\cdot)\mu_*^k(dz), \tag{9.4}$$

where μ_*^k, a measure on E^{k-1}, is the *reduced moment measure* of order k. For $k = 1$, $E^{k-1} = \{0\}$ and $\mu_*^1 = \nu\varepsilon_0$, where ν is the intensity of N.

Cumulant and reduced cumulant measures are defined analogously. If N admits a moment of order k then (cf. Definition 1.62) the measure

$$\gamma^k(\otimes_{j=1}^k f_j) = \sum_{\mathcal{J}=\{J_\ell\}}(-1)^{|\mathcal{J}|-1}(|\mathcal{J}|-1)!\prod_\ell \mu^{|J_\ell|}(\otimes_{j\in J_\ell}f_j) \tag{9.5}$$

is the *cumulant measure* of order k. Here $\otimes f_j(x) = \prod f_j(x_j)$, and the summation is over all partitions $\mathcal{J}$ of $\{1, \ldots, k\}$. The *reduced cumulant measure* γ_*^k, which satisfies

$$\gamma^k(\cdot) = \int_{E^{k-1}} \lambda_z(\cdot) \gamma_*^k(dz), \tag{9.6}$$

is a signed measure in general, but whereas moment measures, reduced moment measures and cumulant measures are not finite except in trivial cases, reduced cumulant measures may have finite total variation. This finiteness (analogous to integrability of the covariance function of a stationary process on $\mathbf{R}$) implies that distant parts of N are nearly independent.

The *covariance measure* ρ is the cumulant measure of order two; for it the disintegration (9.6) becomes $\rho_*(dx) = \mu_*^2(dx) - \nu^2 dx$, with ρ_* known as the *reduced covariance measure*. When N is a Poisson process with intensity ν, then $\rho_* = \nu \varepsilon_0$, a manifestation of independent increments. Estimation procedures applicable to reduced moment measures yield by substitution estimates of the reduced covariance measure; these are applied in Section 4 to the problem of combined inference and linear state estimation.

Reduced moment measures are ordinary moment measures with respect to Palm measures; this property, given as well in Proposition 1.68, is restated here for convenience. The proof appears in Krickeberg (1982).

Lemma 9.2. Let N be a stationary point process admitting a moment of order $k \geq 2$ with respect to P. Then $\mu_*^k = E^*[N^{k-1}]$. $\square$

We now describe the nonparametric estimators whose properties are developed here and in Section 2. Let N be a stationary point process with unknown probability law P and suppose that a single realization of N has been observed over a compact, convex subset K of $\mathbf{R}^d$. We assume throughout that $\Omega = \mathbf{M}_p$.

The fundamental estimators are those of the Palm measure. Given the observations $\mathcal{F}^N(K)$, the integral

$$P^*(H) = \int H \, dP^* = E\left[\int H(N \circ \theta_x) N(dx)\right],$$

is estimated by

$$\hat{P}^*(H) = (1/\lambda(K)) \int_K H(N\tau_x^{-1}) N(dx). \tag{9.7}$$

In (9.7), as customary, we suppress dependence on the "sample size" K. Indeed, the interpretation of K as sample size is rather loose because $\hat{P}^*(H)$ may not be measurable with respect to $\mathcal{F}^N(K)$. In many specific cases, there is a bounded set A (depending on H) such that $H(\mu) = H(\mu_A)$, where μ_A is the restriction of μ to A, which renders $\hat{P}^*(H)$ measurable with respect to

$\mathcal{F}^N(K + A)$. Even this is not true, however, of the estimators of the spectral measure and spectral density introduced in Section 2.

These estimators are unbiased:

$$
\begin{aligned}
E[\hat{P}^*(H)] &= (1/\lambda(K))E[\int_K H(N \circ \theta_x)N(dx)] \\
&= (1/\lambda(K))E^*[H(N)\int_K dx] \\
&= E^*[H(N)] = P^*(H).
\end{aligned}
$$

However, a more compelling justification is based on the conditioning interpretation of P^*. For each $x \in K$ that is a point of N — and only such x contribute to the integral in (9.7) — $N\tau_x^{-1}$ has a point at the origin, and therefore $\hat{P}^*(H)$ is simply a weighted average of H-values of translations of N placing each point in turn at the origin.

Other estimators are derived from $\hat{P}^*$ by substitution. For estimation of the intensity, taking $H \equiv 1$ in (9.7) yields the (obvious) estimator $\hat{\nu} = N(K)/\lambda(K)$, which is $\mathcal{F}^N(K)$-measurable, and whose asymptotic properties were discussed in Section 1.8.

We present now the next simplest and most important special case.

Example 9.3. (Reduced second moment measure). For estimation of the reduced second moment measure, choosing $H(N) = N(f)$, where $f \in C^+(E)$, gives

$$
\hat{\mu}_*^2(f) = (1/\lambda(K))\int_K N(dx) \int f(y - x)N(dy). \tag{9.8}
$$

as $\mathcal{F}^N((\mathrm{supp}\, f) - K)$-measurable estimator of $\mu_*^2(f)$. $\square$

Other reduced moment and cumulant measures may be estimated similarly.

We require additional notation used in connection with Theorem 1.72. For K bounded and convex, let $\delta(K)$ be the supremum of the radii of Euclidean balls contained in K. In order to have the "infinitely much" data necessary for consistent estimation we require that $\delta(K) \to \infty$. The crucial geometric property is that convex sets grow more rapidly than their boundaries: for every $\varepsilon > 0$, $\lambda((\partial K)^\varepsilon)/\lambda(K) \to 0$, where $(\partial K)^\varepsilon$ is the set of points within distance ε of the boundary of K.

Here is our main consistency theorem, for the estimators $\hat{P}^*$ of (9.7).

Theorem 9.4. Assume that P is ergodic and that the intensity ν is positive and finite, and let $\mathcal{H}$ be a uniformly bounded set of continuous functions on $\mathbf{M}_p$ that is compact in the topology of uniform convergence on compact subsets. Then almost surely with respect to P,

$$
\lim_{\delta(K) \to \infty} \sup_{H \in \mathcal{H}} |\hat{P}^*(H) - P^*(H)| = 0. \tag{9.9}
$$

Proof: We combine appeal to Theorem 1.72 with arguments adapted from Karr (1987b). First, for H a fixed element of $C(\mathbf{M}_p)$, Theorem 1.72 applies to the random measure

$$M(A) = \int_A H(N\tau_x^{-1})N(dx) = \int_A H(N \circ \theta_x)N(dx),$$

with the consequence that

$$\lim_{\delta(K)\to\infty} \hat{P}^*(H) = \lim_{\delta(K)\to\infty} M(K)/\lambda(K) = E[M([0,1]^d)] = P^*(H) \quad (9.10)$$

almost surely (and in $L^1(P)$ as well). In particular, from the choice $H \equiv 1$ we infer strong consistency of the estimators $\hat{\nu} = N(K)/\lambda(K)$:

$$\lim_{\delta(K)\to\infty} \hat{\nu} = \nu \quad (9.11)$$

almost surely.

Turning to $\mathcal{H}$, given $\epsilon > 0$ there exists by finiteness of P^* (recall that $P^*(\Omega) = \nu$) a compact subset Γ of $\mathbf{M}_p$ such that $P^*(\mathbf{M}_p \backslash \Gamma) < \epsilon$; we suppose without loss of generality that Γ is a P^*-continuity set, i.e., $P^*(\partial\Gamma) = 0$. Moreover, there exist $\tilde{H}_1, \ldots, \tilde{H}_L \in \mathcal{H}$ such that to each $H \in \mathcal{H}$ there corresponds $\ell(H) \in \{1, \ldots, L\}$ for which $\|H - \tilde{H}_{\ell(H)}\|_\Gamma$ ($= \sup_{\omega \in \Gamma} |H(\omega) - \tilde{H}_{\ell(H)}(\omega)|$) is at most $< \epsilon$. Then assuming that (9.10) holds for $\tilde{H}_1, \ldots, \tilde{H}_L$ and that (9.11) holds as well, in the decomposition

$$\begin{aligned}
\sup_{H \in \mathcal{H}} |\hat{P}^*(H) - P^*(H)| \leq{} & \sup_{H \in \mathcal{H}} |\hat{P}^*(H) - \hat{P}^*(\tilde{H}_{\ell(H)})| \\
& + \max_L |\hat{P}^*(\tilde{H}_\ell) - P^*(\tilde{H}_\ell)| \\
& + \sup_{H \in \mathcal{H}} |P^*(\tilde{H}_{\ell(H)}) - P^*(H)| \,(9.12)
\end{aligned}$$

the second term converges to zero almost surely.

Concerning the third, for each H,

$$\begin{aligned}
|P^*(\tilde{H}_{\ell(H)}) &- P^*(H)| \\
&\leq |P^*(\tilde{H}_{\ell(H)}\mathbf{1}_\Gamma) - P^*(H\mathbf{1}_\Gamma)| + |P^*(\tilde{H}_{\ell(H)}\mathbf{1}_{\Gamma^c}) - P^*(H\mathbf{1}_{\Gamma^c})| \\
&\leq \epsilon P^*(\Gamma) + P^*(\Gamma^c)\sup_{H \in \mathcal{H}}\|H\|_\infty \\
&\leq \epsilon(\nu + \sup_{H \in \mathcal{H}}\|H\|_\infty), \quad (9.13)
\end{aligned}$$

so that this term can be made arbitrarily small by proper choice of ϵ.

Finally, by straightforward arguments (cf. Theorem 4.2), (9.10) implies that almost surely $\hat{P}^* \to P^*$ vaguely as Radon measures on Ω, and hence also weakly, by (9.11). Because Γ is a P^*-continuity set, almost surely $\hat{P}^*(\Gamma) < \varepsilon$ for all sufficiently large K; consequently

$$\sup_{H \in \mathcal{H}} |\hat{P}^*(H) - P^*(\tilde{H}_{\ell(H)})| \leq \varepsilon \hat{P}^*(\Gamma) + \hat{P}^*(\Gamma^c) \sup_{H \in \mathcal{H}} \|H\|_\infty$$
$$\leq \varepsilon(2\nu + \sup_{H \in \mathcal{H}} \|H\|_\infty)$$

once $\delta(K)$ is large enough, which completes the proof. ∎

Strong uniform consistency of the estimators (9.8) of the reduced second moment measure will be shown, ultimately, in two forms, the first more intrinsically useful for estimation of μ_*^2, and the second, in Section 2, directed at estimation of the spectral measure.

Theorem 9.5. Assume that P is ergodic and that under P, N admits a moment of order two. Let $\mathcal{K}$ be a compact, uniformly bounded subset of $C^+(E)$, each element of which is supported in the same compact subset K_0 of E. Then almost surely,

$$\lim_{\delta(K) \to \infty} \sup_{f \in \mathcal{K}} |\hat{\mu}_*^2(f) - \mu_*^2(f)| = 0. \tag{9.14}$$

Proof: Given $f \in \mathcal{K}$, define $H_f : \mathbf{M}_p \to \mathbf{R}$ by $H_f(\mu) = \mu(f)$. Then H_f is continuous, and the proof reduces to showing that the mapping $f \to H_f$ of $C^+(E)$ into the set of bounded, continuous functions on $\mathbf{M}_p$ is itself continuous, for then $\mathcal{H} = \{H_f : f \in \mathcal{K}\}$ is the continuous image of a compact set, and is hence compact. At this point, (9.14) follows from (9.9).

For Γ a compact subset of $\mathbf{M}_p$, $a = \sup_{\mu \in \Gamma} \mu(K_0)$ is finite, and consequently for f and $g \in \mathcal{K}$,

$$\sup_{\mu \in \Gamma} |H_f(\mu) - H_g(\mu)| = \sup_{\mu \in \Gamma} |\mu(f) - \mu(g)| \leq a\|f - g\|_\infty. \blacksquare$$

Finally we consider estimation of the probability law P itself. Even though P is uniquely determined by the Palm measure P^* and even though, as the preceding development confirms, many functionals of interest in inference are easily expressed as functionals of P^*, estimation of P remains an important problem.

Our estimators, given by

$$\hat{P}(H) = (1/\nu\lambda(K)^2) \int_K N(dx) \int_K H(N\tau_{x-y}^{-1}) dy, \tag{9.15}$$

are motivated by the identity

$$E[N(K)H(N)] = E^* \left[\int_K H(N\tau_{-y}^{-1}) dy \right], \tag{9.16}$$

which follows at once from (9.2). In the following theorem we establish strong — but not uniform — consistency of these estimators; indeed, strong consistency in Theorem 9.6 requires the full force of uniformity in Theorem 9.4.

Theorem 9.6. Assume that P is ergodic and that the intensity is positive and finite, and let H be a bounded, continuous function on $\mathbf{M}_p$. Then $\hat{P}(H) \to P(H) = \int H \, dP$ almost surely as $\delta(K) \to \infty$.

Proof: For each K, by (9.16),

$$
\begin{aligned}
|\hat{P}^*(H) - P(H)| \leq \ & \Big| P(H) - (1/\nu) E[H(N) N(K)/\lambda(K)] \Big| \\
& + (1/\nu) \Big| (1/\lambda(K)) \int_K E^*[H(N\tau_y^{-1})] dy \\
& - (1/\lambda(K)^2) \int_K N(dx) \int_K H(N\tau_{x-y}^{-1}) dy \Big| \quad (9.17)
\end{aligned}
$$

Since $N(K)/\lambda(K) \to \nu$, in L^1 (see discussion in the proof of Theorem 9.4), $(1/\nu) E[H(N) N(K)/\lambda(K)] \to E[H(N)] = P(H)$, so that the first term in (9.17) converges to zero. By Theorem 9.4 applied to the family $\mathcal{H}$ of functionals $H_y(\mu) = H(\mu\tau_{-y}^{-1})$ (we omit the straightforward verification of the hypotheses) we infer that

$$
(1/\lambda(K)) \int_K H(N\tau_{x-y}^{-1}) N(dx) \to E^*[H(N\tau_y^{-1})]
$$

uniformly in y. By an analytical argument, for which uniformity in here is crucial, this implies that the second term in (9.17), whose components are $1/\lambda(K)$ times the dy-integrals of the two sides over K, converges to zero almost surely. ∎

We next discuss a central limit theorem and Poisson approximation theorem complementing the consistency theorems. For simplicity we work only in the context of Theorem 9.4.

Our central limit theorem extends that of Jolivet (1981) for estimators of reduced moment measures [the estimators (9.8) correspond to the reduced second moment measure], but does not weaken the hypotheses.

Theorem 9.7. Suppose that P is ergodic, that under P moments of N of every order exist, and that each reduced cumulant measure γ_*^k of order $k \geq 2$ has finite total variation. Then there exists a mean zero Gaussian process $\{G(H) : H \in C(\mathbf{M}_p)\}$ such that for each H, as $\delta(K) \to \infty$

$$
\sqrt{\lambda(K)}[\hat{P}^*(H) - P^*(H)] \xrightarrow{d} G(H). \quad (9.18)
$$

Proof: Consider the class $\mathcal{A}$ of functions $H(\mu) = \sum_{j=1}^{n} c_j \mu^{k_j-1}(f_j)^{d_j}$, where the c_j are real constants, k_j and d_j are positive integers and $f_j \in C(E^{k_j-1})$. This class of "polynomials" is a vector space and an algebra (i.e., is closed under pointwise multiplication) and evidently separates the points of M_p; consequently by the Stone-Weierstrass theorem, the uniform closure of $\mathcal{A}$ is $C(M_p)$. It suffices, therefore, to show that (9.18) holds whenever $H \in \mathcal{A}$. By the continuous mapping theorem, this last assertion holds if for each n, $k_1, \ldots, k_n$ and $f_1, \ldots, f_n$,

$$\sqrt{\lambda(K)} \left[\left(\hat{\mu}_*^{k_1}(f_1), \ldots, \hat{\mu}_*^{k_n}(f_n) \right) - \left(\mu_*^{k_1}(f_1), \ldots, \mu_*^{k_n}(f_n) \right) \right]$$

$$\xrightarrow{d} (G(H_1), \ldots, G(H_n)), \tag{9.19}$$

where $H_j(\mu) = \mu^{k_j-1}(f_j)$ and where $\hat{\mu}_*^k(f) = \int_K N^{k-1} \circ \theta_x(f) N(dx)$, as in (9.8). Using the Cramér-Wold device we can reduce (9.19) to showing that for each k and f,

$$\sqrt{\lambda(K)}[\hat{\mu}_*^k(f) - \mu_*^k(f)] \xrightarrow{d} G(H_f), \tag{9.20}$$

where $H_f(\mu) = \mu^{k-1}(f)$, but (9.20) holds, in the presence of our hypotheses, by Jolivet (1981, Theorem 1), as we now describe briefly.

For a random variable Y admitting moment-generating function $m(t) = E[e^{tY}]$, the coefficients c_ℓ in the expansion $\log m(t) = \sum_{\ell=0}^{\infty} c_\ell t^\ell / \ell!$ are known as the *cumulants* of Y. They are also realized by application of (9.5), with $f_j \equiv 1$, to the random measure $M = Y \varepsilon_0$ on $E^0 = \{0\}$; consequently,

$$c_\ell = \sum (-1)^{|\mathcal{J}|-1} (|\mathcal{J}| - 1)! \prod E[Y^{|\mathcal{J}_q|}],$$

where notation and domain of the summation are as in Definition 1.62. The distribution of Y is normal if and only if $c_2 \neq 0$ but $c_\ell = 0$ for every $\ell > 2$; moreover, convergence of cumulants implies convergence in distribution of random variables (Exercise 9.9).

Given this, the proof is simple conceptually. One shows that as $\delta(K) \to \infty$ the ℓth-order cumulant c_ℓ of $\int_K N^{k-1} \circ \theta_x(f) N(dx)$ satisfies $c_\ell \cong \lambda(K)$ independently of ℓ, from which it follows that the ℓth-order cumulant of $\sqrt{\lambda(K)}[\hat{\mu}_*^k(f) - \mu_*^k(f)]$ behaves asymptotically as $\delta(K)^{1-\ell/2}$, implying not only the convergence (9.20) but also that the limit is normal.

The difficult (albeit mostly tedious) part (Jolivet, 1981) is calculation of the cumulants c_ℓ. With $g_k(x_1, \ldots, x_k) = \mathbf{1}(x_k \in K) f(x_1 - x_k, \ldots, x_{k-1} - x_k)$, we have $\int_K N^{k-1} \circ \theta_x(f) N(dx) = N^k(g_k)$, and therefore for each ℓ,

$$E\left[\left\{ \int_K N^{k-1} \circ \theta_x(f) N(dx) \right\}^\ell \right] = E[N^k(g_k)^\ell] = \mu^{k\ell}(\otimes g_k).$$

By stationarity (and lengthy calculations), $\mu^{k\ell}(\otimes g_k) \cong \lambda(K)$; this is certainly plausible even if it is difficult to verify. The method used by Jolivet (1981) decomposes cumulant measures into factorial moment measures [not introduced here (see Krickeberg, 1982)] and then applies the computational rules of Leonov and Shiryaev (1959). ∎

In addition to Poisson processes, certain Neyman-Scott cluster processes (Section 7.5) are known to fulfill the hypotheses of the theorem. Calculation of the variance σ^2 associated with (9.20) seems difficult. That it depends only on the moment measures $\mu^1, \ldots, \mu^{2k}$ does not appear to help, nor do the calculations in Jolivet (1981).

The hypotheses of Theorem 9.7 are rather severe, in part because of the generality of the convergence condition that $\delta(K) \to \infty$. Another shortcoming is that normal approximations are ineffective for estimation of small probabilities. Poisson approximations, by contrast, can estimate small probabilities. Unfortunately, however, in the following theorem the severity of the assumptions in Theorem 9.7 is not mitigated.

For each $r > 0$ let B_r be the closed ball of radius r centered at the origin.

Theorem 9.8. Let Γ_r, $r > 0$, be decreasing events for which there exists a finite measure ξ on Ω such that as $r \to \infty$ $r^d P^*(\cdot \cap \Gamma_r) \xrightarrow{d} \xi(\cdot)$. For each r, let N_r be the marked point process on $\mathbf{M}_p \times B_1$ defined by

$$N_r = \sum_i \mathbf{1}(N\tau_{X_i}^{-1} \in \Gamma_r)\mathbf{1}(X_i \in B_r)\varepsilon_{(N\tau_{X_i}^{-1}, X_i/r)}. \qquad (9.21)$$

If the hypotheses of Theorem 9.7 are satisfied, then as $r \to \infty$, $N_r \xrightarrow{d} \bar{N}$, where $\bar{N}$ is a Poisson process with mean measure $\eta(\Gamma \times B) = \xi(\Gamma)\lambda(B)$.

Proof: We verify first that $E[N_r] \to \eta$ in the sense of weak convergence. Indeed for Γ a ξ-continuity set and with $rB = \{rx : x \in B\}$,

$$
\begin{aligned}
E[N_r(\Gamma \times B)] &= E\left[\int_{B_r} \mathbf{1}(N\tau_x^{-1} \in \Gamma \cap \Gamma_r))\mathbf{1}(x \in B)N(dx)\right] \\
&= P^*(N \in \Gamma \cap \Gamma_r)\lambda(B_r \cap rB) \\
&= P^*(N \in \Gamma \cap \Gamma_r)r^d\lambda(B_1 \cap B) \\
&= P^*(N \in \Gamma \cap \Gamma_r)r^d\lambda(B) \\
&\to \xi(\Gamma)\lambda(B).
\end{aligned}
$$

It follows from this computation (see Proposition 1.23) that (N_r) is tight, and hence it remains to show that for any "subsequence" $(N_{r'})$ converging in distribution, the limit is Poisson with mean measure η.

For this it suffices by Proposition 1.22 to verify that $P\{N_r(\Gamma \times B) = 0\} \to e^{-\eta(\Gamma \times B)}$ for Γ a ξ-continuity set and B a Borel subset of B_1. Under the

assumptions of Theorem 9.7, in the manner of Jolivet (1981), one may use
(9.21) and the rules of Leonov and Shiryaev (1959) to evaluate the cumulants
of $N_r(\Gamma \times B)$. With details omitted, the result is that for each ℓth-order
cumulant of $N_r(\Gamma \times B)$ converges to $\eta(\Gamma \times B)$, as required ∎

9.2   Spectral Analysis

Although we now treat frequency-domain analysis of stationary point processes, as opposed to the time/space-domain analysis of Sections 1 and 3,
estimation techniques from Section 1 retain central roles. We begin with a
spectral representation theorem for stationary point processes; related structural aspects for stationary point processes on **R**, in particular a Wold decomposition fundamental to linear prediction, are presented in Section 4.
In the remainder of this section we deduce from the spectral representation
corresponding representations for reduced second moment and covariance
measures, and thereafter examine estimation of frequency-domain descriptors. Even though we deal only with second moment structure, we retain the
assumption of stationarity because it permits application of limit theorems
from Section 1.

Let $\mathcal{D}$ denote the space of infinitely differentiable functions on $\mathbf{R}^d$ whose
support is compact ("test functions") and let $\mathcal{S}$ be the set of "rapidly decreasing" functions on $\mathbf{R}^d$. [For topology and functional analysis background
and details, consult Rudin (1973, Chap. 7).] Given a function ψ in $\mathcal{S}$, denote
its Fourier transform by $\tilde{\psi}(v) = \int e^{i\langle v,x \rangle} \psi(x) dx$.

Here is the key result of the section.

Theorem 9.9. Let N be a stationary point process on $\mathbf{R}^d$ admitting
a moment of order 2. Then there exists a unique complex-valued random
measure Z on $\mathbf{R}^d$ with orthogonal increments such that for each $\psi \in \mathcal{D}$,

$$N(\psi) = \int \tilde{\psi}(v) Z(dv). \tag{9.22}$$

Proof: Consider the centered process $M(\psi) = N(\psi) - \nu\lambda(\psi)$, where ν
is the intensity of N, viewed as a *random distribution* on $\mathcal{D}$ (i.e., a random
continuous linear functional). Let G be the spectral measure (Itô, 1953,
Theorem 3.1) induced by M:

$$E[M(\psi)M(\varphi)] = G(\tilde{\psi}\tilde{\varphi}) \tag{9.23}$$

for ψ and $\varphi \in \mathcal{D}$. The Fourier transform $\psi \to \tilde{\psi}$ is a homeomorphism of
$\mathcal{S}$, endowed with the "Schwartz" topology, onto itself. Moreover, $\tilde{\mathcal{D}} = \{\tilde{\psi} :$

$\psi \in \mathcal{D}\}$ is dense in $\mathcal{S}$, and $\mathcal{S}$ is dense in $L^2(G)$; consequently, $\tilde{\mathcal{D}}$ is dense in $L^2(G)$ in *its usual* topology. Given $\psi \in \mathcal{D}$, define $Z_0(\tilde{\psi}) = M(\psi)$. Then as a mapping of $\tilde{\mathcal{D}} \subseteq L^2(G)$ into $L^2(P)$, Z_0 is a linear isometry, the latter since $\|Z_0(\psi)\|^2_{L^2(P)} = E[|M(\psi)|^2] = \|\tilde{\psi}\|^2_{L^2(G)}$. Because $\tilde{\mathcal{D}}$ is dense in $L^2(G)$, Z_0 can be extended to become a linear isometry from $L^2(G)$ into $L^2(P)$, and the formula $Z_0(B) = Z_0(\mathbf{1}_B)$ then defines a random measure with orthogonal increments satisfying

$$N(B) - \nu\lambda(B) = \int Z_0(dv)\int_B e^{i\langle v,x\rangle}dx.$$

Conversion to a random measure Z fulfilling (9.22) is immediate. ∎

We term (9.22) the *spectral representation* of N; there and elsewhere below integrals with respect to Z are in the L^2-sense. In particular, for $u < t$,

$$N_t - N_u = \int \frac{e^{i\langle t,v\rangle} - e^{i\langle u,v\rangle}}{iv} Z(dv).$$

The spectral measure is the fundamental object of interest in frequency-domain analysis of stationary point processes even though, but in other senses precisely because, it depends on the distribution of the process only through the reduced second moment measure (as shown in Proposition 9.11).

Definition 9.10. Let N be a stationary point process with spectral representation (9.22). The measure $F(dv) = E[|Z(dv)|^2]$ is the *spectral measure* of N. The measure G satisfying (9.23) is the *covariance spectral measure* of N. If it exists, the density g_ρ of G is termed the (covariance) *spectral density function* of N.

It is important to work with the covariance spectral density function because it exists under reasonably broad assumptions, whereas the spectral measure F is typically not absolutely continuous. Moreover, the spectral density function can be estimated within our setting.

The crucial relationship linking second moment measures and spectral measures is the following. For $\psi \in \mathcal{D}$, let $\hat{\psi}(x) = (1/2\pi)\int e^{-i\langle x,v\rangle}\psi(v)dv$ denote its inverse Fourier transform.

Proposition 9.11. Let N be a stationary point process on $\mathbf{R}$ admitting moment of order 2. Then the reduced second moment measure μ_*^2 and spectral measure F satisfy, for $\psi \in \mathcal{D}$, the Parseval relations

$$\mu_*^2(\psi) = F(\tilde{\psi}) \tag{9.24}$$
$$F(\psi) = \mu_*^2(\hat{\psi}). \tag{9.25}$$

Proof: It is enough to establish (9.24). Given φ_1 and $\varphi_2 \in \mathcal{D}$, define $\psi \in \mathcal{D}$ by

$$\psi(z) = \int \varphi_1(z + x)\varphi_2(x)dx. \tag{9.26}$$

Then by straightforward calculations, with Z the spectral measure of (9.22),

$$
\begin{aligned}
\mu_*^2(\psi) &= E[N(\varphi_1)N(\varphi_2)] \\
&= E[N(\varphi_1)\overline{N(\varphi_2)}] \\
&= E[Z(\tilde{\varphi}_1)\overline{Z(\tilde{\varphi}_2)}] \\
&= E\left[\left(\int Z(dv)\int e^{i\langle v,x\rangle}\varphi_1(x)dx\right)\left(\int \overline{Z(du)}\int e^{-i\langle u,y\rangle}\varphi_2(y)dy\right)\right] \\
&= \int\int E[Z(dv)\overline{Z(du)}\tilde{\varphi}_1(v)\tilde{\varphi}_2(-u)] \\
&= \int F(dv)\tilde{\varphi}_1(v)\tilde{\varphi}_2(-v) = F(\tilde{\psi}).
\end{aligned}
$$

Hence (9.24) holds for $\psi \in \mathcal{D}$ of the form (9.26); since the set of such ψ is dense, the proof is complete. ∎

For the reduced covariance measure $\rho_*(dx) = \mu_*^2(dx) - \nu^2 dx$, the following version of Proposition 9.11 holds.

Proposition 9.12. For N a stationary point process on $\mathbf{R}$ admitting moment of order 2, the covariance spectral measure G and reduced covariance measure ρ_* satisfy the Parseval relations

$$
\begin{aligned}
\rho_*(\psi) &= G(\tilde{\psi}) \\
G(\psi) &= \rho_*(\hat{\psi})
\end{aligned}
$$

for all $\psi \in \mathcal{D}$. If the spectral density function g_ρ exists, then

$$g_\rho(v) = (1/2\pi)E^*\left[\int e^{-i\langle v,y\rangle}(N - \nu\lambda)(dy)\right] \tag{9.27}$$

for each v. □

Turning now to inference, our statistical model is the following. With the sample space and the point process N fixed, let $\mathcal{P}$ consist of all probability measures P with respect to which

- N is stationary and ergodic;

- N admits a moment of order 2, and hence reduced second moment measure μ_*^2 and spectral measure F. (When it exists the spectral density function is denoted by g_ρ.)

For estimation of the spectral measure we employ techniques developed in Section 1. From (9.8) and (9.25), given a compact subset K of $\mathbf{R}^d$ we obtain the substitution estimator

$$\hat{F}(\psi) = \hat{\mu}_*^2(\hat{\psi}) = (1/\lambda(K))\textstyle\int_K N(dx)\int\hat{\psi}(y-x)N(dy),$$

where $\hat{\psi}$ is the inverse Fourier transform of ψ.

These estimators are unbiased and strongly uniformly consistent.

Theorem 9.13. Assume that P is ergodic and admits a moment of order 2. Let $\mathcal{K}$ be a compact subset of $\mathcal{D}$, all of whose elements are supported in the same compact subset K_0 of $\mathbf{R}^d$. Then almost surely

$$\lim_{\delta(K)\to\infty}\sup_{\psi\in\mathcal{K}}|\hat{F}(\psi) - F(\psi)| = 0. \tag{9.28}$$

Proof: Since $\hat{\psi}$ does not have compact support, (9.28) *does not* ensue immediately from (9.14). Instead, we follow with minor modifications the *reasoning* used to prove Theorem 9.5, and show that

$$\lim_{\delta(K)\to\infty}\sup_{\psi\in\mathcal{K}}|\hat{\mu}_*^2(\hat{\psi}) - \mu_*^2(\hat{\psi})| = 0,$$

which gives (9.28) at once via (9.25).

In view of (9.11) we first replace $\mathbf{M}_p$ in the proof of Theorem 9.4 by $\mathbf{M}_p(\nu) = \{\mu : \lim\mu(K)/\lambda(K) = \nu\}$. As in the proof of Theorem 9.5, define $H_\psi(\mu) = \mu(\hat{\psi})$; it suffices to show that the mapping $\psi \to H_\psi$ is continuous, for then we may appeal to a minor alteration of Theorem 9.4 in order to conclude the proof. By compactness of $\mathcal{K}$ and continuity of the inverse Fourier transform (Rudin 1973, Theorem 7.7), given $\varepsilon > 0$, there is a compact subset K_1 of $\mathbf{R}^d$ such that $\sup_{\psi\in\mathcal{K}}\int_{K_1^c}|\hat{\psi}(x)|dx < \varepsilon$. Given a compact subset Γ of $\mathbf{M}_p(\nu)$, for all ψ and $\varphi \in \mathcal{K}$

$$\sup_{\mu\in\Gamma}|H_\psi(\mu) - H_\varphi(\mu)|$$
$$= \sup_{\mu\in\Gamma}|\mu(\hat{\psi}) - \mu(\hat{\varphi})|$$
$$\leq \sup_{\mu\in\Gamma}\left|\textstyle\int_{K_1}\hat{\psi}d\mu - \int_{K_1}\hat{\varphi}d\mu\right| + \sup_{\mu\in\Gamma}\left|\textstyle\int_{K_1^c}\hat{\psi}d\mu - \int_{K_1^c}\hat{\varphi}d\mu\right|$$
$$\leq \|\hat{\psi} - \hat{\varphi}\|_\infty \sup_{\mu\in\Gamma}\mu(K_1) + 2\varepsilon$$
$$\leq \|\psi - \varphi\|_1 \sup_{\mu\in\Gamma}\mu(K_1) + 2\varepsilon$$
$$\leq \|\psi - \varphi\|_\infty \lambda(K_0) \sup_{\mu\in\Gamma}\mu(K_1) + 2\varepsilon,$$

giving the necessary continuity since K_1 depends on neither ψ nor φ. ∎

By the same pattern of reasoning, since the mapping of $v \in \mathbf{R}^d$ into the functional

$$H_v(\mu) = \int e^{-i\langle v, y\rangle}(\mu - \nu\lambda)(dy) \qquad (9.29)$$

on $\mathbf{M}_p(\nu)$ is continuous, we obtain a consistency theorem for the estimators of the spectral density function given by

$$\hat{g}_\rho(v) = (1/2\pi\lambda(K))\int_K N(dx) \int e^{-i\langle v, y-x\rangle}(N - \nu\lambda)(dy). \qquad (9.30)$$

Theorem 9.14. Assume that P is ergodic and that under P, N admits covariance spectral density function g_ρ. Then for each compact subset K_0 of $\mathbf{R}^d$, almost surely

$$\lim_{\delta(K)\to\infty} \sup_{v\in K_0} |\hat{g}_\rho(v) - g_\rho(v)| = 0. \,\square \qquad (9.31)$$

It is instructive to compare the estimators $\hat{g}_\rho$ with periodogram estimators often used in statistical analysis of stationary point processes [see for example Brillinger (1975b) or Cox and Lewis (1966)]. In our setting the periodogram is given by

$$\hat{g}_\rho^*(v) = (1/2\pi\lambda(K)) \left| \int_K e^{-i\langle v, x\rangle} N(dx) \right|^2. \qquad (9.32)$$

Even for $d = 1$ and $K = [0, t]$ with $t \to \infty$, the periodogram is not a consistent estimator of the spectral density function, which is related to its second moment properties. By contrast, the estimators of (9.30) are strongly uniformly consistent, but at the price (even if truncation is imposed) that their computation is quadratic in $N(K)$, rather than linear.

9.3   Intensity-Based Inference on R

In this section we restrict attention to stationary point processes on $\mathbf{R}$ observed over intervals $[0, t]$, with asymptotics arising as $t \to \infty$. The principal results, Theorems 9.22 and 9.24, pertain to maximum likelihood estimation for finite-dimensional statistical models. While the formulation and hypotheses are reminiscent of the classical case of finite-dimensional models and i.i.d. data, the techniques, based in part on stochastic intensities and martingale methods, are nonclassical. The model is not a multiplicative intensity model, and results such as Theorem 5.13 would not apply even if it were, since we are concerned with single-realization inference. As a preliminary step we describe the structure of the stochastic intensity of a stationary point process on **R**; the conjecture that it is a stationary process is confirmed in Theorem 9.16. Thereafter we take up the main material of the section.

To begin, we present a general singularity theorem for ergodic, stationary point processes.

Theorem 9.15. Let N be a point process on $\mathbf{R}^d$ and let P_1 and P_2 be probability measures with respect to which N is stationary and ergodic. Then the following assertions are equivalent:
a) $P_1 \neq P_2$ on $\mathcal{F}^N$;
b) $P_1 \perp P_2$ on $\mathcal{F}^N$;
c) $P_1 \perp P_2$ on the invariant σ-algebra $\mathcal{I}$.

Proof: That c) implies b), which in turn implies a), is obvious, so it suffices to show that a) implies c). If a) holds, then because the Palm measure P^* determines the probability P, there is $\Gamma \in \mathcal{F}^N$ such that $P_1^*(\Gamma) \neq P_2^*(\Gamma)$. Taking $H = 1_\Gamma$ in Theorem 1.72, we conclude that

$$\lim_{r \to \infty} (1/\lambda(B_r)) \int_{B_r} 1(N \circ \theta_x \in \Gamma) N(dx) = P_i^*(\Gamma)$$

almost surely with respect to each P_i, where B_r is the ball of radius r centered at the origin. The limit is an invariant random variable (Exercise 9.17) and therefore c) holds. ∎

Hence perfect discrimination is possible in principle within the model of ergodic probabilities. Optimality considerations manifested in the Neyman-Pearson lemma dictate that simple hypotheses $H_0 : P = P_0$; $H_1 : P = P_1$ ($\neq P_0$) be tested using likelihood ratios. Unfortunately, few results of generality seem known except for stationary point processes on $\mathbf{R}_+$ that admit stochastic intensities, for which Theorems 2.31 and 5.2 provide likelihood functions and ratios. Issues such as computation of power remain unresolved, in part because stationarity is not a hypothesis conducive to distributional calculations.

Nevertheless, the structure of the stochastic intensity can be described quite precisely. Let N be a stationary point process on $\mathbf{R}$ and for each t let

$$\bar{\mathcal{F}}_t^N = \mathcal{F}^N((-\infty, t]) \tag{9.33}$$

be the entire history, including the "infinitely remote" past, of N at time t. For $t > 0$, let $N_t = N([0, t])$ and, as usual, put $\mathcal{F}_t^N = \mathcal{F}^N([0, t])$. We then have the following characterization.

Theorem 9.16. Suppose that N admits $(P, \bar{\mathcal{F}}^N)$-stochastic intensity $(\bar{\lambda}_t)_{t \in \mathbf{R}_+}$. Then almost surely for each t and s,

$$\bar{\lambda}_t \circ \theta_s = \bar{\lambda}_{t+s}. \tag{9.34}$$

Proof: For fixed $s > 0$, consider the point process $\tilde{N}_t = N_{s+t} - N_s$, together with the filtration $\tilde{\mathcal{F}}_t = \bar{\mathcal{F}}^N_{s+t}$. On the one hand, $\tilde{N}$ has $\tilde{\mathcal{F}}$-stochastic intensity $\tilde{\lambda}_t = \bar{\lambda}_{s+t}$, but since the stochastic intensity is unique, if it also has stochastic intensity $(\bar{\lambda}_t \circ \theta_s)$, then (9.34) is established. For $0 < u < t$ and $\Gamma \in \tilde{\mathcal{F}}_u$,

$$
\begin{aligned}
E[\tilde{N}_t - \tilde{N}_u; \Gamma] &= E[(N_t - N_u) \circ \theta_s; \Gamma] \\
&= E[N_t - N_u; \theta_s^{-1}(\Gamma)] \\
&= E\left[\int_0^t \bar{\lambda}_v dv; \theta_s^{-1}(\Gamma)\right] = E\left[\int_0^t \bar{\lambda}_v \circ \theta_s dv; \Gamma\right],
\end{aligned}
$$

where the second equality holds because $P\theta_s^{-1} = P$ and the third because $\theta_s^{-1}(\Gamma) \in \tilde{\mathcal{F}}_u^N$. ∎

In particular, the process $\bar{\lambda}$ is stationary in law: for each $s > 0$

$$(\bar{\lambda}_{s+t})_{t \geq 0} \overset{d}{=} (\bar{\lambda}_t)_{t \geq 0}. \tag{9.35}$$

Under the hypothesis of Theorem 9.16 there exists a $(P, \mathcal{F}^N)$-stochastic intensity $\lambda_t = E[\bar{\lambda}_t | \mathcal{F}_t^N]$, but it need satisfy neither (9.34) nor (9.35) (Exercise 9.18). This complicates analysis of log-likelihood functions

$$L_t(P) = \int_0^t (1 - \lambda_s) ds + \int_0^t (\log \lambda_s) dN_s$$

corresponding to the observations $\mathcal{F}_t^N$, which would be used to construct likelihood ratio tests. The "pseudo"-log-likelihood function

$$\bar{L}_t(P) = \int_0^t (1 - \bar{\lambda}_s) ds + \int_0^t (\log \bar{\lambda}_s) dN_s,$$

on the other hand, is neither computable from the observations $\mathcal{F}_t^N$ nor the likelihood function associated with $\bar{\mathcal{F}}_t^N$. However, $\bar{L}_t(P)$ is the more amenable to analysis because of stationarity of $\bar{\lambda}$ and is hence a useful tool for deducing properties of $L_t(P)$ under assumptions (as below) that the difference between the two converges to zero as $t \to \infty$.

We move on to maximum likelihood estimation for statistical models of stationary point processes admitting a finite-dimensional parametrization. Let N be a simple point process on $\mathbf{R}$ defined over a measurable space $(\Omega, \mathcal{G})$ supporting shift operators (θ_t) such that $N\tau_t^{-1} = N \circ \theta_t$ identically in t and ω, and let $\bar{\mathcal{F}}_t^N$ and $\mathcal{F}_t^N$ be as defined above. The parameter space is a compact subset I of $\mathbf{R}$, and we suppose given for each $\alpha \in I$ a probability P_α and an $\bar{\mathcal{F}}^N$-predictable process $\bar{\lambda}_t(\alpha)$ such that

- $\bar{\lambda}_{s+t}(\alpha) = \bar{\lambda}_t(\alpha) \circ \theta_s$ identically in s, t and ω [see (9.34)];

- $P_\alpha \theta_t^{-1} = P_\alpha$ for each $t \in \mathbf{R}$ (so that N is P_α-stationary);

- N has $(P_\alpha, \bar{\mathcal{F}}^N)$-stochastic intensity $\bar{\lambda}(\alpha)$;

- N has finite P_α-intensity;

- P_α is ergodic.

Our goals are maximum likelihood estimation of α given the observations $\mathcal{F}_t^N$ and analysis of asymptotic properties of maximum likelihood estimators. The log-likelihood function (with a constant term deleted) is

$$L_t(\alpha) = -\int_0^t \lambda_s(\alpha)ds + \int_0^t \log \lambda_s(\alpha)dN_s, \qquad (9.36)$$

where $\lambda(\alpha)$ is the $(P_\alpha, \mathcal{F}^N)$-stochastic intensity of N.

For analysis of maximum likelihood estimators several assumptions are necessary; they are conveniently segregated into regularity conditions concerning dependence of the $\bar{\lambda}(\alpha)$ on α and approximation hypotheses to the effect that for large values of t, $\lambda_t(\alpha) \cong \bar{\lambda}_t(\alpha)$ for each α (and similarly for derivatives). The latter permit replacement of the log-likelihood functions $L_t(\alpha)$ by the more tractable functions

$$\bar{L}_t(\alpha) = -\int_0^t \bar{\lambda}_s(\alpha)ds + \int_0^t \log \bar{\lambda}_s(\alpha)dN_s. \qquad (9.37)$$

Assumptions 9.17. (Regularity conditions). a) For each t and ω the function $\alpha \to \bar{\lambda}_t(\alpha, \omega)$ is continuous;

b) $P_\alpha\{\bar{\lambda}_0(\alpha) > 0\} = 1$ for each α;

c) For α_1 and $\alpha_2 \in I$, $\bar{\lambda}_0(\alpha_1) = \bar{\lambda}_0(\alpha_2)$ almost surely if and only if $\alpha_1 = \alpha_2$;

d) For each t and ω the function $\alpha \to \bar{\lambda}_t(\alpha, \omega)$ is three times continuously differentiable on (the interior of) I, and for each t and α the random variables $\bar{\lambda}_t'(\alpha)$ and $\bar{\lambda}_t''(\alpha)$ have finite second moments;

e) For each α there exists a neighborhood $U(\alpha)$ of α in I such that

$$E_\alpha\left[\sup_{\beta \in U(\alpha)}\bar{\lambda}_0(\beta)\right] < \infty$$

and

$$E_\alpha\left[\sup_{\beta \in U(\alpha)} |\log \bar{\lambda}_0(\beta)|\right] < \infty;$$

f) For each α

$$0 < I(\alpha) = E_\alpha[\bar{\lambda}_0''(\alpha)^2/\bar{\lambda}_0] < \infty; \qquad (9.38)$$

g) For each α there exists a neighborhood $U(\alpha)$ such that

$$E_\alpha\left[\sup_{\beta \in U(\alpha)} |\bar{\lambda}_0'''(\beta)|\right] < \infty$$

and

$$E_\alpha \left[\bar{\lambda}_0(\alpha)^2 \{\sup_{\beta \in U(\alpha)} |(\log \bar{\lambda}_0)'''|\}^2 \right] < \infty.$$

Each of these implies a corresponding property of the $\mathcal{F}^N$-stochastic intensities $\lambda(\alpha)$. Many of them, which are in some cases substitutes rather than complements, are apparent analogues of conditions imposed in analysis of maximum likelihood estimators given i.i.d. data and a finite-dimensional parameter set. They apply to the model as a whole, whereas the approximation hypotheses apply individually to each α.

Assumptions 9.18. (Approximation hypotheses). a) There is a neighborhood $U(\alpha)$ such that

$$\sup_{\beta \in U(\alpha)} |\lambda_t(\beta) - \bar{\lambda}_t(\beta)| \to 0$$

in P_α-probability, and such that for some $\delta > 0$,

$$\sup_{t \in \mathbf{R}_+} E_\alpha \left[\sup_{\beta \in U(\alpha)} |\log \lambda_t(\beta)|^{2+\delta} \right] < \infty;$$

b) As $t \to \infty$, $\lambda_t(\alpha) - \bar{\lambda}_t(\alpha) \to 0$, $\lambda_t'(\alpha) - \bar{\lambda}_t'(\alpha) \to 0$ and $\lambda_t''(\alpha) - \bar{\lambda}_t''(\alpha) \to 0$ in P_α-probability; there exists $\delta > 0$ such that

$$\sup_{t \in \mathbf{R}_+} E_\alpha \left[|\bar{\lambda}_t(\alpha)/\lambda_t(\alpha)|^{2+\delta} \right] < \infty$$

$$\sup_{t \in \mathbf{R}_+} E_\alpha \left[|\bar{\lambda}_t'(\alpha)/\bar{\lambda}_t(\alpha)|^{2+\delta} \right] < \infty$$

$$\sup_{t \in \mathbf{R}_+} E_\alpha \left[|\bar{\lambda}_t''(\alpha)|^{2+\delta} \right] < \infty;$$

c) As $t \to \infty$,

$$(1/\sqrt{t}) E_\alpha \left[\int_0^t |\lambda_u'(\alpha) - \bar{\lambda}_u'(\alpha)| \, du \right] \to 0$$

$$(1/\sqrt{t}) E_\alpha \left[\int_0^t |\lambda_u(\alpha) - \bar{\lambda}_u(\alpha)| \, [\lambda_u'(\alpha)/\lambda_u(\alpha)] du \right] \to 0;$$

d) There is a neighborhood $U(\alpha)$ such that

$$\sup_{\beta \in U(\alpha)} |\lambda_t'''(\beta) - \bar{\lambda}_t'''(\beta)| \to 0$$

in P_α-probability as $t \to \infty$; there exists $\delta > 0$ satisfying

$$\sup_{t \in \mathbf{R}_+} E_\alpha \left[\sup_{\beta \in U(\alpha)} |\lambda_t'''(\alpha)|^{2+\delta} \right] < \infty$$

$$\sup_{t \in \mathbf{R}_+} E_\alpha \left[\sup_{\beta \in U(\alpha)} |(\log \lambda_t)'''(\alpha)|^{2+\delta} \right] < \infty.$$

Evidently, it is no small matter to verify these assumptions; Ogata (1978) does so for Poisson processes, stationary renewal processes (Exercise 1.31 and Example 9.25), and the self-exciting point process of Hawkes and Oakes (1974) (Exercise 9.19).

In view of compactness of I, given Assumption 9.17a) there exists for each t a maximum likelihood estimator $\hat{\alpha}$ satisfying $L_t(\hat{\alpha}) \geq L_t(\alpha)$ for all $\alpha \in I$, although not necessarily fulfilling the likelihood equation

$$L'_t(\hat{\alpha}) = 0. \tag{9.39}$$

With sufficiently many more of the assumptions above, consistency of maximum likelihood estimators can be established without recourse to (9.39), but asymptotic normality cannot. We begin with three preliminary results, included because they are the principal means through which the special context of stationary point processes enters.

Lemma 9.19. Under Assumptions 9.17a), and d)–f), and with $\bar{L}_t$ given by (9.37), for each t and α

$$E_\alpha[\bar{L}'_t(\alpha)] = 0 \tag{9.40}$$

and

$$E_\alpha[\bar{L}'_t(\alpha)^2] = -E_\alpha[\bar{L}''_t(\alpha)] = E_\alpha\left[\bar{\lambda}'_0(\alpha)/\bar{\lambda}_0(\alpha)\right]. \tag{9.41}$$

Proof: Given these assumptions, derivatives, integrals, and expectations can be interchanged; consequently,

$$
\begin{aligned}
E_\alpha[\bar{L}'_t(\alpha)] &= E_\alpha\left[-\int_0^t \bar{\lambda}'_u(\alpha)du + \int_0^t[\bar{\lambda}'_u(\alpha)/\bar{\lambda}_u(\alpha)]dN_u\right] \\
&= E_\alpha\left[\int_0^t[\bar{\lambda}'_u(\alpha)/\bar{\lambda}_u(\alpha)][dN_u - \bar{\lambda}_u(\alpha)du]\right] = 0,
\end{aligned}
$$

giving (9.40). Derivation of (9.41) is similar. ∎

Lemma 9.20. Let α be fixed and suppose that X is $\bar{\mathcal{F}}^N$-predictable and P_α-stationary with $E_\alpha[X_t^2] < \infty$. Then almost surely

$$\lim_{t \to \infty}(1/t)\int_0^t X_u du = E_\alpha[X_0] = \lim_{t \to \infty}(1/t)\int_0^t[X_u/\bar{\lambda}_u(\alpha)]dN_u. \tag{9.42}$$

Proof: Because the first equality in (9.42) is a standard Césaro limit theorem for stationary processes (see, e.g., Karlin and Taylor, 1975), in order to establish the second it suffices to show that

$$\lim_{t \to \infty}\int_0^t[X_u/\bar{\lambda}_u(\alpha)][dN_u - \bar{\lambda}_u(\alpha)] = 0$$

almost surely. By Theorem 2.9, the stochastic integral process is a mean zero martingale under P_α and consequently (9.42) ensues by a strong law of large numbers for martingales (see, e.g., Hall and Heyde, 1980, Theorem 2.19). ■

If the definition of stochastic intensity is interpreted as "$dN_t = \bar{\lambda}_t(\alpha)dt$ in the small," then (9.42) is a statement that "$dN_t = \bar{\lambda}_t(\alpha)dt$ in the large."

Lemma 9.21. Given α_0 nad α, define

$$\Lambda_1(\alpha_0, \alpha) = \int_0^1 [\bar{\lambda}_u(\alpha) - \bar{\lambda}_u(\alpha_0)]du + \int_0^1 \log[\bar{\lambda}_u(\alpha_0)/\bar{\lambda}_u(\alpha)]dN_u.$$

Then

$$E_{\alpha_0}[\Lambda_1(\alpha_0, \alpha)] \geq 0 \tag{9.43}$$

with equality if and only if $\bar{\lambda}_0(\alpha) = \bar{\lambda}_0(\alpha_0)$ almost surely with respect to P_{α_0}.

Proof: Since $\bar{\lambda}(\alpha)$ is the $(P_\alpha, \bar{\mathcal{F}}^N)$-stochastic intensity of N,

$$
\begin{aligned}
&E_{\alpha_0}[\Lambda_1(\alpha_0, \alpha)] \\
&= \int_0^1 E_{\alpha_0}[\bar{\lambda}_u(\alpha) - \bar{\lambda}_u(\alpha_0)]du + \int_0^1 E_{\alpha_0}\left[\log(\bar{\lambda}_u(\alpha_0)/\bar{\lambda}_u(\alpha))\bar{\lambda}_u(\alpha_0)\right]du \\
&= E_{\alpha_0}[\bar{\lambda}_0(\alpha_0)\{\bar{\lambda}_0(\alpha)/\bar{\lambda}_0(\alpha_0) - 1 + \log(\bar{\lambda}_0(\alpha_0)/\bar{\lambda}_0(\alpha))\}],
\end{aligned}
$$

from which the lemma follows by application of the inequality $\log x - 1 + 1/x \geq 0$, which holds for $x > 0$ with equality if and only if $x = 1$. ■

The expression in (9.43) is analogous to the Kullback-Liebler information in classical statistics; together with Assumption 9.17c), Lemma 9.21 furnishes an identifiability criterion: $\alpha = \alpha_0$ if and only if $E_{\alpha_0}[\Lambda_1(\alpha_0, \alpha)] = 0$.

We come now to the main limit theorems of the section.

Theorem 9.22. If Assumptions 9.17a)–d) hold and if the "true" parameter value α satisfies Assumption 9.18a), then the maximum likelihood estimators $\hat{\alpha}$ are weakly consistent: under P_α, $\hat{\alpha} \to \alpha$ in probability.

Proof: Given a neighborhood U_0 of α, by Lemma 9.21 and Assumption 9.17c) there is $\epsilon > 0$ such that

$$E_\alpha[\Lambda_1(\alpha, \gamma)] \geq 3\epsilon \tag{9.44}$$

for all $\gamma \in I \setminus U_0$. By Assumptions 9.17a) and d), for each such γ there is a neighborhood $U(\gamma)$ satisfying

$$E_\alpha\left[\{\inf_{U(\gamma)}\bar{\lambda}_0(\beta)\} - \bar{\lambda}_0(\alpha) + \bar{\lambda}_0(\alpha)\log\frac{\bar{\lambda}_0(\alpha)}{\sup_{U(\gamma)}\bar{\lambda}_0(\beta)}\right] \geq E_\alpha[\Lambda_1(\alpha, \gamma)] - \epsilon,$$

and by compactness $I \setminus U_0$ can be covered by finitely many such neighborhoods, say $U(\gamma_1), \ldots, U(\gamma_k)$. Predictability of stochastic intensities shows in conjunction with Lemma 9.20 that given $\delta > 0$ there is a (random) time T_0 such that for $t > T_0$ and $i = 1, \ldots, k$,

$$\bar{L}_t(\alpha)/t - \sup_{U(\gamma_i)} \bar{L}_t(\beta)/t \geq E_\alpha[\Lambda_1(\alpha, \gamma)] - 2\varepsilon \geq \varepsilon,$$

the last inequality in consequence of (9.44). Thus there exists a random time T_1, depending on U_0 and ε, such that

$$\sup_{\beta \in U_0} \bar{L}_t(\beta) \geq \sup_{\beta \in I \setminus U_0} \bar{L}_t(\beta) + \varepsilon t$$

for all $t > T_1$. Finally, Assumption 9.18a) implies that the same holds also for $L_t(\cdot)$ with probability converging to 1 as $t \to \infty$, from which we infer that $P_\alpha\{\hat{\alpha} \in U_0\} \to 1$. Since U_0 was arbitrary, weak consistency holds. ∎

Asymptotic normality of the estimation error $\sqrt{t}[\hat{\alpha} - \alpha]$ is demonstrated by the classical argument of showing that the first derivative of the log-likelihood function (L_t in this case) is asymptotically normally distributed, then expanding the log-likelihood function in a Taylor series about a solution to the likelihood equation (9.39).

Theorem 9.23. Suppose that Assumptions 9.17a)–c), and e) and 9.18c) are satisfied. Then under P_α, $L_t'(\alpha)/\sqrt{t} \overset{d}{\to} N(0, I(\alpha))$, where $I(\alpha)$ is the Fisher information given by (9.38).

Proof: By Assumptions 9.17,

$$\bar{L}_t'(\alpha) = \int_0^t [\bar{\lambda}_u'(\alpha)/\bar{\lambda}_u(\alpha)][dN_u - \bar{\lambda}_u(\alpha)du],$$

so by Theorem 2.9, $(\bar{L}_t'(\alpha))$ is a mean zero P_α-martingale. Let $dM_u(\alpha) = dN_u - \bar{\lambda}_u(\alpha)$ be the $(P_\alpha, \bar{\mathcal{F}}^N)$-innovation martingale; then by stationarity of $\bar{\lambda}(\alpha)$ and (9.42),

$$(1/t)\int_0^t [\bar{\lambda}_u'(\alpha)/\bar{\lambda}_u(\alpha)]^2 d\langle M(\alpha)\rangle_u = (1/t)\int_0^t [\bar{\lambda}_u'(\alpha)^2/\bar{\lambda}_u(\alpha)]du$$
$$\to E_\alpha[\bar{\lambda}_0'(\alpha)^2/\bar{\lambda}_0(\alpha)] = I(\alpha). \quad (9.45)$$

In view of ergodicity of N and (9.45), it follows by appeal to Theorem B.21 that $\bar{L}_t(\alpha)/\sqrt{t} \overset{d}{\to} N(0, I(\alpha))$ and Assumption 9-17c) suffices to complete the proof. ∎

Only standard reasoning is required in order to derive the final result of the section.

Theorem 9.24. Suppose that the maximum likelihood estimators $\hat{\alpha}$ satisfy the likelihood equation (9.39) and that Assumptions 9.17a) – d), f), and g), and 9.18b)–d) are fulfilled. Then under P_α,

$$\sqrt{t}[\hat{\alpha} - \alpha] \;\overset{d}{\to}\; N(0, 1/I(\alpha)) \tag{9.46}$$

$$2[L_t(\hat{\alpha}) - L_t(\alpha)] \;\overset{d}{\to}\; \chi_1^2, \tag{9.47}$$

where χ_1^2 is the chi-squared distribution with 1 degree of freedom.

Proof: Within error converging to zero in probability, there holds the Taylor expansion

$$0 = L_t'(\hat{\alpha})/\sqrt{t} \cong L_t'(\alpha)/\sqrt{t} + \sqrt{t}[\hat{\alpha} - \alpha]/t.$$

The first term converges in distribution to $N(0, 1/I(\alpha))$ by Theorem 9.23, while by arguments analogous to those employed to prove Theorem 9.22, $L_t''(\alpha)/t \to I(\alpha)$; consequently, (9.46) holds. Similarly,

$$2[L_t(\hat{\alpha}) - L_t(\alpha)] \cong 2L_t'(\alpha)[\hat{\alpha} - \alpha] + [L_t''(\alpha)/t]\{\sqrt{t}[\hat{\alpha} - \alpha]\}^2,$$

so (9.47) follows from (9.46), Theorem 9.23 and Theorem 9.22. ∎

We illustrate with stationary renewal processes.

Example 9.25. (Stationary renewal process). Let $\{F(\alpha, \cdot) : \alpha \in I\}$ be a finite-dimensional family of distribution functions on $\mathbf{R}_+$, and let

$$H(\alpha, t) = \int_0^t [1 - F(\alpha, x)]dx \,/ \int_0^\infty [1 - F(\alpha, x)]dx,$$

which for a stationary renewal process is the distribution of the forward recurrence time at t for every t (Exercise 9.3). The log-likelihood function is given by Theorem 8.17. Assumptions 9.17a)–c) are satisfied if $\alpha \to F(\alpha)$ is continuous and if $\{F(\alpha)\}$ is identifiable, while d)–g) must be imposed by fiat in the form of existence of derivatives of $\alpha \to F(\alpha, x)$ and moments of the $F(\alpha)$. Integrability in Assumptions 9.18 is ensured by stipulating the existence of moments of $F(\alpha)$ or $H(\alpha)$, but the convergence statements there are satisfied automatically. □

9.4   Linear State Estimation

Because of their tractable second moment structure stationary point processes are natural candidates for linear prediction, that is, calculation of minimum mean squared error *linear* state estimators of unobserved portions of the process. Suppose that N is a stationary point process observed over a bounded set A; then we seek for each function f (assumed without loss of generality to vanish on A) a function $\hat{f}$ on A and a constant a_f such that the linear state estimator $\hat{N}(f) = N(\hat{f}) + a_f$ satisfies

$$E[(N(f) - \hat{N}(f))^2] \le E[(N(f) - N(g) - a)^2] \qquad (9.48)$$

for every function g on A and every $a \in \mathbf{R}$.

This section concentrates on linear state estimation because there exist no results of generality pertaining to nonlinear state estimation for stationary point processes, even on $\mathbf{R}$. Stationarity, although exceedingly useful for some purposes, is less suited to nonlinear state estimation because distributional characteristics other than moments are involved.

We examine two approaches to linear state estimation, a time/space-domain approach applicable on general spaces and a frequency domain/spectral analysis approach for stationary point processes on $\mathbf{R}$. For the latter observations are $\bar{\mathcal{F}}_t^N$ from (9.33), so that there are "infinitely many" data; hence it is natural to consider "moving" problems of prediction of the point process over an interval $(t, t + u]$ extending a fixed length into the future. Martingale methods do not apply because a proper filtration cannot be specified; however, spectral analysis provides very usable techniques. As part of their development we derive a Wold decomposition of a stationary point process on $\mathbf{R}$.

We first analyze linear prediction in the time/space domain. Let us introduce the centered process

$$M(f) = N(f) - \nu\lambda(f), \qquad (9.49)$$

where ν is the intensity of N. Given observations $\mathcal{F}^N(A)$ of N over a bounded set A with $\lambda(A) > 0$ and a function f vanishing on A, we wish to calculate that function $\hat{f}$ on A for which

$$E[(M(f) - M(\hat{f}))^2] \le E[(M(f) - M(g))^2] \qquad (9.50)$$

for every function g on A. Thus $M(\hat{f})$ is the minimum mean squared error linear predictor of $M(f) = N(f) - \nu\lambda(f)$ given $\mathcal{F}^N(A)$, and $M(\hat{f}) + \nu\lambda(f) = N(\hat{f}) - \nu\lambda(f - \hat{f})$ the optimal linear state estimator of $N(f)$.

With P known, solution of (9.50) is straightforward.

Proposition 9.26. Assume that N admits a moment of order 2 with respect to P and let ρ_* be the reduced covariance measure. Given f vanishing on A, the function $\hat{f}$ satisfying (9.50) is the unique function on A such that

$$\int_A \hat{f}(y)\rho_*(dy - x) = \int f(y)\rho_*(dy - x) \qquad (9.51)$$

for almost all $x \in A$ (with respect to Lebesgue measure).

Proof: By standard Hilbert space theory, $M(\hat{f})$ is the projection of $M(f)$ onto the linear space spanned by $\{M(g) : g \in L^2(A)\}$, and hence the unique solution to the normal equations

$$E[\{M(\hat{f}) - M(f)\}M(g)] = 0, \quad g \in L^2(A). \tag{9.52}$$

But

$$E\,[M(f)M(g)] = \int_A g(x)[\int f(y)\rho_*(dy - x)]dx;$$

consequently (9.52) is equivalent to

$$0 = \int_A g(x)[\int_A \hat{f}(y)\rho_*(dy - x) - \int f(y)\rho_*(dy - x)]dx,$$

confirming (9.51). ∎

We next describe the spectral approach to linear prediction for stationary point processes on $\mathbf{R}$. It should not be dismissed as merely an application of the Parseval identity (9.24) to replace the reduced second moment measure μ_*^2 by the spectral measure F because in particular it permits rather explicit solution of "moving" linear prediction problems.

As a preliminary, we state a Wold decomposition for a stationary point process on $\mathbf{R}$ and a backward moving average representation for the purely nondeterministic part.

Let N be a stationary point process on $\mathbf{R}$ and let $\bar{\mathcal{F}}_{-\infty}^N = \cap_{t \in \mathbf{R}} \bar{\mathcal{F}}_t^N$ be the infinitely remote past of N.

Definition 9.27. The stationary point process N is *deterministic* if $\bar{\mathcal{F}}_{-\infty}^N = \mathcal{F}^N$ and *purely nondeterministic* if $\bar{\mathcal{F}}_{-\infty}^N = \{\emptyset, \Omega\}$.

These are the extreme cases; in one, all knowledge about realizations of N is part of the infinitely remote past and in the other no nondeterministic information belongs to the infinitely remote past. The lattice process $\sum_{n=-\infty}^{\infty} \varepsilon_{X+n}$ of Example 1.58 is deterministic because the offset X can be determined from $\bar{\mathcal{F}}_t^N$ regardless of $t \in \mathbf{R}$. By contrast, a Poisson process (more generally, any ergodic stationary point process) is purely nondeterministic (Exercise 9.25).

Stationarity suffices to eliminate intermediate cases, although of course both components may be present simultaneously. Thus we obtain the *Wold decomposition* for a stationary point process.

Theorem 9.28. Let N be a stationary point process on $\mathbf{R}$ admitting moment of order 2. Then there exists a unique decomposition $N = N^1 + N^2$ in which N^1 and N^2 are stationary point processes, N^1 is deterministic, N^2

is purely nondeterministic, and the centered processes $N^i - E[N^i]$, $i = 1, 2$, are orthogonal. The deterministic part N^1 is zero if and only if there exists a spectral density function g_ρ [see (9.27)] satisfying

$$\int_{-\infty}^{\infty} [\log g_\rho(v)/(1 + v^2)]dv > -\infty. \; \square$$

Neither Theorem 9.28 nor Theorem 9.30, the latter providing a backward moving average representation for purely nondeterministic point processes, is proved in detail. However, we do sketch a smoothing approach that permits direct application of results for mean square continuous stationary stochastic processes (see Daley, 1971, and Vere-Jones, 1974, for details).

Lemma 9.29. Given a stationary point process N on $\mathbf{R}$ with intensity ν and moment of order 2, the smoothed process $X_t = \int_{-\infty}^{t} e^{-(t-s)} dN_s$ is strictly stationary with $E[X_t] = \nu$ for all t. The spectral measure F_X of X is related to that of N via $F_X(dv) = 1/(1 + v^2)F(dv)$. $\square$

The centered process M of (9.49) and $\tilde{X}_t = X_t - \nu$ generate the same history and are hence together deterministic or purely nondeterministic, $\tilde{X}$ in the time series sense (see, e.g., Brillinger, 1981). The well-known Wold decomposition $\tilde{X} = \tilde{X}^1 + \tilde{X}^2$, with $\tilde{X}^1$ deterministic and $\tilde{X}^2$ purely nondeterministic, translates at once to a decomposition for M and thence to that for N.

Prediction for a deterministic stationary point process given observations $\bar{\mathcal{F}}_t^N$ is completely trivial, so by Theorem 9.28 we hereafter suppose that N is purely nondeterministic. The spectral representation is complemented in this case by a backward moving average representation.

Theorem 9.30. A stationary point process N on $\mathbf{R}$ is purely nondeterministic if and only if there exists a representation

$$M(\psi) = \int H * \psi(s)\tilde{Z}(ds) \tag{9.53}$$

for the centered process $M(\cdot) = N(\cdot) - \nu\lambda(\cdot)$, where H is a distribution on S with support contained in $(-\infty, 0]$, $\tilde{Z}$ is a random measure with orthogonal increments, and $E[|\tilde{Z}(dt)|^2] = \nu dt$. [The asterisk denotes convolution (see Rudin, 1973).] $\square$

For linear prediction the key points are that M and $\tilde{Z}$ engender the same filtration, because the distribution H in (9.53) is supported in $(-\infty, 0]$, and hence the representation is invertible, and that linear prediction of $\tilde{Z}$, by virtue of its having orthogonal, stationary increments, is essentially trivial. Thus one obtains the following solution to the "moving" linear prediction problem.

Theorem 9.31. Given the setting of Theorem 9.30, there exists a distribution K supported in $(-\infty, 0]$ such that

$$\tilde{Z}(\varphi) = \int (K * \varphi) dM \tag{9.54}$$

for each $\varphi \in \mathcal{D}$. Hence given $u > 0$ and $\psi \in \mathcal{D}$ vanishing outside $[0, u]$, if for each t we define $\psi_t(x) = \psi(t - x)$, then the MMSE linear predictor of

$$M(\psi_t) = \int_t^{t+u} \psi(s - t) dN_s,$$

given the observations $\bar{\mathcal{F}}_t^N$, is

$$\hat{M}(\psi_t) = \int_{-\infty}^t K * H(\psi_{t-y}) dM_y. \tag{9.55}$$

Proof: Invertibility of (9.53) implies that (9.54) holds.

Given this, we have $M(\psi_t) = \int_{-\infty}^{t+u}(H * \psi_t) d\tilde{Z}$, which together with orthogonality of the increments of $\tilde{Z}$ and the property that M and $\tilde{Z}$ generate the same filtration, gives that the optimal linear predictor is

$$\hat{M}(\psi_t) = \int_{-\infty}^t (H * \psi_t) d\tilde{Z}.$$

Substitution of (9.54) into this expression verifies (9.55). ∎

The Fourier transform of the distribution $K * H$ appearing in (9.55) has been termed the *spectral characteristic for prediction* (see Yaglom, 1962); one can formulate a frequency-domain analogue of (9.55) using it. Note that the measure (in ψ_t) defined by (9.55) may be absolutely continuous even though N is a point process; this happens, for example, if N is Poisson.

We conclude with brief analysis of a Poisson cluster process.

Example 9.32. Let N be a Neyman-Scott cluster process (see Section 7.5) in which the cluster center process is Poisson with rate η and cluster center-cluster member distances are independent and identically exponentially distributed with parameter α; we do not restrict the distribution of the numbers of cluster members except to the extent that it admit finite variance σ^2 and mean μ. Then the smoothed process X of Lemma 9.29 has rational spectral density function $f_X(v) = \eta[\alpha^2(\sigma + \mu) + \mu v^2]/(\alpha^2 + v^2)$, and consequently N has rational spectral density function of the generic form $f_\rho(v) = |a(b + iv)/(c + iv)|^2$. In the notation of Theorem 9.31, it follows that for each t

$$M(\psi_t) = (b - c)\int_0^t e^{-cs}\psi(s)ds + \int_{-\infty}^t e^{b(t-s)}dN_s. \;\square$$

The final topic of the section is combined inference and linear state estimation.

Suppose that P is not known, specifically that ρ_* is unknown. We construct pseudo-state estimators that approximate the true state estimators $M(\hat{f})$ arising from (9.51) and describe their asymptotic behavior. More precisely, let $f \in C(E)$ be fixed with compact support, and suppose that N is observed over compact, convex sets K such that $K \cap \text{supp } f = \emptyset$. We then construct, using the estimators $\hat{\rho}_* = \hat{\mu}_*^2 - \hat{\nu}^2\lambda$, estimators $\tilde{f}$ of the solution to (9.51) with $A = K$, and establish that as $\delta(K) \to \infty$, $M(\hat{f} - \tilde{f}) \to 0$, in an appropriate sense.

Given $\hat{\rho}_*$, let $\tilde{f} = \tilde{f}_K$ be the unique function on K minimizing

$$\left\| \int \tilde{f}(y)\hat{\rho}_*(dy - \cdot) - \int f(y)\hat{\rho}_*(dy - \cdot) \right\|_2 .$$

As pseudo-state estimator we then take

$$\hat{M}(f) = \int_K \tilde{f}(x)N(dx) - \hat{\nu}\int_K f(x)dx.$$

While the function f in Proposition 9.26 seems to depend on A, in fact it does not: there exists a single function $\hat{f} \in L^2(E)$ such that for each A, $\hat{f}_A = \hat{f}1_A$ is the solution to (9.51). Consequently $\hat{M}(f)$ is an approximation to the true state estimator.

Theorem 9.33. Assume that P is ergodic and that ρ_* has finite total variation. Then as $\delta(K) \to \infty$, $\hat{M}(f) - M(\hat{f}) \to 0$ almost surely.

Proof: We begin with the decomposition

$$\begin{aligned} \hat{M}(f) - M(\hat{f}) &= \int_K(\tilde{f} - \hat{f})dN + (\nu - \hat{\nu})\int_K\hat{f}(x)dx \\ &\quad + \hat{\nu}[\int_K\hat{f}(x)dx - {}_K\tilde{f}(x)dx]. \end{aligned} \tag{9.56}$$

The second term is dealt with most easily; by ergodicity of P, $\hat{\nu} \to \nu$ almost surely, and hence since $\hat{f} \in L^2(E)$, this term, converges to zero as $\delta(K) \to \infty$
We next show that almost surely $\tilde{f} - \hat{f}$ converges to zero in L^2. Indeed,

$$\begin{aligned} &\left\| \int \tilde{f}(y)\hat{\rho}_*(dy - \cdot) - \int f(y)\rho_*(dy - \cdot) \right\|_2 \\ &\leq \left\| \int \tilde{f}(y)\hat{\rho}_*(dy - \cdot) - \int f(y)\hat{\rho}_*(dy - \cdot) \right\|_2 \\ &\quad + \| \int f(y)\hat{\rho}_*(dy - \cdot) - \int f(y)\rho_*(dy - \cdot)\|_2 \\ &\leq \left\| \int \hat{f}(y)\hat{\rho}_*(dy - \cdot) - \int f(y)\hat{\rho}_*(dy - \cdot) \right\|_2 \\ &\quad + \| \int f(y)\hat{\rho}_*(dy - \cdot) - \int f(y)\rho_*(dy - \cdot)\|_2 \\ &\leq \left\| \int \hat{f}(y)\hat{\rho}_*(dy - \cdot) - \int f(y)\rho_*(dy - \cdot) \right\|_2 \end{aligned}$$

$$+ 2\left\| \int f(y)\hat{\rho}_*(dy - \cdot) - \int f(y)\rho_*(dy - \cdot)\right\|_2$$
$$= \left\| \int \tilde{f}(y)\hat{\rho}_*(dy - \cdot) - \int \hat{f}(y)\hat{\rho}_*(dy - \cdot)\right\|_2$$
$$+ 2\left\| \int f(y)\hat{\rho}_*(dy - \cdot) - \int f(y)\rho_*(dy - \cdot)\right\|_2,$$

which converges to zero by Theorem 9.4 applied in the manner of Theorem 9.5 to the estimators $\hat{\rho}_*$

In view of the preceding paragraph, the third term in (9.56) converges to zero almost surely, and so does the first, which completes the proof. ∎

EXERCISES

9.1 Let N be a stationary point process on $\mathbf{R}$ with intensity $\nu > 0$.
a) Prove that $P\{N(-\infty, 0) = N(0, \infty) = \infty\} + P\{N \equiv 0\} = 1$.
b) Construct an example in which both probabilities are nonzero.

9.2 Let N be a stationary renewal process on $\mathbf{R}$. Prove that if N has independent increments, then it is a Poisson process.

9.3 Let N be a stationary renewal process on $\mathbf{R}$ with interarrival distribution F and let $H(t) = (1/m)\int_0^t [1 - F(u)]du$, where m is the mean of F. Prove that for every t the forward recurrence time $W_t = \sup\{s : N((t, t+s]) = 0\}$ has distribution H.

9.4 Prove Lemma 9.2.

9.5 Let N be a stationary Poisson process on $\mathbf{R}$. Calculate the second moment measure and covariance measure, and interpret the results in view of the independent increments property of N.

9.6 Show that all moment and cumulant measures of a stationary point process are invariant under appropriate diagonal groups. (See Section 1.8.)

9.7 Let N be a stationary point process with moment of order k. Devise analogues of the estimators (9.8) for the reduced kth-order moment measure μ_*^k and verify that they are unbiased.

9.8 Let N be a stationary point process admitting moment of order 2. Devise unbiased estimators of the reduced covariance measure and establish their asymptotic properties.

9.9 Let X_n, X be random variables for which there exist cumulant sequences $(c_{n\ell})$, (c_n). Prove that $X_n \overset{d}{\to} X$ if and only if $c_{n\ell} \to c_\ell$ for each ℓ.

9.10 Let N be a stationary Poisson process on $\mathbf{R}^d$ that under P_ν has intensity ν. Compare asymptotic behavior or the estimators $\hat{\nu} = N(K)/\lambda(K)$ with that of the alternative estimators

$$\hat{\nu}^* = \sqrt{(1/\lambda(K))\int_K N(J + x)N(dx)},$$

where $J = (0, 1]^d$. [The latter arise from the role of ν in connection with the reduced second moment measure: $\mu_*^2(J) = \nu^2$, so that $(\hat{\nu}^*)^2 = \hat{\mu}_*^2(J)$.]

9.11 Let N be a stationary point process on $\mathbf{R}$ with respect to the probability P, and suppose that observation is initiated at the time of an event of N.
a) Argue that one can take that time to be T_0 and that the law of the observations is not P but rather the Palm distribution P_0.
b) Show that the (P-) intensity of N satisfies $\nu = 1/E_0[U_1]$.
c) Imposing suitable hypotheses, describe the asymptotic behavior of the estimators $\hat{\nu}^* = [(1/n) \sum_{i=1}^{n} U_i]^{-1}$.

9.12 Let N be a stationary point process on $\mathbf{R}$, observed asynchronously, and suppose that the parameter of interest is $m = E_0[U_i]$, the mean interarrival time under the Palm distribution.
a) Calculate the bias of the $\mathcal{F}_t^N$-estimators $\hat{m} = (1/N_t) \sum_{i=1}^{N_t} U_i$.
b) Describe large-sample properties of these estimators.

9.13 Suppose that under the probability P_p, $0 < p < 1$, N' is the p-thinning of a stationary point process N on $\mathbf{R}^d$ with known intensity ν, whose law does not depend on p. Only N' is observable.
a) Prove that N' is stationary with P_p-intensity $p\nu$.
b) Establish — given appropriate assumptions — asymptotic behavior of the $\mathcal{F}^{N'}(K)$-estimators $\hat{p} = N'(K)/[\nu\lambda(K)]$.

9.14 (Continuation of Exercise 9.13) Continue to assume that the law of N does not depend on p, but suppose now that ν is unknown and that N is observable. Describe joint asymptotic properties of the $\mathcal{F}^N(K) \bigvee \mathcal{F}^{N'}(K)$-estimators $\hat{p} = N'(K)/N(K)$.

9.15 Let $\tilde{N}$ be a stationary Poisson process on $\mathbf{R}^2$ with intensity ν, and for each r and x let $B_r(x)$ be the open disk with radius r centered at x. Introduce the *interpoint distance process* (with K fixed, a point process on $\mathbf{R}_+$) given by

$$N_K(r) = (1/2)\int_K \tilde{N}(B_r(x) \cap K)\tilde{N}(dx),$$

which represents the number of interpoint distances, among pairs of points in K, not exceeding r. Assume that K is compact and convex. Prove that as $\delta(K) \to \infty$, $\left\{ N_k\left(r/\sqrt{\lambda(K)}\right) : r \geq 0 \right\} \xrightarrow{d} N$, where N is a Poisson process on $\mathbf{R}_+$ with $E[N_r] = \nu\pi r^2/2$.

9.16 Let N be a stationary point process and introduce the *aggregation* functional $a(C) = \mu^2(C \times C)/\nu(C)$, where μ^2 is the unreduced second moment measure.
a) Prove that $a(C) = E_0[\int \lambda(C \cap (C - x))N(dx)]/\lambda(C)$. [Interpretation: $a(C)$ is large if there are many other points nearby given that there is one at the origin, and hence measures aggregation (clustering) at scale size C.]
b) Prove that if N is Poisson with intensity ν, then $a(C) = 1 + \nu\lambda(C)$.
c) Construct unbiased estimators of $a(\cdot)$ given that the intensity of N is known.

9.17 In the context and notation of Theorem 9.15, prove that the random variable $Z = \lim_{r \to \infty}(1/\lambda(B_r)) \int_{B_r} 1(N \circ \theta_x \in \Gamma) N(dx)$ is invariant.

9.18 Let N be a stationary point process on $\mathbf{R}$ with $\bar{\mathcal{F}}^N$-stochastic intensity $(\bar{\lambda}_t)_{t \in \mathbf{R}}$, where $\bar{\mathcal{F}}^N_t = \mathcal{F}^N((-\infty, t])$.
a) Prove that there exists an $\mathcal{F}^N$-stochastic intensity (λ_t) satisfying $\lambda_t = E[\bar{\lambda}_t | \mathcal{F}^N_t]$ almost surely for each t
b) Prove (by example) that λ need not fulfill either (9.34) or (9.35).

9.19 Let N be a stationary point process on $\mathbf{R}$ that under P_α, admits $\bar{\mathcal{F}}^N$-stochastic intensity $\bar{\lambda}_t(\alpha) = \nu + \beta \int_{(-\infty, t)} e^{-\gamma(t-u)} dN_u$, where $\alpha = (\nu, \beta, \gamma)$ with $\nu > 0$ and $0 < \beta < \gamma$. Note that the interval of integration *excludes* t.
a) Prove that N is a Poisson cluster process.
b) Verify that Assumptions 9.17 and 9.18 are fulfilled.
c) Calculate the Fisher information (9.38).

9.20 Let N be a stationary point process such that under the Palm distribution P_0 there exists a decomposition $E_0[N(dx)] = \varepsilon_0(dx) + \nu_0(x) dx$, with ν_0, the *Palm intensity function*, a positive function on $\mathbf{R}^d$. Prove that the spectral density function then satisfies

$$g_\rho(v) = (1/2\pi)\{\nu + \int [\nu \nu_0(x) - \nu^2] e^{-i\langle v, x \rangle} dx\}.$$

9.21 Calculate the spectral measure of a stationary Poisson process N on $\mathbf{R}^d$.

9.22 Calculate the spectral measure of a stationary renewal process N on $\mathbf{R}$ with interarrival distribution F.

9.23 Let N be a stationary point process on $\mathbf{R}$ stipulated to admit covariance spectral density function g_ρ, and introduce the periodogram

$$J_t(v) = (1/t) \left| (1/\sqrt{2\pi}) \int_0^t e^{ivu} dN_u \right|^2.$$

a) Prove that under the hypotheses of Theorem 9.7, $J_t(v) \xrightarrow{d} g_\rho(v)^2 X$ for each v as $t \to \infty$, where X has χ^2-distribution with 1 degree of freedom.
b) Discuss implications concerning use $J_t(v)$ as estimator of $g_\rho(v)$.

9.24 Let $N = \sum_{n=-\infty}^{\infty} \varepsilon_{(T_n, X_n)}$ be a marked point process on $\mathbf{R}$ with mark space $\{1, 2\}$ such that the processes $N_i = \sum 1(X_n = i)\varepsilon_{T_n}$ are stationary and admit spectral density functions f_1 and f_2.
a) Show that the *(cross-)covariance measure* $\eta(A \times B) = \text{Cov}(N_1(A), N_2(B))$ admits a representation $\eta(dt \times du) = h(t - u) dt\, du$ for some function h.
b) Define the cross-spectral density $f_{12}(v) = (1/2\pi) \int_{-\infty}^{\infty} e^{-iuv} h(u) du$, with h as in a), and the *coherence* between N_1 and N_2 as the function $C(v) = |f_{12}(v)|^2 / f_1(v) f_2(v)$. Show that $0 \le C(v) \le 1$.
c) Discuss estimation of $C(v)$ from observations $\mathcal{F}_t^{N_1} \vee \mathcal{F}_t^{N_2}$.

9.25 Prove that a stationary Poisson process on $\mathbf{R}$ is purely nondeterministic.

9.26 Prove Lemma 9.29.

9.27 Let $\tilde{N} = \sum_{n=-\infty}^{\infty} \varepsilon_{T_n}$ be a Poisson process on $\mathbf{R}$ with intensity α and consider the point process $N = \tilde{N} + \sum_{n=-\infty}^{\infty} \varepsilon_{T_n + h}$, where $h \in \mathbf{R}$. [Interpretation: N is a process of two-point clusters, the two members of each cluster separated by h.]

a) Prove that N is stationary and calculate its intensity, reduced second moment measure, and reduced covariance measure.

b) Calculate the spectral measure of N.

c) Derive a backward moving-average representation for N.

9.28 Verify unproved assertions in Example 9.32.

NOTES

The pervasive influences on the chapter are Neveu (1977) on the theory of stationary point processes and Krickeberg (1982) on inference in a general setting. Concerning theory, Matthes *et al.* (1978), despite its being sometimes difficult to read, contains much on Palm distributions and some on ergodic theory, with emphasis on Poisson cluster processes; along these lines seminal papers are Matthes (1963a), Kerstan and Matthes (1964a), and Franken *et al.* (1965). Cox and Isham (1980) is a more elementary exposition of selected aspects of the theory.

Other expository treatments of inference, albeit only in one dimension, are Brillinger (1975b, 1978); the latter, an extended and occasionally forced analogy between point processes and time series, yields numerous insights.

We have omitted from this chapter the discrete case $E = \mathbf{Z}^d$, typically dealt with by random field rather than point process methods. That is, one analyzes directly the stationary random field $\{N(z) : z \in \mathbf{Z}^d\}$, where $N(z) = N(\{z\})$; point process structure of N plays only an incidental role.

Applications as diverse as cosmology (Neyman and Scott, 1972), earthquakes (Vere-Jones, 1970), queueing systems (Franken *et al.*, 1981, which treats in addition marked point processes that are stationary only in one coordinate. (See also Krickeberg, 1982, where other versions of stationarity and isotropy are considered), stereology (Mecke and Stoyan, 1983), and others have been examined. Analysis of *real* data is the focus of Diggle (1983) and Ripley (1981).

Section 9.1

In Krickeberg (1982) the fundamental role of Palm measures in inference for stationary point processes was recognized and first explored systematically. The estimators (9.7) are formulated there; as described in the text, they are averages of translations of N placing each observed point at the origin. Our development of moment measures is incomplete: factorial moment measures, which figure in the proof of Theorem 9.7 and through which, for example, one can characterize simple stationary point processes, have been omitted. They are treated in Krickeberg (1974, 1981, 1982), with additional statistical aspects in Hanisch (1983) and Osher (1983). Theorem 9.4, the basic consistency theorem, is a consequence of Theorem 1.72, and is due to Karr (1987b), as are Theorems 9.5, 9.6, 9.7 and 9.8. Jolivet (1981) is the ultimate source of Theorem 9.7, whose proof is accomplished by time-series techniques, at whose heart are the computational rules of Leonov and Shiryaev (1959).

The "clean" moment theory for stationary point process is counterbalanced by a difficult distribution theory; typical consequences are that linear state estimation is well developed (Section 4) while little is known about nonlinear state estimation and that distributions of test statistics often cannot be calculated. Results that are known pertain mainly to interarrival distributions for stationary point processes on $\mathbf{R}$; Cramér et al. (1971), Leadbetter (1969b), and Eriksson (1978) are representative.

Section 9.2

The relative brevity of this section reveals our preference for time/space-domain over frequency-domain methods. More complete presentations are Brillinger (1972, 1975b, 1978), Daley (1971), Daley and Vere-Jones (1988) and Vere-Jones (1974); the key early works are those of Bartlett (1963, 1964). Theorem 9.9 is a special case of results in Itô (1953). Taken from Vere-Jones (1974), Proposition 9.11 is but one of several inversion formulas. Theorems 9.13 and 9.14 are due to Karr (1987b). The estimators $\hat{g}_\rho$ of (9.30) are, however, of limited practical value because they are not computable from observation over a compact set. Nor is the periodogram (9.32) free from flaws: the spectral density function is related to its second moment rather than the first (Exercise 9.23). Properties of periodograms are described in Brillinger (1975b) and Cox and Lewis (1966); behavior in the Poisson case has been known since Bartlett (1963). Pham (1981) treats estimation of parametrically specified spectral density functions, while Rice (1975) considers estimation of the spectral density function in factored form.

Section 9.3

This section could equally well be entitled "Likelihood-Based Inference." Although obviously true, Theorem 9.16 seems not to have appeared before; there are some subtleties involved since $\mathcal{F}^N$-stochastic intensities need not be stationary. Beneveniste and Jacod (1975) treats additional aspects. The remainder of the section — Assumptions 9.17 and 9.18, Lemmas 9.19–9.21, Theorems 9.22–9.24, and Example 9.25 — is from Ogata (1978). Although martingale limit theorems are used, the structure of the argument is very much that from the classical case of i.i.d. data; see Ibragimov and Haśminskii (1981), the most complete source, or Wilks (1962), which is more elementary. Intensity-based estimation and model comparison for stationary point process models of precipitation is carried out in Smith and Karr (1985).

The results in this section extend without difficulty to the case of multi-dimensional index sets.

The dearth of hypothesis tests stems from lack of distribution theory. Problems that have been addressed include testing whether a stationary point process is Poisson (Brown and Silverman, 1978; Cox and Lewis, 1966; Davies, 1977) and discrimination between Poisson and particular Poisson cluster processes (Kryscio and Saunders, 1983).

Section 9.4

Jowett and Vere-Jones (1972) is a nice development of linear prediction for stationary point processes on **R** and from it are taken Theorems 9.28, 9.30 and 9.31, the main results of the section. Definition 9.27 introduces properties well known for time series (see, e.g., Brillinger, 1981). Ibragimov and Rozanov (1978) and, from a more elementary perspective, Karlin and Taylor (1975) discuss the analogous problem of linear prediction for stationary processes on **R**.

Exercises

9.2 See Daley (1971b).

9.7 See Theorems 9.4 and 9.5.

9.10 Similar effects arise, for example, depending on whether one views the parameter of a Poisson distribution as the mean or the variance.

9.11 Failure to use the Palm distribution to describe synchronous observations has led to confusion and errors in the past.

9.15 Kryscio and Saunders (1983) contains analogous features of the nearest neighbor indicator process $W_t(x) = \int_{K_1} 1(N(B_t(x) \cap K) = 1)N(dx)$, not only for Poisson but also for certain Poisson cluster processes. Roberts (1969) contains partial results on the unresolved problem of the distribution of the number of nearest neighbors of a point in a planar Poisson process.

9.16 This functional is proposed in Jolivet (1978); "aggregation" lacks possibly misleading connotations of "clustering."

9.18 See Exercise 2.16.

9.19 The process is described in Hawkes and Oakes (1974).

9.23 See Brillinger (1975b).

9.24 Coherence is used by Brillinger (1974) to study certain queueing systems. It can be interpreted as a measure of linear predictability of one process given observation of the other.

9.28 See Jowett and Vere-Jones (1972) for these and other examples.

10

Inference for Stochastic Processes Based on Poisson Process Samples

This chapter focuses not on inference for point processes but rather on inference for other stochastic processes given observations resulting from a point process sampling mechanism. It is also a sample in itself: it examines only Markov processes, stationary processes on $\mathbf{R}$ and stationary random fields, and only Poisson process samples. We use marked point processes to unify the problem formulation, but this produces a coherent statement, not broadly applicable methods. Nevertheless, problems that arise in point process sampling *are* of practical interest and do lead to challenging mathematics.

To impart structure and harmony, we concentrate on *Poisson sampling* of stochastic processes. One facet, Poisson sampling of renewal processes, has been treated in Section 8.1. In more general terms the problem is as follows. Given a process $X = (X_t)$ on $\mathbf{R}$ with unknown probability law and a Poisson process $N = \sum \varepsilon_{T_n}$, often but not invariably presumed homogeneous, such that X and N are independent, the goals are to effect statistical inference for the law of X and state estimation concerning unobserved past values or not yet (and possibly never to be) observed future values, given as observations single realizations of the Poisson samples $X(T_n)$, the values of X at the points of N, together possibly with N itself. Observable aspects are encapsulated in the marked point process $\bar{N} = \sum_n \varepsilon_{(T_n, X(T_n))}$, in which each point of N is marked by the current value of X. As in Chapter 9, asymptotics arise given observations over $[0, t]$ as $t \to \infty$.

Before proceeding we consider the foundation issue of whether inference concerning the law of X is even possible. Is $\mathcal{L}_X$ uniquely determined by the law of the Poisson sample sequence? Under the assumption that N is homogeneous the answer is affirmative with no restriction on X other than that the $X(T_n)$ be random variables. A spatial version appears in Section 3.

Proposition 10.1. Let X_1 and X_2 be measurable stochastic processes, each continuous in probability, and let N be a Poisson process with rate one independent of X_1 and X_2. If $(X_1(T_n))_{n\geq 0} \stackrel{d}{=} (X_2(T_n))_{n\geq 0}$, then $X_1 \stackrel{d}{=} X_2$.

Proof: It is important to remember that $X_1 \stackrel{d}{=} X_2$ means only equality of finite-dimensional distributions. With n and $x_1,\ldots,x_n$ fixed, define functions

$$F_i(t_1,\ldots,t_n) = P\{X_i(t_j) \leq x_j, j = 1,\ldots,n\}, \quad i = 1,2.$$

Given positive integers $k_1,\ldots,k_n$, let $K_j = \sum_1^j k_\ell$ and let $S_j = T_{K_j}$. Then the integrals

$$\int_0^\infty \cdots \int_0^\infty F_i(u_1, u_1 + u_2, \ldots, \textstyle\sum_{\ell=1}^n u_\ell)[\prod_{j=1}^n u_j^{k_j-1} e^{-u_j}/(k_j - 1)!]du_1 \cdots du_n$$

$$= E[F_i(S_1,\ldots,S_n)] = P\{X_i(S_j) \leq x_j, j = 1,\ldots,n\}$$

are equal by assumption. Multiplication by $\prod_{j=1}^n (1 - \theta_j)^{k_j}$, where $|\theta_j| \leq 1$, followed by summation over $k_1,\ldots,k_n$, confirms that

$$\int_0^\infty \cdots \int_0^\infty F_1(u_1,\ldots,\textstyle\sum_{\ell=1}^n u_\ell) \exp\{-\textstyle\sum_{j=1}^n \theta_j u_j\}du_1 \cdots du_n$$
$$= \int_0^\infty \cdots \int_0^\infty F_2(u_1,\ldots,\textstyle\sum_{\ell=1}^n u_\ell) \exp\{-\textstyle\sum_{j=1}^n \theta_j u_j\}du_1 \cdots du_n.$$

By the assumption of continuity in probability, F_1 and F_2 are continuous; therefore, this last expression implies that $F_1 = F_2$. Since n and the x_j were arbitrary, $X_1 \stackrel{d}{=} X_2$. ∎

For the Markov process model of Section 1, the Poisson sample sequence is a Markov chain, to which, given assumptions of irreducibility and recurrence, well-known methods apply. Indeed, there is a converse, Theorem 10.3, to Proposition 10.1: if the Poisson sample sequence is Markov, then so must be X itself. For binary (0-1 valued) Markov processes, consistent estimation is possible even if the rate of N is unknown and the observations are only the thinned point process $N' = \sum X(T_n)\varepsilon_{T_n}$. This situation has been discussed, from the viewpoint of partially observed Poisson processes, in Section 6.3. In Section 1 we also consider combined inference and state estimation, for which, in the finite state space case, a satisfactory solution is available.

In Section 2 we examine Poisson sampling of stationary stochastic processes on **R**. Although the law of X may not be recoverable from single realizations, consistent estimation of the mean and the covariance function is possible from asynchronous and synchronous observations of the Poisson sample sequence. In fact, Poisson sampling is superior to sampling at regularly spaced instants $t_n = n\Delta$, for because of "aliasing," estimation of the covariance function given regular samples is impossible in general.

Finally, in Section 3 we consider Poisson sampling of stationary random fields. Results mimic those in Section 2 but are more fragmentary.

10.1   Markov Processes

In this section we examine Poisson sampling for Markov processes. To this end, let X be a Markov process with finite state space E, let P_i be the law of X under the initial condition $X_0 = i$, let $P_t(i, j) = P_i\{X_t = j\}$ be the transition function, let $A = P_0'$ be the generator matrix and for $\alpha > 0$ let

$$U^\alpha(i,j) = \int_0^\infty e^{-\alpha t} P_t(i,j) dt$$

be the α-potential matrix. Then $U^\alpha = (\alpha I - A)^{-1}$, so that (P_t) is determined by each U^α.

Suppose that $N = \sum \epsilon_{T_n}$ is a Poisson process with rate λ, and is independent of X. Then the following result is immediate.

Proposition 10.2. The Poisson sample process $(X(T_n))$ is a Markov chain with transition matrix $Q = \lambda U^\lambda$. $\square$

Less apparent is the converse: a Markov Poisson sample process can arise only from a Markov process.

Theorem 10.3. Let X be a measurable process, continuous in probability, with finite state space E, and let N be a unit rate Poisson process independent of X. If the Poisson sample process is a (homogeneous) Markov chain, then there exists a Markov process $\tilde{X}$ such $X \stackrel{d}{=} \tilde{X}$.

Proof: Let $R_t(i,j) = P\{X_t = j | X_0 = i\}$; these are continuous functions of t, and the transition matrix Q of $(X(T_n))$ satisfies

$$Q^n(i,j) = \int_0^\infty R_u(i,j)(e^{-u} u^{n-1}/(n-1)!) du$$

for each n. From the Chapman-Kolmogorov equation for $(X(T_n))$ we infer that for each n and m,

$$\int_0^\infty \int_0^\infty [R_{u+v}(i,j) - \sum_{k \in E} R_u(i,k) R_v(k,j)] u^{n-1} v^{m-1} e^{-(u+v)} du dv = 0$$

and consequently (by transform or Hilbert space arguments)

$$R_{u+v}(i,j) = \sum_{k \in E} R_u(i,k) R_v(k,j)$$

for all u, v, i and j. There exists a Markov process $\tilde{X}$ with transition function (R_t), and it is immediate that $(\tilde{X}(T_n)) \stackrel{d}{=} (X(T_n))$, from which the theorem follows by appeal to Proposition 10.1. ∎

We now consider statistical inference, for which we impose the complication that the rate of the Poisson process is unknown, but permit N to be observed. Our goal is nonparametric estimation of λ and the generator A from observations $\mathcal{F}_t^{\tilde{N}}$ of the marked point process $\tilde{N} = \sum_n \varepsilon_{(T_n, X(T_n))}$ over sets $[0,t] \times E$, and with asymptotics as $t \to \infty$. More explicitly, $\mathcal{F}_t^{\tilde{N}}$ corresponds to observation of N over $[0,t]$ and the Poisson samples $X(T_1), \ldots, X(T_{N_t})$.

The transition function satisfies

$$P_t = e^{tA} = \sum_{n=0}^{\infty} t^n A^n / n!, \tag{10.1}$$

and hence the $\mathcal{F}_t^{\tilde{N}}$-likelihood function is

$$L_t(\lambda, A) = [e^{-\lambda t} \lambda^{N_t} / N_t!] \prod_{i,j \in E} \{ e^{U_t(i,j)A}(i,j) \}^{N_t(i,j)}, \tag{10.2}$$

where

$$N_t(i,j) = \sum_{k=1}^{N_t} \mathbf{1}(X(T_{k-1}) = i, X(T_k) = j) \tag{10.3}$$

is the number of i-to-j transitions for $(X(T_n))$ during $[0,t]$ and

$$U_t(i,j) = \sum_{k=1}^{N_t} (T_k - T_{k-1}) \mathbf{1}(X(T_{k-1}) = i, X(T_k) = j),$$

which is the time spent in state i during sojourns that terminate with a jump to j.

Maximization of L_t with respect to A and λ seems impossible. Even in the two-state case (see Karr, 1984a), dependence of L_t on A is too complicated to permit explicit solution of the likelihood equations. Of course, since L_t factors into a function of λ alone and a function of A alone, we can easily derive that the maximum likelihood estimator of λ is $\hat{\lambda} = N_t/t$.

For estimation of A we resort to a simplification: $(X(T_n))$ has transition matrix $Q = \lambda U^\lambda = \lambda(\lambda I - A)^{-1}$; consequently,

$$A = \lambda(I - Q^{-1}). \tag{10.4}$$

If the Markov chain $(X(T_n))$ alone were observable, the nonparametric maximum likelihood estimator of Q would be (Billingsley, 1961a, 1961b)

$$\hat{Q}(i,j) = N_t(i,j) / \sum_{k=1}^{N_t} \mathbf{1}(X(T_{k-1}) = i), \tag{10.5}$$

where $N_t(i, j)$ is given by (10.3).

Our estimators amalgamate (10.4)–(10.5):

$$\hat{A} = \hat{\lambda}(I - \hat{Q}^{-1}). \tag{10.6}$$

While not maximum likelihood estimators, they are strongly consistent and asymptotically normal notwithstanding, and inferior only to the extent that they are inefficient.

Theorem 10.4. Assume that X is irreducible (hence positive recurrent). Then as $t \to \infty$, $\hat{A} \to A$ almost surely.

Proof: That $\hat{\lambda} \to \lambda$ almost surely was established in Proposition 3.23, while by the strong law of large numbers for irreducible Markov chains (Doob, 1953)

$$\lim_{n \to \infty} \frac{\sum_{k=1}^n \mathbf{1}(X(T_{k-1}) = i, X(T_k) = j)}{\sum_{k=1}^n \mathbf{1}(X(T_{k-1}) = i)} = \frac{\nu(i)Q(i,j)}{\nu(i)} = Q(i,j),$$

almost surely, where ν is the unique limit distribution of $(X(T_n))$. Since $N_t \to \infty$ almost surely, it follows, with $\hat{Q}$ given by (10.5), that $\hat{Q} \to Q$ almost surely. In particular, since Q is nonsingular, $\hat{Q}$ is nonsingular for all sufficiently large t, removing any doubts that the estimators $\hat{A}$ are well defined. Since matrix inversion is continuous, consistency is confirmed. $\blacksquare$

Theorem 10.5. Assume that X is irreducible. Then

$$\sqrt{t}[\hat{A} - A] \overset{d}{\to} N(0, R), \tag{10.7}$$

on $\mathbf{R}^d$, where $d = |E|^2$ and the covariance matrix R is given by (10.12) below.

Proof: Consider the random elements $(\hat{\lambda}, \hat{A})$ of $\mathbf{R}^{d+1}$. By Proposition 3.21,

$$\sqrt{n}[n/T_n - \lambda] \overset{d}{\to} N(0, \lambda^2), \tag{10.8}$$

while by the central limit theorem for Markov chains (Doob, 1953; see also Basawa and Prakasa Rao, 1980 and Billingsley, 1961a, 1961b), with

$$\hat{Q}^* = \frac{\sum_{k=1}^n \mathbf{1}(X(T_{k-1}) = i, X(T_k) = j)}{\sum_{k=1}^n \mathbf{1}(X(T_{k-1}) = i)},$$

we have

$$\sqrt{n}[\hat{Q}^* - Q] \overset{d}{\to} N(0, R_0), \tag{10.9}$$

where R_0 is the $d \times d$ matrix

$$R_0((i,j),(i',j')) = \frac{\mathbf{1}(i=i')}{\nu(i)}[\mathbf{1}(j=j')Q(i,j) - Q(i,j)Q(i',j')]. \quad (10.10)$$

(Again ν is the unique invariant distribution for Q.) Independence of X and N allows us to combine (10.8) and (10.9):

$$\sqrt{n}[(n/T_n, \hat{Q}^*) - (\lambda, Q)] \overset{d}{\to} N(0, R_1)$$

on $\mathbf{R}^{1+d}$, where

$$
\begin{aligned}
R_1(0,0) &= \lambda^2 \\
R_1(0,(i,j)) &= 0 \\
R_1((i,j),(i',j')) &= R_0((i,j),(i',j')).
\end{aligned}
$$

By appeal to Serfozo (1975, Theorem 8.1), we then have

$$\sqrt{t}[(\hat{\lambda}, \hat{Q}) - (\lambda, Q)] \overset{d}{\to} N(0, R_2), \quad (10.11)$$

where $R_2 = (1/\lambda)R_1$.

Now let G be the open set of points (μ, V) in $\mathbf{R}^{1+d}$ such that the matrix V is nonsingular, and let $H : G \to \mathbf{R}$ be the function $H(\mu, V) = \mu(I - V^{-1})$. According to (10.4), $A = H(\lambda, Q)$, while Theorem 10.4 implies that almost surely $(\hat{\lambda}, \hat{Q}) \in G$ for all sufficiently large t, so that (10.6) gives $\hat{A} = H(\hat{\lambda}, \hat{Q})$. With $J_H(\lambda, Q)$ the Jacobian of H evaluated at (λ, Q), it follows from (10.11) and multivariate transformation theory that (10.7) holds with

$$R = J_H(\lambda, Q)R_2 J_H(\lambda, Q)^T. \ \blacksquare \quad (10.12)$$

State estimation for X is easy. Given observations $\mathcal{F}_t^{\tilde{N}} = \mathcal{F}^{\tilde{N}}([0,t] \times E)$, in order to calculate the conditional distribution $P\{X_u = j | \mathcal{F}_t^{\tilde{N}}\}$, we notice that by independence of X and N, for $u \in (T_{k-1}, T_k)$,

$$P\{X_u = j | \mathcal{F}_t^{\tilde{N}}\} = \frac{P_{u-T_{k-1}}(X(T_{k-1}), j)P_{T_k-u}(j, X(T_k))}{P_{U_k}(X(T_{k-1}), X(T_k))}, \quad (10.13)$$

where (P_t) is the transition function and $U_k = T_k - T_{k-1}$ is the kth interarrival time of N. For $u \geq T_{N_t}$,

$$P\{X_u = j | \mathcal{F}_t^{\tilde{N}}\} = P_{u-T_{N_t}}(Z_t, j), \quad (10.14)$$

where $Z_t = X(T_{N_t})$ is the most recently observed value of X. In particular, choosing $u = t$ solves the filtering problem:

$$P\{X_t = j | \mathcal{F}_t^N\} = P_{V_t}(Z_t, j),$$

(10.15)

where V_t is the backward recurrence time of N at t (see Chapter 8). Nowhere in (10.13)–(10.15) have we utilized the Poisson nature of N; these formulas are valid for any sampling process independent of X. However, in the context of combined statistical inference and state estimation the Poisson assumption will be important again.

We now examine a more restricted form of observation, but under the assumption that $E = \{0, 1\}$, so that X is a *binary Markov process* with transition rates a $(0 \to 1)$ and b $(1 \to 0)$. Suppose that X is observable at t if and only if $dN_t = X_t = 1$, so that the observations are the point process

$$N' = \sum_{n=0}^{\infty} 1(X(T_n) = 1)\varepsilon_{T_n} = \sum_{n=0}^{\infty} X(T_n)\varepsilon_{T_n},$$

(10.16)

that is, $N_t' = \int_0^t X \, dN$. This process is a thinning of N, albeit not in the sense of Definition 1.38, and has already appeared several times, viewed mainly as partial observation of N forced by X, rather than *vice versa*. From Theorem 8.24, N' is a renewal process whose interarrival distribution F satisfies

$$\ell_{1-F}(\alpha) = \int_0^\infty e^{-\alpha t}[1 - F(t)]dt = \frac{a + b + \alpha}{\alpha^2 + \alpha(a + b + \lambda) + \lambda a}.$$

(10.17)

We assume that a, b and λ are all positive.

From observations $\mathcal{F}_t^{N'}$ consistent estimation of a, b and λ is possible in the following manner. There exists (Exercise 10.4) an invertible function H satisfying

$$(a, b, \lambda) = H(\ell_{1-F}(0), \ell_{1-F}(1), \ell_{1-F}(2)).$$

(10.18)

With $\hat{F}(x) = (1/N_t') \sum_{k=1}^{N_t'} 1(U_k' \leq x)$ the $\mathcal{F}_t^{N'}$-estimator of F from Section 8.1, (a, b, λ) is then estimated by substitution:

$$(\hat{a}, \hat{b}, \hat{\lambda}) = H\left(\int_0^\infty [1 - \hat{F}], \int_0^\infty e^{-t}[1 - \hat{F}], \int_0^\infty e^{-2t}[1 - \hat{F}]\right).$$

These estimators (see Karr, 1984a, for proof) are strongly consistent and asymptotically normal.

Proposition 10.6. a) $(\hat{a}, \hat{b}, \hat{\lambda}) \to (a, b, \lambda)$ almost surely;

b) $\sqrt{t}[(\hat{a}, \hat{b}, \hat{\lambda}) - (a, b, \lambda)] \xrightarrow{d} N(0, R)$, where R is a covariance matrix not calculated here. $\square$

The principal state estimation problem, the filtering problem of calculating $P\{X_t = 1|\mathcal{F}_t^{N'}\}$, is solved in Theorem 6.22; variants are treated in Karr (1984a).

Reverting to Poisson sampling with observations $\bar{N} = \sum \varepsilon_{(T_n, X(T_n))}$, we conclude the section with analysis of combined statistical inference and state estimation. The state space E is again an arbitrary finite set. Since the main state estimation issue is filtering, we confine attention to it (for extension to prediction, see Exercise 10.5). By (10.1) and (10.15), for each t,

$$P\{X_t = j|\mathcal{F}_t^{\bar{N}}\} = e^{V_t A}(Z_t, j), \tag{10.19}$$

where V_t is the backward recurrence time at t in the sampling process N and $Z_t = X(T_{N_t})$ is the most recent observation of X.

Following our method in Sections 7.4 and 8.3, we construct pseudo-state estimators by substituting the $\mathcal{F}_t^{\bar{N}}$-estimator $\hat{A}$ into (10.19):

$$\hat{P}\{X_t = j|\mathcal{F}_t^{\bar{N}}\} = e^{V_t \hat{A}}(Z_t, j).$$

The main feature is asymptotic normality of the difference between the pseudo- and the true state estimators.

Theorem 10.7. Assume that X is irreducible, and let π be the limit distribution of X_t as $t \to \infty$. Then as random elements of $\mathbf{R}^{|E|}$,

$$\sqrt{t}\left[\hat{P}\{X_t \in (\cdot)|\mathcal{F}_t^{\bar{N}}\} - P\{X_t \in (\cdot)|\mathcal{F}_t^{\bar{N}}\}\right] \xrightarrow{d} Y(\cdot),$$

where $Y(\cdot) = e^{V_\infty A_\infty}(Z_\infty, \cdot)$, and
a) $A_\infty \sim N(0, R)$, with R is the covariance matrix of (10.12);
b) V_∞ is exponentially distributed with parameter λ;
c) Z_∞ is a random element of E with distribution π;
d) A_∞, V_∞ and Z_∞ are mutually independent.

Proof: It suffices to show that $(\sqrt{t}[\hat{A} - A], V_t, Z_t) \xrightarrow{d} (A_\infty, V_\infty, Z_\infty)$ with a)–d) fulfilled. That $\sqrt{t}[\hat{A} - A] \xrightarrow{d} A_\infty$ satisfying a) is established in Theorem 10.5; $\hat{A}$ is asymptotically independent of (V_t, Z_t) by the mixing character of X and independence of X and N. Concerning (V_t, Z_t), let F be the exponential distribution with parameter λ and $R(t) = 1 + \lambda t$ the associated renewal function [see Chapter 8]. Then for each j and for $y < t$,

$$
\begin{aligned}
P\{V_t > y, Z_t = j\} &= \sum_{k=0}^{\infty} P\{T_k < t - y, X(T_k) = j, T_{k+1} > t\} \\
&= \sum_{k=0}^{\infty} \int_{[0, t-y)} F^k(du) P\{X_u = j\}[1 - F(t - u)] \\
&= P\{X_0 = j\}e^{-\lambda t} + \lambda \int_0^{t-y} P\{X_u = j\}e^{-\lambda(t-u)}du \\
&\cong P\{X_0 = j\}e^{-\lambda t} + P\{X_t = j\}e^{-\lambda y} \\
&\to \pi(j)e^{-\lambda y}.
\end{aligned}
$$

Consequently, $(V_t, Z_t) \overset{d}{\to} (V_\infty, Z_\infty)$ such that b)–d) are fulfilled. ∎

This argument does not depend crucially on the Poisson nature of N. Provided that N is a renewal process with nonarithmetic interarrival distribution, Theorem 10.7 can be extended (see Karr, 1984a).

10.2   Stationary Processes on R

The question addressed in this section is statistical estimation for stationary stochastic processes on **R** given Poisson samples. Since "stationary" will be used in two senses, we begin by introducing them.

A real-valued process $(X_t)_{t \in \mathbf{R}}$ is *stationary* if its probability law is invariant under translations of the time axis: $(X_t)_{t \in \mathbf{R}} \overset{d}{=} (X_{t+s})_{t \in \mathbf{R}}$ for every $s \in \mathbf{R}$. This relatively strong form neither requires nor entails existence of moments, but is too restrictive in some applications, not so much because it is believed to fail as because it is difficult to confirm, for example by statistical hypothesis tests. Nonetheless, it is a potent assumption in comparison with the more commonly stipulated L^2-stationarity.

The process X is L^2-*stationary* if its first and second moment structure is translation invariant; that is, if $E[X_t^2] < \infty$ for each t, if the *mean* $m = E[X_t]$ is independent of t, and if the *covariance function* $R(t) = \text{Cov}(X_u, X_{u+t})$ depends, as indicated, only on $|t|$. In particular (Exercise 10.8), R is symmetric and positive definite with $R(0) = \text{Var}(X_t)$. The law of X is not specified by m and R except in the Gaussian case, so that they do not define an identifiable statistical model, but of course one may perform inference for them nonetheless.

For L^2-stationary processes there holds an analogue of the spectral theory for stationary point processes developed in Section 9.2. Its cornerstone is a spectral representation theorem (see Brillinger, 1981, or Rozanov, 1967): there is a complex-valued random measure Z (in the L^2-sense), with mean zero and orthogonal increments, such that for each t, $X_t = m + \int_{-\infty}^{\infty} e^{itv} Z(dv)$.

It follows that $R(t) = \int_{-\infty}^{\infty} e^{itv} E[|Z(dv)|^2]$, and the measure $F(dv) = E[|Z(dv)|^2]$ is called the *spectral measure* of X. If the covariance function is integrable, i.e., $\int_{-\infty}^{\infty} |R(t)| dt < \infty$, then F admits *spectral density function* f satisfying

$$f(v) = (1/2\pi) \int_{-\infty}^{\infty} e^{-ivt} R(t) dt.$$

Our concern is with estimation of the mean, the covariance function and the spectral density function, given Poisson samples of X. Throughout, X is assumed to be continuous in probability.

Let $N = \sum_{n=0}^{\infty} \varepsilon_{T_n}$, a Poisson process on $\mathbf{R}_+$ independent of X and with *known* intensity ν, be the sampling process. We presume that observation commences at $t = 0$ and that N is observable, so that the observations constitute the marked point process $\bar{N} = \sum_{n=0}^{\infty} \varepsilon_{(T_n, X(T_n))}$. Estimation can be based on asynchronous (real-time) data $\mathcal{H}_t = \mathcal{F}^{\bar{N}}([0,t] \times \mathbf{R})$ or synchronous data $\mathcal{H}_{T_n} = \sigma(T_1, \ldots, T_n; X(T_1), \ldots, X(T_n))$. The former not only is more important in applications but also can be generalized to higher dimensions (Section 3); however the latter is also interesting, especially since the two differ in mathematical details.

We begin with estimation of the mean from asynchronous data. Even though the problem is solved easily it illustrates the differing implications of the two forms of stationarity. Given that the intensity ν is known, the obvious $\mathcal{H}_t$-estimator of m is

$$\hat{m} = (1/\nu t)\int_0^t X_u dN_u. \tag{10.20}$$

Elementary properties valid under L^2-stationarity are provided by the following result, whose proof is left as Exercise 10.9.

Proposition 10.8. Suppose that X is L^2-stationary and let $\hat{m}$ be given by (10.20). Then $E[\hat{m}] = m$ for each t and

$$\text{Var}(\hat{m}) = (R(0) + m^2)/\nu t + (1/t)\int_{-t}^t (1 - |s|/t)R(s)ds. \ \square$$

Thus the estimators $\hat{m}$ are unbiased, and if the covariance function is integrable they are mean square consistent: $\hat{m} \to m$ in L^2.

Stationarity implies strong consistency, and if X satisfies an appropriate central limit theorem, then the $\hat{m}$ are asymptotically normal.

Proposition 10.9. Assume that X is stationary and that $m = E[X_t]$ exists. Then the estimators $\hat{m}$ of (10.20) are strongly consistent.

Proof: Relative to the filtration $\mathcal{G}_t = \mathcal{F}^N([0,t]) \bigvee \mathcal{F}_\infty^X$, by independence of X and N the stochastic intensity of N remains the $\mathcal{F}^N$-stochastic intensity, namely the deterministic process identically equal to ν. Lemma 9.20 applies, and hence $\lim_{t \to \infty} \hat{m} = \lim_{t \to \infty} (1/t) \int_0^t (X_u/\nu)dN_u = E[X_0] = m$. ∎

Proposition 10.10. Assume that X is stationary with $E[X_t^2] < \infty$ for each t. Suppose that the covariance function R is integrable and that

$$(1/\sqrt{t})\int_0^t (X_u - m)du \xrightarrow{d} N(0, \int R(s)ds). \tag{10.21}$$

Then $\sqrt{t}[\hat{m} - m] \xrightarrow{d} N(0, \sigma^2)$, where $\sigma^2 = [(R(0) + m^2)/\nu] \int R(s)ds$.

Proof: Conditions for validity of (10.21) are given, for example, in Rozanov (1965); the asymptotic variance is necessarily $\int R(s)ds = 2\pi f(0)$, where f is the spectral density function. For $\alpha \in \mathbf{R}$ and $\tilde{X} = X - m$,

$$
\begin{aligned}
E&\left[\exp\left\{\{(i\alpha/\sqrt{t})\int_0^t \tilde{X}dN\}\right\}\right] \\
&= E\left[E[\exp\{(i\alpha/\sqrt{t})\int_0^t \tilde{X}dN\}|X]\right] \\
&= E\left[\exp\{\nu\int_0^t[e^{(i\alpha/\sqrt{t})\tilde{X}} - 1]du\}\right] \\
&\cong E\left[\exp\left\{(i\nu\alpha/\sqrt{t})\int_0^t \tilde{X}du - (\nu\alpha^2/2t)\int_0^t \tilde{X}^2 du\right\}\right] \\
&\rightarrow \exp\left\{-(\nu^2\alpha^2/2)\int R(s)ds - (\nu\alpha^2/2)[R(0)+m^2]\right\},
\end{aligned}
$$

where the error converges in probability to zero by Proposition 10.9, and where we have applied (10.21) as well as Proposition 10.9 to the stationary process X^2. ∎

One can even permit the intensity of the sampling process to be unknown. The estimators $\hat{\nu} = N_t/t$ are strongly consistent and asymptotically normal by Proposition 3.23, and $\hat{m}$ can be replaced by

$$\hat{m}^* = (1/N_t)\int_0^t X_u dN_u = \int_0^t X_u dN_u / \int_0^t dN_u; \qquad (10.22)$$

some properties are examined in Exercises 10.10 and 10.11.

For estimation of the covariance function from asynchronous data we employ kernel methods. Assume for simplicity that $m = 0$. Let w be a symmetric, positive, bounded and continuous function on $\mathbf{R}$ satisfying $\int w(x)dx = 1$ and let k_t be positive numbers such that $k_t \rightarrow 0$ and $tk_t \rightarrow \infty$ as $t \rightarrow \infty$. As $\mathcal{H}_t$-estimator of R we take

$$\hat{R}(s) = (1/\nu^2 t)\int_0^t \int_0^t w_t(s - s_1 + s_2)X(s_1)X(s_2)N^{(2)}(ds_1, ds_2), \qquad (10.23)$$

where $w_t(x) = (1/k_t)w(x/k_t)$ and (see Section 1.2)

$$N^{(2)}(ds_1, ds_2) = N(ds_1)(N - \varepsilon_{s_1})(ds_2) = 1(s_1 \neq s_2)N(ds_1)N(ds_2).$$

The rationale is that the probability measures associated with the density functions $w_t(\cdot)$ converge weakly to ε_0 as $t \rightarrow \infty$, giving the approximation

$$\hat{R}(s) \cong (1/\nu^2 t)\int_0^t \int_0^t 1(s_1 - s_2 = s)X(s_1)X(s_2)N^{(2)}(ds_1, ds_2),$$

whose role as estimator of $R(s)$ is apparent.

The estimators $\hat{R}$ are only asymptotically unbiased.

Weak consistency can be established under rather broad hypotheses, which is done in the following analogue of Proposition 10.9. In part b) of Theorem 10.11 we employ the fourth-order cumulant function Q of X, which exists under the assumption that $E[X_t^4] < \infty$ and — because $m = 0$ — satisfies

$$
\begin{aligned}
Q(s_1, s_2, s_3) \;=\; & E\left[X(t+s_1)X(t+s_2)X(t+s_3)\right] - R(s_1)R(s_3 - s_2) \\
& - R(s_2)R(s_3 - s_1) - R(s_3)R(s_2 - s_1) \qquad (10.24)
\end{aligned}
$$

for all t, s_1, s_2 and s_3.

Theorem 10.11. Assume that X is L^2-stationary. Then
a) For each s,

$$
E[\hat{R}(s)] = \int_{-t}^{t}(1 - |v|/t)w_t(s - v)R(v)dv; \qquad (10.25)
$$

b) If in addition R is integrable and the fourth-order cumulant function Q is bounded and continuous and satisfies

$$
\sup_{s_1, s_2} \int |Q(u + s_1, u, s_2)|du < \infty,
$$

then

$$
\begin{aligned}
\mathrm{Cov}(\hat{R}(s_1), \hat{R}(s_2)) \;=\; & (1/\nu^2 t)\int_{-t}^{t}\Big\{(1 - |u|/t)[Q(0, u, u) + 2R(u)^2 + R(0)^2] \\
& \times w_t(s_2 - u)[w_t(s_1 - u) + w_t(s_1 + u)]\Big\}du \\
& + O(1/t); \qquad (10.26)
\end{aligned}
$$

in particular,

$$
\lim_{t\to\infty}(tk_t)\mathrm{Cov}(\hat{R}(s_1), \hat{R}(s_2)) = \sigma^2(s_1)\mathbf{1}(s_1 = \pm s_2), \qquad (10.27)
$$

where

$$
\sigma^2(s) = [\textstyle\int w(x)^2 dx/\nu^2][Q(0, s, s) + 2R(s)^2 + R(0)^2]. \qquad (10.28)
$$

Proof: (Sketch). The arguments are primarily computational, so we do not present them in detail. The expression (10.25) follows essentially immediately from (10.23): since $E[N^{(2)}(ds_1, ds_2)] = \nu^2 ds_1 ds_2$,

$$
E[\hat{R}(s)] = (1/\nu)\int_0^t\int_0^t w_t(s - s_1 + s_2)R(s_1 - s_2)ds_1 ds_2,
$$

which becomes (10.25) after change of variables.

Independence of X and N implies that

$$
\begin{aligned}
& \mathrm{Cov}(\hat{R}(s_2), \hat{R}(s_2)) \\
& = (1/\nu^2 t)\int_0^t\int_0^t\int_0^t\int_0^t w_t(s_1 - u_1 + r_1)w_t(s_2 - u_2 + r_2) \\
& \quad \times E[X(u_1)X(r_1)X(u_2)X(r_2)]E[N^{(2)}(du_1, dr_2)N^{(2)}(du_2, dr_2)] \\
& \quad - E[\hat{R}(s_1)]E[\hat{R}(s_2)].
\end{aligned}
$$

The latter term is calculated using (10.25), while from (10.24),

$$E[X(u_1)X(r_1)X(u_2)X(r_2)]$$
$$= Q(u_1 - r_1, u_1 - u_2, u_1 - r_2) + R(u_1 - u_2)R(r_1 - r_2)$$
$$+ R(u_1 - r_1)R(u_2 - r_2) + R(u_1 - r_2)R(r_1 - u_2),$$

and consequently,

$$\text{Cov}(\hat{R}(s_1), \hat{R}(s_2))$$
$$= (1/\nu^2 t)\int_0^t \int_0^t \int_0^t \int_0^t w_t(s_1 - u_1 + r_1)w_t(s_2 - u_2 + r_2)$$
$$\times Q(u_1 - r_1, u_1 - u_2, u_1 - r_2)E[N^{(2)}(du_1, dr_1)N^{(2)}(du_2, dr_2)]$$
$$+ (1/\nu^2 t)\int_0^t \int_0^t \int_0^t \int_0^t w_t(s_1 - u_1 + r_1)w_t(s_2 - u_2 + r_2)R(u_1 - u_2)R(r_1 - r_2)$$
$$\times \left\{ E[N^{(2)}(du_1, dr_1)N^{(2)}(du_2, dr_2)] - \nu^4 du_1 dr_1 du_2 dr_2 \right\}$$
$$+ (1/\nu^2 t)\int_0^t \int_0^t \int_0^t \int_0^t w_t(s_1 - u_1 + r_1)w_t(s_2 - u_2 + r_2)R(u_1 - r_1)R(u_2 - r_2)$$
$$\times E[N^{(2)}(du_1, dr_1)N^{(2)}(du_2, dr_2)]$$
$$+ (1/\nu^2 t)\int_0^t \int_0^t \int_0^t \int_0^t w_t(s_1 - u_1 + r_1)w_t(s_2 - u_2 + r_2)R(u_1 - r_2)R(r_1 - u_2)$$
$$\times E[N^{(2)}(du_1, dr_1)N^{(2)}(du_2, dr_2)] \tag{10.29}$$

Exercise 10.12 implies that

$$E[N^{(2)}(du_1, dr_1)N^{(2)}(du_2, dr_2)]$$
$$= \nu^4 du_1 du_2 dr_1 dr_2$$
$$+ \nu^3 du_1 dr_1 \epsilon_{u_1}(du_2)dr_2 + \nu^3 du_1 dr_1 du_2 \epsilon_{u_1}(dr_2)$$
$$+ \nu^3 du_1 dr_1 \epsilon_{r_1}(du_2)dr_2 + \nu^3 du_1 dr_1 du_2 \epsilon_{r_1}(dr_2)$$
$$+ \nu^2 du_1 dr_1 \epsilon_{u_1}(du_2)\epsilon_{r_1}(dr_2)$$
$$+ \nu^2 du_1 dr_1 \epsilon_{r_1}(du_2)\epsilon_{u_1}(dr_2), \tag{10.30}$$

which one then substitutes into (10.29). The dominant components of each term arise from $\nu^2 du_1 dr_1 \epsilon_{u_1}(du_2)\epsilon_{r_1}(dr_2)$ and $\nu^2 du_1 dr_1 \epsilon_{r_1}(du_2)\epsilon_{u_1}(dr_2)$ (see Masry, 1983, for details; the integrability assumption on Q is used within these omitted calculations) and therefore within $O(1/t)$

$$\text{Cov}(\hat{R}(s_1), \hat{R}(s_2))$$
$$= (1/\nu^2 t)\left[\int_{-t}^t (1 - |u|/t)w_t(s_1 - u)w_t(s_2 - u)Q(u, 0, u)du \right.$$
$$\left. + \int_{-t}^t (1 - |u|/t)w_t(s_1 + u)w_t(s_2 - u)Q(0, u, u)du \right]$$
$$+ (1/\nu^2 t)\left[\int_{-t}^t (1 - |u|/t)w_t(s_1 - u)w_t(s_2 - u)R(u)^2 du \right.$$
$$\left. + \int_{-t}^t (1 - |u|/t)w_t(s_1 - u)w_t(s_2 + u)R(u)^2 du \right]$$

$$+ (1/\nu^2 t) \left[\int_{-t}^{t} (1 - |u|/t) w_t(s_1 - u) w_t(s_2 - u) R(u)^2 du \right.$$
$$+ R(0)^2 \int_{-t}^{t} (1 - |u|/t) w_t(s_1 + u) w_t(s_2 - u) du \right]$$
$$+ (R(0)^2/\nu^2 t) \left[\int_{-t}^{t} (1 - |u|/t) w_t(s_1 - u) w_t(s_2 - u) du \right.$$
$$+ \int_{-t}^{t} (1 - |u|/t) w_t(s_1 + u) w_t(s_2 - u) R(u)^2 du \right].$$

Since $Q(u, 0, u) = Q(0, u, u)$, (10.26) holds; derivation of (10.27) from it is computational. ∎

The asymptotic covariance between $\hat{R}(s_1)$ and $\hat{R}(s_2)$ is zero unless $s_1 = \pm s_2$, which precludes a functional central limit theorem for the $\hat{R}$ as stochastic processes. Nevertheless, given additional assumptions asymptotic normality can be established for single values $\hat{R}(s)$.

Theorem 10.12. Assume that X admits moments of all orders, that for every $k \geq 2$ the cumulant function of order k satisfies

$$\int_{\mathbf{R}^{k-1}} |Q^{(k)}(u_1, \ldots, u_{k-1})| du_1 \cdots du_{k-1} < \infty, \qquad (10.31)$$

that the covariance function is twice continuously differentiable, and that $\int x^2 w(x) dx < \infty$. Then for each s, $\sqrt{tk_t}[\hat{R}(s) - R(s)] \overset{d}{\to} N(0, \sigma^2(s))$, where $\sigma^2(s)$ is given by (10.28).

Proof: By stationarity of X and N individually and their independence, the process $M(dt) = X(t)N(dt)$ has stationary increments. The main idea is to apply the reasoning used to prove Theorem 9.7 to M, suitably centered. The kth-order cumulant measure of M (see Definition 1.62) has reduced-form representation $\gamma^k(dt_1, \ldots, dt_k) = \gamma_*^k(dt_1 - t_k, \ldots, dt_{k-1} - t_k) dt_k$, where γ_*^k, the reduced cumulant measure of order k, has finite variation by (10.31) and properties of N. As in the proof of Theorem 9.7, to establish that

$$\sqrt{tk_t}\{\hat{R}(s) - E[\hat{R}(s)]\} \overset{d}{\to} N(0, \sigma^2(s)),$$

it suffices to show that for each $\ell \geq 3$ the ℓth-order cumulant of $\hat{R}(s) - E[\hat{R}(s)]$ is of order $(tk_t)^{1-\ell}$, for then the same cumulant of $\sqrt{tk_t}\{\hat{R}(s) - E[\hat{R}(s)]\}$ is of order $(tk_t)^{1-\ell/2}$ and hence converges to zero. Directly from (10.23),

$$\hat{R}(s)^\ell = (1/[\nu^2 tk_t^\ell]) \prod_{j=1}^{\ell} \int_{-t}^{t} \int_{-t}^{t} w([s - s_{2j-1} + s_{2j}]/k_t)$$
$$\times \mathbf{1}(s_{2j-1} \neq s_{2j}) M(ds_{2j-1}) M(ds_{2j}),$$

from which the asserted behavior of cumulants is at least intuitively clear. Details are given in Masry (1983) (see also Jolivet, 1981).

Given that (10.27) holds, in order to complete the proof it remains to show that $\sqrt{tk_t}\{E[\hat{R}(s)] - R(s)\} \to 0$, which follows from (10.25) by a Taylor series expansion of R. ∎

Rather than exhaust all cases, we move on to estimation of the spectral density function given synchronous data. Finite sample covariances differ from those for asynchronous data; moreover, asymptotic normality of the estimators $\hat{f}$ defined in (10.32) has not been established. We suppose that $E[X_t] = 0$ and that $E[X_t^2] < \infty$.

The observations $\mathcal{H}_{T_n}$ comprise the sampled values $X(T_1), \ldots, X(T_n)$ and the observation times $T_1, \ldots, T_n$. The spectral density function, by symmetry of the covariance function, satisfies

$$f(v) = (1/\pi)\int_0^\infty \cos(vt)R(t)dt,$$

so we take as its $\mathcal{H}_{T_n}$-estimator the function

$$\begin{aligned}
\hat{f}(v) &= (1/\pi\nu n)\sum_{\ell=1}^{n-1}\sum_{k=1}^{n-\ell}X(T_k)X(T_{k+\ell}) \\
&\times w_n(T_{k+\ell} - T_k)\cos(v(T_{k+\ell} - T_k)),
\end{aligned} \tag{10.32}$$

where the kernel w_n is constructed in the following manner. Let h be a symmetric density function on $\mathbf{R}$ (a "spectral window"), let k_n be positive numbers with $k_n \to 0$ and $nk_n \to \infty$, let $h_n(x) = (1/k_n)h(x/k_n)$, and finally let $w_n(t) = \tilde{h}(tk_n)$, where $\tilde{h}$ is the Fourier transform of h.

One can express (10.32) in an alternative form more amenable to computation and analysis:

$$\hat{f}(v) = \int_{-\infty}^\infty h_n(v - u)\hat{I}(u)du - (1/2\pi\nu n)\sum_{k=1}^n X(T_k)^2, \tag{10.33}$$

where

$$\hat{I}(u) = (1/2\pi\nu n)\left|\sum_{k=1}^n X(T_k)e^{ivT_k}\right|^2,$$

which is analogous to periodogram estimators.

We begin with a computational result, whose proof, along with those of the results that follow, is given in Masry (1983).

Proposition 10.13. a) If R is integrable, then as $n \to \infty$

$$E[\hat{f}(v)] = \int h_n(v - u)f(u)du + o(1)$$

uniformly in v;

b) If $\int |tR(t)|dt < \infty$, then, also uniformly in v

$$E[\hat{f}(v)] = \int h_n(v - u)f(u)du + O(1/n).\,\square$$

The final results of the section describe the covariance structure of the estimators $\hat{f}$. We introduce the notation $\hat{J}(u) = \hat{I}(u) - (1/2\pi\nu n)\sum_{k=1}^{n} X(T_k)^2$, and re-write (10.33) as

$$\hat{f}(v) = \int_{-\infty}^{\infty} h_n(v - u)\hat{J}(u)du.$$

The asymptotic covariance of $\hat{J}$ is the key to that of $\hat{f}$.

Proposition 10.14. Suppose that $\int |tR(t)|dt < \infty$. Then

$\mathrm{Cov}(\hat{J}(v_1), \hat{J}(v_2))$
$$= (1/n)[f(v_1) + R(0)/2\pi\nu]^2[\Gamma_n(v_1 + v_2) - \Gamma_n(v_1 - v_2)] + O(1/n)$$

uniformly in v_1 and v_2, where $\Gamma_n(v) = 2\sum_{\ell=1}^{n}(1 - \ell/n)\Re[\nu/(\nu - iv)]^\ell$. $\square$

Properties for $\hat{f}$ are then as follows.

Theorem 10.15. Suppose that the hypotheses of Proposition 10.14 hold and that in addition h is bounded and differentiable with bounded derivative. Then

$$\mathrm{Cov}(\hat{f}(v_1), \hat{f}(v_2)) = (2\pi\nu/n)\int_{-\infty}^{\infty}\left\{[f(u) - R(0)/2\pi\nu]^2 h_n(v_1 - u)\right.$$
$$\times[h_n(v_2 + u) - h_n(v_2 - u)]\Big\} du$$
$$+ O(\log(1/nk_n)/n^2k_n^2),$$

and consequently

$$\lim_{n\to\infty}(nk_n)\mathrm{Cov}(\hat{f}(v_1), \hat{f}(v_2))$$
$$= \left(2\pi\nu[f(v_1) + R(0)/\nu^2]^2\int h(x)^2 dx\right)\mathbf{1}(v_1 = \pm v_2).\ \square \qquad (10.34)$$

It follows that the estimators $\hat{f}(v)$ are mean square consistent for each v. In addition, they are asymptotically uncorrelated for different values of $|v|$. If the sampling rate (of the Poisson process N) is controllable, it can be optimized to produce minimal asymptotic variance for a given value of v (Exercise 10.14).

The covariance structure of the estimators $\hat{f}$ differs from that of the asynchronous data estimators [note the analogy to (10.33)] given by

$$f^*(v) = \int_{-\infty}^{\infty}\frac{1}{k_t}h\left(\frac{v - u}{k_t}\right)\hat{H}(u)du - \frac{1}{2\pi\nu t}\int_0^t X_u^2 dN_u, \qquad (10.35)$$

where $k_t \to 0$, $tk_t \to \infty$ and $\hat{H}$ is the periodogram

$$\hat{H}(u) = (1/2\pi\nu t)\left|\int_0^t e^{-ius}X(s)dN_s\right|^2.$$

In particular, if $\int |tR(t)|\,dt < \infty$, then

$$E[\hat{H}(u)] = \nu^2 f(u) + \nu R(0)/2\pi + O(1/t) \qquad (10.36)$$

and

$$
\begin{aligned}
\mathrm{Cov}(\hat{H}(v_1), \hat{H}(v_2)) = \; & (1/t)[\nu^2 f(v_1) + \nu R(0)/2\pi] \\
& \times [\Delta_t(v_1 + v_2) - \Delta_t(v_1 - v_2)] \\
& + O(1/t), \qquad\qquad\qquad (10.37)
\end{aligned}
$$

where Δ_t is the Féjer kernel $\Delta_t(v) = (1/t)[2\sin(vt/2)/v]^2$, which also arises in time series analysis (see Masry, 1978a, or Exercise 10.15).

    Notwithstanding finite sample differences, asymptotics are the same as for synchronous data.

10.3   Stationary Random Fields

This section is a spatial version of Section 2, but is incomplete, and ends with problems, not solutions. Before proceeding to the main results, we introduce random fields and present a multidimensional analogue of Proposition 10.1.

    A random field is a stochastic process with a multidimensional (Euclidean) parameter set, and stationarity is translation invariance in the obvious sense.

    Definition 10.16. a) A *random field* on $\mathbf{R}^d$ is a measurable stochastic process $(Y(x))_{x \in \mathbf{R}^d}$ taking values in $\mathbf{R}$;

    b) The random field Y is *stationary* if $(Y(x+y))_{x\in\mathbf{R}^d} \stackrel{d}{=} (Y(x))_{x\in\mathbf{R}^d}$ for each $y \in \mathbf{R}^d$.

    As in the one-dimensional case, if Y is stationary and if $E[|Y(x)|^2] < \infty$ for each x, then Y is L^2-stationary and there exist a covariance function $R(y) = \mathrm{Cov}(Y(x), Y(x+y))$ and spectral measure F satisfying

$$R(y) = \int_{\mathbf{R}^d} e^{i\langle y,v\rangle} F(dv), \qquad y \in \mathbf{R}^d,$$

where $\langle \cdot, \cdot \rangle$ denotes the inner product. However, whereas in one dimension the covariance function is symmetric and hence a function only of $|t|$, in the multidimensional case the stronger condition of isotropy is not implied by stationarity.

    Definition 10.17. An L^2-stationary random field Y is *isotropic* if for every rotation γ of $\mathbf{R}^d$, $\mathrm{Cov}(Y(\gamma x), Y(\gamma y)) = \mathrm{Cov}(Y(x), Y(y))$ for all x and $y \in \mathbf{R}^d$.

For an isotropic random field the covariance function reduces to a function R on $\mathbf{R}_+$: $\text{Cov}(Y(x), Y(y)) = R(|x - y|)$, and the spectral measure satisfies $F\gamma^{-1} = F$ for each rotation γ.

For simplicity we often assume that $E[Y(x)] = 0$, in which case there exists a spectral representation analogous to that in Theorem 9.9: $Y(x) = \int_{\mathbf{R}^d} e^{i\langle v, x\rangle} Z(dv)$, where Z is a mean zero random measure with orthogonal increments and $E[Z(A_1)\overline{Z(A_2)}] = F(A_1 \cap A_2)$. For isotropic random fields more detailed spectral representations can be derived; these and additional material can be found in Yadrenko (1983).

Our main topic in the section is estimation of means and covariance functions for stationary random fields given Poisson samples. That is, let Y be an L^2-stationary random field on $\mathbf{R}^d$ and let $N = \sum \epsilon_{X_n}$ be a stationary Poisson process on $\mathbf{R}^d$, independent of Y, with intensity ν. The data are the marked point process $\bar{N} = \sum \epsilon_{(X_n, Y(X_n))}$, with each point X_n of N marked by the value of Y at X_n. We assume that the sampling process is observable, and consider "asynchronous" data $\mathcal{H}(K) = \mathcal{F}^{\bar{N}}(K \times \mathbf{R})$, where K is a compact, convex subset of $\mathbf{R}^d$.

Before examining estimation of the mean and covariance function we present a generalization of Proposition 10.1 to the effect that the law of the marked point process determines that of the random field. Indeed, in the following result neither process need be stationary; however, minimal regularity conditions are required: Y must be continuous in probability and the mean measure of N must place positive mass in every open set.

Theorem 10.18. Let Y be a random field on $\mathbf{R}^d$ that is continuous in probability, let N be a Poisson process on $\mathbf{R}^d$ with diffuse mean measure μ satisfying $\mu(G) > 0$ for every open set G, assume that Y and N are independent. Then the law of the marked point process $\bar{N} = \sum \epsilon_{(X_n, Y(X_n))}$ determines that of Y.

Proof: Let Y_1 and Y_2 be random fields fulfilling the hypotheses of the theorem, with associated marked point processes $\bar{N}_1$ and $\bar{N}_2$, and suppose without loss of generality that $|Y_1| \leq 1$ and $|Y_2| \leq 1$. Then for h a function on $\mathbf{R}^d$ with $0 \leq h < 1$ and $f(x, u) = -\log[1 - uh(x)]$, $|u| \leq 1$,

$$
\begin{aligned}
E\left[\exp\{-\int h(x)Y_1(x)\mu(dx)\}\right] &= E\left[\exp\{-\int[1 - e^{-f(x, Y_1(x))}]\mu(dx)\}\right] \\
&= E\left[E[\exp\{-\int f(x, Y_1(x))N(dx)\}|Y_1]\right] \\
&= E[\exp\{-\int f d\bar{N}_1\}] \\
&= E[\exp\{-\int h(x)Y_2(x)\mu(dx)\}];
\end{aligned}
$$

the last equality is a reversal of the first three. Consequently, by Theorem 1.12 the random measures $M_1(dx) = Y_1(dx)\mu(dx)$ and $M_2(dx) = Y_2(x)\mu(dx)$

are identically distributed. Given $x_1, \ldots, x_k$, in $\mathbf{R}^d$, for each i choose open sets G_{in} such that as $n \to \infty$, $G_{in} \downarrow \{x_i\}$. Then $\mu(G_{in}) > 0$ for each i and n, while $\mu(G_{in}) \to 0$ for each i. Thus continuity in probability of Y_1 and Y_2

$$
\begin{aligned}
(Y_1(x_1), \ldots, Y_1(x_k)) &= \lim_{n \to \infty} (M_1(G_{1n})/\mu(G_{1n}), \ldots, M_1(G_{kn})/\mu(G_{kn})) \\
&\overset{d}{=} \lim_{n \to \infty} (M_2(G_{1n})/\mu(G_{1n}), \ldots, M_2(G_{kn})/\mu(G_{kn})) \\
&= (Y_2(x_1), \ldots, Y_2(x_k)),
\end{aligned}
$$

where the limits are in the sense of convergence in distribution. Hence $Y_1 \overset{d}{=} Y_2$. ∎

Suppose now that Y is an L^2-stationary random field on $\mathbf{R}^d$ with unknown mean and covariance function to be estimated from observations $\mathcal{H}(K)$ of the Poisson sample process $\bar{N}$. (Once again N is a stationary Poisson process with known intensity ν.) By analogy with (10.20) we introduce $\mathcal{H}(K)$-estimators

$$\hat{m} = (1/\nu\lambda(K))\int_K Y \, dN, \tag{10.38}$$

where K is compact and convex. As customary, the "sample size" K is suppressed.

Were ν unknown, one could use the estimators $\hat{m}^* = (1/N(K)) \int_K Y \, dN$, whose properties are superior in some ways even when ν is known (see Karr, 1986a).

Mean square consistency of the $\hat{m}$ is one consequence of the following analogue of Proposition 10.8, whose computational proof is omitted.

Proposition 10.19. Let $\hat{m}$ be given by (10.38). Then for each K, $E[\hat{M}] = m$ and

$$
\begin{aligned}
\text{Var}(\hat{m}) &= (R(0) + m^2)/\nu\lambda(K) \\
&\quad + (1/\lambda(K))\int_{\mathbf{R}^d} R(y)[\lambda(K \cap (K - y))/\lambda(K)] \, dy, \tag{10.39}
\end{aligned}
$$

where $K - y = \{x - y : x \in K\}$. □

To establish consistency not only of the $\hat{m}$ but also of estimators of the covariance function, we need the following. Recall from Chapters 1 and 9 that for a compact, convex set K, $\delta(K)$ denotes the supremum of radii of Euclidean balls contained in K. Then, for each $y \in \mathbf{R}^d$,

$$\lim_{\delta(K) \to \infty} \lambda(K \cap (K - y))/\lambda(K) = 1. \tag{10.40}$$

We can now show consistency of the $\hat{m}$.

Proposition 10.20. Assume that the covariance function is integrable: $\int_{\mathbf{R}^d} |R(y)| dy < \infty$. Then $\lim_{\delta(K)\to\infty} \mathrm{Var}(\hat{m}) = 0$, and hence the estimators $\hat{m}$ are mean square consistent.

Proof: Evidently, the first term in (10.39) converges to zero since $\delta(K) \to \infty$ implies that $\lambda(K) \to \infty$. By (10.40), integrability of R and the dominated convergence theorem,

$$\lim_{\delta(K)\to\infty} \int_{\mathbf{R}^d} R(y)[\lambda(K \cap (K-y))/\lambda(K)] dy = \int_{\mathbf{R}^d} R(y) dy;$$

therefore the second term converges to zero as well. ∎

For estimation of the covariance function we suppose for simplicity that $m = 0$. By analogy to the one-dimensional case, the $\mathcal{H}(K)$-estimator is

$$\hat{R}(x) = (1/\nu^2 \lambda(K)) \int_K \int_K w_K(x - x_1 + x_2) Y(x_1) Y(x_2) N^{(2)}(dx_1, dx_2), \tag{10.41}$$

where $N^{(2)}(dx_1, dx_2) = N(dx_1)(N - \varepsilon_{x_1})(dx_2)$, where the kernel w is a positive, bounded, isotropic density function on $\mathbf{R}^d$, and where $w_K(x) = (1/\alpha_K^d) w(x/\alpha_K)$ with $\alpha_K \to 0$ and $\alpha_K^d \lambda(K) \to \infty$ as $\delta(K) \to \infty$. Motivation is the same as for the one-dimensional version (10.23) and properties are substantially similar.

Theorem 10.21. Assume that X is L^2-stationary, that the covariance function is continuous and integrable, and that the fourth-order cumulant function Q [see (10.24)] exists and satisfies

$$\sup_{x_1, x_2} \int_{\mathbf{R}^d} |Q(x + x_1, x, x_2)| dx < \infty.$$

Then
    a) For each x,

$$E[\hat{R}(x)] = \int_{K-K} w(y) R(x - \alpha_K y)[\lambda(K \cap (K - x + \alpha_K y))/\lambda(K)] dy, \tag{10.42}$$

and hence $E[\hat{R}(x)] \to R(x)$ as $\delta(K) \to \infty$;
    b) For each x_1 and x_2,

$$\lim_{\delta(K)\to\infty} \lambda(K)\alpha_K^d \mathrm{Cov}(\hat{R}(x_1), \hat{R}(x_2)) = \sigma^2(x_1) \mathbf{1}(x_1 = \pm x_2), \tag{10.43}$$

where

$$\sigma^2(x) = [\int w(y)^2 dy/\nu^2][Q(0, x, x) + 2R(x)^2 + R(0)^2]. \tag{10.44}$$

Note the exact correspondence between (10.43) and (10.27).

Proof: The calculations are nearly the same as those used to prove Theorem 10.11. As when $d = 1$, $E[N^{(2)}(dx_1, dx_2)] = \nu^2 dx_1 dx_2$, so that

$$
\begin{aligned}
E[\hat{R}(x)] &= (1/\lambda(K)) \int_K \int_K w_K(x - x_1 + x_2) R(x_1 - x_2) dx_1 dx_2 \\
&= \int_K w(y) R(x - \alpha_K y)[\lambda(K \cap (K - x + \alpha_K y))/\lambda(K)] dy,
\end{aligned}
$$

which is the first part of (10.42). As $\delta(K) \to \infty$, $\alpha_K \to 0$ by assumption, while $\lambda(K \cap (K - x + \alpha_K y))/\lambda(K) \cong \lambda(K \cap (K - x))/\lambda(K)$, which converges to 1 by (10.40), giving the second part of (10.42).

Concerning b),

$$
\begin{aligned}
E[\hat{R}(x)\hat{R}(y)] &= (1/\nu^2\lambda(K)) \int_K \int_K w_K(x - x_1 + x_2) w_K(y - y_1 - y_2) \\
&\quad \times E[Y(x_1)Y(x_2)Y(y_1)Y(y_2)] \\
&\quad \times E[N^{(2)}(dx_1, dx_2)N^{(2)}(dy_1, dy_2)],
\end{aligned}
$$

which is expanded as in the proof of Theorem 10.11. First of all,

$$
\begin{aligned}
E[Y(x_1)&Y(x_2)Y(y_1)Y(y_2)] \\
&= Q(x_2 - x_1, y_1 - x_1, y_2 - x_1) + R(x_2 - x_1)R(y_2 - y_1) \\
&\quad + R(y_1 - x_1)R(y_2 - x_2) + R(y_2 - x_1)R(x_2 - y_1).
\end{aligned}
$$

In addition, (10.30) remains valid with only notational changes. The dominant contributions to $\text{Cov}(\hat{R}(x), \hat{R}(y))$ are analogous to those in the one-dimensional case, with the result (Karr, 1986a) that within $O(1/\lambda(K))$,

$$
\begin{aligned}
\text{Cov}&(\hat{R}(x), \hat{R}(y)) \\
&= (1/\nu^2\lambda(K)) \int_K w_K(x - z) \Big\{[w_K(y - z) + w_K(y + z)] \\
&\quad \times [Q(0, z, z) + 2R(z)^2 + R(0)^2][\lambda(K \cap (K - z))/\lambda(K)]\Big\} dz \\
&= (1/\alpha_K^d \nu^2 \lambda(K)) \int_K w(z) \Big\{[w(z + (y - x)/\alpha_K) + w(z + (yu + x)/\alpha_K)] \\
&\quad \times [Q(0, x - \alpha_K z, x - \alpha_K z) + 2R(x - \alpha_K z)^2 + R(0)^2] \\
&\quad \times [\lambda(K \cap (K - x + \alpha_K z))/\lambda(K)]\Big\} dz, \quad\quad (10.45)
\end{aligned}
$$

from which (10.43) follows by another application of (10.40) in conjunction with the remaining assumptions. (It is also necessary to observe that $\int_{K-K} w(z)^2 dz \to \int_{\mathbf{R}^d} w(z)^2 dz$, but this is shown easily.) ∎

Hence in particular, $\text{Var}(\hat{R}(x)) \to 0$ for each x, so that the estimators $\hat{R}$ are mean square consistent pointwise in x. Moreover, for $x_1 \neq \pm x_2$, $\hat{R}(x_1)$ and $\hat{R}(x_2)$ are asymptotically uncorrelated; while pleasant in some respects this conclusion is disappointing in others: it precludes, for example, a functional central limit theorem for the $\hat{R}$ as random processes, since a Gaussian limit process would have independent values at distinct points.

Although the global error measure $E[\int_K (\hat{R} - R)^2]$ does not converge to 0, its rate of growth can be ascertained rather precisely under mild additional assumptions.

Theorem 10.22. Suppose that the hypotheses of Theorem 10.21 are satisfied, that the covariance function R is bounded and twice continuously differentiable, that

$$\int_{\mathbf{R}^d} |y|^2 w(y)dy < \infty \tag{10.46}$$

and that $\lambda(K)\alpha_K^{4+d} \to 0$. Then

$$\lim_{\delta(K)\to\infty} \alpha_K^d E\left[\int_K \{\hat{R}(x) - R(x)\}^2 dx\right] = R(0)^2. \tag{10.47}$$

Proof: From (10.45), for each x, within error $O(1/\lambda(K))$,

$$\begin{aligned}
\text{Var}(\hat{R}(x)) &= (1/\alpha_K^d \lambda(K))\int_{K-K} w(z)^2 \Big\{ [Q(0, x - \alpha_K z, x - \alpha_K z) \\
&\quad + R(x - \alpha_K z)^2 + R(0)^2] \\
&\quad \times [\lambda(K \cap (K - x - \alpha_K z))/\lambda(K)] \Big\} dz;
\end{aligned}$$

consequently

$$\begin{aligned}
\int_K \text{Var}(\hat{R}(x))dx &= (1/\alpha_K^d \lambda(K))(\int_{K-K} w(z)^2 dz) \\
&\quad \times \Big\{ \int_K \{Q(0, x, x) + R(x)^2\} dx + \lambda(K)R(0)^2 \Big\} \\
&\quad + O(1/\alpha_K^d \lambda(K)) \\
&= R(0)/\alpha_K^d + o(1). \tag{10.48}
\end{aligned}$$

The differentiability assumption on R and Taylor's theorem combine with (10.42) to yield (subscripts denote partial derivatives)

$$\begin{aligned}
E[\hat{R}(x)] &= \int_{K-K} w(z) \Big\{ \Big[R(x) + \alpha_K \sum_i R_i(x)z_i + (\alpha_K^2/2)\sum_{i,j} R_{ij}(x)z_i z_j + O(\alpha_K^3) \Big] \\
&\quad \times [\lambda(K \cap (K - x + \alpha_K z))/\lambda(K)] \Big\} dz \\
&= R(x)\int_{K-K} w(z)[\lambda(K \cap (K - x + \alpha_K z))/\lambda(K)]dz + O(\alpha_K^2) \\
&= R(x) + O(1/\lambda(K)) + O(\alpha_K^2).
\end{aligned}$$

The integrated first-derivative terms vanish individually by isotropy of w, while the second-derivative term is $O(\alpha_K^2)$ by (10.46). Therefore

$$\int_K \{E[\hat{R}(x)] - R(x)\}^2 dx = O(1/\lambda(K) + \alpha_K^2 + \lambda(K)\alpha_K^4).$$

Together (10.48) and this expression imply (10.47). ∎

Asymptotic normality of $\hat{R}(x)$ for individual values of x can be established under suitably stringent hypotheses.

Theorem 10.23. Assume that Y admits finite moments of all orders and that for every $k \geq 2$ the cumulant function of order k satisfies

$$\int_{\mathbf{R}^{k-1}} |Q^{(k)}(y_1, \ldots, y_{k-1})| dy_1 \cdots dy_{k-1} < \infty. \tag{10.49}$$

Assume in addition that R is twice continuously differentiable and that (10.46) holds. Then for each x, $\sqrt{\lambda(K)\alpha_K^d}[\hat{R}(x) - R(x)] \xrightarrow{d} N(0, \sigma^2(x))$, where $\sigma^2(x)$ is given by (10.44). ☐

The argument used to prove Theorem 10.12 applies with essentially only notational changes.

We conclude our treatment of estimation with a brief discussion of alternative estimators, to which results concerning estimation for stationary point processes are germane. The signed measure $S(A) = \int_A R(x)dx$ can be estimated as follows. Assume that A is bounded; then with $M(dx) = Y(x)N(dx)$, the estimator

$$\hat{S}(A) = (1/\nu^2\lambda(K))\int_K M(dx_1) \int \mathbf{1}(x_2 - x_1 \in A)M(dx_2)$$

is $\mathcal{H}(K \cup (A + K))$-measurable. Provided that $0 \notin A$, this estimator is unbiased.

Moreover, strong consistency holds under ergodicity hypotheses as an immediate consequence of Theorem 1.72.

Theorem 10.24. Suppose that the random measure M is ergodic. Then for each set A with $0 \notin A$, $\hat{S}(A) \to S(A)$ almost surely as $\delta(K) \to \infty$. ☐

Asymptotic normality for $\hat{S}$ follows from the central limit theorem of Jolivet (1981), whose ultimate basis is the same as that of Theorems 9.4 and 10.12; however, for the usual technical reasons we require a smooth integrand. Thus, for f a continuous function with compact support, we estimate $S(f) = \int R(x)f(x)dx$ by

$$\hat{S}(f) = (1/\nu^2\lambda(K))\int_K M(dx_1) \int f(x_2 - x_1)M(dx_2). \tag{10.50}$$

Provided that $f(0) = 0$, $\hat{S}(f)$ is unbiased for each K. Moreover Theorem 10.24 remains valid and in addition the following central limit theorem obtains.

Theorem 10.25. Suppose that M is ergodic, that (10.49) is fulfilled and that f is continuous with compact support not containing 0. Then $\sqrt{\lambda(K)}[\hat{S}(f) - S(f)] \xrightarrow{d} N(0, \sigma^2(f))$, where $\sigma^2(f)$ is a variance not calculated here. $\square$

Comparison of (10.50) with estimators in Chapter 9 reveals that the $\hat{S}$ estimate the reduced second moment measure of M.

State estimation for random fields given point process samples has a significant applications component. For example, mineral and oil reserves must be estimated and mapped from a small number of test drillings, and areal precipitation must be inferred from raingage measurements.

We consider only linear state estimation. Let Y be an L^2-stationary random field on $\mathbf{R}^d$ with known mean $m = 0$ and known covariance function R. We retain the assumption that the sampling process N is Poisson, independent of X and observable. Given observations of the marked point process $\bar{N}$ over a bounded set A in $\mathbf{R}^d$, we wish to calculate for each x the MMSE linear state estimator $\hat{Y}^*(x)$ of $Y(x)$. "Linear" means that we restrict attention to state estimators

$$\hat{Y}(x) = \int_A h(x, z) Y(z) N(dz), \qquad (10.51)$$

with h a function on $\mathbf{R}^d \times A$.

The optimal function h^* can be characterized as the solution to an integral equation.

Proposition 10.26. With A fixed, the optimal linear state estimator corresponds to any function h^* satisfying the equation

$$R(x - y) = \nu \int_A h^*(x, z) R(y - z) dz + h^*(x, y) R(0) \qquad (10.52)$$

for $x \in \mathbf{R}^d$ and $y \in A$. For each x,

$$E[(\hat{Y}^*(x) - Y(x))^2] = h^*(x, x) R(0). \qquad (10.53)$$

Proof: The following argument can be made more rigorous by putting it in integrated form, but is clearer as it stands. With x fixed, by Hilbert space theory the optimal linear state estimator $\hat{Y}^*(x)$ fulfills the normal equations

$$E[\{\hat{Y}^*(x) - Y(x)\} Y(y) N(dy)] = 0, \qquad y \in A$$

therefore h^* must satisfy

$$
\begin{aligned}
\nu R(x-y)dy &= E[\hat{Y}^*(x)Y(y)N(dy)] \\
&= E[\{\int_A h^*(x,z)Y(z)N(dz)\}Y(y)N(dy)] \\
&= \int_A h^*(x,z)R(y-z)E[N(dz)N(dy)] \\
&= \int_A h^*(x,z)R(y-z)[\nu^2 dz + \nu \varepsilon_y(dz)]dy,
\end{aligned}
$$

which is the same as (10.52).

Derivation of (10.53) from (10.52) is a straightforward calculation, given here in skeletal form:

$$
\begin{aligned}
E[(\hat{Y}^*(x) &- Y(x))^2] \\
&= E[\int h^*(x,y)Y(y)N(dy)\int h^*(x,z)Y(z)N(dz)] \\
&\quad - 2\nu \int h^*(x,y)R(x-y)dy + R(0) \\
&= \nu^2 \int \int h^*(x,y)h^*(x,z)R(y-z)dydz \\
&\quad + \nu \int h^*(x,y)^2 R(0)dy - 2\nu \int h^*(x,y)R(x-y)dy + R(0) \\
&= \nu \int h^*(x,y)R(x-y)dy - 2\nu \int h^*(x,y)R(x-y)dy + R(0) \\
&= h^*(x,x)R(0),
\end{aligned}
$$

where the last two equalities both hold by virtue of (10.52). ∎

The formal solution of (10.52) is

$$
h^*(x,y) = \sum_{k=0}^{\infty}(-\nu)^k R(0)^{-(k+1)} R_A^k(x,y), \tag{10.54}
$$

where $R_A^0(x,y) = R(x-y)$ and $R_A^{k+1}(x,y) = \int_A R_A^k(x,z)R(z-y)dz$.

The optimal estimators $\hat{Y}^*(x)$ have one glaring shortcoming, engendered by our using an averaged (via the expectation) error criterion: if $x \in A$ is a point X_i of N, then even though $Y(x) = Y(X_i)$ is known exactly, we do not have $\hat{Y}^*(x) = Y(x)$. This occurs, of course, because $P\{N(\{x\}) \neq 0\} = 0$, but even so is unsatisfactory. The estimator

$$
\hat{Y}'(x) = \hat{Y}^*(x) + \int_A \{Y(z) - \hat{Y}^*(z)\}\mathbf{1}(z=x)N(dz)
$$

satisfies $\hat{Y}'(X_i) = Y(X_i)$ for each point X_i of N and — albeit not linear — is as easily calculated as $\hat{Y}^*$. Moreover, since $P\{\hat{Y}'(x) = \hat{Y}^*(x)\} = 1$ for each x, (10.53) remains valid; thus pointwise behavior is improved without impairing the mean-squared error.

Yet another approach is to allow estimators that are linear functionals of the marked point process $\bar{N}$, i.e.,

$$
\hat{Y}(x) = \int_{A \times \mathbb{R}} h(x,y,u)\bar{N}(dy,du), \tag{10.55}
$$

although this fails to alleviate the difficulty. An argument analogous to the proof of Proposition 10.26 identifies the optimal h^*.

Proposition 10.27. The MMSE linear state estimator of the form (10.55) corresponds to the function h^* satisfying

$$\frac{dE[Y(x)\mathbf{1}(Y(y) \in (\cdot))]}{dP\{Y(y) \in (\cdot)\}}(u)$$

$$= h^*(x, y, u)$$

$$+ \int_{A \times \mathbf{R}} h^*(x, y', u') \frac{dP\{Y(y') \in du', Y(y) \in (\cdot)\}}{dP\{Y(y) \in (\cdot)\}}(u)dy' \quad (10.56)$$

for all x, y and u. □

This procedure resembles the "disjunctive kriging" of Matheron (1976). Solution of (10.56) requires more than just the covariance function of Y, but not full knowledge of the distribution: only bivariate distributions appear in (10.56), which is usable if these can be calculated.

Combined inference and state estimation remains to be treated. Given Poisson samples of Y observed over K, it is clear what one *might* do. First, estimate R using (10.41), then — assuming that $m = 0$ and that linear state estimators of the form (10.51) are sought — substitute $\hat{R}$ for R in (10.54) to obtain an estimator $\hat{h}^*$ and finally form pseudo-state estimators

$$\hat{Y}(x) = \int_K \hat{h}^*(x, z)Y(z)N(dz).$$

A myriad of complications ensues, including whether there is convergence in (10.54) when $\hat{R}$ replaces R, the dual role of the observations, and, for asymptotics, the dependence of the domain of integration (as well as the estimators) on A.

As a mirror of reality, we end here, with unresolved questions and vagueness rather than answers and certainty.

EXERCISES

10.1 Prove Proposition 10.2.

10.2 Derive the likelihood function $L_t(\lambda, A)$ of (10.2).

10.3 Verify that (10.13)–(10.14) obtain for any observable sampling process N independent of X.

10.4 Let N' be the renewal process given by (10.16), where X is a binary Markov process with (positive) transition rates a ($0 \to 1$) and b ($1 \to 0$) and $N = \sum \varepsilon_{T_n}$ is Poisson with rate λ and independent of X.
a) Prove that the interarrival distribution F of N' satisfies (10.17).
b) Prove that there is a function H such that (10.18) holds.

10.5 After defining suitable pseudo-state estimators, extend Theorem 10.7 to state estimators $P\{X_{t+s} = j|\mathcal{F}_t^N\}$, where $s > 0$ is fixed, used for prediction of X.

10.6 Let X be a binary Markov process with transition rates a $(0 \to 1)$ and b $(1 \to 0)$, observed at the "jittered" random times $T_n = n\Delta + \delta_n$, where $\Delta > 0$ and the δ_n are i.i.d. random variables independent of X with $P\{|\delta_n| < \Delta/2\} = 1$. (The latter condition prevents "crossed" observations.) Assume that the distribution G of the δ_n is known and that observations are the marked point process $\bar{N} = \sum \varepsilon_{(T_n, X(T_n))}$.
a) Prove that $(X(T_n))$ is a Markov chain and calculate its transition matrix.
b) Devise $\mathcal{F}_t^N$-estimators of a and b that are strongly consistent and asymptotically normal, and verify these properties.

10.7 Let X and N be as in Section 1, let (L_t) be the *left*-continuous backward recurrence time process of N, and let $Z_t = X(T_{N_t})$ be the most recently observed value of X.
a) Show that the $(P, \mathcal{F}^N)$-stochastic intensity of the point process $N_t(i, j)$ of (10.3) is $\lambda_t(A, i, j) = P_{L_t}(Z_{t-}, i)A(i, j)$, where (P_t) and A are the transition function and generator of X.
b) Describe large sample, as $t \to \infty$, behavior of the martingale estimators $\hat{A}(i, j) = \int_0^t \hat{P}_{L_t}(Z_{s-}, i)dN_s(i, j)$, where $\hat{P}_u(i, j) = \exp[u\hat{A}](i, j)$.

10.8 Let R be the covariance function of an L^2-stationary process X on R. Prove that R is symmetric and positive definite with $R(0) = \text{Var}(X_t)$ for each t.

10.9 Prove Proposition 10.8.

10.10 Show that under the hypotheses of Propositions 10.9 and 10.10 the estimators $\hat{m}^*$ of (10.22) are strongly consistent and asymptotically normal.

10.11 (Continuation of Exercise 10.10) Calculate the asymptotic efficiency of the estimators $\hat{m}$ of (10.20) relative to that of the estimators $\hat{m}^*$ of (10.22).

10.12 Verify (10.30).

10.13 Prove that under the assumptions and notation of Theorem 10.12, for each k the reduced kth-order cumulant measure γ_*^k of the process $M(dt) = X(t)N(dt)$ has finite total variation.

10.14 Let $\hat{f}(v)$ be the $\mathcal{H}_{T_n}$-estimators of the spectral density function given by (10.32) and let v be fixed. Determine the value of the sampling rate ν that minimizes the asymptotic variance in (10.34).

10.15 Confirm that the $\mathcal{H}_t$-estimators $\hat{f}^*(v)$ of the spectral density function given by (10.35) fulfill (10.36) and (10.37) provided that $\int |tR(t)|dt < \infty$.

10.16 Verify (10.40).

10.17 a) Formulate analogues of the estimators (10.41) under the stipulation that Y is a stationary, isotropic random field on R^d with mean zero and covariance function $\text{Cov}(Y(x), Y(y)) = R(|x - y|)$.
b) Establish a corresponding analogue of Theorem 10.21.

c) Determine the efficiency that would be lost if instead of the estimators in a) one were to ignore isotropy of Y and use instead the estimators $\hat{R}$ in (10.23).

NOTES

Even in one dimension there is as yet no general theory of point process sampling of stochastic processes. This chapter is somewhat unified but hardly complete: it omits more specific models than it presents. Of course, when a "field" has yet to progress beyond isolated special cases, little else is possible. Two forms of point process sampling have been investigated in some breadth: deterministic — usually regular — and Poisson sampling; both assume independence of the sampling process and the process being sampled.

Regular sampling has been studied for birth/death processes (Keiding, 1975), binary Markov processes (Brown *et al.*, 1977, 1979), and stationary processes (time series). For the latter there arises (see Masry, 1978b) the phenomenon of "aliasing:" from regular observations spaced Δ apart one cannot recover the spectral density function at frequencies above the Nyquist frequency $2\pi/\Delta$.

By contrast, Poisson sampling has been known at least since Shapiro and Silverman (1960) to be alias-free for a very broad class of stationary processes. (Proposition 10.1 and Theorem 10.18 have similar implications.) Masry (1978b, 1983) examines other alias-free sampling methods.

Deterministic but not necessarily regular sampling is considered for diffusion processes in Le Breton (1975) and for time series by Robinson (1977).

The sole result in the introduction, Proposition 10.1, is due to Kingman (1963); it is generalized to random fields in Theorem 10.18.

Section 10.1

Theorem 10.3, the converse to the elementary Proposition 10.2, comes from Kingman (1963). The substitution estimators $\hat{A}$ of (10.6), and Theorems 10.4 and 10.5 concerning them, are new here but based on ideas in Karr (1984a), the source as well of Proposition 10.6 and the estimators to which it applies. More primitive versions of the latter appear in Karr (1982). Although formally new, Theorem 10.7 is a fairly straightforward extension of Karr (1984a, Theorem 10.1), which pertains only to binary Markov processes; the proof here is less intricate.

Statistical inference for Markov chains and processes with finite state space given complete observations is developed in Billingsley (1961a, 1961b) and less completely in Basawa and Prakasa Rao (1980); identification of the estimator $\hat{Q}$ of (10.5) as nonparametric maximum likelihood estimator is noted in all. Doob (1953) remains a lucid presentation of limit theorems underlying the asymptotics.

Section 10.2

The main sources are Masry (1978a, 1983) on synchronous and asynchronous Poisson sampling of stationary processes on **R**. The latter provides the estimators $\hat{R}$ of (10.23) and Theorems 10.11 and 10.12 concerning their properties, while from the former come the estimators $\hat{f}$ in (10.32), Propositions 10.13 and 10.14, and Theorem 10.15. Some details of proofs are given in these two papers; Masry and Lui (1976) contains related material. Central limit theorems once again rest ultimately on Leonov and Shiryaev (1959) (see also Brillinger, 1972).

Background on stationary processes can be found in Cramér and Leadbetter (1967) and Rozanov (1965); Karlin and Taylor (1975) is more elementary.

Section 10.3

All the results are taken from Karr (1986a). Theorem 10.18 generalizes, even in one dimension, Proposition 10.1, while each remaining aspect — estimators and properties alike — has an apparent analogue in Section 2.

The whole topic of random fields has developed only rather recently. Problems on which research has centered are local behavior, extreme values, Markov properties, asymptotics and linear extrapolation; intricacies and subtleties are legion [central limit theorems are especially difficult (see, e.g., Deo, 1975)]. General sources are Adler (1981), Rozanov (1982) and Yadrenko (1983).

Exercises

10.4 See Karr (1982, 1984a), Theorem 8.24, and Exercise 3.15.

10.6 See Karr (1984a).

10.7 See also Section 8.4.

10.13 See also Theorem 9.7.

10.14 Proofs are given in Masry (1978a).

10.17 See Karr (1986a).

Appendix A:
Spaces of Measures

This appendix should be regarded only as a very selective compilation of definitions and results; motivation and proof are equally absent. Cohn (1980), Dunford and Schwartz (1958), Matthes *et al.* (1978), and Rudin (1975) are more substantial treatments; Kallenberg (1983) contains an appendix at an intermediate level of detail.

Our setting is a locally compact Hausdorff space E with Borel σ-algebra $\mathcal{E}$. Let $\mathcal{B}$ be the family of bounded (relatively compact) sets in $\mathcal{E}$. By C, C_0 and C_K we denote the set of bounded, continuous functions on E, the set of bounded, continuous functions vanishing at infinity ($f \in C_0$ if and only if f is continuous and for each $\varepsilon > 0$ there is a compact set K such that $|f(x)| < \varepsilon$ for $x \notin K$) and the set of continuous functions on E with compact support. Both C and C_0 are Banach spaces under the uniform norm, the latter is separable, and C_K is dense in it. The Baire σ-algebra $\sigma(C_K)$ coincides with $\mathcal{E}$.

Except at the end of the ensuing discussion we restrict attention to positive measures.

Definition A.1. A measure μ on E is a *Radon measure* if $\mu(B) < \infty$ for each $B \in \mathcal{B}$.

Evidently, μ is a Radon measure if and only if $|\mu(f)| = |\int f d\mu| < \infty$ for every $f \in C_K$. Some other basic properties merit mention.

Proposition A.2. If μ is a Radon measure, then
a) μ is σ-finite;
b) μ is regular: for $A \in \mathcal{E}$, $\mu(A) = \sup\{\mu(K) : K \subseteq A,\ K$ is compact$\} = \inf\{\mu(G) : A \subseteq G,\ G$ is open$\}$. $\square$

Among Radon measures we distinguish several key subclasses.

Definition A.3. A Radon measure μ is

a) A *point* (or *counting*) *measure* if $\mu(A) \in \mathbf{N}$ for every $A \in \mathcal{B}$;

b) A *simple point measure* if μ is a point measure and if $\mu(\{x\}) \leq 1$ for each $x \in E$;

c) *Diffuse* if $\mu(\{x\}) = 0$ for every $x \in E$.

The fundamental point measures are the *point masses* ϵ_x, defined by

$$\epsilon_x(A) = \mathbf{1}(x \in A) \begin{array}{ll} = 1 & \text{if } x \in A \\ = 0 & \text{if } x \notin A. \end{array}$$

The alternative notation δ_x and terminology *Dirac measure* are used commonly, especially for $E = \mathbf{R}$. Theorem A.4 exhibits a point measure as a countable sum of point masses.

A point x is an *atom* of μ if $\mu(\{x\}) > 0$; the set of atoms is countable.

At the opposite extreme, diffuse measures have no atoms at all and hence are analogous to continuous distribution functions on $\mathbf{R}$. Corresponding to the familiar decomposition of a distribution function into discrete and continuous (but not necessarily absolutely continuous) parts is the following representation.

Theorem A.4. A Radon measure μ can be decomposed as

$$\mu = \mu_d + \sum_{i=1}^{K} a_i \epsilon_{x_i}, \qquad (A.1)$$

where μ_d is diffuse, $0 \leq K \leq \infty$, $a_i > 0$ for each i and the x_i are distinct points in E. The decomposition is unique up to reordering of the (x_i, a_i). $\square$

The Radon measure μ is *purely atomic* if the diffuse component is zero. A purely atomic measure is a point measure if and only if $a_i \in \mathbf{N}$ for each i and in this case $\{x_i\}$ can have no accumulation points in E. For a simple point measure, $a_i = 1$ for each i, giving $\mu(A) = \sum_{i=1}^{K} \epsilon_{x_i}(A) = \sum_{i=1}^{K} \mathbf{1}(x_i \in A)$ as the number of points in A.

We denote by $\mathbf{M}$ the set of Radon measures on E and by $\mathbf{M}_p$, $\mathbf{M}_s$, $\mathbf{M}_a$ and $\mathbf{M}_d$ the subsets of point measures, simple point measures, purely atomic measures and diffuse measures. Once measurability structure is established, a *random measure* (purely atomic random measure, diffuse random measure) is by definition a random element of $\mathbf{M}$ ($\mathbf{M}_a$, $\mathbf{M}_d$,) and a *point process* (simple point process) a random element of $\mathbf{M}_p$ ($\mathbf{M}_s$).

Unlike function spaces, $\mathbf{M}$ possesses refreshingly simple measurability and topological structure. Essentially all functionals of interest are measurable with respect to the σ-algebra we are about to introduce, and many are continuous with respect to a Polish topology whose Borel σ-algebra coincides with it.

On **M** we define the σ-algebra

$$\mathcal{M} = \sigma(\mu \to \mu(f) : f \in C_K), \qquad (A.2)$$

the smallest σ-algebra rendering measurable each mapping $\mu \to \mu(f)$, $f \in C_K$.

Characterization in terms of sets is also possible.

Proposition A.5. The σ-algebra $\mathcal{M}$ is generated by the mappings $\mu \to \mu(A)$, $A \in \mathcal{B}$. ∎

In particular, $\mathcal{M}$ is countably generated. Each of the sets $\mathbf{M}_p$, $\mathbf{M}_s$, $\mathbf{M}_a$ and $\mathbf{M}_d$ belongs to $\mathcal{M}$, and we define $\mathcal{M}_p = \mathcal{M} \cap \mathbf{M}_p$, ...

For our purposes only the vague topology on **M** is required, but for completeness we mention also the weak topology as well as the strong topology engendered by the total variation norm.

Definition A.6. Suppose that μ_n and μ are elements of **M**; then (μ_n) converges *vaguely* to μ if $\mu_n(f) \to \mu(f)$ for every $f \in C_K$.

We write vague convergence simply as $\mu_n \to \mu$. In many situations the definition is adequate for direct verification, but alternative forms are available.

Theorem A.7. For μ_n, $\mu \in \mathbf{M}$, the following assertions are equivalent:
a) $\mu_n \to \mu$;
b) For every compact set K, $\limsup_n \mu_n(K) \leq \mu(K)$ and for every bounded open set G, $\liminf_n \mu_n(G) \geq \mu(G)$;
c) $\lim_n \mu_n(A) = \mu(A)$ for every $A \in \mathcal{B}$ with $\mu(\partial A) = 0$;
d) $\lim_n \mu_n(f) = \mu(f)$ for every bounded, measurable function f on E vanishing outside some compact set and satisfying $\mu(D_f) = 0$, where D_f is the discontinuity set of f. ∎

In fact, if $\mu_n \to \mu$ and f_n, f are such that (f_n) is uniformly bounded, if the f_n are all zero outside a common compact set, and if $f_n(x_n) \to f(x)$ whenever $x_n \to x$ in E, then $\mu_n(f_n) \to \mu(f)$.

The mapping $x \to \varepsilon_x$ of E into $\mathbf{M}_p$ is one-to-one and vaguely continuous; thus we sometimes regard E as a subset of $\mathbf{M}_p$.

Measurability structure and the vague topology are intimately related.

Theorem A.8. The σ-algebra $\mathcal{M}$ is the Borel (and Baire) σ-algebra relative to the vague topology. ∎

While $\mathbf{M}_p$ is closed, none of $\mathbf{M}_s$, $\mathbf{M}_a$ or $\mathbf{M}_d$ is.

For random measures and point processes, defined (Definition 1.1) as random elements of $(\mathbf{M}, \mathcal{M})$ and $(\mathbf{M}_p, \mathcal{M}_p)$, respectively, convergence in distribution can be investigated with standard methods (set forth, e.g., in Billingsley, 1968) by virtue of the next result.

Theorem A.9. The spaces $\mathbf{M}$ and $\mathbf{M}_p$ are Polish spaces (i.e., metrizable as complete, separable metric spaces) in the vague topology. ☐

In establishing convergence in distribution the key step is usually to show tightness, which is ordinarily done by direct or indirect appeal to Prohorov's theorem (see Billingsley, 1968), necessitating, in turn, a tractable characterization of relatively compact subsets, in this case of $\mathbf{M}$ or $\mathbf{M}_p$. For random measures and point processes this step is rarely difficult because the following result (see Lemma 1.20) reduces tightness to that of one-dimensional distributions.

Theorem A.10. For a subset Γ of $\mathbf{M}$ the following assertions are equivalent:

a) Γ is relatively compact in the vague topology;
b) $\sup_{\mu \in \Gamma} \mu(A) < \infty$ for each $A \in \mathcal{B}$;
c) $\sup_{\mu \in \Gamma} |\mu(f)| < \infty$ for each $f \in C_K$. ☐

The set of *finite* elements of $\mathbf{M}$ can also be endowed with the *weak* topology, which despite the name is stronger than the vague topology: (μ_n) converges weakly to μ, written $\mu_n \xrightarrow{w} \mu$, if $\mu_n(f) \to \mu(f)$ for every $f \in C_K$, (i.e., $\mu_n \to \mu$ vaguely) and in addition $\mu_n(E) \to \mu(E)$ or, equivalently, if $\mu_n(f) \to \mu(f)$ for every $f \in C$. Analogues of the results above hold (see Kallenberg, 1983).

Duality considerations also lead to the weak topology; they stem from the Riesz representation theorem.

Theorem A.11. Given a positive linear functional L on C_K, there is a unique element μ of $\mathbf{M}$ such that $L(f) = \mu(f)$ for all f.

The vector space V of differences $\mu - \lambda$, where μ and λ are *finite* elements of $\mathbf{M}$, is normed by the *total variation norm*

$$\|\mu\| = \sup \left\{ \sum_i |\mu(A_i)| : (A_i) \text{ is a finite partition of } E \right\}.$$

This norm, applied to probability measures, is utilized in Sections 1.6 and 2.4. Its properties on V include the following.

Theorem A.12. a) V is a Banach space under the total variation norm;
b) V is the Banach space dual of C_0;
c) The weak*-topology induced on V by C_0 is the weak topology as defined above. ☐

Appendix B:
Continuous Time Martingales

More so than Appendix A, this appendix is selective and incomplete. All of the subtlety and much of the beauty of the theory are missing from our presentation. With one exception no proofs are given, but some heuristic arguments germane to point processes are sketched.

The setting, save in Theorem B.21, where the filtration and martingale have particular form, is a probability space $(\Omega, \mathcal{F}, P)$ together with a filtration $\mathcal{H} = (\mathcal{H}_t)_{t \geq 0}$ satisfying the *"conditions habituelles"* ($\mathcal{H}$ is right-continuous and $\mathcal{H}_0$ contains all null sets in $\mathcal{H}_\infty = \bigvee_{t \geq 0} \mathcal{H}_t$). Important concepts: adapted processes, stopping times, predictable processes,... are mentioned, albeit briefly, in Section 2.1. The qualification "with respect to $\mathcal{H}$" is omitted throughout.

We recall that a process X is càdlàg (*continue à droite, limité à gauche*) if almost surely the path $t \to X_t(\omega)$ is right-continuous on $[0, \infty)$ and admits left-hand limits on $(0, \infty)$. For martingales, this can be proved rather than stipulated, but since in this book "càdlàguity" holds by construction (point processes are right-continuous), we incorporate it in the definition. The reader is warned, however, that several of the results below are not valid without right-continuity.

Definition B.1. a) An adapted càdlàg process M is a *martingale* if $E[|M_t|] < \infty$ for each t and if whenever $s < t$, $E[M_t | \mathcal{H}_s] = M_s$.

b) An adapted càdlàg process Z is a *submartingale* if $E[|Z_t|] < \infty$ for each t and if $s < t$ implies that $E[Z_t | \mathcal{H}_s] \geq Z_s$.

c) An adapted càdlàg process Z is a *supermartingale* if the process $-Z = (-Z_t)$ is a submartingale.

415

In connection not only with point processes but also Wiener processes (see, e.g., Kallianpur, 1980, or Liptser and Shiryaev, 1978) square integrable martingales are particularly important.

Definition B.2. A martingale M is *square integrable* if $\sup_{t \geq 0} E[M_t^2] < \infty$.

If M is a square integrable martingale, then by Jensen's inequality for conditional expectations, the process M^2 is a submartingale.

The important concept of the predictable variation of a square integrable martingale M stems from a particular decomposition of the submartingale M^2. Before presenting it, we introduce further terminology.

Definition B.3. An adapted process A is an *increasing process* if $A_0 = 0$ and if each sample path $t \to A_t$ is nondecreasing and right-continuous. If in addition, $\sup_t E[A_t] < \infty$, then A is *integrable*.

Predictable increasing processes, one component of the Doob-Meyer decomposition (B.1), can be characterized via martingales.

Theorem B.4. An integrable increasing process A is predictable if and only if for every positive martingale M and each $t \geq 0$,

$$E[\textstyle\int_0^t M_u dA_u] = E[\textstyle\int_0^t M_{u-} dA_u]. \;\square$$

We do not pursue here the rather profound analogy between supermartingales and superharmonic functions (see, e.g., Neveu, 1975), but within this context one distinguishes supermartingales analogous to potentials (i.e., superharmonic functions vanishing "at the boundary").

Definition B.5. A supermartingale Z is a *potential* if $Z \geq 0$ and if $\lim_{t \to \infty} E[Z_t] = 0$.

We come now to the fundamental decomposition theorem, due to Doob for discrete-time supermartingales and Meyer (1965) in the continuous-time case.

Theorem B.6. Given a potential Z, the following statements are equivalent:

a) There exists a predictable increasing process A (necessarily integrable) such that for all t,

$$Z_t = E[A_\infty | \mathcal{H}_t] - A_t; \tag{B.1}$$

b) Z is of class (D): the family $\{Z_T : T \text{ is a finite stopping time}\}$ is uniformly integrable. $\square$

When it exists, the *Doob-Meyer decomposition* (B.1) is unique within an evanescent process (satisfying $P\{X_t = 0 \text{ identically in } t\} = 1$). Note that the process $M_t = E[A_\infty|\mathcal{H}_t]$ is a uniformly integrable martingale, so that one interpretation of (B.1) is that the nonmartingale part of a supermartingale is predictable.

A point process N is evidently a submartingale and hence given sufficient integrability admits a Doob-Meyer decomposition $N = M + A$, with M a martingale and A, the *compensator* of N, a predictable increasing process. (In Chapter 2, the compensator was derived using dual predictable projections, an approach at once more general and more amenable to computation.)

The fundamental tool for analysis of the innovation martingale M is the predictable variation process $\langle M \rangle$, which results from the Doob-Meyer decomposition of the submartingale M^2.

Theorem B.7. Let M be a square integrable martingale. Then there exists a unique predictable increasing process $\langle M \rangle$ such that $\langle M \rangle_0 = M_0$ and such that $M^2 - \langle M \rangle$ is a martingale. $\square$

Definition B.8. The process $\langle M \rangle$ is the *predictable variation* of M.

Since a square integrable martingale has orthogonal increments, for $s < t$, we have
$$E[(M_t - M_s)^2|\mathcal{H}_s] = E[\langle M \rangle_t - \langle M \rangle_s|\mathcal{H}_s].$$
By analogy with Theorem 2.17, if t is fixed, then almost surely
$$\langle M \rangle_t = M_0^2 + \lim_{n\to\infty} \sum_{k=1}^{2^n} E[(M_{tk/2^n} - M_{t(k-1)/2^n})^2|\mathcal{H}_{t(k-1)/2^n}]. \qquad (B.2)$$

For example, if N is a point process with stochastic intensity λ (Definition 2.28), then

$$
\begin{aligned}
\langle M \rangle_t &= \lim_{n\to\infty} \sum_{k=1}^{2^n} E\left[\left\{N_{tk/2^n} - N_{t(k-1)/2^n} - \int_{t(k-1)/2^n}^{tk/2^n}\lambda_s ds\right\}^2 \middle| \mathcal{H}_{t(k-1)/2^n}\right] \\
&= \lim_{n\to\infty} \sum_{k=1}^{2^n} E[(N_{tk/2^n} - N_{t(k-1)/2^n})^2|\mathcal{H}_{t(k-1)/2^n}] \\
&\quad - \lim_{n\to\infty} \sum_{k=1}^{2^n} \left(\int_{t(k-1)/2^n}^{tk/2^n}\lambda_s ds\right) E[N_{tk/2^n} - N_{t(k-1)/2^n}|\mathcal{H}_{t(k-1)/2^n}] \\
&\quad + \lim_{n\to\infty} \sum_{k=1}^{2^n} \left(\int_{t(k-1)/2^n}^{tk/2^n}\lambda_s ds\right)^2 \\
&= \lim_{n\to\infty} E[N_{tk/2^n} - N_{t(k-1)/2^n}|\mathcal{H}_{t(k-1)/2^n}] \\
&= \int_0^t \lambda_s ds,
\end{aligned}
$$

where the second equality holds by predictability of λ and the last is by Theorem 2.17.

The space of square integrable martingales becomes a Hilbert space under the norm $\|M\| = E[\langle M \rangle_\infty]$; the inner product is constructed under the assumption that the polar identity holds, leading to predictable covariation processes.

Definition B.9. For square integrable martingales M and N, the process $\langle M, N \rangle = (1/4)[\langle M + N \rangle - \langle M - N \rangle]$ is the *predictable covariation* of M and N.

Orthogonality is then defined in the obvious manner.

Definition B.10. Square integrable martingales M and M are *orthogonal*, written $M \perp N$, if $\langle M, N \rangle \equiv 0$.

Evidently, $\langle M, M \rangle = \langle M \rangle$. The characterization that $M^2 - \langle M \rangle$ is a martingale extends: $\langle M, N \rangle$ is the unique predictable càdlàg process such that $\langle M, N \rangle_0 = 0$ and such that $MN - \langle M, N \rangle$ is a martingale.

Orthogonality is then characterized easily.

Theorem B.11. Square integrable martingales M and N are orthogonal if and only if MN is a martingale. □

Stochastic integration with respect to martingales is treated in depth and generality in (among others) Elliott (1982), Jacod (1979), Kallianpur (1980), Metivier and Pellaumial (1980) and Protter (1990). We do not require such generality, because integrals with respect to innovation martingales for point processes exist pathwise as Lebesgue-Stieltjes integrals.

Given a process C and martingale M for which the stochastic integral process $\int_0^t C\,dM$ exists, whether by general theory or by specialized construction, we denote it by $C * M$. Here as throughout the book integrals $\int_0^t$ are over the closed interval $[0, t]$. The central property underlying martingale inference for point processes is that the stochastic integral $C * M$ of a predictable process with respect to a martingale is itself a martingale; heuristically,

$$
\begin{aligned}
E[(C * M)_{t+\Delta t} - (C * M)_t | \mathcal{H}_t] &\cong E[C_t(M_{t+\Delta t} - M_t)|\mathcal{H}_t] \\
&= C_t E[M_{t+\Delta t} - M_t|\mathcal{H}_t] = 0
\end{aligned}
$$

by predictability of C followed by the property that M is a martingale. Theorem 2.9 contains a formal proof based on elementary predictable processes and the monotone class theorem. A similarly heuristic argument to the effect that

$$
\begin{aligned}
\Delta\langle C * M \rangle_t &\cong E[\{(C * M)_{t+\Delta t} - (C * M)_t\}^2|\mathcal{H}_t] \\
&\cong C_t^2 E[(M_{t+\Delta t} - M_t)^2|\mathcal{H}_t]
\end{aligned}
$$

provides motivation for the second part of the next result.

Theorem B.12. If M is a square integrable martingale and C is a bounded, predictable process, then the stochastic integral $C * M$ is a square integrable martingale with $\langle C * M \rangle_t = \int_0^t C_s^2 d\langle M \rangle_s$. $\Box$

More generally, we have the following property.

Theorem B.13. Suppose that $M(1)$ and $M(2)$ are square integrable martingales, that $C(1)$ and $C(2)$ are bounded, predictable processes, and that $\tilde{M}_t(i) = \int_0^t C_s(i) dM_s(i)$, $i = 1, 2$. Then

$$\langle \tilde{M}(1), \tilde{M}(2) \rangle_t = \int_0^t C_u(1) C_u(2) d\langle M(1), M(2) \rangle_u. \Box$$

Especially, if $M(1)$ and $M(2)$ are orthogonal, then so are $\tilde{M}(1)$ and $\tilde{M}(2)$. This key property is used, for example, in the proof of Proposition 5.4.

Only a minor role is played in this book by the quadratic variation of a square integrable martingale, but for completeness we mention it briefly. The quadratic variation of M is the process

$$[M]_t = \langle M^c \rangle_t + \sum_{s \leq t} (\Delta M_s)^2,$$

where M^c is the continuous martingale part of M, realized from an orthogonal decomposition $M = M^c + M^d$, where M^c is a continuous martingale, M^d is a "compensated sum of jumps," and $M^c \perp M^d$ (see Jacod, 1979, for formalities). The more easily understood approach is via a counterpart of (B.2):

$$[M]_t = M_0^2 + \lim_{n \to \infty} \sum_{k=1}^{2^n} (M_{tk/2^n} - M_{t(k-1)/2^n})^2.$$

Thus if $M_t = N_t - \int_0^t \lambda_s ds$ is the innovation martingale of a point process N with stochastic intensity λ, then $[M]_t = N_t$ for each t. The dual predictable projection of the quadratic variation is the predictable variation.

None of the three grand classes of martingale results

- Inequalities

- Convergence theorems

- Optional sampling theorems,

is central to this book, but no presentation can ignore them. We present parallel versions of each, for supermartingales and for martingales.

Martingale inequalities are important because they lead to the convergence theorems, but are also the most important link between martingale theory and classical analysis.

Theorem B.14. If Z is a supermartingale, then for each t and $a > 0$,
$P\{\sup_{s \leq t} Z_s \geq a\} \leq (1/a)(E[Z_0] - E[Z_t^-]).\ \Box$

Theorem B.15. For each $p > 1$ there exist universal constants c_p and
C_p such that for each martingale M and each $t \geq 0$,

$$c_p E[[M]_t^{p/2}] \leq E[\{\sup_{s \leq t}|M_s|^p\}] \leq C_p E[[M]_t^{p/2}].\ \Box$$

Of the sometimes bewildering variety of martingale convergence theorems, the following are two of the most useful.

Theorem B.16. If Z is a supermartingale with $\sup_t E[Z_t^-] < \infty$, then
there exists a random variable $Z_\infty \in L^1(P)$ such that $Z_t \to Z_\infty$ almost
surely as $t \to \infty$. $\Box$

Theorem B.16 applies in particular to nonnegative martingales and supermartingales, among them likelihood ratios; in general, the convergence
$Z_t \to Z_\infty$ is not realized in $L^1(P)$. For martingales, L^1-convergence is equivalent to uniform integrability and in this case the martingale has very simple
structure.

Theorem B.17. For a martingale M the following assertions are equivalent:
   a) M is uniformly integrable;
   b) There is a random variable $M_\infty \in L^1(P)$ such that $M_t \to M_\infty$ almost
surely and in $L^1(P)$;
   c) There is a random variable $M_\infty \in L^1(P) \cap \mathcal{H}_\infty$ such that $M_t = E[M_\infty|\mathcal{H}_t]$ for each t. $\Box$

Optional sampling theorems pertain to preservation of martingale or supermartingale properties when the constant times in Definition B.1 are replaced by stopping times.

Theorem B.18. Let Z be a supermartingale for which there exists a
random variable $Y \in L^1(P)$ satisfying $Z_t \geq E[Y|\mathcal{H}_t]$ for each t (i.e., Z is
bounded below by a uniformly integrable martingale, and hence fulfills the
hypotheses of Theorem B.16). Then for all stopping times $T_1 \leq T_2 \leq \infty$,
Z_{T_1} and Z_{T_2} belong to $L^1(P)$ and $E[Z_{T_2}|\mathcal{H}_{T_1}] \leq Z_{T_1}.\ \Box$

Theorem B.19. Let $M_t = E[M_\infty|\mathcal{H}_t]$ be a uniformly integrable mar-
tingale. Then for each stopping time $T \leq \infty$, $M_T = E[M_\infty|\mathcal{H}_T].\ \Box$

Often a process fails to be, for example, a martingale only because integrability or square integrability requirements fail, yet the property in ques-

tion holds for certain "stopped" processes. Localization makes this idea precise and usable.

Definition B.20. A process X is said to possess a property (P) (e.g., that of being a martingale) *locally* if there exist stopping times $T_1 \leq T_2 \leq \cdots$ converging to ∞ almost surely such that for each n the stopped process $(X_{t \wedge T_n})_{t \geq 0}$ has property (P). The T_n are termed *localizing times*.

For example, M is a *local martingale* if there are localizing times T_n such that each process $(M_{t \wedge T_n})$ is a martingale. In general, the most obvious exception being the convergence theorems, which degenerate, results above carry over if both hypotheses and conclusions are localized correspondingly.

Ordinarily, the localizing times depend on the property (P) but for a simple point process $N = \sum \varepsilon_{T_n}$ and $\mathcal{H} = \mathcal{F}^N$, the points T_n serve universally. For example, if $E[N_t] < \infty$ for each t, then the compensator exists, the innovation martingale M is locally square integrable, and hence there exists a locally integrable, predictable, increasing process $\langle M \rangle$ such that $M^2 - \langle M \rangle$ is a local martingale. For clarity and ease of exposition, in the text we have imposed hypotheses sufficiently strong to avoid localization; most of the results extend, however. When the underlying space is $[0, 1]$, as in Chapter 5, no harm ensues, but for processes on $[0, \infty)$, localization is mandatory in nearly all interesting cases.

We conclude with a central limit theorem specific to point process martingales. Among its applications is Theorem 9.23.

Theorem B.21. Let N be a point process with continuous compensator A and $\mathcal{F}^N$-innovation martingale M, and suppose that Y is a predictable process satisfying

$$\lim_{t \to \infty} (1/t) \int_0^t Y_s^2 dA_s = \sigma^2 \qquad (B.3)$$

in the sense of convergence in probability, where $0 < \sigma^2 < \infty$. Assume also that the Lindeberg condition that for every $\varepsilon > 0$

$$\lim_{t \to \infty} (1/t) E \left[\int_0^t Y_s^2 \mathbf{1}(|Y_s| > \varepsilon \sqrt{t}) dA_s \right] = 0$$

is fulfilled. Then

$$(1/\sqrt{t}) \int_0^t Y_s dM_s \xrightarrow{d} N(0, \sigma^2). \qquad (B.4)$$

Proof: (Sketch). The argument is substantially that employed by Kutoyants (1975) to establish a similar result for integrals with respect to diffusion processes. See also Kutoyants (1984, Theorem 4.5.4).

For each predictable process B, the process

$$M_t(B) = \exp\{-\int_0^t B dN - \int_0^t (e^B - 1) dA\}$$

is locally a square integrable martingale: it is the likelihood ratio associated with the probability measure under which N has compensator $d\tilde{A} = e^B dA$ (see Theorem 2.31). Glossing over technicalities we infer via the optional sampling theorem that

$$E[M_T(B)] = E[M_0(B)] = 1 \qquad (B.5)$$

for each stopping time T.

By appeal to (B.3), define a random time change (T_t) by

$$\int_0^{T_t} Y^2 dA = t\sigma^2.$$

Strictly speaking, $t\sigma^2$ must be replaced by $(1 - \epsilon)t\sigma^2$, but these details can be dealt with using the methods of Kutoyants (1975) and we do not dwell on them. With t fixed and (B.5) applied to $B_s = iuY^s/\sqrt{t}$, where $u \in \mathbf{R}$,

$$
\begin{aligned}
1 &= E[M_{T_t}(iuY/\sqrt{t})] \\
&= E\left[\exp\left\{(iu/\sqrt{t})\int_0^{T_t} Y dN - \int_0^{T_t}(e^{iuY/\sqrt{t}} - 1)dA\right\}\right] \\
&= E\left[\exp\left\{(iu/\sqrt{t})\int_0^{T_t} Y dN - (iu/\sqrt{t})\int_0^{T_t} Y dA + (u^2/2t)\int_0^{T_t} Y^2 dA\right\}\right. \\
&\qquad \times \left. \exp\left\{-\int_0^{T_t}[\textstyle\sum_{k=3}^{\infty}(iuY/\sqrt{t})^k/k!]dA\right\}\right] \\
&= e^{u^2\sigma^2/2} E\left[\exp\left\{(iu/\sqrt{t})\int_0^{T_t} Y dM\right\} R_t\right],
\end{aligned}
$$

where in view of (B.3), $R_t \to 1$ in probability. Consequently,

$$(1/\sqrt{t})\int_0^{T_t} Y dM \xrightarrow{d} N(0,\sigma^2).$$

To complete the proof it remains to show that

$$(1/\sqrt{t})\left[\int_0^t Y dM - \int_0^{T_t} Y dM\right] \to 0$$

in probability, to which once more the reasoning of Kutoyants (1975, 1984) applies with only minor modification. ∎

Bibliography

[1] O. O. Aalen. *Statistical Inference for a Family of Counting Processes.* PhD thesis, University of California, Berkeley, 1975.

[2] O. O. Aalen. Nonparametric inference in connection with multiple decrement models. *Scand. J. Statist.*, 3:15–27, 1976.

[3] O. O. Aalen. Weak convergence of stochastic integrals related to counting processes. *Z. Wahrsch. verw. Geb.*, 38:261–277, 1977.

[4] O. O. Aalen. Nonparametric inference for a family of counting processes. *Ann. Statist.*, 6:701–726, 1978.

[5] O. O. Aalen. A model for nonparametric regression analysis of counting processes. *Lect. Notes Statist.*, 2:1–25, 1980.

[6] O. O. Aalen and J. M. Hoem. Random time changes for multivariate counting processes. *Skand. Akt.*, 1978:81 – 101, 1978.

[7] O. O. Aalen and S. Johansen. An empirical transition matrix for nonhomogeneous Markov chains based on censored observations. *Scand. J. Statist.*, 5:141 – 150, 1978.

[8] E. Abdurachman and H. T. David. Cesàro limits for marked point processes on the line. *Commun. Statist. Stochastic Models*, 4:77 – 98, 1988.

[9] S. R. Adke and E. S. Murphee. Efficient sequential estimation for compound Poisson processes. *Commun. Statist. Theory Methods*, 17:443 – 460, 1988.

[10] R. J. Adler. *The Geometry of Random Fields.* Wiley, New York, 1981.

[11] V. E. Akman and A. E. Raftery. Bayes factors for nonhomogeneous Poisson processes with vague prior information. *J. Roy. Statist. Soc. B*, 48:322 – 329, 1986.

[12] M. G. Akritas and R. A. Johnson. Asymptotic inference in Lévy processes of discontinuous type. *Ann. Statist.*, 9:604 – 614, 1981.

[13] P. Albrecht. Über einige Eigenschaften des gemischten Poissonprozesse. *Mitt. Verein. Schweiz. Versichrungsmath.*, 1981:241 – 250, 1981.

[14] P. Albrecht. Testing the goodness of fit of a mixed Poisson process. *Insur. Math. Econ.*, 1:27 – 33, 1982a.

[15] P. Albrecht. On some statistical methods connected with the mixed Poisson process. *Scand. Actuarial J.*, 1982:1 – 14, 1982b.

[16] D. Aldous. Stopping times and tightness. *Ann. Probab.*, 6:335 – 340, 1978.

[17] S.-I. Amari and M. Kumon. Estimation in the presence of infinitely many nuisance parameters — geometry of estimating functions. *Ann. Statist.*, 16:1044 – 1068, 1988.

[18] D. A. Amato. A generalized Kaplan-Meier estimator for heterogeneous populations. *Commun. Statist. Theory Methods*, 17:263 – 286, 1988.

[19] R. V. Ambartzumian. Two inverse problems concerning the superposition of recurrent point processes. *J. Appl. Probab.*, 2:449 – 454, 1965.

[20] R. V. Ambartzumian. Palm distributions and superpositions of independent point processes in r^n. In P. A. W. Lewis, editor, *Stochastic Point Processes*. Wiley, New York, 1972.

[21] R. V. Ambartzumian. *Combinatorial Integral Geometry: With Applications to Mathematical Stereology*. Wiley, New York, 1982.

[22] R. V. Ambartzunian and B. S. Nahapetian. The Palm distribution and limit theorems for random point processes. *Akad. Nauk. Armjan. SSR Dokl.*, 71:87 – 90, 1980.

[23] L. P. Ammann and P. F. Thall. Count distributions, orderliness and invariance of Poisson cluster processes. *J. Appl. Probab.*, 16:261 – 273, 1979.

[24] P. K. Andersen. Testing goodness of fit of Cox's regression and life model. *Biometrics*, 38:67 – 77, 1982.

[25] P. K. Andersen. Comparing survival distributions via hazard ratio estimates. *Scand. J. Statist.*, 10:77 – 86, 1983.

[26] P.K. Andersen and Ø. Borgan. Counting process models for life history data: a review. *Scand. J. Statist.*, 12:97 – 158, 1985.

[27] P. K. Andersen, Ø. Borgan, R. D. Gill, and N. Keiding. Linear nonparametric tests for comparison of counting processes, with application to censored survival data. *Internat. Statist. Rev.*, 50:219 – 244, 1982.

[28] P. K. Andersen, Ø. Borgan, R. D. Gill, and N. Keiding. Censoring, truncation and filtering in statistical models based on counting processes. *Contemp. Math.*, 80:19 – 60, 1988.

[29] P. K. Andersen, Ø. Borgan, R. D. Gill, and N. Keiding. *Statistical Models for Counting Processes*. Springer-Verlag, New York, 1990. (To appear).

[30] P. K. Andersen and R. D. Gill. Cox's regression model for counting processes: a large sample study. *Ann. Statist.*, 10:1100 – 1120, 1982.

[31] T. W. Anderson. *An Introduction to Multivariate Statistical Analysis*. Wiley, New York, 1958.

[32] T. W. Anderson and D. A. Darling. Asymptotic theory of certain 'goodness of fit' criteria based on stochastic processes. *Ann. Math. Statist.*, 23:193 – 212, 1952.

[33] V. V. Anisimov. Estimation of the parameters of switched Poisson processes. *Theor. Probab. Appl.*, 31:1 – 11, 1985.

[34] G. T. Apoyan and Yu. A. Kutoyants. On compensator estimation of inhomogeneous Poisson process. *Problems Control Inform. Theory*, pages 135 – 142, 1987.

[35] A. Araujo and E. Giné. *The Central Limit Theorem for Real and Banach-Valued Random Variables*. Wiley, New York, 1980.

[36] E. Arjas and P. Haara. A marked point process approach to censored failure data with complicated covariates. *Scand. J. Statist.*, 11:193 – 209, 1984.

[37] E. Arjas and P. Haara. A logistic regression model for hazard: asymptotic results. *Scand. J. Statist.*, 14:1 – 18, 1987.

[38] E. Arjas and P. Haara. A note on asymptotic normality in the Cox regression model. *Ann. Statist.*, 16:1133 – 1140, 1988a.

[39] E. Arjas and P. Haara. A note on the exponentiality of total hazards before failure. *J. Multivariate Anal.*, 26:207 – 218, 1988b.

[40] E. Arjas and I. Norros. Change of life distribution via a hazard transformation: an inequality with application to minimal repair. *Math. Oper. Res.*, 14:355 – 361, 1988.

[41] R. B. Asher and D. G. Lainiotis. Adaptive estimation of doubly stochastic Poisson processes. *Inform. Sci.*, 12:245 – 261, 1977.

[42] K. B. Athreya, R. L. Tweedie, and D. Vere-Jones. Asymptotic behaviour of point processes with Markov dependent intervals. *Math. Nachr.*, 99:301 – 313, 1980.

[43] T. Aven. A theorem for determining the compensator of a counting process. *Scand. J. Statist.*, 12:69 – 72, 1985.

[44] T. Aven. Bayesian inference in a parametric counting process model. *Scand. J. Statist.*, 13:87 – 97, 1986.

[45] T. Aven. A counting process approach to replacement models. *Optimization*, 18:285 – 296, 1987.

[46] Y. Baba. Maximum likelihood estimation of parameters in birth and death processes by Poisson sampling. *J. Oper. Res. Soc. Japan*, 25:99 – 112, 1982.

[47] F. Bacelli and P. Brémaud. *Palm Probabilities and Stationary Queues*, volume 41 of *Lecture Notes in Statistics*. Springer-Verlag, New York, 1987.

[48] A. Baddeley. A limit theorem for statistics of spatial data. *Adv. Appl. Probab.*, 12:447 – 461, 1980.

[49] A. Baddeley and J. Møller. Nearest-neighbour Markov point processes and random sets. *Internat. Statist. Rev.*, 57:89 – 122, 1989.

[50] K. R. Bailey. The asymptotic joint distribution of regression and survival parameter estimates in the Cox regression model. *Ann. Statist.*, 11:39 – 48, 1983.

[51] A. V. Balakrishnan. A martingale approach to linear recursive state estimation. *SIAM J. Control*, 10:754 – 766, 1972.

[52] R. Banys. The convergence of sums of dependent point processes to Poisson processes. *Litovsk. Mat. Sb.*, 15:11 – 23, 1975.

[53] R. Banys. The convergence of superpositions of integer-valued random measures. *Lithuanian Math. J.*, 19:1 – 15, 1979a.

[54] R. Banys. Limit theorems for superpositions of multi-dimensional integer-valued random processes. *Lithuanian Math. J.*, 19:15 – 24, 1979b.

[55] R. Banys. On superposition of random measures and point processes. *Lect. Notes Statist.*, 2:26 – 37, 1980.

[56] R. Banys. On the convergence of superpositions of point processes in the space $D[0, 1]^2$. *Litovsk. Mat. Sb.*, 22:3 – 7, 1982.

[57] R. Banys. Convergence of random measures and point processes in the plane. *Probab. Math. Statist.*, 5:211 – 219, 1985.

[58] R. Banys. Convergence of random measures in the space $D[0, \infty)$. *Litovsk. Mat. Sb.*, 26:3 – 9, 1986.

[59] A. D. Barbour and G. K. Eagleson. An improved Poisson limit theorem for sums of dissociated random variables. *J. Appl. Probab.*, 24:586 – 599, 1987.

[60] R. E. Barlow. Likelihood ratio tests for restricted families of probability distributions. *Ann. Math. Statist.*, 39:547 – 560, 1968.

[61] R. E. Barlow, D. J. Bartholomew, J. M. Bremner, and H. D. Brunk. *Statistical Inference under Order Restrictions*. Wiley, New York, 1972.

[62] R. E. Barlow and F. Proschan. A note on tests for monotone failure rate based on incomplete data. *Ann. Math. Statist.*, 40:595 – 600, 1969.

[63] R. E. Barlow and F. Proschan. *Statistical Theory of Reliability and Life Testing: Probability Models*. Holt, Rinehart & Winston, New York, 1975.

[64] O. Barndorff-Nielson. Identifiability of mixtures of exponential families. *J. Math. Anal. Appl.*, 12:115 – 121, 1965.

[65] O. Barndorff-Nielson and G. F. Yeo. Negative binomial processes. *J. Appl. Probab.*, 6:633 – 647, 1969.

[66] P. Bártfai and J. Tomkó. *Point Processes and Queueing Problems*. North-Holland, Amsterdam, 1981.

[67] D. J. Bartholomew. Tests for randomness in a series of events when the alternative is trend. *J. Roy. Statist. Soc. B*, 18:234 – 239, 1956.

[68] M. S. Bartlett. Processus stochastiques ponctuels. *Ann. Inst. H. Poincaré*, 14:35 – 60, 1954.

[69] M. S. Bartlett. The spectral analysis of point processes. *J. Roy. Statist. Soc. B*, 25:264 – 296, 1963.

[70] M. S. Bartlett. The spectral analysis of two-dimensional point processes. *Biometrika*, 51:299 – 311, 1964.

[71] M. S. Bartlett. The spectral analysis of line processes. *Proc. Fifth Berkeley Symp. Math. Statist. Prob.*, III:135 – 153, 1967.

[72] M. S. Bartlett. The statistical analysis of spatial pattern. *Adv. Appl. Probab.*, 6:336 – 358, 1974.

[73] M. S. Bartlett. *The Statistical Analysis of Spatial Pattern*. Chapman & Hall, London, 1975.

[74] R. Bartoszyński, B. W. Brown, C. McBride, and J. R. Thompson. Some nonparametric techniques for estimating the intensity function of a cancer related nonstationary Poisson process. *Ann. Statist.*, 9:1050 – 1060, 1981.

[75] I. V. Basawa and B. L. S. Prakasa Rao. *Statistical Inference for Stochastic Processes*. Academic Press, New York, 1980.

[76] I. W. Basawa and D. J. Scott. *Asymptotic Optimal Inference for some Non-ergodic Models*, volume 17 of *Lecture Notes in Statistics*. Springer-Verlag, New York, 1983.

[77] M. Baudin. Likelihood and nearest-neighbor distance properties of multidimensional Poisson cluster processes. *J. Appl. Probab.*, 18:879 – 888, 1981.

[78] N. Becker and P. Yip. Nonparametric inference for a partially observed compartmental process. *Austral. J. Statist.*, 29:143 – 150, 1987.

[79] J. M. Begun, W. J. Hall, W.-M. Huang, and J. A. Wellner. Information and asymptotic efficiency in parametric-nonparametric models. *Ann. Statist.*, 11:432 – 452, 1983.

[80] B. Belkin. First passage to a general threshold for a process corresponding to sampling at Poisson times. *J. Appl. Probab.*, 8:573 – 588, 1971.

[81] A. Beneveniste and J. Jacod. Intensité stochastique du processus ponctuel stationnaire, reconstruction du processus ponctuel à partir de son intensité stochastique. *C. R. Acad. Sci. Paris*, 280:A821 – A824, 1975.

[82] M. Berman. Some multivariate generalizations of results in univariate stationary point processes. *J. Appl. Probab.*, 14:748 – 757, 1977.

[83] M. Berman. Regenerative multivariate point processes. *Adv. Appl. Probab.*, 10:411 – 430, 1978.

[84] M. Berman. Inhomogeneous and modulated gamma processes. *Biometrika*, 68:143 – 152, 1981.

[85] M. Berman. Testing for spatial association between a point process and another stochastic process. *Appl. Statist.*, 35:54 – 62, 1986.

[86] M. Berman and P. J. Diggle. Estimating weighted integrals of the second-order intensity of a spatial point process. *J. Roy. Statist. Soc. B*, 51:81 – 92, 1989.

[87] J. E. Besag. Spatial interaction and the statistical analysis of lattice systems. *J. Roy. Statist. Soc. B*, 36:192 – 236, 1974.

[88] J. E. Besag. On the statistical analysis of dirty pictures. *J. Roy. Statist. Soc. B*, 48:259 – 302, 1986.

[89] J. E. Besag, R. K. Milne, and S. Zachary. Point process limits of lattice processes. *J. Appl. Probab.*, 19:210 – 216, 1982.

[90] F. J. Beutler and F. B. Dolivo. Recursive integral equations for the detection of counting processes. *J. Appl. Math. and Opt.*, 3:65 – 72, 1976.

[91] F. J. Beutler and O. A. Z. Leneman. The theory of stationary point processes. *Acta. Math.*, 116:159 – 197, 1966a.

[92] F. J. Beutler and O. A. Z. Leneman. Random sampling of random processes: stationary point processes. *Inform. and Control*, 9:325 – 346, 1966b.

[93] P. J. Bickel. Tests for monotone failure rate, II. *Ann. Math. Statist.*, 40:1250 – 1260, 1969.

[94] P. J. Bickel and K. A. Doksum. Tests for monotone failure rate based on normalized spacings. *Ann. Math. Statist.*, 40:1216 – 1235, 1968.

[95] P. J. Bickel and K. A. Doksum. *Mathematical Statistics: Basic Ideas and Selected Topics*. Holden-Day, San Francisco, 1977.

[96] O. Bie, Ø. Borgan, and K. Liestøl. Confidence intervals and confidence bands for the cumulative hazard rate and their small sample properties. *Scand. J. Statist.*, 14:221 – 233, 1987.

[97] P. Billingsley. *Statistical Inference for Markov Processes*. University of Chicago Press, Chicago, 1961a.

[98] P. Billingsley. Statistical methods in Markov chains. *Ann. Math. Statist.*, 32:12 – 40, 1961b.

[99] P. Billingsley. *Convergence of Probability Measures*. Wiley, New York, 1968.

[100] P. Billingsley and F. Tøpsøe. Uniformity in weak convergence. *Z. Wahrsch. verw. Geb.*, 7:1 – 16, 1967.

[101] A. Birnbaum. Statistical methods for Poisson processes and exponential populations. *J. Amer. Statist. Assoc.*, 49:254 – 266, 1954.

[102] T. Björk and J. Grandell. Exponential inequalities for ruin probabilities in the Cox case. *Scand. Actuar. J.*, 1988:77 – 111, 1988.

[103] D. Blackwell. A renewal theorem. *Duke Math. J.*, 15:145 – 150, 1948.

[104] A. Blanc-Pierre and R. Sultan. Sur deux modèles de fonctions aléatoires dérivées d'un processus de Poisson. *C. R. Acad. Sci. Paris*, 269:A326 – A328, 1969.

[105] R. M. Blumenthal and R. K. Getoor. *Markov Processes and Potential Theory*. Academic Press, New York, 1968.

[106] S. Blumenthal, J. A. Greenwood, and L. Herbach. Superimposed nonstationary renewal processes. *J. Appl. Probab.*, 8:184 – 192, 1971.

[107] R. K. Boel and V. Beñes. Recursive nonlinear estimation of a diffusion acting as the rate of an observed Poisson process. *IEEE Trans. Inform. Theory*, IT-26:561 – 575, 1980.

[108] R. Boel, P. Varaiya, and E. Wong. Martingales on jump processes, I: Representation results, and II: Applications. *SIAM J. Control*, 13:999 – 1061, 1975.

[109] F. Böker. Limits of generalized compound point processes. *Statistics*, 16:97 – 105, 1985.

[110] F. Böker. Convergence of thinning processes using compensators. *Stochastic Process. Appl.*, 23:143 – 152, 1986.

[111] F. Böker. *Über Statistiche Methoden bei Punktprozessen*. Verlag Otto Schwartz, Göttingen, 1987.

[112] F. Böker and R. F. Serfozo. Ordered thinnings of point processes and random measures. *Stochastic Process. Appl.*, 15:113 – 132, 1983.

[113] Ø. Borgan. Maximum likelihood estimation in a parametric counting process model, with application to censored failure time data and multiplicative models. *Scand. J. Statist.*, 11:1 – 16, 1984.

[114] Ø. Borgan and H. Ramlau-Hansen. Demographic incidence rates and estimation of intensities with incomplete information. *Ann. Statist.*, 13:564 – 582, 1985.

[115] M. T. Boswell. Estimating and testing trend in a stochastic process of Poisson type. *Ann. Math. Statist.*, 37:1564 – 1573, 1966.

[116] M. T. Boswell and H. D. Brunk. Distribution of likelihood ratio in testing against trend. *Ann. Math. Statist.*, 40:371 – 380, 1969.

[117] L. Breiman. The Poisson tendency in traffic distribution. *Ann. Math. Statist.*, 34:308 – 311, 1963.

[118] P. Brémaud. *A Martingale Approach to Point Processes*. PhD thesis, The University of California, Berkeley, 1972.

[119] P. Brémaud. An extension of Watanabe's characterization theorem. *J. Appl. Probab.*, 8:396 – 399, 1975a.

[120] P. Brémaud. La méthode des semi-martingales en filtrage lorsque l'observation est un processus ponctuel marqué. *Lect. Notes Math.*, 511:1 – 18, 1975b.

[121] P. Brémaud. On the information carried by a stochastic point process. *Cahiers du CETHEDEC*, 43:43 – 70, 1975c.

[122] P. Brémaud. Bang-bang controls of point processes. *Adv. Appl. Probab.*, 8:385 – 394, 1976.

[123] P. Brémaud. Optimal thinning of a point process. *SIAM J. Control and Optimization*, 17:222 – 230, 1979.

[124] P. Brémaud. *Point Processes and Queues: Martingale Dynmaics*. Springer-Verlag, Berlin, 1981.

[125] P. Brémaud. An averaging principle for filtering a jump process with point process observations. *IEEE Trans. Inform. Theory*, IT-34:582 – 586, 1988.

[126] P. Brémaud and J. Jacod. Processus ponctuels et martingales: résultats récents sur la modèlisation et le filtrage. *Adv. Appl. Probab.*, 9:362 – 416, 1977.

[127] N. Breslow. Covariance analysis of censored survival data. *Biometrics*, 30:89 – 99, 1974.

[128] N. Breslow. Analysis of survival data under the proportional hazards model. *Internat. Statist. Rev.*, 43:45 – 58, 1975.

[129] N. Breslow and J. Crowley. A large sample study of the life table and product limit estimators under random censorship. *Ann. Statist.*, 2:437 – 453, 1974.

[130] J. Bretagnolle and C. Huber-Carol. Effects of omitting covariates in Cox's model for survival data. *Scand. J. Statist.*, 15:125 – 138, 1988.

[131] D. R. Brillinger. The asymptotic representation of the sample distribution function. *Bull. Amer. Math. Soc.*, 75:545 – 547, 1969.

[132] D. R. Brillinger. The spectral analysis of stationary interval functions. *Proc. Sixth Berkeley Symp. Math. Statist. Prob.*, 1:483 – 513, 1972.

[133] D. R. Brillinger. Cross-spectral analysis of processes with stationary increments, including the $G/G/\infty$ queue. *Ann. Probab.*, 2:815 – 827, 1974.

[134] D. R. Brillinger. Identification of point process systems. *Ann. Probab.*, 3:909 – 929, 1975a.

[135] D. R. Brillinger. Statistical inference for stationary point processes. In M. L. Puri, editor, *Stochastic Processes and Related Topics*. Academic Press, New York, 1975b.

[136] D. R. Brillinger. Comparative aspects of the study of ordinary time series and point processes. In P. R. Krishnaiah, editor, *Developments in Statistics*. Academic Press, New York, 1978.

[137] D. R. Brillinger. Analyzing point processes subjected to random deletions. *Canad. J. Statist.*, 7:21 – 27, 1979.

[138] D. R. Brillinger. *Time Series: Data Analysis and Theory*. Holden-Day, San Francisco, 1981.

[139] D. R. Brillinger. Some statistical methods for random process data from seismology and neurophysiology. *Ann. Statist.*, 16:1 – 54, 1988.

[140] M. Brown. An invariance property of Poisson processes. *J. Appl. Probab.*, 6:453 – 458, 1969.

[141] M. Brown. A property of Poisson processes and its application to equilibrium of particle systems. *Ann. Math. Statist.*, 41:1935 – 1941, 1970.

[142] M. Brown. Discrimination of Poisson processes. *Ann. Math. Statist.*, 42:773 – 776, 1971.

[143] M. Brown. Statistical analysis of nonhomogeneous Poisson processes. In P. A. W. Lewis, editor, *Stochastic Point Processes*. Wiley, New York, 1972.

[144] M. Brown. Bounds, inequalities and monotonicity properties for some specialized renewal processes. *Ann. Probab.*, 8:227 – 240, 1980.

[145] M. Brown and H. Solomon. Some results for secondary processes generated by a Poisson process. *Stochastic Process. Appl.*, 2:337 – 348, 1974.

[146] M. Brown, H. Solomon, and M. A. Stephens. Estimation of parameters of zero-one processes by interval sampling. *Oper. Res.*, 25:493 – 505, 1977.

[147] M. Brown, H. Solomon, and M. A. Stephens. Estimation of parameters of zero-one processes by interval sampling: an adaptive approach. *Oper. Res.*, 27:606 – 615, 1979.

[148] T. C. Brown. A martingale approach to the Poisson convergence of simple point processes. *Ann. Probab.*, 6:615 – 628, 1978.

[149] T. C. Brown. Position dependent and stochastic thinnings of point processes. *Stochastic Process. Appl.*, 9:189 – 193, 1979.

[150] T. C. Brown. Compensators and Cox convergence. *Proc. Cambridge Philos. Soc.*, 90:305 – 319, 1981.

[151] T. C. Brown. Some Poisson approximations using compensators. *Ann. Probab.*, 11:726 – 744, 1983.

[152] T. C. Brown, B. G. Ivanoff, and N. C. Weber. Poisson convergence in two dimensions with application to row and column exchangeable arrays. *Stochastic Process. Appl.*, 23:307 – 318, 1986.

[153] T. C. Brown and J. Kupka. Ramsey's theorem and Poisson random measures. *Ann. Probab.*, 11:904 – 908, 1983.

[154] T. C. Brown and M. G. Nair. Poisson approximations for time-changed point processes. *Stochastic Process. Appl.*, 29:247 – 256, 1988a.

[155] T. C. Brown and M. G. Nair. A simple proof of the multivariate random time change theorem for point processes. *J. Appl. Probab.*, 25:210 – 214, 1988b.

[156] T. C. Brown and B. W. Silverman. Short distances, flat triangles, and Poisson limits. *J. Appl. Probab.*, 15:815 – 825, 1978.

[157] T. C. Brown, B. W. Silverman, and R. K. Milne. A class of 2-type point processes. *Z. Wahrsch. verw. Geb.*, 58:299 – 308, 1981.

[158] W. M. Brown. Sampling with random jitter. *J. SIAM*, 11:460 – 473, 1963.

[159] W. Bryc. A characterization of the Poisson process by conditional moments. *Stochastics*, 20:17 – 26, 1987.

[160] J. A. Bucklew and S. Cambanis. Estimating random integrals from noisy observations: sampling designs and their performance. *IEEE Trans. Inform. Theory*, IT-34:111 – 127, 1988.

[161] R. S. Bucy and R. Kalman. New results in linear filtering and prediction theory. *J. Basic Engg.*, 83:95 – 108, 1961.

[162] M. D. Burke, S. Csörgő, and L. Horváth. Strong approximation of some biometric estimators under random censorship. *Z. Wahrsch. verw. Geb.*, 56:87 – 112, 1981.

[163] R. M. Burton and M. M. Franzosa. Positive dependence properties of point processes. *Ann. Probab.*, 18:359 – 377, 1990.

[164] R. M. Burton and T.-S. Kim. An invariance principle for associated random fields. *Pacific J. Math.*, 132:11 – 19, 1988.

[165] R. M. Burton and E. C. Waymire. Scaling limits for point random fields. *J. Multivariate Anal.*, 15:237 – 251, 1984.

[166] R. C. Burton and E. C. Waymire. Scaling limits for associated random measures. *Ann. Probab.*, 13:1267 – 1278, 1985.

[167] K. Byth. θ-stationary point processes and their second-order analysis. *J. Appl. Probab.*, 18:864 – 878, 1981.

[168] V. R. Cane. A class of nonidentifiable stochastic models. *J. Appl. Probab.*, 14:475 – 482, 1977.

[169] D. S. Carter and P. M. Prenter. Exponential spaces and counting processes. *Z. Wahrsch. verw. Geb.*, 21:1 – 19, 1972.

[170] C. Chandramodan, R. D. Foley, and R. L. Disney. Thinning of point processes — covariance analysis. *Adv. Appl. Probab.*, 17:127 – 146, 1985.

[171] I. S. Chang and C. A. Hsuing. Finite sample optimality of maximum partial likelihood estimation in Cox's model for counting processes. *J. Statist. Plann. Inference*, 25:35 – 42, 1990.

[172] M. N. Chang. Weak convergence of a self-consistent estimator of the survival function with doubly censored data. *Ann. Statist.*, 18:391 – 404, 1990.

[173] M. N. Chang and P. V. Rao. Berry-Esséen bounds for the Kaplan-Meier estimator. *Commun. Statist. Theory Methods*, 17:4647 – 4661, 1988.

[174] M. N. Chang and G. L. Yang. Strong consistency of a nonparametric estimator of the survival function with doubly censored data. *Ann. Statist.*, 15:1536 – 1547, 1987.

[175] Y. Y. Chen, M. Hollander, and N. A. Langberg. Testing whether new is better than used with randomly censored data. *Ann. Statist.*, 11:267 – 274, 1983.

[176] F. S. Chong. A point process with second order Markov dependent intervals. *Math. Nachr.*, 103:155 – 163, 1981.

[177] A. Chouinard and D. McDonald. A characterization of nonhomogeneous Poisson processes. *Stochastics*, 15:113 – 119, 1985.

[178] K. L. Chung. The Poisson process as renewal process. *Period. Math. Hungar.*, 2:41 – 48, 1972.

[179] K. L. Chung. *A Course in Probability Theory*. Academic Press, New York, 1974.

[180] K. L. Chung and J. L. Doob. Fields, optionality and measurability. *Amer. J. Math.*, 87:397 – 424, 1965.

[181] E. Çinlar. On the superposition of m-dimensional point processes. *J. Appl. Probab.*, 5:169 – 176, 1968.

[182] E. Çinlar. Superposition of point processes. In P. A. W. Lewis, editor, *Stochastic Point Processes*. Wiley, New York, 1972.

[183] E. Çinlar. *Introduction to Stochastic Processes*. Prentice-Hall, Englewood Cliffs, NJ, 1975a.

[184] E. Çinlar. Markov renewal theory: a survey. *Management Sci.*, 21:727 – 752, 1975b.

[185] E. Çinlar and R. A. Agnew. On the superposition of point processes. *J. Roy. Statist. Soc. B*, 30:576 – 581, 1968.

[186] E. Çinlar and J. Jacod. Representation of semimartingale Markov processes in terms of Wiener processes and Poisson random measures. In E. Çinlar, K. L. Chung, and R. K. Getoor, editors, *Seminar on Stochastic Processes 1981*. Birkhauser, Boston, 1981.

[187] E. Çinlar and P. Jagers. Two mean values which characterize the Poisson process. *J. Appl. Probab.*, 10:678 – 681, 1973.

[188] D. B. Clarkson and D. B. Wolfson. An application of a displaced Poisson process. *Statist. Neerl.*, 37:21 – 28, 1983.

[189] G. Clayton and T. F. Cox. Some robust density estimators for spatial point processes. *Biometrics*, 42:753 – 768, 1986.

[190] M. L. Clevenson and J. W. Zidek. Bayes linear estimators of the intensity function of the nonstationary Poisson process. *J. Amer. Statist. Assoc.*, 72:112 – 120, 1977.

[191] A. D. Cliff and J. K. Ord. Model building and the analysis of spatial pattern in human geography. *J. Roy. Statist. Soc. B*, 37:297 – 348, 1975.

[192] A. D. Cliff and J. K. Ord. *Spatial Processes: Models and Applications*. Pion, London, 1981.

[193] P. Clifford, S. Richardson, and D. Hémon. Estimating the significance of the correlation between two spatial processes. *Biometrics*, 45:123 – 134, 1989.

[194] D. L. Cohn. *Measure Theory*. Birkhauser, Boston, 1980.

[195] R. Cowan. Further results on single-lane traffic flow. *J. Appl. Probab.*, 16:523 – 531, 1980.

[196] D. R. Cox. Some statistical models connected with series of events. *J. Roy. Statist. Soc. B*, 17:129 – 164, 1955.

[197] D. R. Cox. *Renewal Theory*. Methuen, London, 1962.

[198] D. R. Cox. On the estimation of the intensity function of a stationary point process. *J. Roy. Statist. Soc. B*, 27:332 – 337, 1965.

[199] D. R. Cox. The statistical analysis of dependencies in point processes. In P. A. W. Lewis, editor, *Stochastic Point Processes*. Wiley, New York, 1972a.

[200] D. R. Cox. Regression models and life tables. *J. Roy. Statist. Soc. B*, 34:187 – 220, 1972b.

[201] D. R. Cox. Partial likelihood. *Biometrika*, 62:269 – 276, 1975.

[202] D. R. Cox and D. V. Hinkley. *Theoretical Statistics*. Chapman & Hall, London, 1974.

[203] D. R. Cox and V. Isham. *Point Processes*. Chapman & Hall, London, 1980.

[204] D. R. Cox and P. A. W. Lewis. *The Statistical Analysis of Series of Events*. Chapman & Hall, London, 1966.

[205] D. R. Cox and P. A. W. Lewis. Multivariate point processes. *Proc. Sixth Berkeley Symp. Math. Statist. Prob.*, 3:401 – 448, 1972.

[206] D. R. Cox and D. Oakes. *Analysis of Survival Data*. Chapman & Hall, London, 1984.

[207] H. Cramér and M. R. Leadbetter. *Stationary and Related Stochastic Processes*. Wiley, New York, 1967.

[208] H. Cramér, M. R. Leadbetter, and R. J. Serfling. On distribution function-moment relationships in a stationary point process. *Z. Wahrsch. verw. Geb.*, 18:1 – 8, 1971.

[209] N. Cressie. Spatial prediction and ordinary kriging. *Mathematical Geol.*, 20:405 – 421, 1988.

[210] K. S. Crump. On point processes having an order statistic structure. *Sankhyā A*, 37:396 – 404, 1975.

[211] S. Csörgő. Estimation in the proportional hazards model of random censorship. *Statistics*, 19:437 – 463, 1988.

[212] M. Csörgő, S. Csörgő, and L. Horváth. Estimation of total time on test transforms and Lorenz curves under random censorship. *Statistics*, 18:77 – 97, 1987.

[213] M. Csörgő, L. Horváth, and J. Steinebach. Invariance principles for renewal processes. *Ann. Probab.*, 15:1441 – 1460, 1987.

[214] M. Csörgő and P. Révész. *Strong Approximation in Probability and Statistics*. Academic Press, New York, 1981.

[215] D. M. Dąbrowska. Nonparametric regression with censored survival time data. *Scand. J. Statist.*, 14:181 – 197, 1987.

[216] D. M. Dąbrowska. Kaplan-Meier estimate in the plane. *Ann. Statist.*, 16:1475 – 1489, 1988.

[217] D. M. Dąbrowska and K. A. Doksum. Estimates and confidence intervals for the median and mean life in the proportional hazards model. *Biometrika*, 74:799 – 807, 1987.

[218] A. R. Dabrowski. Extremal point processes and intermediate quantile functions. *Probab. Th. Rel. Fields*, 85:365 – 386, 1990.

[219] D. J. Daley. Weakly stationary point processes and random measures. *J. Roy. Statist. Soc. B*, 33:406 – 428, 1971.

[220] D. J. Daley. Asymptotic properties of stationary point processes with generalized clusters. *Z. Wahrsch. verw. Geb.*, 21:65 – 76, 1972.

[221] D. J. Daley. Poisson and alternating renewal processes with superposition a renewal process. *Math. Nachr.*, 57:359 – 369, 1973.

[222] D. J. Daley. Various concepts of orderliness for point processes. In E. J. Harding and D. G. Kendall, editors, *Stochastic Geometry*. Wiley, New York, 1974.

[223] D. J. Daley. Queueing output processes. *Adv. Appl. Probab.*, 8:395 – 415, 1976.

[224] D. J. Daley. Stationary point processes with Markov-dependent intervals and infinite intensity. *J. Appl. Probab.*, 19A:313 – 320, 1982a.

[225] D. J. Daley. Infinite intensity mixtures of point processes. *Math. Proc. Cambridge Philos. Soc.*, 92:109 – 114, 1982b.

[226] D. J. Daley and R. K. Milne. Theory of point processes: a bibliography. *Internat. Statist. Rev.*, 41:183 – 201, 1973.

[227] D. J. Daley and R. K. Milne. Orderliness, intensities and Palm-Khinchin equations for multivariate point processes. *J. Appl. Probab.*, 12:383 – 389, 1975.

[228] D. J. Daley and D. Oakes. Random walk point processes. *Z. Wahrsch. verw. Geb.*, 30:1 – 16, 1974.

[229] D. J. Daley and D. Vere-Jones. A summary of the theory of point processes. In P. A. W. Lewis, editor, *Stochastic Point Processes*. Wiley, New York, 1972.

[230] D. J. Daley and D Vere-Jones. The extended probability generating functional, with application to mixing properties of cluster point processes. *Math. Nachr.*, 131:311 – 319, 1987.

[231] D. J. Daley and D. Vere-Jones. *An Introduction to the Theory of Point Processes.* Springer-Verlag, New York, 1988.

[232] R. Dalhaus and W. Künsch. Edge effects and efficient parameter estimation for stationary random fields. *Biometrika*, 74:887 – 892, 1987.

[233] H. E. Daniels. The Poisson process with a curved absorbing boundary. *Bull. Internat. Statist. Inst.*, 40:994 – 1008, 1963.

[234] D. A. Darling. The Kolmogorov-Smirnov, Cramér-von Mises tests. *Ann. Math. Statist.*, 28:823 – 838, 1953.

[235] R. Davidson. Exchangeble point-processes. In E. J. Harding and D. G. Kendall, editors, *Stochastic Geometry*. Wiley, New York, 1974.

[236] R. B. Davies. Testing the hypothesis that a point process is Poisson. *Adv. Appl. Probab.*, 9:724 – 746, 1977.

[237] D. J. Davis. An analysis of some failure data. *J. Amer. Statist. Assoc.*, 47:113–150, 1952.

[238] M. H. A. Davis. The representation of martingales of jump processes. *SIAM J. Control*, 14:623 – 638, 1976.

[239] M. H. A. Davis, T. Kailath, and A. Segall. Nonlinear filtering with counting observations. *IEEE Trans. Inform. Theory*, IT-21:143 – 150, 1975.

[240] H. Debes, J. Kerstan, A. Liemant, and K. Matthes. Verallgemeinerungen eines Satzes von Dobruschin I. *Math. Nachr.*, 47:183 – 244, 1970.

[241] H. Debes, J. Kerstan, A. Liemant, and K. Matthes. Verallgemeinerungen eines Satzes von Dobruschin III. *Math. Nachr.*, 50:299 – 383, 1971.

[242] A. Deffner and E. Hæusler. A characterization of order statistic point processes that are mixed Poisson processes and mixed sample processes simultaneously. *J. Appl. Probab.*, 22:314 – 323, 1985.

[243] V. DeGruttola and S. W. Lagakos. Analysis of doubly-censored survival data, with application to AIDS. *Biometrics*, 45:1 – 12, 1989.

[244] P. Deheuvels. Point processes and multivariate extreme values. *J. Multivariate Anal.*, 13:257 – 272, 1983.

[245] P. Deheuvels, A. F. Karr, D. Pfeifer, and R. J. Serfling. Poisson approximation in selected metrics by coupling and semigroup methods with applications. *J. Statist. Plann. Inference*, 20:1 – 22, 1988.

[246] P. Deheuvels and D. Pfeifer. Poisson approximations of multinomial distributions and point processes. *J. Multivariate Anal.*, 25:65 – 89, 1988.

[247] C. Dellacherie *Capacités et Processus Stochastiques*. Springer-Verlag, Berlin, 1972.

[248] C. Dellacherie and P.-A. Meyer. *Potentiel et Probabilités*. North-Holland, Amsterdam, 1980.

[249] A. P. Dempster, N. M. Laird, and D. B. Rubin. Maximum likelihood from incomplete data via the EM algorithm. *J. Roy. Statist. Soc. B*, 39:1 – 38, 1977.

[250] Y. L. Deng. On the comparison of point processes. *J. Appl. Probab.*, 22:300 – 313, 1985.

[251] C. M. Deo. A functional central limit theorem for stationary random fields. *Ann. Probab.*, 3:708 – 715, 1975.

[252] G. Dia. Estimation of a regression function of a Poisson process. *Statist. Prob. Lett.*, 6:47 – 54, 1987a.

[253] G. Dia. Étude d'un estimateur de la fonction de régression pour un processus ponctuel à valeurs dans $r_+^s \times r$ $(s \geq 1)$. *Serdica*, 13:382 – 395, 1987b.

[254] P. J. Diggle. On parameter estimation and goodness-of-fit testing for spatial point patterns. *Biometrics*, 35:87 – 101, 1979.

[255] P. J. Diggle. *Statistical Analysis of Spatial Point Patterns*. Academic Press, New York, 1983.

[256] P. J. Diggle. A kernel method for smoothing point process data. *Appl. Statist.*, 34:138 – 147, 1985.

[257] P. J. Diggle, J. E. Besag, and J. T. Gleaves. Statistical analysis of spatial point patterns by distance methods. *Biometrics*, 32:659 – 667, 1976.

[258] P. J. Diggle, D. J. Gates, and A. Stibbard. A nonparametric estimator for pairwise-interaction point processes. *Biometrika*, 74:763 – 770, 1987.

[259] P. J. Diggle and R. K. Milne. Negative binomial quadrat counts and point processes. *Scand. J. Statist.*, 10:257 – 267, 1983.

[260] P. J. Diggle and R. K. Milne. Bivariate Cox processes: some models for bivariate spatial point patterns. *J. Roy. Statist. Soc. B*, 45:11 – 21, 1983b.

[261] R. L. Disney and P. C. Kiessler. Equivalence and reversibility of point processes. *Lect. Notes in Control and Inform. Sci.*, 91:30 – 41, 1987.

[262] R. L. Dobrushin. On the Poisson law for distributions of particles in space. *Ukrain. Mat. Z.*, 8:127 – 134, 1956.

[263] R. L. Dobrushin. The description of a random field by means of its conditional distributions. *Theor. Probab. Appl.*, 13:197 – 224, 1968.

[264] K. A. Doksum. An extension of partial likelihood methods for proportional hazards models to general transformation models. *Ann. Statist.*, 15:325 – 345, 1987.

[265] C. Doléans-Dade. Quelques applications de la formule de changement de variables pour les semimartingales. *Z. Wahrsch. verw. Geb.*, 16:181 – 194, 1970.

[266] J. D. Dollard and C. N. Friedman. *Product Integration with Applications to Differential Equations*. Addison-Wesley, Reading, MA, 1979.

[267] M. Donsker. Justification and extension of Doob's heuristic approach to the Kolmogorov-Smirnov theorems. *Ann. Math. Statist.*, 23:277 – 283, 1952.

[268] J. L. Doob. Heuristic approach to the Kolmogorov-Smirnov theorems. *Ann. Math. Statist.*, 20:393 – 403, 1949.

[269] J. L. Doob. *Stochastic Processes*. Wiley, New York, 1953.

[270] H. Doss. On estimating the dependence between two point processes. *Ann. Statist.*, 17:749 – 763, 1989.

[271] C. Driancourt and F. Streit. Mise en evidence de tendances pour les processus ponctuels de Poisson à l'aide de tests statistiques. *Publ. Inst. Statist. Univ. Paris*, 28:1 – 20, 1983.

[272] M. F. Driscoll and N. A. Weiss. Random translations of stationary point processes. *J. Math. Anal. Appl.*, 48:423 – 433, 1974.

[273] L. E. Dubins and D. A. Freedman. Random distribution functions. *Proc. Fifth Berkeley Symp. Math. Statist. Prob.*, 2:183–214, 1967.

[274] D. M. Dudley. Distances of probability measures and random variables. *Ann. Math. Statist.*, 39:1563 – 1572, 1968.

[275] R. M. Dudley. Speeds of metric probability convergence. *Z. Wahrsch. verw. Geb.*, 22:323 – 332, 1972.

[276] R. M. Dudley. Sample functions of Gaussian processes. *Ann. Probab.*, 1:66 – 103, 1973.

[277] R. M. Dudley. Metric entropy and the central limit theorem in $C(S)$. *Ann. Inst. Fourier*, 242:49 – 60, 1974.

[278] R. M. Dudley. Central limit theorems for empirical measures. *Ann. Probab.*, 6:899 – 929, 1978.

[279] N. Dunford and J. T. Schwartz. *Linear Operators I: General Theory*. Interscience, New York, 1958.

[280] J. Durbin. Some methods for constructing exact tests. *Biometrika*, 48:41 – 55, 1961.

[281] J. Durbin. Boundary-crossing problems for the Brownian motion and Poisson processes and techniques for computing the power of the Kolmogorov-Smirnov test. *J. Appl. Probab.*, 8:431 – 453, 1971.

[282] J. Durbin. *Distribution Theory for Tests Based on the Sample Distribution Function*. SIAM, Philadelphia, 1973a.

[283] J. Durbin. Weak convergence of the sample distribution function when parameters are estimated. *Ann. Statist.*, 1:279 – 290, 1973b.

[284] A. Dvoretsky, J. Kiefer, and J. Wolfowitz. Sequential decision problems for processes with continuous time parameter: testing hypotheses. *Ann. Math. Statist.*, 24:254 – 264, 1953a.

[285] A. Dvoretsky, J. Kiefer, and J. Wolfowitz. Sequential decision problems for processes with continuous time parameter: problems of estimation. *Ann. Math. Statist.*, 24:403 – 415, 1953b.

[286] A. Dvoretsky, J. Kiefer, and J. Wolfowitz. Asymptotic minimax character of the sample distribution function and of the classical multinomial estimator. *Ann. Math. Statist.*, 27:642 – 669, 1956.

[287] R. L. Dykstra and C. J. Feltz. Nonparametric maximum likelihood estimation of survival functions with a general stochastic ordering and its dual. *Biometrika*, 76:331 – 341, 1989.

[288] M. Ebe. On sets of motion of point processes. *J. Oper. Res. Soc. Japan*, 11:97 – 113, 1969.

[289] B. Efron. The two-sample problem with censored data. *Proc. Fifth Berkeley Symp. Math. Statist. Prob.*, 4:831 – 852, 1967.

[290] B. Efron. Efficiency of Cox's likelihood function for censored data. *J. Amer. Statist. Assoc.*, 72:557 – 565, 1977.

[291] B. Efron. Logistic regression, survival analysis and the Kaplan-Meier curve. *J. Amer. Statist. Assoc.*, 83:414 – 425, 1988.

[292] B. Efron and I. M. Johnstone. Fisher's information in terms of the hazard rate. *Ann. Statist.*, 18:38 – 62, 1990.

[293] J. H. J. Einmahl and F. H. Ruymgaart. The order of magnitude of the moments of the modulus of continuity of multiparameter Poisson and empirical processes. *J. Multivariate Anal.*, 21:263 – 273, 1987.

[294] N. El-Karoui and J.-P. Lepeltier. Représentation des processus ponctuels multivariés à l'aide d'un processus de Poisson. *Z. Wahrsch. verw. Geb.*, 39:111 – 133, 1977.

[295] R. J. Elliot. Stochastic integrals for martingales of a jump process with partially accessible jump times. *Z. Wahrsch. verw. Geb.*, 36:213 – 226, 1976.

[296] R. J. Elliott. *Stochastic Calculus and Applications.* Springer-Verlag, New York, 1982.

[297] S. P. Ellis. A limit theorem for spatial point processes. *Adv. Appl. Probab.*, 18:646 – 659, 1986.

[298] S. P. Ellis. Second-order approximations to the characteristic function of certain point process integrals. *Adv. Appl. Probab.*, 19:546– 559, 1987.

[299] B. Epstein. Tests for the validity of the assumption that the underlying distribution of life is exponential. *Technometrics*, 2:83 – 101, 1960.

[300] B. Eriksson. An approximation of the variance of counts for a stationary point process. *Scand. J. Statist.*, 5:111 – 115, 1978.

[301] J. Etezadi-Amoli and A. Ciampi. A general model for testing the proportional hazards and accelerated failure time hypotheses in the analysis of censored survival data with covariates. *Commun. Statist. Theory Methods*, 14:651 – 667, 1985.

[302] J. Etezadi-Amoli and A. Ciampi. Extended hazard regression for censored survival data with covariates: A spline approximation for the baseline hazard function. *Biometrics*, 43:181 – 192, 1987.

[303] V. I. Fedorseev and F. P. Shirokov. Spatial-temporal nonlinear filtration for Poisson random fields. *Problems Inform. Transmission*, 12:20 – 28, 1976.

[304] P. D. Feigin. On the characterization of point processes with the order statistic property. *J. Appl. Probab.*, 16:297 – 304, 1979.

[305] P. D. Feigin. Maximum likelihood estimation for continuous time stochastic processes. *Adv. Appl. Probab.*, 8:712 – 736, 1976.

[306] P. D. Feigin. Conditional exponential families and a representation theorem for asymptotic inference. *Ann. Statist.*, 9:597 – 603, 1981.

[307] W. Feller. *An Introduction to Probability Theory and its Applications, II.* Wiley, New York, 2nd edition, 1971.

[308] A. Fellows and J. Granara. Une caractérisation de processus de Poisson. *Z. Wahrsch. verw. Geb.*, 39:71 – 79, 1977.

[309] T. S. Ferguson. *Mathematical Statistics: A Decision-Theoretic Approach.* Academic Press, New York, 1967.

[310] A. Feuerverger and R. A. Mureika. The empirical characteristic function and its applications. *Ann. Statist.*, 5:88 – 97, 1977.

[311] K. H. Fichtner. Charakterisierung Poissonscher zufälliger Punktfolgen und infinitesimale Verdunnungsschemata. *Math. Nachr.*, 68:93 – 104, 1975.

[312] T. Fiksel. Estimation of interaction potentials of Gibbsian point processes. *Statistics*, 19:77 – 86, 1988.

[313] L. Fisher. A survey of the mathematical theory of multidimensional point processes. In P. A. W. Lewis, editor, *Stochastic Point Processes*. Wiley, New York, 1972.

[314] P. H. Fishman and D. L. Snyder. The statistical analysis of space-time point processes. *IEEE Trans. Inform. Theory*, IT-22:257 – 274, 1976.

[315] K. Fleischmann. Ergodicity properties of infinitely divisible stochastic point processes. *Math. Nachr.*, 102:127 – 135, 1981.

[316] T. R. Fleming. Asymptotic distribution results in competing risks estimation. *Ann. Statist.*, 6:1071 – 1079, 1978a.

[317] T. R. Fleming. Nonparametric estimation for nonhomogeneous Markov processes in the problem of competing risks. *Ann. Statist.*, 6:1057 – 1070, 1978b.

[318] A. Foldes and L. Rejto. Strong uniform consistency for nonparametric survival curve estimators from randomly censored data. *Ann. Statist.*, 9:122 – 129, 1981a.

[319] A. Foldes and L. Rejto. A LIL type result for the product limit estimator. *Z. Wahrsch. verw. Geb.*, 56:75 – 86, 1981b.

[320] R. Fortet. Random functions from a Poisson process. *Proc. Second Berkeley Symp. Math. Statist. Prob.*, 1:373 – 385, 1951.

[321] B. Fox and P. Glynn. Estimating time averages via randomly spaced observations. *SIAM J. Appl. Math.*, 47:186 – 200, 1987.

[322] P. Franken. Approximation durch Poissonische Prozesse. *Math. Nachr.*, 26:101 – 114, 1963.

[323] P. Franken. Approximation der Verteilungen von Summen unabhangiger nichtnegativer ganzzahliger Zufällsgrossen durch Poissonische Verteilungen. *Math. Nachr.*, 27:303 – 340, 1964.

[324] P. Franken, B.-M. Kirstein, and A. Streller. Reliability of complex systems with repair. *Elektron. Informationsverarb. Kybernet.*, 20:407 – 422, 1984.

[325] P. Franken, D. König, U. Arndt, and V. Schmidt. *Queues and Point Processes*. Akademie-Verlag, Berlin, 1981.

[326] P. Franken, A. Liemant, and K. Matthes. Stationäre zufällige Punktfolgen III. *Jahresber. Deutsch. Math.-Verein*, 67:183 – 202, 1965.

[327] P. Franken and A. Streller. Reliability analysis of complex repairable systems by means of marked point processes. *J. Appl. Probab.*, 17:154 – 167, 1980.

[328] D. S. Freed and L. A. Shepp. A Poisson process whose rate is a hidden Markov process. *Adv. Appl. Probab.*, 14:21 – 36, 1982.

[329] D. Freedman. Poisson processes with a random arrival rate. *Ann. Math. Statist.*, 33:924 – 929, 1962.

[330] E. W. Frees. Nonparametric renewal function estimation. *Ann. Statist.*, 14:1366 – 1378, 1986.

[331] J. Fritz. Entropy of point processes. *Studia. Sci. Math. Hungar.*, 4:389 – 400, 1969.

[332] J. Fritz. Generalization of McMillan's theorem to random set functions. *Studia. Sci. Math. Hungar.*, 5:369 – 394, 1970.

[333] M. Fujisaki, G. Kallianpur, and H. Kunita. Stochastic differential equations for nonlinear filtering problems. *Osaka J. Math.*, 9:19 – 40, 1972.

[334] P. Gacs and D. Szasz. On a problem of Cox concerning point processes in r^k of 'controlled variability'. *Ann. Probab.*, 3:597 – 607, 1975.

[335] P. Gaenssler. *Empirical Processes: On Some Basic Results from the Probabilistic Point of View.* Institute of Mathematical Statistics, Hayward, CA, 1984.

[336] P. Gaenssler and W. Stute. On uniform convergence of measures with application to uniform convergence of empirical distributions. *Lect. Notes Math.*, 566:45 – 56, 1976.

[337] L. Galtchouk and B. Rozovskii. The disruption problem for a Poisson process. *Theor. Probab. Appl.*, 16:712 – 716, 1971.

[338] S. A. Galun and A. P. Tribnov. Detection and estimation of the instant of change of intensity of a Poisson flow. *Automat. Remote Control*, 43:782 – 790, 1982.

[339] P. Gaenssler and W. Stute. Empirical processes: a survey of results for independent and identically distributed random variables. *Ann. Probab.*, 7:193 – 243, 1979.

[340] D. J. Gates and M. Westcott. Clustering estimates for spatial point distributions with unstable potentials. *Ann. Inst. Statist. Math.*, 38:123 – 135, 1986.

[341] D. J. Gates and M. Westcott. Point processes with a clustering transition. *Austral. J. Statist.*, 30A:107 – 122, 1988.

[342] D. P. Gaver. Random hazard in reliability problems. *Technometrics*, 5:211 – 226, 1963.

[343] D. P. Gaver and P. A. W. Lewis. First order autoregressive gamma sequences and point processes. *Adv. Appl. Probab.*, 12:727 – 745, 1980.

[344] E. A. Gehan. A generalized Wilcoxon test for comparing arbitrarily singly-censored data. *Biometrika*, 52:203 – 233, 1965.

[345] D. Geman and S. Geman. Stochastic relaxation, Gibbs distributions and the Bayesian restoration of images. *IEEE Trans. Pattern Anal. Mach. Intell.*, PAMI-6:721 – 741, 1984.

[346] D. Geman and J. Horowitz. Remarks on Palm measures. *Ann. Inst. H. Poincaré*, 9:215 – 232, 1973.

[347] S. Geman and C.-R. Hwang. Nonparametric maximum likelihood estimation by the method of sieves. *Ann. Statist.*, 10:401 – 414, 1982.

[348] E. A. Geraniotis and H. V. Poor. Minimax discrimination for observed Poisson processes with uncertain rate functions. *IEEE Trans. Inform. Theory*, IT-31:660 – 669, 1985.

[349] B. Gerlach. Testing exponentiality against increasing failure rate with randomly censored data. *Statistics*, 18:275 – 286, 1987.

[350] A. Getis and B. Boots. *Models of Spatial Processes*. Cambridge University Press, Cambridge, 1978.

[351] R. D. Gill. Nonparametric estimation based on censored observations of a Markov renewal process. *Z. Wahrsch. verw. Geb.*, 53:97 – 116, 1980a.

[352] R. D. Gill. *Censoring and Stochastic Integrals*. Mathematisch Centrum, Amsterdam, 1980b.

[353] R. D. Gill. Large sample behaviour of the product-limit estimator on the whole line. *Ann. Statist.*, 11:49 – 58, 1983.

[354] R. D. Gill. Understanding Cox's regression model: a martingale approach. *J. Amer. Statist. Assoc.*, 79:441 – 447, 1984.

[355] R. D. Gill. The total time on test plot and cumulative total time on test statistic for a counting process. *Ann. Statist.*, 14:1234 – 1239, 1986a.

[356] R. D. Gill. On estimating transition intensities of a Markov process with aggregate data of a certain type: 'occurrences but no exposures'. *Scand. J. Statist.*, 13:113 – 134, 1986b.

[357] R. D. Gill. Non- and semi-parametric maximum likelihood estimators and the von Mises method. *Scand. J. Statist.*, 16:97 – 128, 1989.

[358] R. D. Gill and S. Johansen. A survey of product integration with a view towards application in survival analysis. Technical report, Universities of Copenhagen and Utrecht, Copenhagen and Utrecht, 1989.

[359] R. D. Gill, Y. Vardi, and J. A. Wellner. Large sample theory of empirical distribution functions in biased sampling models. *Ann. Statist.*, 16:1069 – 1102, 1988.

[360] I. Girsanov. On transforming a certain class of stochastic processes by absolutely continuous substitution of measure. *Theor. Probab. Appl.*, 5:285 – 301, 1960.

[361] L. Glass and W. R. Tobler. Uniform distribution of objects in a homogeneous field: cities on a plane. *Nature*, 233:67 – 68, 1971.

[362] E. Glotzl. On the singularity of σ point processes. *Math. Nachr.*, 92:211 – 213, 1979.

[363] V. P. Godambe. The foundations of finite sample estimation in stochastic processes. *Biometrika*, 72:419 – 428, 1985.

[364] V. P. Godambe and C. C. Heyde. Quasi-likelihood and optimal estimation. *Internat. Statist. Rev.*, 55:231 – 244, 1987.

[365] J. R. Goldman. Infinitely divisible point processes in r^n. *J. Math. Anal. Appl.*, 17:133 – 146, 1967a.

[366] J. R. Goldman. Stochastic point processes: limit theorems. *Ann. Math. Statist.*, 38:771 – 779, 1967b.

[367] Z. Govindarjulu. *Sequential Statistical Procedures*. Academic Press, New York, 1975.

[368] C. Góźdź and M. Polak. An approximation theorem for Poisson processes defined on an abstract space. *Studia. Sci. Math. Hungar.*, 22:169 – 174, 1987.

[369] J. Grandell. On stochastic processes generated by a stochastic intensity function. *Skand. Aktuar. Tidskrift*, 54:204 – 240, 1971a.

[370] J. Grandell. A note on the linear estimation of the intensity in a doubly stochastic Poisson field. *J. Appl. Probab.*, 8:612 – 614, 1971b.

[371] J. Grandell. On the estimation of intensities in a stochastic process generated by a stochastic intensity sequence. *J. Appl. Probab.*, 9:542 – 556, 1972a.

[372] J. Grandell. Statistical inference for doubly stochastic Poisson processes. In P. A. W. Lewis, editor, *Stochastic Point Processes*. Wiley, New York, 1972b.

[373] J. Grandell. *Doubly Stochastic Poisson Processes*, volume 529 of *Lecture Notes in Mathematics*. Springer-Verlag, Berlin, 1976.

[374] J. Grandell. Point processes and random measures. *Adv. Appl. Probab.*, 9:502 – 526, 1977.

[375] J. Grandell. *Stochastic Models of Air Pollution Concentration*, volume 30 of *Lecture Notes in Statistics*. Springer-Verlag, New York, 1985.

[376] R. J. Gray. Some diagnostic methods for Cox's regression models through hazard smoothing. *Biometrics*, 46:93 – 102, 1990.

[377] P. E. Greenwood and A. N. Shiryaev. *Contiguity and the Statistical Invariance Principle*. Gordon and Breach, London, 1985.

[378] G. Gregoire. Negative binomial point processes. *Stochastic Process. Appl.*, 16:179 – 188, 1984.

[379] U. Grenander. Stochastic processes and statistical inference. *Ark. Math.*, 1:195 – 277, 1950.

[380] U. Grenander. On empirical spectral analysis of stochastic processes. *Ark. Math.*, 1:503 – 531, 1951.

[381] U. Grenander. On the theory of mortality measurement, II. *Skand. Akt.*, 39:125 – 153, 1956.

[382] U. Grenander. *Abstract Inference*. Wiley, New York, 1981.

[383] R. C. Griffiths and R. K. Milne. A class of bivariate Poisson processes. *J. Multivariate Anal.*, 8:380 – 395, 1985.

[384] B. Grigelionis. On the convergence of random step processes to a Poisson process. *Theor. Probab. Appl.*, 8:177 – 182, 1963.

[385] B. Grigelionis. On the representation of integer-valued random measures by means of stochastic integrals with respect to a Poisson measure. *Litovsk. Mat. Sb.*, 11:783 – 794, 1971.

[386] B. Grigelionis. On the nonlinear filtering theory and absolute continuity of measures corresponding to stochastic processes. *Lect. Notes Math.*, 330:80 – 94, 1973.

[387] B. Grigelionis. Random point processes and martingales. *Litovsk. Mat. Sb.*, 15:101 – 114, 1975.

[388] B. Grigelionis. A martingale approach to the statistical problems of point processes. *Scand. J. Statist.*, 7:190 – 196, 1980.

[389] B. Grigelionis and R. Mikulevicius. On stably weak convergence of semi-martingales and of point processes. *Theor. Probab. Appl.*, 28:337 – 350, 1983.

[390] J. H. J. Guerts. On the small-sample performance of Efron's and Gill's version of the product limit estimator under nonproportional hazards. *Biometrics*, 43:683 – 692, 1987.

[391] P. Guttorp and M. L. Thompson. Nonparametric estimation of intensities for sampled counting processes. *J. Roy. Statist. Soc. B*, 52:157 – 173, 1990.

[392] L. Gyorfi. Poisson processes defined on an abstract space. *Studia. Sci. Math. Hungar.*, 7:243 – 248, 1972.

[393] E. Haberland. Infinitely divisible recurrent point processes. *Math. Nachr.*, 70:259 – 264, 1975.

[394] D. I. Hadjiev. On the filtering of semimartingales in case of observation of point processes. *Theor. Probab. Appl.*, 23:169 – 178, 1978.

[395] F. Haight. Counting distributions for renewal processes. *Biometrika*, 52:395 – 403, 1965.

[396] J. Hajek and Z. Sidak. *Theory of Rank Tests*. Academic Press, New York, 1967.

[397] P. Hall. On Starr and Vardi's estimates of the number of transmission sources. *J. Appl. Probab.*, 19:52 – 63, 1982.

[398] P. Hall. *Introduction to the Theory of Coverage Processes*. Wiley, New York, 1988.

[399] P. Hall and C. C. Heyde. *Martingale Limit Theory and its Application*. Academic Press, New York, 1980.

[400] W. J. Hall and J. A. Wellner. Confidence bands for a survival curve from censored data. *Biometrika*, 67:133 – 143, 1980.

[401] A. Hanen. Mesures aléatoires stationnaires et mesure de Palm. *Ann. Inst. H. Poincaré B*, 9:311 – 325, 1973.

[402] K.-H. Hanisch. On inversion formulae for *n*-fold Palm distributions of point processes on LCS-spaces. *Math. Nachr.*, 106:171 – 179, 1982.

[403] K.-H. Hanisch. Reduction of the *n*th moment measures and the special case of the third moment measure of stationary and isotropic planar point processes. *Math. Operationsforsch. Statist.*, 14:421 – 435, 1983.

[404] K.-H. Hanisch. Scattering analysis of point processes and random measures. *Math. Nachr.*, 117:235 – 245, 1984.

[405] K.-H. Hanisch and D. Stoyan. Formulas for the second-order analysis of marked point processes. *Math. Operationsforsch. Statist.*, 10:555 – 560, 1979.

[406] K.-H. Hanisch and D. Stoyan. Remarks on statistical inference and prediction for a hard-core clustering model. *Math. Operationsforsch. Statist.*, 14:559 – 567, 1983.

[407] E. J. Harding and D. G. Kendall, editors. *Stochastic Geometry*. Wiley, New York, 1974.

[408] T. E. Harris. *The Theory of Branching Processes*. Springer-Verlag, Berlin, 1963.

[409] T. E. Harris. Counting measures, monotone random set functions. *Z. Wahrsch. verw. Geb.*, 10:102 – 119, 1968.

[410] T. E. Harris. Random measures and motions of point processes. *Z. Wahrsch. verw. Geb.*, 18:85 – 115, 1971.

[411] A. G. Hawkes. Spectra of some self-exciting and mutually exciting point processes. *Biometrika*, 58:83 – 104, 1971.

[412] A. G. Hawkes and D. Oakes. A cluster process representation of a self-exciting point process. *J. Appl. Probab.*, 11:493 – 503, 1974.

[413] T. Hayashi. Laws of large numbers in self-correcting point processes. *Stochastic Process. Appl.*, 23:319 – 326, 1986.

[414] L. Heinrich. On a test of randomness of spatial point patterns. *Math. Operationsforsch. Statist.*, 15:413 – 420, 1984.

[415] L. Heinrich. Asymptotic behaviour of an empirical nearest-neighbour distance for stationary Poisson cluster processes. *Math. Nachr.*, 136:131 – 148, 1988.

[416] L. Heinrich and V. Schmidt. Normal convergence of multidimensional shot noise and rates of this convergence. *Adv. Appl. Probab.*, 17:709 – 730, 1985.

[417] I. S. Helland. Central limit theorems for martingales with discrete or continuous time. *Scand. J. Statist.*, 9:79 – 94, 1982.

[418] B. E. Helvik and A. R. Swenson. Modelling of clustering effects in point processes. An application to failures in SPC systems. *Scand. J. Statist.*, 14:57 – 66, 1987.

[419] T. Hida. On some asymptotic properties of Poisson processes. *Nagoya J. Math.*, 6:29 – 36, 1953.

[420] N. Y. Hjort. Bayes estimators and asymptotic efficiency in parametric counting process models. *Scand. J. Statist.*, 13:63 – 85, 1986.

[421] P. Holgate. Tests of randomness based on distance methods. *Biometrika*, 52:345 – 353, 1965.

[422] P. Holgate. The use of distance methods for the analysis of spatial distribution of points. In P. A. W. Lewis, editor, *Stochastic Point Processes*. Wiley, New York, 1972.

[423] M. Hollander and F. Proschan. Testing whether new is better than used. *Ann. Math. Statist.*, 43:1136 – 1146, 1972.

[424] M. Hollander and F. Proschan. Testing to determine the underlying distribution using randomly censored data. *Biometrics*, 35:393 – 401, 1979.

[425] M. Hollander and D. A. Wolfe. *Nonparametric Statistical Methods*. Wiley, New York, 1973.

[426] R. Höpfner. Asymptotic inference for continuous time Markov chains. *Probab. Th. Rel. Fields*, 77:537 – 550, 1988.

[427] J. Horowitz. Measure-valued stochastic processes. *Z. Wahrsch. verw. Geb.*, 70:213 – 236, 1985.

[428] J. Horowitz. Gaussian random measures. *Stochastic Process. Appl.*, 22:129 – 133, 1986.

[429] L. Horváth. Strong approximation of renewal processes. *Stochastic Process. Appl.*, 18:127 – 138, 1984.

[430] P. Hougaard. Modelling multivariate survival. *Scand. J. Statist.*, 14:291 – 304, 1987.

[431] T. Hsing. On the characterization of certain point processes. *Stochastic Process. Appl.*, 26:297 – 316, 1987.

[432] T. Hsing, J. Hüsler, and M. R. Leadbetter. On the exceedance point process for a stationary sequence. *Probab. Th. Rel. Fields*, 78:97 – 112, 1988.

[433] P. Huber. *Robust Statistics.* Wiley, New York, 1981.

[434] J. J. Hunter. Renewal theory in two dimensions: basic results. *Adv. Appl. Probab.*, 6:376 – 391, 1974a.

[435] J. J. Hunter. Renewal theory in two dimensions: asymptotic results. *Adv. Appl. Probab.*, 6:546 – 562, 1974b.

[436] W. J. Huster, R. Brookmeyer, and S. G. Self. Modelling paired survival data with covariates. *Biometrics*, 45:145 – 156, 1989.

[437] J. E. Hutton and P. I. Nelson. Quasi-likelihood estimation for semimartingales. *Stochastic Process. Appl.*, 22:245 – 257, 1986.

[438] I. A. Ibragimov and R. Z. Has'minskii. *Statistical Estimation: Asymptotic Theory.* Springer-Verlag, New York, 1981.

[439] I. A. Ibragimov and Yu. A. Rozanov. *Gaussian Random Processes.* Springer-Verlag, New York, 1978.

[440] P. Imkeller. Stochastic integrals of point processes and the decomposition of two-parameter martingales. *J. Multivariate Anal.*, 30:98 – 123, 1989.

[441] V. Isham. Dependent thinning of point processes. *J. Appl. Probab.*, 17:987 – 995, 1980.

[442] V. Isham. An introduction to spatial point processes and random fields. *Internat. Statist. Rev.*, 49:21 – 43, 1981.

[443] V. Isham. Multitype Markov point processes: some applications. *Proc. Royal Soc. London A*, 391:39 – 53, 1984.

[444] V. Isham, D. N. Shanbhag, and M. Westcott. A characterization of the Poisson process using forward recurrence times. *Math. Proc. Cambridge Philos. Soc.*, 78:513 – 516, 1975.

[445] V. Isham and M. Westcott. A self-correcting point process. *Stochastic Process. Appl.*, 8:335 – 347, 1979.

[446] K. Itô. Stationary random distributions. *Mem. Coll. Sci. Univ. Kyoto, Ser. A*, 28:209 – 223, 1953.

[447] K. Itô. Poisson point processes attached to Markov processes. *Proc. Sixth Berkeley Symp. Math. Statist. Prob.*, 3:225 – 239, 1972.

[448] Y. Ito. Superposition of distinguishable point processes. *J. Appl. Probab.*, 14:200 – 204, 1977.

[449] Y. Ito. Superposition and decomposition of stationary point processes. *J. Appl. Probab.*, 15:481 – 493, 1978.

[450] Y. Ito. On renewal processes decomposable into IID components. *Adv. Appl. Probab.*, 12:672 – 688, 1980.

[451] B. G. Ivanoff. Central limit theorems for point processes. *Stochastic Process. Appl.*, 12:171 – 186, 1982.

[452] B. G. Ivanoff. Poisson convergence for point processes in the plane. *J. Austral. Math. Soc. A*, 39:253 – 269, 1985.

[453] B. G. Ivanoff. Compensator approximations for point processes on the plane. *Stochastics*, 28:317 – 342, 1989.

[454] B. G. Ivanoff and E. Merzbach. Characterization of compensators for point processes on the plane. *Stochastics*, 29:395 – 406, 1989.

[455] M. Jacobsen. *Statistical Analysis of Counting Processes*, volume 12 of *Lecture Notes in Statistics*. Springer-Verlag, New York, 1982.

[456] M. Jacobsen. Maximum likelihood estimation in the multiplicative intensity model: a survey. *Internat. Statist. Rev.*, 52:193 – 207, 1984.

[457] M. Jacobsen. Right censoring and the Kaplan-Meier and Nelson-Aalen estimators: summary of results. *Contemp. Math.*, 80:61 – 66, 1988.

[458] M. Jacobsen. Right censoring and martingale methods for failure time data. *Ann. Statist.*, 17:1133 – 1156, 1989.

[459] J. Jacod. Two dependent Poisson processes whose sum is still a Poisson process. *J. Appl. Probab.*, 12:170 – 172, 1975a.

[460] J. Jacod. Multivariate point processes: predictable projection, Radon-Nikodym derivatives, representation of martingales. *Z. Wahrsch. verw. Geb.*, 31:235 – 253, 1975b.

[461] J. Jacod. *Calcul Stochastique et Problèmes de Martingales*, volume 714 of *Lecture Notes in Mathematics*. Springer-Verlag, New York, 1979.

[462] J. Jacod. Partial likelihood and asymptotic normality. *Stochastic Process. Appl.*, pages 47 –71, 1987a.

[463] J. Jacod. Sur la convergence des processus ponctuels. *Probab. Th. Rel. Fields*, 76:439 – 455, 1987b.

[464] J. Jacod. Filtered statistical models and Hellinger processes. *Stochastic Process. Appl.*, 31:3 – 45, 1989a.

[465] J. Jacod. Convergence of filtered statistical models and Hellinger processes. *Stochastic Process. Appl.*, 31:47 – 68, 1989b.

[466] J. Jacod and A. N. Shiryaev. *Limit Theorems for Stochastic Processes*. Springer-Verlag, Berlin, 1987.

[467] P. Jagers. On the weak convergence of superpositions of point processes. *Z. Wahrsch. verw. Geb.*, 22:1 – 7, 1972.

[468] P. Jagers. On Palm probabilities. *Z. Wahrsch. verw. Geb.*, 26:17 – 32, 1973.

[469] P. Jagers and T. Lindvall. Three theorems on the thinning of point processes. *Adv. Appl. Probab.*, 5:14 – 15, 1973.

[470] P. Jagers and T. Lindvall. Thinning and rare events in point processes. *Z. Wahrsch. verw. Geb.*, 28:89 – 98, 1974.

[471] N. Jain and M. B. Marcus. Central limit theorems for $C(S)$-valued random variables. *J. Funct. Anal.*, 19:216 – 231, 1975.

[472] U. Jansen. A generalization of insensitivity results by cyclically marked stationary point processes. *Elektron. Informationsverarb. Kybernet.*, 19:307 – 320, 1983.

[473] S. Janson. Poisson convergence and Poisson processes with applications to random graphs. *Stochastic Process. Appl.*, 26:1 – 30, 1987.

[474] A. Janssen. Local asymptotic normality for randomly censored models with application to rank tests. *Statist. Neerl.*, 43:109 – 125, 1990.

[475] E. B. Jensen, A. J. Baddeley, H. J. G. Gunderson, and R. Sundberg. Recent trends in stereology. *Internat. Statist. Rev.*, 53:99 – 108, 1985.

[476] E. B. Jensen and R. Sundberg. Statistical models for stereological inference about spatial structures: On the applicability of best linear unbiased estimators in stereology. *Biometrics*, 42:735 – 752, 1988.

[477] W. S. Jewell. Properties of recurrent event processes. *Oper. Res.*, 8:446 – 472, 1960.

[478] J. M. Jobe. Error rates for Poisson process discrimination. *Commun. Statist. Theory Methods*, 16:647 – 658, 1987.

[479] S. Johansen. The product limit estimator as maximum likelihood estimator. *Scand. J. Statist.*, 5:195 – 199, 1978.

[480] S. Johansen. An extension of Cox's regression model. *Internat. Statist. Rev.*, 51:165 – 174, 1983.

[481] N. L. Johnson and S. Kotz. A vector multivariate hazard rate. *J. Multivariate Anal.*, 5:53 – 66, 1975.

[482] E. Jolivet. Caractérisation et test du caractère agregatif des processus ponctuels stationnaires sur $\mathbf{R}^2$. *Lect. Notes Math.*, 636:1 – 25, 1978.

[483] E. Jolivet. Central limit theorem and convergence of empirical processes for stationary point processes. In P. Bártfai and J. Tomkó, editors, *Point Processes and Queueing Problems*. North-Holland, Amsterdam, 1981.

[484] E. Jolivet. Parametric estimation of the covariance density for a stationary point process on $\mathbf{r}^d$. *Stochastic Process. Appl.*, 22:111 – 119, 1986.

[485] M. P. Jones and J. Crowley. A general class of nonparametric tests for survival analysis. *Biometrics*, 45:157 – 170, 1989.

[486] D. W. Jorgensen. Multiple regression analysis of a Poisson process. *J. Amer. Statist. Assoc.*, 56:235 – 245, 1961.

[487] J. H. Jowett and D. Vere-Jones. The prediction of stationary point processes. In P. A. W. Lewis, editor, *Stochastic Point Processes*. Wiley, New York, 1972.

[488] Yu. M. Kabanov. The capacity of a channel of Poisson type. *Theor. Probab. Appl.*, 23:143 – 147, 1979.

[489] Yu. M. Kabanov. Contiguity of distributions of multivariate point processes. *Lect. Notes Math.*, 1299:140 – 157, 1988.

[490] Yu. M. Kabanov and R. S. Liptser. On convergence in variation of the distribution of multivariate point processes. *Z. Wahrsch. verw. Geb.*, 63:475 – 485, 1983.

[491] Yu. M. Kabanov, R. S. Liptser, and A. N. Shiryaev. Some limit theorems for simple point processes (martingale approach). *Stochastics*, 3:203 – 216, 1980a.

[492] Yu. M. Kabanov, R. S. Liptser, and A. N. Shiryaev. Representation of integer-valued random measures and local martingales by means of random measures with deterministic compensators. *Mat. Sb.*, 111:293 – 307, 1980b.

[493] Yu. M. Kabanov, R. S. Liptser, and A. N. Shiryaev. Weak and strong convergence of the distributions of counting processes. *Theor. Probab. Appl.*, 28:303 – 336, 1983.

[494] Yu. Kabanov, R. S. Liptser, and A. N. Shiryaev. On the variation distance for probability measures defined on a filtered space. *Probab. Th. Rel. Fields*, 71:19 – 36, 1986.

[495] T. Kailath. The innovations approach to detection and estimation theory. *Proc. IEEE*, 58:680 – 695, 1970.

[496] T. Kailath and A. Segall. Radon-Nikodym derivatives with respect to measures induced by discontinuous independent increment processes. *Ann. Probab.*, 3:449 – 464, 1975.

[497] T. Kailath and A. Segall. The modelling of random modulated jump processes. *IEEE Trans. Inform. Theory*, IT-21:135 – 142, 1975b.

[498] J. D. Kalbfleisch and R. L. Prentice. *The Statistical Analysis of Failure Time Data*. Wiley, New York, 1980.

[499] J. D. Kalbfleisch and R. L. Prentice. Estimation of the average hazard rate. *Biometrika*, 68:105 – 112, 1981.

[500] O. Kallenberg. Characterization and convergence of random measures and point processes. *Z. Wahrsch. verw. Geb.*, 27:9 – 21, 1973.

[501] O. Kallenberg. Extremality of Poisson and sample processes. *Stochastic Process. Appl.*, 2:73 – 83, 1974.

[502] O. Kallenberg. Limits of compound and thinned point processes. *J. Appl. Probab.*, 12:269 – 278, 1975a.

[503] O. Kallenberg. On symmetrically distributed random measures. *Trans. Amer. Math. Soc.*, 202:105 – 121, 1975b.

[504] O. Kallenberg. Stability of critical cluster fields. *Math. Nachr.*, 77:7 – 43, 1977.

[505] O. Kallenberg. L^p intensities of random measures. *Stochastic Process. Appl.*, 9:155 – 161, 1979.

[506] O. Kallenberg. On conditional intensities of point processes. *Z. Wahrsch. verw. Geb.*, 41:205 – 220, 1980.

[507] O. Kallenberg. Conditioning in point processes. Technical Report 82-7, Department of Mathematics, Chalmers Tekniska Högskola and University of Göteborg, Göteborg, Sweden, 1982.

[508] O. Kallenberg. *Random Measures*. Akademie-Verlag, Berlin, 3rd edition, 1983.

[509] O. Kallenberg. An informal guide to the theory of conditioning in point processes. *Internat. Statist. Rev.*, 52:151 – 164, 1984.

[510] O. Kallenberg. Exchangeable random measures in the plane. *J. Theoretical Prob.*, 3:81 – 135, 1990.

[511] G. Kallianpur. *Stochastic Filtering Theory*. Springer-Verlag, New York, 1980.

[512] G. Kallianpur and C. Striebel. Estimation of stochastic systems: arbitrary system process with additive white noise observation errors. *Ann. Math. Statist.*, 39:785 – 801, 1968.

[513] R. E. Kalman. A new approach to linear filtering and prediction problems. *J. Basic Engg.*, 82:35 – 45, 1960.

[514] E. L. Kaplan and P. Meier. Nonparametric estimation from incomplete observations. *J. Amer. Statist. Assoc.*, 53:457 – 481, 1958.

[515] S. Karlin and W. J. Studden. *Tchebycheff Systems: With Applications in Analysis and Statistics*. Interscience, New York, 1966.

[516] S. Karlin and H. M. Taylor. *A First Course in Stochastic Processes*. Academic Press, New York, 2nd edition, 1975.

[517] S. Karp and J. R. Clark. Photon counting: a problem in classical noise theory. *IEEE Trans. Inform. Theory*, IT-16:672 – 680, 1970.

[518] A. F. Karr. Two extreme value processes arising in hydrology. *J. Appl. Probab.*, 13:190 – 194, 1976a.

[519] A. F. Karr. A conditional expectation for random measures. Technical report, Department of Mathematical Sciences, The Johns Hopkins University, Baltimore, 1976b.

[520] A. F. Karr. Derived random measures. *Stochastic Process. Appl.*, 8:159 – 169, 1978.

[521] A. F. Karr. Classical limit theorems for measure-valued Markov processes. *J. Multivariate Anal.*, 9:234 – 247, 1979.

[522] A. F. Karr. A partially observed Poisson process. *Stochastic Process. Appl.*, 12:249 – 269, 1982.

[523] A. F. Karr. State estimation for Cox processes on general spaces. *Stochastic Process. Appl.*, 14:209 – 232, 1983.

[524] A. F. Karr. Estimation and reconstruction for zero-one Markov processes. *Stochastic Process. Appl.*, 16:219 – 255, 1984a.

[525] A. F. Karr. Combined nonparametric inference and state estimation for mixed Poisson processes. *Z. Wahrsch. verw. Geb.*, 66:81 – 96, 1984b.

[526] A. F. Karr. The martingale method: introductory sketch and access to the literature. *Oper. Res. Lett.*, 3:59 – 63, 1984c.

[527] A. F. Karr. State estimation for Cox processes with unknown probability law. *Stochastic Process. Appl.*, 20:115 – 131, 1985a.

[528] A. F. Karr. Inference for thinned point processes, with application to Cox processes. *J. Multivariate Anal.*, 16:368 – 392, 1985b.

[529] A. F. Karr. Inference for stationary random fields given Poisson samples. *Adv. Appl. Probab.*, 18:406 – 422, 1986.

[530] A. F. Karr. Maximum likelihood estimation in the multiplicative intensity model, via sieves. *Ann. Statist.*, 15:473–490, 1987a.

[531] A. F. Karr. Estimation of Palm measures of stationary point processes. *Probab. Th. Rel. Fields*, pages 55 – 69, 1987b.

[532] A. F. Karr. Palm distributions of point processes and their applications to statistical inference. *Contemp. Math.*, 80:331 – 358, 1988.

[533] A. F. Karr, D. L. Snyder, M. I. Miller, and M. J. Appel. Estimation of the intensity functions of Poisson processes, with application to positron emission tomography. Technical report, The Johns Hopkins Univeristy, Baltimore, 1990.

[534] M. L. Kavvas and J. Delleur. A stochastic cluster model of daily rainfall occurrences. *Water Resources Res.*, 17:1151 – 1160, 1981.

[535] N. Keiding. Estimation in the birth process. *Biometrika*, 61:71 – 80, 1974.

[536] N. Keiding. Maximum likelihood estimation in the birth and death process. *Ann. Statist.*, 3:363 – 372, 1975.

[537] N. Keiding and R. D. Gill. Random truncation models and Markov processes. *Ann. Statist.*, 18:582 – 602, 1990.

[538] A. M. Kellerer. The variance of a Poisson process of domains. *J. Appl. Probab.*, 23:307 – 321, 1986.

[539] A. M. Kellerer. Minkowski functionals of Poisson processes. *Z. Wahrsch. verw. Geb.*, 67:63 – 84, 1987b.

[540] H. G. Kellerer. Markov properties of point processes. *Probab. Th. Rel. Fields*, 76:71 – 80, 1987a.

[541] D. G. Kendall. Foundations of a theory of random sets. In E. J. Harding and D. G. Kendall, editors, *Stochastic Geometry*. Wiley, New York, 1974.

[542] J. Kerstan. Teilprozesse Poissonscher Prozesse. *Trans. Third Prague Conf.*, pages 377 – 403, 1962.

[543] J. Kerstan and K. Matthes. Stationare zufällige Punktfolgen II. *Jahresber. Deutsch. Math.-Verein*, 66:106 – 118, 1964a.

[544] J. Kerstan and K. Matthes. Verallgemeinerung eines Satzes von Sliwnjak. *Rev. Roumaine Math. Pures Appl.*, 9:811 – 829, 1964b.

[545] J. Kerstan and K. Matthes. A generalization of the Palm-Khinchin theorem. *Ukrain. Mat. Z.*, 17:29 – 36, 1965.

[546] J. Kerstan, K. Matthes, and U. Prehn. Verallgemeinerungen eines Satzes von Dobruschin II. *Math. Nachr.*, 51:149 – 188, 1971.

[547] A. Ya. Khintchine. Streams of events without aftereffects. *Theor. Probab. Appl.*, 1:1 – 15, 1956a.

[548] A. Ya. Khintchine. On Poisson streams of events. *Theor. Probab. Appl.*, 1:248 – 255, 1956b.

[549] A. Ya. Khintchine. *Mathematical Models in the Theory of Queueing*. Griffin, London, 1960.

[550] J. Kiefer. Skorohod embedding of multivariate random variables and the sample distribution function. *Z. Wahrsch. verw. Geb.*, 24:1 – 35, 1972.

[551] J. Kiefer and J. Wolfowitz. Consistency of the maximum likelihood estimator in the presence of infinitely many incidental parameters. *Ann. Math. Statist.*, 27:887 – 906, 1956.

[552] J. Kiefer and J. Wolfowitz. Sequential tests of hypotheses about the mean occurrence time of a continuous parameter Poisson process. *Naval Res. Logist. Quart.*, 3:205 – 219, 1957.

[553] J. F. C. Kingman. Poisson counts for random sequences. *Ann. Math. Statist.*, 34:1217 – 1232, 1963.

[554] J. F. C. Kingman. Random discrete distributions. *J. Roy. Statist. Soc. B*, 37:1 – 22, 1964a.

[555] J. F. C. Kingman. On doubly stochastic Poisson processes. *Proc. Cambridge Philos. Soc.*, 60:923 – 930, 1964b.

[556] J. F. C. Kingman. Completely random measures. *Pacific J. Math.*, 21:59 – 78, 1967.

[557] A. N. Kolmogorov and V. M. Tihomirov. ε-entropy and ε-capacity of sets in functional spaces. *Amer. Math. Soc. Trans.*, 17:277 – 364, 1961.

[558] T. Komatsu. Statistics of processes with jumps. *Lect. Notes Math.*, 550:276 – 289, 1976.

[559] J. Komlós, P. Major, and G. Tusnády. An approximation of partial sums of independent random variables and the sample distribution function, I. *Z. Wahrsch. verw. Geb.*, 32:111 – 131, 1975.

[560] J. Komlós, P. Major, and G. Tusnády. An approximation of partial sums of independent random variables and the sample distribution function, II. *Z. Wahrsch. verw. Geb.*, 34:33 – 58, 1976.

[561] F. Konency. Parameter estimation for point processes with partial observations: a filtering approach. *Systems Control Lett.*, 4:281 – 286, 1984.

[562] F. Konency. Maximum likelihood estimation for doubly stochastic Poisson processes with partial observations. *Stochastics*, 16:51 – 63, 1986.

[563] F. Konency. The asymptotic properties of maximum likelihood estimators for marked Poisson processes with a cyclic intensity measure. *Metrika*, 34:143 – 155, 1987.

[564] D. König and V. Schmidt. Imbedded and nonimbedded stationary characteristics of queueing systems with varying rates and point processes. *J. Appl. Probab.*, 17:753 – 767, 1980.

[565] J. A. Koziol and S. B. Green. A Cramér-von Mises statistic for randomly censored data. *Biometrika*, 63:465 – 474, 1976.

[566] O. K. Kozlov. Gibbsian description of point random fields. *Theor. Probab. Appl.*, 21:339 – 355, 1976.

[567] W. Kremers and D. S. Robson. Unbiased estimation when sampling from renewal processes: the single sample and k-sample random means cases. *Biometrika*, 74:329 – 336, 1987.

[568] K. Krickeberg. The Cox process. *Symp. Math.*, 9:151 – 167, 1972.

[569] K. Krickeberg. Moments of point processes. In E. J. Harding and D. G. Kendall, editors, *Stochastic Geometry*. Wiley, New York, 1974.

[570] K. Krickeberg. An alternative approach to the Glivenko-Cantelli theorems. *Lect. Notes Math.*, 566:57 – 67, 1976.

[571] K. Krickeberg. Statistical problems on point processes. *Banach Center Publ.*, 6:197 – 223, 1980.

[572] K. Krickeberg. Moment analysis of stationary point processes in r^d. In P. Bártfai and J. Tomkó, editors, *Point Processes and Queueing Problems.* North-Holland, Amsterdam, 1981.

[573] K. Krickeberg. Processus ponctuels en statistique. *Lect. Notes Math.,* 929:205 – 313, 1982.

[574] R. J. Kryscio and R. Saunders. On interpoint distances for planar Poisson processes. *J. Appl. Probab.,* 20:513 – 528, 1983.

[575] U. Küchler and M. Sørensen. Exponential families of stochastic processes: a unifying semimartingale approach. *Internat. Statist. Rev.,* 57:123 – 160, 1989.

[576] Y. Kumazawa. Testing whether new is better than used using randomly censored data. *Ann. Statist.,* 15:420 – 426, 1987.

[577] H. Kunita. Asymptotic distribution of the non-linear filtering errors of Markov processes. *J. Multivariate Anal.,* 1:365 – 393, 1971.

[578] H. Kunita and S. Watanabe. On square integrable martingales. *Nagoya J. Math.,* 30:209 – 245, 1967.

[579] T. G. Kurtz. Point processes and completely monotone set functions. *Z. Wahrsch. verw. Geb.,* 31:57 – 67, 1974.

[580] T. G. Kurtz. Representation and approximation of counting processes. *Lect. Notes in Control and Inform. Sci.,* 42:117 – 191, 1982.

[581] Yu. A. Kutoyants. Local asymptotic normality for the diffusion type processes. *Izv. Akad. Nauk. Armen. SSR,* 10:103 – 112, 1975.

[582] Yu. A. Kutoyants. Intensity parameter estimation of an inhomogeneous Poisson process. *Problems Control Inform. Theory,* 8:137 – 149, 1979.

[583] Yu. A. Kutoyants. *Estimation of Parameters of Random Processes.* Akad. Nauk. Armyan. SSR, Erevan, USSR, 1980.

[584] Yu. A. Kutoyants. Multidimensional parameter estimation of the intensity function of inhomogeneous Poisson processes. *Problems Control Inform. Theory,* 11:325 – 334, 1982.

[585] Yu. A. Kutoyants. *Parameter Estimation for Stochastic Processes.* Helderman Verlag, Berlin, 1984a.

[586] Yu. A. Kutoyants. On nonparametric estimation of intensity function of inhomogeneous Poisson process. *Problems Control Inform. Theory,* 13:253 – 258, 1984b.

[587] S. W. Lagakos. General right censoring and its impact on analysis of survival data. *Biometrics,* 35:139 – 156, 1979.

[588] T. L. Lai and Z. Ying. Stochastic integrals of empirical-type processes with applications to censored regression. *J. Multivariate Anal.,* 27:334 – 358, 1988.

[589] G. Last. Some remarks on conditional distributions for point processes. *Stochastic Process. Appl.*, 34:121 – 135, 1990.

[590] J. F. Lawless. *Statistical Models and Methods for Lifetime Data.* Wiley, New York, 1982.

[591] J. F. Lawless. Regression methods for Poisson process data. *J. Amer. Statist. Assoc.*, 82:808 – 815, 1987.

[592] A. J. Lawrance. Dependency of intervals between events in superposition processes. *J. Roy. Statist. Soc. B*, 35:306 – 315, 1973.

[593] A. Lawson. On tests for spatial trend in a nonhomogeneous Poisson process. *J. Appl. Statist.*, 15:225 – 234, 1988.

[594] M. R. Leadbetter. On streams of events and mixtures of streams. *J. Roy. Statist. Soc. B*, 28:218 – 227, 1966.

[595] M. R. Leadbetter. On three basic results in the theory of stationary point processes. *Proc. Amer. Math. Soc.*, 19:115 – 117, 1968.

[596] M. R. Leadbetter. On certain results for stationary point processes and their application. *Bull. Internat. Statist. Inst.*, 43:309 – 320, 1969a.

[597] M. R. Leadbetter. On the distributions of the times between events in a stationary stream of events. *J. Roy. Statist. Soc. B*, 31:295 – 302, 1969b.

[598] M. R. Leadbetter. On basic results of point process theory. *Proc. Sixth Berkeley Symp. Math. Statist. Prob.*, 3:449 – 462, 1972a.

[599] M. R. Leadbetter. Point processes generated by level crossings. In P. A. W. Lewis, editor, *Stochastic Point Processes.* Wiley, New York, 1972b.

[600] M. R. Leadbetter, G. Lindgren, and H. Rootzén. *Extremes and Related Properties of Random Sequences and Processes.* Springer-Verlag, New York, 1983.

[601] A. Le Breton. On continuous and discrete sampling for parameter estimation in diffusion type processes. *Math. Programming Studies*, 5:124 – 144, 1975.

[602] L. LeCam. Locally asymptotically normal families of distributions. *Univ. Calif. Publ. Statist.*, 3:37 – 98, 1960a.

[603] L. LeCam. An approximation theorem for the Poisson binomial distribution. *Pacific J. Math.*, 10:1181 – 1197, 1960b.

[604] L. LeCam. *Asymptotic Methods in Statistical Decision Theory.* Springer-Verlag, New York, 1986.

[605] L. Lee. Distribution-free tests for comparing trends in Poisson series. *Stochastic Process. Appl.*, 12:107 – 113, 1982.

[606] P. M. Lee. Infinitely divisible stochastic processes. *Z. Wahrsch. verw. Geb.*, 7:147 – 160, 1967.

[607] P. M. Lee. Some examples of infinitely divisible point processes. *Studia. Sci. Math. Hungar.*, 3:219 – 224, 1968.

[608] C. Léger and D. B. Wolfson. Hypothesis tests for a nonhomogeneous Poisson process. *Commun. Statist. Stochastic Models*, 3:439 – 455, 1987.

[609] E. Lehmann. *Testing Statistical Hypotheses*. Wiley, New York, 1959.

[610] E. Lehmann. *Theory of Point Estimation*. Wiley, New York, 1983.

[611] E. Lenglart. Relation de domination entre deux processus. *Ann. Inst. H. Poincaré*, 13:171 – 179, 1977.

[612] N. N. Leonenkov and A. V. Ivanov. *Statistical Analysis of Random Fields*. Kluwer, Dordrecht, 1989.

[613] V. P. Leonov and A. N. Shiryaev. On a method of calculation of semi-invariants. *Theor. Probab. Appl.*, 4:319 – 329, 1959.

[614] J. Leśkow. Estimation of the periodic function in the multiplicative intensity model. *Probab. Math. Statist.*, 8:103 – 110, 1987.

[615] J. Leśkow. A note on kernel smoothing of an estimator of a periodic function in the multivariate intensity model. *Statist. Prob. Lett.*, 7:395 – 400, 1989.

[616] P. Lévy. Processus semi-Markoviens. *Proc. Internat. Cong. Math.*, 3:416 – 426, 1954.

[617] P. A. W. Lewis. A branching Poisson process model for the analysis of computer failure patterns. *J. Roy. Statist. Soc. B*, 26:398 – 456, 1964.

[618] P. A. W. Lewis. Some results on tests for Poisson processes. *Biometrika*, 52:67 – 78, 1965.

[619] P. A. W. Lewis. A computer program for the statistical analysis of series of events. *IBM Systems J.*, 5:202 – 225, 1966.

[620] P. A. W. Lewis. Non-homogeneous branching Poisson processes. *J. Roy. Statist. Soc. B*, 29:343 – 354, 1967.

[621] P. A. W. Lewis. Asymptotic properties and equilibrium conditions for branching Poisson processes. *J. Appl. Probab.*, 6:355 – 371, 1969.

[622] P. A. W. Lewis. Recent results in the statistical analysis of univariate point processes. In P. A. W. Lewis, editor, *Stochastic Point Processes*. Wiley, New York, 1972a.

[623] P. A. W. Lewis, editor. *Stochastic Point Processes: Statistical Analysis, Theory and Applications*. Wiley, New York, 1972b.

[624] P. A. W. Lewis and G. S. Shedler. Analysis and modelling of point processes in computer systems. *Bull. Internat. Statist. Inst.*, 47:193 – 210, 1977.

[625] T. Lewis and L. J. Govier. Some properties of counts for certain types of point processes. *J. Roy. Statist. Soc. B*, 26:325 – 337, 1964.

[626] A. Liemant. Invariante zufällige Punktfolgen. *Wiss. Z. Friedrich-Schiller-Universität Jena Math.-Naturwiss-Reihe.*, 18:361 – 372, 1969.

[627] A. Liemant. Verallgemeinerungen eines Satzes von Dobruschin V. *Math. Nachr.*, 70:387 – 390, 1975.

[628] A. Liemant and K. Matthes. Verallgemeinerungen eines Satzes von Dobruschin IV. *Math. Nachr.*, 59:311 – 317, 1974.

[629] A. Liemant and K. Matthes. Verallgemeinerungen eines Satzes von Dobruschin VI. *Math. Nachr.*, 80:7 – 18, 1977.

[630] A. Liemant, K. Matthes, and A. Wakolbinger. *Equilibrium Distributions of Branching Processes.* Akademie-Verlag, Berlin, 1988.

[631] F. Liese. Uberlagerung verdunnter und schwach abhangiger Punktprozesse. *Math. Nachr.*, 95:177 – 186, 1980.

[632] G. Lindgren. Point processes of exits by bivariate Gaussian processes and extremal theory of the chi-squared process and its concomitants. *J. Multivariate Anal.*, 10:181 – 206, 1980.

[633] B. G. Lindsay. The geometry of mixture likelihoods. I. A general theory. *Ann. Statist.*, 11:86 – 94, 1983a.

[634] B. G. Lindsay. The geometry of mixture likelihoods. II. The exponential family. *Ann. Statist.*, 11:783 – 792, 1983b.

[635] T. Lindvall. An invariance principle for thinned random measures. *Studia. Sci. Math. Hungar.*, 11:269 – 275, 1976.

[636] T. Lindvall. On coupling of discrete renewal processes. *Z. Wahrsch. verw. Geb.*, 48:57 – 70, 1981.

[637] T. Lindvall. On coupling of continuous time renewal processes. *J. Appl. Probab.*, 19:82 – 89, 1982.

[638] T. Lindvall. Ergodicity and inequalities in a class of point processes. *Stochastic Process. Appl.*, 30:121 – 131, 1988.

[639] Yu. N. Linkov. Asymptotic power of statistical criteria for counting processes. *Prob. Inform. Transmission*, 17:196 – 205, 1981.

[640] R. S. Liptser and A. N. Shiryaev. *Statistics of Random Processes*, volume I and II. Springer-Verlag, Berlin, 1978.

[641] R. S. Liptser and A. N. Shiryaev. A functional central limit theorem for semimartingales. *Theor. Probab. Appl.*, 25:667 – 688, 1980.

[642] B. Lisek and M. Lisek. A new method for testing whether a point process is Poisson. *Statistics*, 16:445 – 450, 1985.

[643] R. Y. C. Liu and J. Van Ryzin. A histogram estimator of the hazard rate with censored data. *Ann. Statist.*, 13:592 – 605, 1985.

[644] A. Y. Lo. Bayesian nonparametric statistical inference for Poisson point processes. *Z. Wahrsch. verw. Geb.*, 59:55 – 66, 1982.

[645] S. H. Lo, Y. P. Mack, and J.-L. Wang. Density and hazard rate estimation for censored data via strong representation of the Kaplan-Meier estimator. *Probab. Th. Rel. Fields*, 80:461– 474, 1989.

[646] S. H. Lo and K. Singh. The product-limit estimator and the bootstrap: some asymptotic representations. *Probab. Th. Rel. Fields*, 71:455 – 465, 1986.

[647] M. Loève. *Probability Theory*, volume I. Springer-Verlag, New York, 1977.

[648] G. G. Lorentz. *Approximation of Functions*. Holt, Reinhart & Winston, New York, 1966.

[649] H. W. Lotwick. Some models for multitype spatial point processes, with remarks on analysing multitype patterns. *J. Appl. Probab.*, 21:575 – 582, 1984.

[650] H. W. Lotwick and B. W. Silverman. Methods for analysing spatial processes of several types of points. *J. Roy. Statist. Soc. B*, 44:406 – 413, 1982.

[651] O. Lundberg. *On Random Processes and their Application to Sickness and Accident Statistics*. Almqvist & Wiksells, Uppsala, 1940.

[652] O. Macchi. Stochastic point processes and multicoincidences. *IEEE Trans. Inform. Theory*, IT-17:1 – 7, 1971.

[653] O. Macchi. The coincidence approach to point processes. *Adv. Appl. Probab.*, 7:83 – 122, 1975.

[654] O. Macchi. Stochastic point processes in pure and applied physics. *Bull. Internat. Statist. Inst.*, 47(2):211 – 241, 1977.

[655] O. Macchi and B. Picinbono. Estimation and detection of weak optical signals. *IEEE Trans. Inform. Theory*, IT-18:562 – 573, 1972.

[656] N. R. Mann, R. E. Schafer, and N. D. Singpurwalla. *Methods for Statistical Analysis of Reliability and Life Data*. Wiley, New York, 1974.

[657] R. L. Marcellus. A Markov renewal approach to the Poisson disorder problem. *Commun. Statist. Stochastic Models*, 6:213 – 228, 1990.

[658] K. V. Mardia. Multi-dimensional multivariate Gaussian random fields with application to image processing. *J. Multivariate Anal.*, 24:265 – 284, 1988.

[659] A. W. Marshall and F. Proschan. Maximum likelihood estimation for distributions with monotone failure rate. *Ann. Math. Statist.*, 36:69 – 77, 1965.

[660] E. Masry. Poisson sampling and spectral estimation of continuous-time processes. *IEEE Trans. Inform. Theory*, IT-24:173 – 183, 1978a.

[661] E. Masry. Alias-free sampling: an alternative conceptualization and its applications. *IEEE Trans. Inform. Theory*, IT-24:317 – 324, 1978b.

[662] E. Masry. Nonparametric covariance estimation from irregularly spaced data. *Adv. Appl. Probab.*, 15:113 – 132, 1983.

[663] E. Masry. Random sampling of continuous-parameter stochastic processes: statistical properties of joint density estimators. *J. Multivariate Anal.*, 26:133 – 165, 1988.

[664] E. Masry and M.-C. C. Lui. Discrete spectral estimation of continuous-parameter processes — a new consistent estimate. *IEEE Trans. Inform. Theory*, IT-22:289 – 312, 1976.

[665] P. Massart. Strong approximation for multivariate empirical and related processes, via KMT constructions. *Ann. Probab.*, 17:266 – 291, 1989.

[666] G. Mathéron. A simple substitute for conditional expectation: the disjunctive kriging. In M. Guariscio, M. David, and C. Huijbregts, editors, *Advanced Statistics in the Mining Industry*. Reidel, Boston, 1976.

[667] K. Matthes. Stationäre zufällige Punktfolgen I. *Jahresber. Deutsch. Math.-Verein*, 66:66 – 79, 1963a.

[668] K. Matthes. Unbeschrankt teilbare verteilungsgesetze stationärer zufälliger Punktfolgen. *Wiss. Z. Hochsch. Elektrotech. Ilm*, 9:235 – 238, 1963b.

[669] K. Matthes, J. Kerstan, and J. Mecke. *Infinitely Divisible Point Processes*. Wiley, New York, 1978.

[670] M. Matthes, W. Warmuth, and J. Mecke. Bemerkungen zu einer Arbeit von Nguyen Xuan Xanh und Hans Zessin. *Math. Nachr.*, 88:117 – 127, 1979.

[671] J. Mau. Statistical modeling via partitioned counting processes. *J. Statist. Plann. Inference*, 12:171 – 176, 1985.

[672] J. Mau. Nonparametric estimation of the integrated intensity of an unobservable transition in a Markov illness-death process. *Stochastic Process. Appl.*, 21:275 – 289, 1986.

[673] J. Mau. A generalization of a nonparametric test for stochastically ordered distributions from censored survival data. *J. Roy. Statist. Soc. B*, 50:403 – 412, 1988a.

[674] J. Mau. A comparison of counting process models for complicated life histories. *Appl. Stoch. Models Data Anal.*, 4:283 – 298, 1988b.

[675] G. Mazziotto and J. Szpirglas. Équations du filtrage pour un processus de Poisson mélangé à deux indices. *Stochastics*, 4:89 – 119, 1980.

[676] G. Mazziotto and E. Merzbach. Point processes indexed by directed sets. *Stochastic Process. Appl.*, 30:105 – 120, 1988.

[677] J. A. McFadden. On the lengths of intervals in a stationary point process. *J. Roy. Statist. Soc. B*, 24:364 – 382, 1962.

[678] J. A. McFadden. The entropy of a point process. *SIAM J. Appl. Math.*, 13:988 – 994, 1965.

[679] J. A. McFadden and W. Weissblum. Higher order properties of a stationary point process. *J. Roy. Statist. Soc. B*, 25:413 – 431, 1963.

[680] I. W. McKeague. Estimation for a semimartingale model using the method of sieves. *Ann. Statist.*, 14:579 – 589, 1986.

[681] I. W. McKeague. A counting process approach to the regression analysis of grouped survival data. *Stochastic Process. Appl.*, 28:221 – 239, 1988a.

[682] I. W. McKeague. Asymptotic theory for weighted least squares estimators in Aalen's additive risk model. *Contemp. Math.*, 80:139 – 152, 1988b.

[683] E. McKenzie. Some ARMA models for dependent sequences of Poisson counts. *Adv. Appl. Probab.*, 20:822 – 835, 1988.

[684] J. Mecke. Zum Problem der Zeiligbarkeit stationärer rekurrenter zufälliger Punktfolgen. *Math. Nachr.*, 35:311 – 321, 1967a.

[685] J. Mecke. Stationäre zufällige Masse auf lokalkompakten Abelschen Gruppen. *Z. Wahrsch. verw. Geb.*, 9:36 – 58, 1967b.

[686] J. Mecke. Eine charakteristische Eigenschaft der doppelt stochastischen Poissonschen Prozesse. *Z. Wahrsch. verw. Geb.*, 11:74 – 81, 1968.

[687] J. Mecke. Zufällige Masse auf lokalkompakten Hausdorffschen Raumen. *Beiträge Anal.*, 3:7 – 30, 1972.

[688] J. Mecke. A characterization of the mixed Poisson processes. *Rev. Roumaine Math. Pures Appl.*, 21:1355 – 1360, 1976.

[689] B. Melamed and J. Walrand. On the one-dimensional distributions of counting processes with stochastic intensities. *Stochastics*, 19:1 – 9, 1986.

[690] N. Mendes Lopes. Covergence et optimisation d'un estimateur de la répartition locale des coleurs d'un processus ponctuel chromatique. *Publ. Inst. Statist. Univ. Paris*, 29:49 – 68, 1984.

[691] E. Merzbach. Point processes in the plane. *Acta. Appl. Math.*, 12:79 – 101, 1988.

[692] E. Merzbach and D. Nualart. A characterization of the spatial Poisson process and changing time. *Ann. Probab.*, 14:1380 – 1390, 1986.

[693] E. Merzbach and D. Nualart. A martingale approach to point processes in the plane. *Ann. Probab.*, 16:265 – 274, 1988.

[694] E. Merzbach and D. Nualart. Markov properties for point processes in the plane. *Ann. Probab.*, 18:342 – 358, 1990.

[695] M. Metivier and J. Pellaumail. *Stochastic Integration*. Academic Press, New York, 1980.

[696] P.-A. Meyer. *Probability and Potentials*. Ginn-Blaisdell, Waltham, MA, 1965.

[697] P.-A. Meyer. Guide détaillé de la théorie 'générale' des processus. *Lect. Notes Math.*, 51:140 – 165, 1969.

[698] P.-A. Meyer. Processus de Poisson ponctuels, d'apres K. Itô. *Lect. Notes Math.*, 191:177 – 180, 1971.

[699] P.-A. Meyer. Un cours sur les intégrales stochastiques. *Lect. Notes Math.*, 511:245 – 400, 1976.

[700] R. E. Miles. On the homogeneous planar Poisson process. *Math. Biosci.*, 6:85 – 127, 1970.

[701] P. W. Millar. Asymptotic minimax theorems for the sample distribution function. *Z. Wahrsch. verw. Geb.*, 48:233 – 252, 1979.

[702] R. G. Miller. *Survival Analysis*. Wiley, New York, 1981.

[703] C. Milne. Transient behaviour of the interrupted Poisson process. *J. Roy. Statist. Soc. B*, 44:398 – 405, 1982.

[704] R. K. Milne. Identifiability for random translations of Poisson processes. *Z. Wahrsch. verw. Geb.*, 15:195 – 201, 1970.

[705] R. K. Milne. Simple proofs of some theorems on point processes. *Ann. Math. Statist.*, 42:368 – 372, 1971.

[706] N. M. Mirasol. The output of an $M/G/\infty$ queueing system is Poisson. *Oper. Res.*, 11:282 – 284, 1963.

[707] M. L. Moeschberger and J. P. Klein. A comparison of several methods of estimating the survival function when there is extreme right censoring. *Biometrics*, 41:253 – 259, 1985.

[708] J. Mogyorodi. Some remarks on the rarefaction of the renewal processes. *Litovsk. Mat. Sb.*, 11:303 – 315, 1971.

[709] J. Mogyorodi. Rarefaction of renewal processes, I and II. *Studia. Sci. Math. Hungar.*, 7:285 – 305, 1972.

[710] J. Mogyorodi. Rarefaction of renewal processes, III and IV. *Studia. Sci. Math. Hungar.*, 8:21 – 38, 1973a.

[711] J. Mogyorodi. Rarefaction of renewal processes, V and VI. *Studia. Sci. Math. Hungar.*, 8:193 – 209, 1973b.

[712] E. Moore and R. Pyke. Estimation of transition distributions of a Markov renewal process. *Ann. Inst. Statist. Math.*, 20:411 – 424, 1968.

[713] P. A. P. Moran. A note on the serial coefficients. *Biometrika*, 57:670 – 673, 1970.

[714] T. Mori. On random translations of point processes. *Yokohama Math. J.*, 19:119 – 139, 1971.

[715] T. Mori. Stationary random measures and renewal theory. *Yokohama Math. J.*, 23:31 – 54, 1975.

[716] T. Mori. A generalization of Poisson point processes with application to a classical limit theorem. *Z. Wahrsch. verw. Geb.*, 54:331 – 340, 1980.

[717] J. E. Moyal. The general theory of stochastic population processes. *Acta Math.*, 108:1 – 31, 1962.

[718] H.-G. Müller and J.-L. Wang. Nonparametric analysis of changes in hazard rates for censored survival data: an alternative to change-point models. *Biometrika*, 77:305 – 314, 1990.

[719] S. A. Murphy. *Time-Dependent Coefficients in a Cox-Type Regression Model.* PhD thesis, University of North Carolina at Chapel Hill, 1989.

[720] S. A. Murphy and P. K. Sen. Time-dependent coefficients in a Cox-type regression model. Technical report, University of North Carolina, Chapel Hill, 1990.

[721] T. Naes. The asymptotic distribution of the estimator of the regression parameter in Cox's regression model. *Scand. J. Statist.*, 9:107 – 115, 1982.

[722] B. S. Nahapetian. Palm distribution and limit theorems for random point processes in r^n. *Probab. Math. Statist.*, 4:123 – 132, 1984.

[723] K. Nawrotzki. Ein Grenzwertsatz fur homogene zufällige Punktfolgen. *Math. Nachr.*, 24:201 – 217, 1962.

[724] W. B. Nelson. Hazard plotting methods for analysis of life data with different failure modes. *J. Qual. Technology*, 2:126 – 141, 1970.

[725] W. B. Nelson. Theory and applications of hazard plotting for censored failure data. *Technometrics*, 14:945 – 966, 1972.

[726] W. B. Nelson. *Applied Life Data Analysis.* Wiley, New York, 1982.

[727] G. Neuhaus. Asymptotically optimal rank tests for the two-sample problem with randomly censored data. *Commun. Statist. Theory Methods*, 17:2037 – 2058, 1988.

[728] J. Neveu. *Mathematical Foundations of the Calculus of Probability.* Holden-Day, San Francisco, 1965.

[729] J. Neveu. Sur la structure des processus ponctuels stationnaires. *C. R. Acad. Sci. Paris*, A267:561 – 564, 1968.

[730] J. Neveu. *Discrete-Parameter Martingales.* North-Holland, Amsterdam, 1975.

[731] J. Neveu. Sur les mesures de Palm de deux processus ponctuels stationnaires. *Z. Wahrsch. verw. Geb.*, 34:199–203, 1976.

[732] J. Neveu. Processus ponctuels. *Lect. Notes Math.*, 598:249 – 447, 1977.

[733] D. S. Newman. A new family of point processes which are characterized by their second moment properties. *J. Appl. Probab.*, 7:338 – 358, 1970.

[734] C. M. Newman and Y. Rinott. Nearest neighbors and Voronoï volumes in high-dimensional point processes with various distance functions. *Adv. Appl. Probab.*, 17:794 – 809, 1985.

[735] J. Neyman and E. L. Scott. Statistical approach to problems of cosmology. *J. Roy. Statist. Soc. B*, 20:1 – 43, 1958.

[736] J. Neyman and E. L. Scott. Processes of clustering and applications. In P. A. W. Lewis, editor, *Stochastic Point Processes*. Wiley, New York, 1972.

[737] J. Neyman, E. L. Scott, and C. D. Shane. Statistics of images of galaxies with particular reference to clustering. *Proc. Third Berkeley Symp. Math. Statist. Prob.*, 3:75 – 111, 1956.

[738] H. T. Nguyen. On point process sampling of continuous-time stochastic models. *Publ. Inst. Statist. Univ. Paris*, 30:73 – 95, 1985.

[739] Nguyen van Huu. A limit theorem for functionals of a Poisson process. *Comment. Math. Univ. Carolin.*, 20:547–564, 1979.

[740] X. X. Nguyen. Ergodic theorems for subadditive spatial processes. *Z. Wahrsch. verw. Geb.*, 48:159 – 176, 1979.

[741] X. X. Nguyen and H. Zessin. Punktprozesse mit Wechselwirkung. *Z. Wahrsch. verw. Geb.*, 37:91 – 126, 1976a.

[742] X. X. Nguyen and H. Zessin. Martin-Dynkin boundary of mixed Poisson processes. *Z. Wahrsch. verw. Geb.*, 37:191 – 200, 1976b.

[743] X. X. Nguyen and H. Zessin. Ergodic theorems for spatial processes. *Z. Wahrsch. verw. Geb.*, 48:133 – 158, 1979a.

[744] X. X. Nguyen and H. Zessin. Integral and differential characterizations of the Gibbs process. *Math. Nachr.*, 88:105 – 115, 1979b.

[745] G. Nieuwenhuis. Equivalence of limit theorems for stationary point processes and their palm distributions. *Probab. Th. Rel. Fields*, 81:593 – 608, 1989.

[746] I. Norros. A compensator representation of multivariate life length distributions, with applications. *Scand. J. Statist.*, 13:99 – 112, 1986.

[747] A. A. Noura and K. L. Q. Read. Proportional hazards changepoint models in survival analysis. *Appl. Statist.*, 39:241 – 254, 1900.

[748] D. Oakes. Survival times: aspects of partial likelihood. *Internat. Statist. Rev.*, 49:235 – 264, 1981.

[749] D. Oakes. Partial likelihood: applications, ramifications, generalizations. *Contemp. Math.*, 80:67 – 78, 1988.

[750] Y. Ogata. Asymptotic behavior of maximum likelihood estimators for stationary point processes. *Ann. Inst. Statist. Math.*, 30:243 – 251, 1978.

[751] Y. Ogata. Likelihood analysis of point processes and its application to seismological data. *Bull. Internat. Statist. Inst.*, 50(2):943 – 961, 1983.

[752] Y. Ogata. Statistical models for earthquake occurrences and residual analysis for point processes. *J. Amer. Statist. Assoc.*, 83:9 – 27, 1988.

[753] Y. Ogata and H. Akaike. On linear intensity models for mixed doubly stochastic Poisson and self-exciting point processes. *J. Roy. Statist. Soc. B*, 44:102 – 107, 1982.

[754] Y. Ogata and K. Katsura. Point-process models with linearly parametrized intensity for the application to earthquake catalogue. *J. Appl. Probab.*, 23A:231 – 240, 1986.

[755] Y. Ogata and M. Tanamura. Estimation of interaction potentials of spatial point patterns through the maximum likelihood procedure. *Ann. Inst. Statist. Math.*, 33B:315 – 338, 1981.

[756] Y. Ogata and M. Tanamura. Likelihood analysis of spatial point patterns. *J. Roy. Statist. Soc. B*, 46:496 – 518, 1984.

[757] Y. Ogata and D. Vere-Jones. Inference for earthquake models: a self-correcting model. *Stochastic Process. Appl.*, 17:337 – 347, 1984.

[758] J. Osher. On estimators for the reduced second moment measure of point processes. *Math. Operationsforsch. Statist.*, 14:63 – 71, 1983.

[759] J. Osher and D. Stoyan. On the second-order and orientation analysis of planar stationary point processes. *Biometrics*, 23:523 – 533, 1981.

[760] G. A. Ososkov. A limit theorem for flows of similar events. *Theor. Probab. Appl.*, 1:248 – 255, 1956.

[761] T. Ozaki. Maximum likelihood estimation of Hawkes' self-exciting point process. *Ann. Inst. Statist. Math.*, 31:144 – 155, 1979.

[762] C. Palm. Intensitätsschwankungen in Fernsprechverkehr. *Ericsson Technics*, 44:1 – 189, 1943.

[763] F. Papangelou. The Ambrose-Kakutani theorem and the Poisson process. *Lect. Notes Math.*, 160:234 – 240, 1970.

[764] F. Papangelou. Integrability of expected increments of point processes and a related change of scale. *Trans. Amer. Math. Soc.*, 165:483 – 506, 1972.

[765] F. Papangelou. On the Palm probabilities of processes of points and processes of lines. In E. J. Harding and D. G. Kendall, editors, *Stochastic Geometry*. Wiley, New York, 1974a.

[766] F. Papangelou. The conditional intensity of general point processes and an application to line processes. *Z. Wahrsch. verw. Geb.*, 28:207 – 226, 1974b.

[767] F. Papangelou. Point processes on spaces of flats and other homogeneous spaces. *Proc. Cambridge Philos. Soc.*, 80:297 – 314, 1976.

[768] F. Papangelou. On the entropy rate of stationary point processes and its discrete approximation. *Z. Wahrsch. verw. Geb.*, 44:191 – 211, 1978.

[769] M. S. Pepe and T. R. Fleming. Weighted Kaplan-Meier statistics: A class of distance tests for censored survival data. *Biometrics*, 45:497 – 508, 1989.

[770] P. Peruničić and Z. Glišić. Estimation of the rarefying function. *Mat. Vesnik*, 39:447 – 453, 1987.

[771] D. Pfeifer and U. Heller. A martingale characterization of mixed Poisson processes. *J. Appl. Probab.*, 24:246 – 251, 1987.

[772] T. D. Pham. Estimation of the spectral parameters of a stationary point process. *Ann. Statist.*, 9:615 – 627, 1981.

[773] M. J. Phelan. Inference from censored Markov chains with applications to multiwave data. *Stochastic Process. Appl.*, 29:85 – 102, 1988.

[774] M. J. Phelan. Estimating the transition probabilities from censored Markov renewal processes. *Statist. Prob. Lett.*, 10:43 – 47, 1990.

[775] J. Pickands III. The two-dimensional Poisson process and extremal processes. *J. Appl. Probab.*, 8:745 – 756, 1971.

[776] M. Polak and D. Szynd. An approximation for Poisson processes defined on an abstract space. *Bull. Acad. Polon. Sci.*, 25:1037 – 1043, 1977.

[777] D. Pollard. *Convergence of Stochastic Processes*. Springer-Verlag, New York, 1984.

[778] D. Pollard and J. Strobel. On the construction of random measures. *Bull. Soc. Math. Grèce*, 20:67 – 80, 1979.

[779] O. Pons. Test sur la loi d'un processus ponctuel. *C. R. Acad. Sci. Paris*, 292:91 – 94, 1981.

[780] O. Pons. A test of independence between two censored survival times. *Scand. J. Statist.*, 13:173 – 185, 1986a.

[781] O. Pons. Vitesse de convergence des estimateurs à noyau pour l'intensité d'un processus ponctuel. *Statistics*, 17:577 – 584, 1986b.

[782] O. Pons and E. de Turckheim. Estimation in Cox's periodic model with a histogram-type estimator for the underlying intensity. *Scand. J. Statist.*, 14:329 – 345, 1987.

[783] O. Pons and E. de Turckheim. On Cox's periodic regression model. *Ann. Statist.*, 16:678 – 693, 1988.

[784] S. C. Port and C. J. Stone. Infinite particle systems. *Trans. Amer. Math. Soc.*, 178:307 – 340, 1973.

[785] B. L. S. Prakasa Rao. Statistical inference from sampled data for stochastic processes. *Contemp. Math.*, 80:249 – 284, 1988.

[786] A. Prékopa. On stochastic set functions, I. *Acta Math. Hungar.*, 8:215 – 256, 1956a.

[787] A. Prékopa. On stochastic set functions, II and III. *Acta Math. Hungar.*, 8:337 – 400, 1956b.

[788] A. Prékopa. On Poisson and compound Poisson stochastic set functions. *Studia. Math.*, 16:142 – 155, 1957.

[789] A. Prékopa. On secondary processes generated by a random point distribution of Poisson type. *Ann. Univ. Sci. Budapest Sect. Math.*, 1:153 – 170, 1958.

[790] R. L. Prentice and J. D. Kalbfleisch. Hazard rate models with covariates. *Biometrics*, 35:25 – 39, 1979.

[791] R. L. Prentice and S. G. Self. Asymptotic distribution theory of Cox-type regression models with general relative risk form. *Ann. Statist.*, 11:803 – 813, 1983.

[792] F. Proschan and R. Pyke. Tests for monotone failure rate. *Proc. Fifth Berkeley Symp. Math. Statist. Prob.*, 3:293 – 312, 1967.

[793] P. Protter. *Stochastic Integration and Differential Equations: A New Approach.* Springer-Verlag, New York, 1990.

[794] H. Pruscha. Estimating a parametric trend component in a continuous time stochastic process. *Stochastic Process. Appl.*, 28:241 – 257, 1988.

[795] J. Przyborowski and H. Wilenski. Homogeneity of results in testing samples from a Poisson series. *Biometrika*, 31:313 – 323, 1940.

[796] M. L. Puri and P. K. Sen. *Nonparametric Methods in Multivariate Analysis.* Wiley, New York, 1971.

[797] P. S. Puri. On the characterization of point processes with the order statistic structure without the moment condition. *J. Appl. Probab.*, 19:39 – 51, 1982.

[798] R. Pyke. The supremum and infimum of the Poisson process. *Ann. Math. Statist.*, 30:568 – 576, 1959.

[799] R. Pyke. Markov renewal processes: definition and preliminary properties. *Ann. Math. Statist.*, 32:1231 – 1242, 1961a.

[800] R. Pyke. Markov renewal processes with finitely many states. *Ann. Math. Statist.*, 32:1243 – 1259, 1961b.

[801] R. Pyke. Spacings. *J. Roy. Statist. Soc. B*, 27:395 – 449, 1965.

[802] R. Pyke. The weak convergence of the empirical process with random sample size. *Proc. Cambridge Philos. Soc.*, 64:155 – 160, 1968.

[803] R. Pyke and R. Schaufele. Limit theorems for Markov renewal processes. *Ann. Math. Statist.*, 35:1746 – 1764, 1964.

[804] M. H. Quenouille. Problems in plane sampling. *Ann. Math. Statist.*, 20:355 – 375, 1949.

[805] A. S. Qureishi. The discrimination between two Weibull processes. *Technometrics*, 6:57 – 76, 1964.

[806] R. L. Racicot. Fitting a filtered Poisson process. *J. Inst. Appl. Math.*, 7:260 – 272, 1971.

[807] L. Räde. Limit theorems for thinning of renewal processes. *J. Appl. Probab.*, 9:847 – 851, 1972b.

[808] L. Räde. Thinning of renewal processes: A flow graph study. Göteborg, Sweden, 1972a.

[809] R. M. Radecke. Entwicklungsgesetze stationärer markierter Punktprozesse. *Math. Nachr.*, 81:83 – 167, 1977.

[810] R. M. Radecke. Uber eine spezielle Klasse abhangiger Verschiebungen Poissonscher Punktprozesse. *Math. Nachr.*, 92:129 – 138, 1979.

[811] A. E. Raftery and V. E. Akman. Asymptotic inference for a change-point Poisson process. *Ann. Statist.*, 14:1583 – 1590, 1986.

[812] H. Ramlau-Hansen. Smoothing counting process intensities by means of kernel functions. *Ann. Statist.*, 11:453 – 466, 1983a.

[813] H. Ramlau-Hansen. The choice of a kernel function in the graduation of counting process intensities. *Scand. Actuar. J.*, 1983:165 – 182, 1983b.

[814] D. H. Randles and D. A. Wolfe. *An Introduction to the Theory of Nonparametric Statistics*. Wiley, New York, 1971.

[815] M. Rao and R. Wedel. Poisson processes as renewal processes invariant under translations. *Ark. Mat.*, 8:539 – 541, 1968.

[816] R. Rebolledo. Remarques sur la convergence en loi des martingales vers martingales continues. *C. R. Acad. Sci. Paris*, 285:465 – 468, 1977a.

[817] R. Rebolledo. Remarques sur la convergence en loi des martingales vers martingales continues. *C. R. Acad. Sci. Paris*, 285:517–520, 1977b.

[818] R. Rebolledo. Sur les applications de la théorie des martingales à l'étude statistique d'une famille de processus ponctuels. *Lect. Notes Math.*, 636:27 – 70, 1978.

[819] R. Rebolledo. Central limit theorem for local martingales. *Z. Wahrsch. verw. Geb.*, 51:269 – 286, 1980.

[820] R. A. Redner and H. F. Walker. Mixture densities, maximum likelihood and the EM algorithm. *SIAM Rev.*, 26:195 – 239, 1984.

[821] N. Reid. Influence functions for censored data. *Ann. Statist.*, 9:78 – 92, 1981.

[822] A. Rényi. On an extremal property of the Poisson process. *Ann. Inst. Statist. Math.*, 16:129 – 133, 1964a.

[823] A. Rényi. On two mathematical models of the traffic on a divided highway. *J. Appl. Probab.*, 1:311 – 320, 1964b.

[824] A. Rényi. Remarks on the Poisson process. *Studia. Sci. Math. Hungar.*, 2:119 – 123, 1967.

[825] S. I. Resnick. Point processes, regular variation and weak convergence. *Adv. Appl. Probab.*, 18:66 – 138, 1986.

[826] S. I. Resnick. *Extreme Values, Regular Variation and Point Processes.* Springer-Verlag, New York, 1987.

[827] J. Rice. Estimated factorization of the spectral density of a stationary point process. *J. Appl. Probab.*, 7:801 – 817, 1975.

[828] B. D. Ripley and J. P. Rasson. Finding the edge of a Poisson forest. *J. Appl. Probab.*, 14:483 – 491, 1977.

[829] B. D. Ripley. Locally finite random sets: foundations for point process theory. *Ann. Probab.*, 4:983 – 994, 1976a.

[830] B. D. Ripley. The foundations of stochastic geometry. *Ann. Probab.*, 4:995 – 998, 1976b.

[831] B. D. Ripley. On stationarity and superposition of point processes. *Ann. Probab.*, 4:999 – 1005, 1976c.

[832] B. D. Ripley. The second-order analysis of stationary point processes. *J. Appl. Probab.*, 13:255 – 266, 1976d.

[833] B. D. Ripley. Modelling spatial patterns. *J. Roy. Statist. Soc. B*, 39:172 – 212, 1977.

[834] B. D. Ripley. *Spatial Statistics.* Wiley, New York, 1981.

[835] B. D. Ripley. *Statistical Inference for Spatial Processes.* Cambridge University Press, Cambridge, 1988.

[836] B. D. Ripley and F. P. Kelly. Markov point processes. *J. London Math. Soc. II Ser.*, 15:188 – 192, 1977.

[837] Y. Ritov and J. A. Wellner. Censoring, martingales and the Cox model. *Contemp. Math.*, 80:191 – 220, 1988.

[838] F. D. K. Roberts. Nearest neighbors in a Poisson ensemble. *Biometrika*, 56:401 – 406, 1969.

[839] P. M. Robinson. Estimation of a time series from irregularly spaced data. *Stochastic Process. Appl.*, 6:9 – 24, 1977.

[840] H. Rohde and J. Grandell. On the removal time of aerosol particles from the atmosphere by precipitation scavenging. *Tellus*, 24:443 – 454, 1972.

[841] H. Rohde and J. Grandell. Estimates of characteristic times for precipitation scavenging. *J. Atmospheric Sci.*, 38:370 – 386, 1981.

[842] T. Rolski. *Stationary Random Processes Associated with Point Processes*, volume 5 of *Lecture Notes in Statistics.* Springer-Verlag, New York, 1981.

[843] T. Rolski. Ergodic properties of Poisson processes with almost periodic intensity. *Probab. Th. Rel. Fields*, 84:27 – 38, 1990.

[844] H. Rootzén. A note on the central limit theorem for doubly stochastic Poisson processes. *J. Appl. Probab.*, 13:809 – 813, 1976.

[845] S. M. Ross. Optimal dispatching of a Poisson process. *J. Appl. Probab.*, 6:692 – 699, 1969.

[846] S. M. Ross. *Stochastic Processes*. Wiley, New York, 1982.

[847] G. G. Roussas. Nonparametric estimation of the transition distribution function of a Markov process. *Ann. Math. Statist.*, 40:1386 – 1400, 1969.

[848] G. G. Roussas. *Contiguity of Probability Measures: Some Applications in Statistics*. Cambridge University Press, Cambridge, 1972.

[849] Yu. A. Rozanov. *Stationary Random Processes*. Holden-Day, San Francisco, 1967.

[850] Yu. A. Rozanov. *Markov Random Fields*. Springer-Verlag, New York, 1982.

[851] I. Rubin. Regular point processes and their detection. *IEEE Trans. Inform. Theory*, IT-18:547 – 557, 1972.

[852] I. Rubin. Regular jump processes and information processing. *IEEE Trans. Inform. Theory*, IT-2O:617 – 624, 1974.

[853] M. Rudemo. Doubly stochastic Poisson processes and process control. *Adv. Appl. Probab.*, 4:318 – 338, 1972.

[854] M. Rudemo. State estimation for partially observed Markov chains. *J. Math. Anal. Appl.*, 44:581 – 611, 1973a.

[855] M. Rudemo. Point processes generated by transitions of Markov chains. *Adv. Appl. Probab.*, 5:262 – 286, 1973b.

[856] M. Rudemo. On a random transformation of a point process to a Poisson process. In *Mathematics and Statistics*. Chalmers Tekniska Högskola, Göteborg, Sweden, 1973c.

[857] M. Rudemo. Prediction and smoothing for partially observed Markov chains. *J. Math. Anal. Appl.*, 49:1 – 23, 1975.

[858] W. Rudin. *Functional Analysis*. McGraw-Hill, New York, 1973.

[859] W. Rudin. *Real and Complex Analysis*. McGraw-Hill, New York, 2nd edition, 1975.

[860] L. Rüschendorf. Inference for random sampling processes. *Stochastic Process. Appl.*, 32:129 – 140, 1989.

[861] C. Ryll-Nardzewski. Remarks on processes of calls. *Proc. Fourth Berkeley Symp. Math. Statist. Prob.*, 2:455 – 465, 1961.

[862] S. O. Samuelson. Nonparametric estimation of the cumulative intensity from doubly censored data. *Scand. J. Statist.*, 16:1 – 21, 1989.

[863] R. Saunders and G. M. Funk. Poisson limits for a clustering model of Strauss. *J. Appl. Probab.*, 14:776 – 784, 1977.

[864] R. Saunders, R. J. Kryscio, and G. M. Funk. Poisson limits for a hard-core clustering model. *Stochastic Process. Appl.*, 12:97 – 116, 1982.

[865] I. R. Savage. Contributions to the theory of rank order statistics — the two sample case. *Ann. Math. Statist.*, 27:590 – 615, 1956.

[866] J. G. Saw. Tests on the intensity of a Poisson process. *Commun. Statist.*, 4:777 – 782, 1975.

[867] V. Schmidt. Qualitative and asymptotic properties of stochastic integrals related to random marked point processes. *Optimization*, 18:737 – 759, 1987.

[868] T. Sellke and D. Siegmund. Sequential analysis of the proportional hazards model. *Biometrika*, 70:315 – 326, 1983.

[869] P. K. Sen. The Cox regression model, invariance principle for some induced quantile processes and some repeated significance tests. *Ann. Statist.*, 9:109 – 121, 1981.

[870] B. Sendov. Some questions of the theory of approximation of functions and sets in the Hausdorff metric. *Russian Math. Surveys*, 24:143 – 183, 1969.

[871] R. J. Serfling. Some elementary results on Poisson approximation in a sequence of Bernoulli trials. *SIAM Rev.*, 20:567 – 579, 1978.

[872] R. J. Serfling. *Approximation Theorems of Mathematical Statistics.* Wiley, New York, 1980.

[873] R. F. Serfozo. Conditional Poisson processes. *J. Appl. Probab.*, 9:288 – 302, 1972a.

[874] R. F. Serfozo. Processes with conditional stationary independent increments. *J. Appl. Probab.*, 9:303 – 315, 1972b.

[875] R. F. Serfozo. Functional limit theorems for stochastic processes based on embedded processes. *Adv. Appl. Probab.*, 7:123 – 139, 1975.

[876] R. F. Serfozo. Compositions, inverses and thinnings of random measures. *Z. Wahrsch. verw. Geb.*, 37:253 – 265, 1977.

[877] R. F. Serfozo. Thinning of cluster processes: convergence of sums of thinned point processes. *Math. Oper. Res.*, 9:522 – 533, 1984.

[878] R. F. Serfozo. Partitions of point processes: multivariate Poisson approximations. *Stochastic Process. Appl.*, 20:281 – 294, 1985.

[879] R. F. Serfozo. Poisson functionals of Markov processes and queueing networks. *Adv. Appl. Probab.*, 21:595 – 611, 1989.

[880] J. Serra. *Image Analysis and Mathematical Morphology.* Academic Press, New York, 1982.

[881] D. N. Shanbhag and M. Westcott. A note on infinitely divisible point processes. *J. Roy. Statist. Soc. B*, 39:331 – 332, 1977.

[882] H. S. Shapiro and R. A. Silverman. Alias-free sampling of random noise. *J. SIAM*, 8:225 – 248, 1960.

[883] A. N. Shiryaev. Martingales: recent developments, results and applications. *Internat. Statist. Rev.*, 49:199 – 233, 1981.

[884] G. R. Shorack and J. A. Wellner. *Empirical Processes with Applications to Statistics.* Wiley, New York, 1986.

[885] R. W. Shorrock. Extremal processes and random measures. *J. Appl. Probab.*, 12:316 – 323, 1975.

[886] B. W. Silverman. On the estimation of a probability density function by the maximum penalized likelihood method. *Ann. Statist.*, 10:795 – 810, 1982.

[887] B. W. Silverman. *Density Estimation for Statistics and Data Analysis.* Chapman & Hall, London, 1986.

[888] B. W. Silverman, M. C. Jones, J. D. Wilson, and D. W. Nychka. A smoothed EM approach to indirect estimation problems with particular reference to stereology and emission tomography. *J. Roy. Statist. Soc. B*, 52:271 – 324, 1990.

[889] L. Simar. Maximum likelihood estimation of a compound Poisson process. *Ann. Statist.*, 4:1200 – 1209, 1976.

[890] I. M. Slivnjak. Some properties of stationary flows of homogeneous random events. *Theor. Probab. Appl.*, 7:347 – 352, 1962.

[891] E. V. Slud. Sequential linear rank tests for 2-sample censored survival data. *Ann. Statist.*, 12:551 – 571, 1984.

[892] W. L. Smith. Regenerative stochastic processes. *Proc. Roy. Soc. A*, 232:6 – 31, 1955.

[893] W. L. Smith. Renewal theory and its ramifications. *J. Roy. Statist. Soc. B*, 20:243 – 284, 1958.

[894] J. A. Smith. *Point Process Models of Rainfall.* PhD thesis, The Johns Hopkins University, Baltimore, 1981.

[895] J. A. Smith and A. F. Karr. A point process model of summer season rainfall occurrences. *Water Resources Res.*, 19:95 – 103, 1983.

[896] J. A. Smith and A. F. Karr. Statistical inference for point process models of rainfall. *Water Resources Res.*, 21:73 – 79, 1985.

[897] J. A. Smith and A. F. Karr. Flood frequency analysis using the Cox regression model. *Water Resources Res.*, 22:890 – 896, 1986.

[898] R. T. Smythe. Ergodic properties of marked point processes in r$^\nu$. *Ann. Inst. H. Poincaré*, 11:109 – 125, 1975.

[899] D. L. Snyder. Filtering and detection for doubly stochastic Poisson processes. *IEEE Trans. Inform. Theory*, IT-18:91 – 102, 1972a.

[900] D. L. Snyder. Smoothing for doubly stochastic Poisson processes. *IEEE Trans. Inform. Theory*, IT-18:558 – 562, 1972b.

[901] D. L. Snyder. Information processing for observed jump processes. *Inform. and Control*, 22:69 – 78, 1973.

[902] D. L. Snyder. *Random Point Processes*. Wiley-Interscience, New York, 1975.

[903] H. Solomon and P. C. C. Wang. Nonhomogeneous Poisson fields of random lines with applications to traffic flow. *Proc. Sixth Berkeley Symp. Math. Statist. Prob.*, 3:383 – 400, 1972.

[904] W. Solomon. Representation and approximation of large population age distribution using Poisson random measures. *Stochastic Process. Appl.*, 26:237 – 255, 1987.

[905] N. Starr. Optimal and adaptive stopping based on capture rates. *J. Appl. Probab.*, 11:294 – 301, 1974.

[906] V. Stefanov. Efficient sequential estimation in exponential-type processes. *Ann. Statist.*, 14:1606 – 1611, 1986.

[907] M. Stein. Asymptotically efficient prediction of a random field with a misspecified covariance function. *Ann. Statist.*, 16:55 – 63, 1988.

[908] M. Stein. Uniform asymptotic optimality of linear prediction of a random field using an incorrect second-order structure. *Ann. Statist.*, 18:850 – 872, 1990.

[909] C. J. Stone. On a theorem of Dobrushin. *Ann. Math. Statist.*, 39:1391 – 1401, 1968a.

[910] C. J. Stone. Infinite particle systems and multidimensional renewal theory. *J. Math. Mech.*, 18:201 – 227, 1968b.

[911] H. Störmer. Zur überlagerung von Erneurrungsprozessen. *Z. Wahrsch. verw. Geb.*, 13:9 – 24, 1969.

[912] D. Stoyan. Interrupted point processes. *Bimoetrical J.*, 21:607 – 610, 1979.

[913] D. Stoyan. *Comparison Methods for Queues and Other Stochastic Processes*. Wiley-Interscience, New York, 1983a. English edition edited by D. J. Daley.

[914] D. Stoyan. Inequalities and bounds for variances of point processes and fiber processes. *Math. Operationsforsch. Statist.*, 14:409 – 419, 1983b.

[915] D. Stoyan. On correlations of marked point processes. *Math. Nachr.*, 116:197 – 207, 1984a.

[916] D. Stoyan. Correlations of the marks of marked point processes — statistical inference and simple models. *Elektron. Informationsverarb. Kybernet.*, 20:285 – 294, 1984b.

[917] D. Stoyan. Thinnings of point processes and their use in the statistical analysis of a settlement pattern with deserted villages. *Statistics*, 19:45 – 56, 1988.

[918] D. Stoyan and J. Mecke. *Stochastische Geometrie.* Akademie-Verlag, Berlin, 1983.

[919] D. Stoyan and H. Stoyan. On one of Matérn's hard-core point process models. *Math. Nachr.*, 122:205 – 214, 1985.

[920] J. M. Stoyanov and D. I. Vladeva. Estimation of unknown parameters of continuous time stochastic processes by observations at random moments. *C. R. Acad. Bulgare Sci.*, 35:153   156, 1982.

[921] V. Strassen. An invariance principle for the LIL. *Z. Wahrsch. verw. Geb.*, 3:211 – 226, 1964.

[922] V. Strassen. The existence of probability measures with given marginals. *Ann. Math. Statist.*, 36:423 – 439, 1965.

[923] D. J. Strauss. A model for clustering. *Biometrika*, 62:467 – 475, 1972.

[924] F. Streit. Optimality properties of statistical tests for Poisson processes. *Adv. Appl. Probab.*, 10:330 – 331, 1978.

[925] F. Stulajter. Some nonlinear statistical problems of a Poisson process. *Kybernitka (Prague)*, 18:397 – 407, 1982.

[926] W. Stute. Empirische Prozesse in der Datenanalyse. *Jahresber. Deutsch. Math.-Verein*, 90:129 – 144, 1988.

[927] U. Sumita and J. G. Shanthikumar. An age-dependent counting process generated from a renewal process. *Adv. Appl. Probab.*, 20:739 – 755, 1988.

[928] B. Sundt. On the problem of testing whether a mixed Poisson process is homogeneous. *Insurance Math. Econ.*, 1:253 – 254, 1982.

[929] V. Susarla and J. Van Ryzin. Large sample theory for an estimator of the true survival time from censored samples. *Ann. Statist.*, 8:1002 – 1016, 1980.

[930] A. Svennson. Asymptotic estimation in counting processes with parametric intensities based on one realization. *Scand. J. Statist.*, 17:23 – 34, 1990.

[931] L. Takacs. On secondary stochastic processes generated by a multidimensional Poisson process. *Publ. Math. Inst. Hungar. Acad. Sci.*, 2:71 – 79, 1957.

[932] R. Takacs. Estimation for the pair potential of a Gibbsian point process. *Statistics*, 17:429 – 433, 1986.

[933] M. A. Tanner and W. H. Wong. The estimation of the hazard function from randomly censored data by the kernel method. *Ann. Statist.*, 11:989 – 993, 1983.

[934] R. A. Tapia and J. R. Thompson. *Nonparametric Probability Density Estimation.* The Johns Hopkins University Press, Baltimore, 1978.

[935] P. Tautu. Stochastic spatial processes in biology: a concise historical survey. *Lect. Notes Math.*, 1212:1 – 41, 1986.

[936] H. Teicher. Identifiability of mixtures. *Ann. Math. Statist.*, 32:244 – 248, 1961.

[937] L. Telksnys, editor. *Detection of Changes in Random Processes.* Optimization Software, New York, 1986.

[938] P. F. Thall. A theorem on regular infinitely divisible counting processes. *Stochastic Process. Appl.*, 16:205 – 210, 1984.

[939] P. F. Thall. Mixed Poisson regression models for longitudinal interval count data. *Biometrics*, 44:197 – 210, 1988.

[940] P. F. Thall and J. M. Lachin. Analysis of recurrent events: nonparametric methods for random-interval count data. *J. Amer. Statist. Assoc.*, 83:339 – 347, 1988.

[941] A. Thavaneswaran and M. E. Thompson. Optimal estimation for semi-martingales. *J. Appl. Probab.*, 23:409 – 417, 1986.

[942] T. Thédeen. A note on the Poisson tendency in traffic distribution. *Ann. Math. Statist.*, 35:1823 – 1824, 1964.

[943] T. Thédeen. Convergence and invariance questions for point systems in r^1 under random motion. *Ark. Math.*, 7:211 – 239, 1967a.

[944] T. Thédeen. On stochastic stationarity of renewal processes. *Ark. Math.*, 7:249 – 263, 1967b.

[945] T. Thédeen. On road traffic with free overtaking. *J. Appl. Probab.*, 6:524 – 549, 1969.

[946] H. R. Thompson. Spatial point processes, with application to ecology. *Biometrika*, 42:102 – 115, 1955.

[947] S. K. Thompson and F. L. Ramsey. Detectability functions in observing spatial point processes. *Biometrics*, 43:355 – 362, 1987.

[948] W. A. Thompson. *Point Process Models with Application to Safety and Reliability.* Chapman & Hall, London, 1988.

[949] P. J. Thomson. Signal estimation using an array of recorders. *Stochastic Process. Appl.*, 13:201 – 214, 1982.

[950] H. Thorisson. A complete coupling proof of Blackwell's renewal theorem. *Stochastic Process. Appl.*, 26:87 – 97, 1987.

[951] M. L. Thornett. A class of second-order stationary random measures. *Stochastic Process. Appl.*, 8:323 – 334, 1979.

[952] K. D. Tocher. Extension of the Neyman-Pearson theory to the discrete case. *Biometrika*, 37:130 – 144, 1950.

[953] F. Tøpsøe. Uniformity in convergence of measures. *Z. Wahrsch. verw. Geb.*, 39:1 – 30, 1977.

[954] W. Y. Tsai and J. Crowley. A large sample study of generalized maximum likelihood estimators from incomplete data via self-consistency. *Ann. Statist.*, 13:1317 – 1334, 1985.

[955] W. J. Tsai, S. Leurgans, and J. Crowley. Nonparametric estimation of a bivariate survival function in the presence of censoring. *Ann. Statist.*, 14:1351 – 1365, 1986.

[956] A. A. Tsiatis. An example of nonidentifiability in competing risks. *Scand. Actuar. J.*, 1978:235 – 239, 1978.

[957] A. A. Tsiatis. A large sample study of Cox's regression model. *Ann. Statist.*, 9:93 – 108, 1981.

[958] H. G. Tucker. An estimate of the compounding distribution of a compound Poisson distribution. *Theor. Probab. Appl.*, 8:195 – 200, 1964.

[959] S. Tulya-Muhika. A characterization of E-processes and Poisson processes in $\mathbf{r}^n$. *Z. Wahrsch. verw. Geb.*, 20:199–216, 1971.

[960] G. J. G. Upton and B. Fingleton. *Spatial Data Analysis by Example, I.* Wiley, Chichester, 1985.

[961] E. Valkeila. A general Poisson approximation theorem. *Stochastics*, 7:159 – 171, 1982.

[962] P. C. T. van der Hoeven. Une projection de processus ponctuels. *Z. Wahrsch. verw. Geb.*, 61:483 – 499, 1982.

[963] P. C. T. van der Hoeven. *On Point Processes*, volume 165 of *Mathematics Centre Tracts*. Mathematisch Centrum, Amsterdam, 1983.

[964] A. P. Van der Plas. On the estimation of the parameters of Markov probability models using macro data. *Ann. Statist.*, 11:78 – 85, 1983.

[965] A. W. van der Vaart. *Statistical Estimation in Large Parameter Spaces.* Mathematisch Centrum, Amsterdam, 1988.

[966] J. Van Schuppen. Filtering, prediction and smoothing for counting observations: a martingale approach. *SIAM J. Appl. Math.*, 32:552 – 570, 1977.

[967] M. C. A. Van Zuijlen. Properties of the empirical distribution function for independent nonidentically distributed random variables. *Ann. Probab.*, 6:250 – 266, 1978.

[968] V. N. Vapnik and A. Ya. Chervonenkis. On the uniform convergence of relative frequencies of events to their probabilities. *Theor. Probab. Appl.*, 16:264 – 280, 1971.

[969] P. Varaiya. The martingale theory of jump processes. *IEEE Trans. Automat. Control*, AC-20:34 – 42, 1975.

[970] Y. Vardi. Asymptotically optimal sequential estimation: the Poisson case. *Ann. Statist.*, 7:1040 – 1051, 1979.

[971] Y. Vardi. On a stopping time of Starr and its use in estimating the number of transmission sources. *J. Appl. Probab.*, 17:235 – 242, 1980.

[972] Y. Vardi. Nonparametric estimation in renewal processes. *Ann. Statist.*, 10:772 – 785, 1982a.

[973] Y. Vardi. Nonparametric estimation in the presence of length bias. *Ann. Statist.*, 10:616 – 620, 1982b.

[974] Y. Vardi, L. A. Shepp, and L. Kaufman. A statistical model for positron emission tomography. *J. Amer. Statist. Assoc.*, 80:8 – 20, 1975.

[975] D. Vere-Jones. Some applications of probability generating functionals to the study of input-output systems. *J. Roy. Statist. Soc. B*, 30:321 – 333, 1968.

[976] D. Vere-Jones. Stochastic models for earthquake occurrence. *J. Roy. Statist. Soc. B*, 32:1 – 62, 1970.

[977] D. Vere-Jones. An elementary approach to the spectral theory of stationary random measures. In E. J. Harding and D.G. Kendall, editors, *Stochastic Geometry*. Wiley, New York, 1974.

[978] D. Vere-Jones. A renewal equation for point processes with Markov-dependent intervals. *Math. Nachr.*, 68:133 – 139, 1975a.

[979] D. Vere-Jones. On updating algorithms for inference for stochastic processes. In J. Gani, editor, *Perspectives in Probability and Statistics*. Academic Press, New York, 1975b.

[980] D. Vere-Jones. Space-time correlations for micro-earthquakes — a pilot study. *Suppl. Adv. Appl. Prob.*, 10:73 – 87, 1978.

[981] D. Vere-Jones. On the estimation of frequency in point process data. *J. Appl. Probab.*, 19A:383 – 394, 1982.

[982] V. Vervaat. Thinned and stretched Poisson processes. *Statist. Neerl.*, 21:245 – 268, 1967.

[983] J. G. Voekel and J. Crowley. Nonparametric inference for a class of semi-Markov processes with censored observations. *Ann. Statist.*, 12:142 – 160, 1984.

[984] V. A. Volkonskii. An ergodic theorem for the distribution of the duration of fades. *Theor. Probab. Appl.*, 5:323 – 326, 1960.

[985] W. Von Waldenfels. Charakteristische Funktionale zufälliger Masse. *Z. Wahrsch. verw. Geb.*, 10:279 – 283, 1968.

[986] L. Yu. Vostrikova. On necessary and sufficient conditions for convergence of probability measures in variation. *Stochastic Process. Appl.*, 18:99 – 112, 1984.

[987] A. Wald. *Sequential Analysis*. Wiley, New York, 1947.

[988] C. B. Wan and M. H. A. Davis. The general point process disorder problem. *IEEE Trans. Inform. Theory*, IT-23:538 – 540, 1977.

[989] J.-L. Wang. Estimators of a distribution function with increasing failure rate average. *J. Statist. Plann. Inference*, 16:415 – 427, 1987a.

[990] J.-L. Wang. Estimating IFRA and NBU survival curves based on censored data. *Scand. J. Statist.*, 14:199 – 210, 1987b.

[991] M.-C. Wang. Product-limit estimates: a generalized maximum likelihood study. *Commun. Statist. Theory Methods*, 16:3117 – 3132, 1987.

[992] M.-C. Wang. A semiparametric model for randomly truncated data. *J. Amer. Statist. Assoc.*, 84:742 – 748, 1989.

[993] M.-C. Wang, N. P. Jewell, and W.-Y. Tsai. Asymptotic properties of the product limit estimate under random truncation. *Ann. Statist.*, 14:1597 – 1605, 1986.

[994] S. Watanabe. On discontinuous additive functionals and Lévy measures of a Markov process. *Japan J. Math.*, 34:53 – 70, 1964.

[995] S. Watanabe. Poisson point process of Brownian excursions and its applications to diffusion processes. *Proc. Symp. Pure Math.*, 31:153 – 164, 1977.

[996] G. S. Watson and M. R. Leadbetter. Hazard analysis I. *Biometrika*, 51:175 – 184, 1964a.

[997] G. S. Watson and M. R. Leadbetter. Hazard analysis II. *Sankhyā*, 26A:110 – 116, 1964b.

[998] E. C. Waymire and V. K. Gupta. The mathematical structure of rainfall representations, 2: A review of the theory of point processes. *Water Resources Res.*, 17:1273 – 1286, 1981.

[999] E. C. Waymire and V. K. Gupta. An analysis of the Pólya point process. *Adv. Appl. Probab.*, 15:39 – 53, 1983.

[1000] H. Wegmann. Characterization of Palm distributions and infinitely divisible random measures. *Z. Wahrsch. verw. Geb.*, 39:257 – 262, 1977.

[1001] W. Weil. Point processes of cylinders, particles and flats. *Acta Appl. Math.*, 9:103 – 136, 1987.

[1002] P. Weiss. On the singularity of Poisson processes. *Math. Nachr.*, 92:111 – 115, 1979.

[1003] J. A. Wellner. A martingale inequality for the empirical process. *Ann. Probab.*, 5:303 – 308, 1977.

[1004] J. A. Wellner. Limit theorems for the ratio of the empirical distribution function to the true distribution function. *Z. Wahrsch. verw. Geb.*, 45:73 – 88, 1978.

[1005] J. A. Wellner. Asymptotic optimality of the product limit estimator. *Ann. Statist.*, 10:595 – 602, 1982.

[1006] J. A. Wellner. A heavy censoring limit theorem for the product limit estimator. *Ann. Statist.*, 13:150 – 162, 1985.

[1007] S. A. West. Upper and lower probability inferences pertaining to Poisson processes. *J. Amer. Statist. Assoc.*, 72:448 – 452, 1977.

[1008] M. Westcott. On existence and mixing results for cluster point processes. *J. Roy. Statist. Soc. B*, 33:290 – 300, 1971.

[1009] M. Westcott. The probability generating functional. *J. Austral. Math. Soc.*, 14:448 – 466, 1972.

[1010] W. Whitt. Limits for the superposition of m-dimensional point processes. *J. Appl. Probab.*, 9:462 – 465, 1972.

[1011] W. Whitt. On the quality of Poisson approximations. *Z. Wahrsch. verw. Geb.*, 28:23 – 36, 1973.

[1012] W. Whitt. Approximating a point process by a renewal process. I. Two basic methods. *Oper. Res.*, 30:125 – 147, 1982.

[1013] M. J. Wichura. On the construction of almost uniformly convergent random variables with given weakly convergent laws. *Ann. Math. Statist.*, 41:284 – 291, 1970.

[1014] K. H. Wickwire. Optimal stopping problems for differential equations perturbed by Poisson processes. *Naval Logist. Res. Quart.*, 28:423 – 429, 1981.

[1015] S. S. Wilks. *Mathematical Statistics*. Wiley, New York, 1962.

[1016] J. S. Williams and S. Lagakos. Models for censored survival analysis: constant-sum and variable-sum models. *Biometrika*, 64:215 – 224, 1977.

[1017] J. S. Willie. Measuring the association of a time series and a point process. *J. Appl. Probab.*, 19:597 – 608, 1982a.

[1018] J. S. Willie. Covariation of a time series and a point process. *J. Appl. Probab.*, 19:609 – 618, 1982b.

[1019] B. B. Winter. Nonparametric estimation with censored data from a distribution function with nondecreasing density. *Commun. Statist. Theory Methods*, 16:93 – 120, 1987.

[1020] T. K. M. Wisniewski. Bivariate stationary point processes: fundamental relations and first recurrence times. *Adv. Appl. Probab.*, 4:296 – 317, 1972.

[1021] A. Wold. On stationary point processes and Markov chains. *Skand. Aktuar.*, 31:229 – 240, 1948.

[1022] W. H. Wong. Theory of partial likelihood. *Ann. Statist.*, 14:88 – 123, 1986.

[1023] M. Woodroofe. Estimating a distribution function with truncated data. *Ann. Statist.*, 13:163 – 177, 1985.

[1024] C. F. J. Wu. On the convergence properties of the EM algorithm. *Ann. Statist.*, 11:95 – 103, 1983.

[1025] M. Yadin and S. Zacks. The visibility of stationary and moving targets in the plane subject to a Poisson field of shadowing elements. *J. Appl. Probab.*, 22:776 – 786, 1985.

[1026] M. I. Yadrenko. *Spectral Theory of Random Fields*. Optimization Software, Inc., New York, 1983.

[1027] A. M. Yaglom. *An Introduction to the Theory of Stationary Random Functions*. Prentice-Hall, Englewood Cliffs, NJ, 1962.

[1028] B. S. Yandell. Nonparametric inference for rates and densities with censored serial data. *Ann. Statist.*, 11:1119 – 1135, 1983.

[1029] B. S. Yandell and L. Horváth. Bootstrapped multidimensional product limit process. *Austral. J. Staitst.*, 30:342 – 358, 1988.

[1030] N. Yannaros. On Cox processes and gamma renewal processes. *J. Appl. Probab.*, 25:423 – 427, 1988a.

[1031] N. Yannaros. The inverses of thinned renewal processes. *J. Appl. Probab.*, 25:822 – 828, 1988b.

[1032] A. Yashin and E. Arjas. A note on random intensities and conditional survivor functions. *J. Appl. Probab.*, 25:630 – 635, 1988.

[1033] M. Yor. Représentation des martingales de carré integrable relative aux processus de Wiener et de Poisson à n parametres. *Z. Wahrsch. verw. Geb.*, 39:121 – 129, 1976.

[1034] L. Younes. Parametric inference for imperfectly observed Gibbsian fields. *Probab. Th. Rel. Fields*, 82:625 – 645, 1989.

[1035] J. C. A. Zaat. Interrupted Poisson processes, a model for the mechanical thinning of beet plants. *Statist. Neerl.*, 18:311 – 324, 1964.

[1036] S. Zacks. *Theory of Statistical Inference*. Wiley, New York, 1971.

[1037] U. Zähle. Asymptotic properties of point processes of rare events in Gaussian fields, I. *Theor. Probab. Appl.*, 28:703 – 713, 1983.

[1038] U. Zähle. Asymptotic properties of point processes of rare events in Gaussian fields, II. *Theor. Probab. Appl.*, 30:66 – 74, 1985.

[1039] M. Zakai. On the optimal filtering of diffusion processes. *Z. Wahrsch. verw. Geb.*, 11:230 – 243, 1969.

[1040] P. Zeephongsekul. Laplace functional approach to point processes occurring in a traffic model. *Ann. Probab.*, 9:1034 – 1040, 1981.

[1041] C.-H. Zhang. Fourier methods for estimating mixing densities and distributions. *Ann. Statist.*, 18:806 – 831, 1990.

[1042] M. Zhou. The-sided bias bound of the Kaplan-Meier estimator. *Probab. Th. Rel. Fields*, 79:165– 174, 1988.

[1043] D. M. Zucker and A. F. Karr. Nonparametric survival analysis with time-dependent covariate effects: a penalized partial likelihood approach. *Ann. Statist.*, 18:329 – 353, 1990.

Index